Reaction Mechanisms in Carbon Dioxide Conversion

Michele Aresta • Angela Dibenedetto •
Eugenio Quaranta

Reaction Mechanisms in Carbon Dioxide Conversion

Springer

Michele Aresta
Chemical and Biomolecular
Engineering Department
NUS
Singapore, Singapore

CIRCC
Pisa, Italy

Angela Dibenedetto
Department of Chemistry and CIRCC
University of Bari
Bari, Italy

Eugenio Quaranta
Department of Chemistry and CIRCC
University of Bari
Bari, Italy

ISBN 978-3-662-46830-2 ISBN 978-3-662-46831-9 (eBook)
DOI 10.1007/978-3-662-46831-9

Library of Congress Control Number: 2015950373

Springer Heidelberg New York Dordrecht London

Printed on acid-free paper

Springer-Verlag GmbH Berlin Heidelberg is part of Springer Science+Business Media (www.springer.com)

Lella, Gianluca, Brunella, Manuela, Mattia,
Nicolo', Federica, Gabriele, Lorenzo
Gianluca, Rosangela
Milena, Chiara

Preface

Carbon dioxide (CO_2) is today at the centre of attention of the scientific and technological world because of its potential role in climate change, which is supposed to be, directly or indirectly, caused by the growing level of CO_2 in the atmosphere. Policy-makers work out solutions ranging over a large number of options, including carbon capture and disposal or sequestration (CCS) or carbon capture and utilization (CCU).

The conversion of CO_2 into added-value products (chemicals, materials, fuels) is today one of the most widespread research themes, and the scientific literature presents a large number of reviews, papers and books that disclose the CO_2 chemistry or, in general, utilization. Industry is also much involved, as witnessed by the ever growing number of industrial patents and new processes.

Despite the large number of volumes published, an organized collection of information about the mechanisms concerning the many reactions in which CO_2 is involved does not exist. However, this book has been designed with the precise idea of filling the gap and collecting the knowledge we have of the *reaction mechanism* of CO_2 conversion when it is used either as a building block for the production of chemicals and materials, or as a carbon source for C1 and Cn energy-rich molecules.

This book covers 40 years of research efforts in understanding the reactivity of CO_2. Its aim is not to present an exhaustive list of all reactions in which CO_2 can be involved, even if it shows quite a large number of them, but to analyze the thermodynamics and kinetics of its conversion.

The book is organized in ten chapters, each devoted to a particular subject. References point the reader to more specific issues or more general applications.

Chapter 1 presents the CO_2 molecule and gives fundamental information for understanding the chemistry of the heterocumulene. It also discusses the excited states of CO_2 and its radical anion and radical cation. Such species are often encountered in other chapters. The spectroscopic techniques (infrared, ultraviolet, nuclear magnetic resonance) used for collecting information about the state of the

CO_2 moiety in the products are introduced here, and applied in the following chapters.

Chapter 2 discusses the co-ordination of CO_2 to metal centres in various temperature conditions and the effects of such interaction on the heterocumulene structure. It addresses the question of whether preliminary co-ordination of CO_2 to a metal centre is an essential prerequisite for CO_2 conversion.

In Chap. 3 the interaction of CO_2 with selected electron-rich moieties such as: H^-, OH^-, NR_2^-, R_3C^-, RO^- and NRR′R″ is presented. This chapter discusses reactions not mediated by a metal centre, which are discussed extensively in the following chapter.

Chapter 4 is dedicated to the analysis of the interaction of CO_2 with E–X bonds, where E–X=M–H, M–R, M–OH, M–OR, M–O_2, M–M, C–C, C–O, N–H, N–C, N–P and Si–H. The analysis of such elementary steps is a bridge to the more general aspect of reactivity of CO_2. Such reactions illustrate the role of metals in initiating the reactions and give the fundamentals for understanding the use of CO_2 in synthetic chemistry under catalytic conditions.

Chapter 5 covers the topic of CO_2 interaction with multiple C–C bonds (alkenes, alkynes, cumulenes and conjugated dienes) relevant to carboxylation reactions with formation of new C–C bonds and to the synthesis of fine chemicals or intermediates.

Chapter 6 is dedicated to the synthesis of organic carbonates, linear and cyclic, molecular compounds and polymeric materials, and of polyurethanes. Organic carbonates (monomeric) find a wide use in various sectors of the chemical and energy industries. Polymers, which can be considered as chemical sinks of CO_2, find a continuously growing application in architecture and are used for making CDs, sanitary tools and laboratory glassware, among others. Polyurethanes play a key role as insulators and packaging materials.

Chapter 7 introduces the high-temperature processes of CO_2 conversion: Dry Reforming of Methane (DRM) and the relevant general use of CO_2 as oxidant or dehydrogenating (DH) agent. The conversion of CO_2 into methanol is discussed here as it has similarities to the other processes. Such applications deal with the conversion of large volumes of CO_2 into fuels or energy-rich molecules.

Chapter 8 deals with the $1e^-{-}2e^-$ transfer to CO_2 in electrochemical, photochemical, photo-electrochemical processes, a topic of great importance for the conversion of CO_2 into energy-rich molecules. This chapter is relevant to the use of perennial energy sources, such as solar, wind, geothermal, hydro-energy, in the conversion of large volumes of CO_2. The use of solar energy brings about the man-made photosynthesis (direct and indirect) which is the future technology for the conversion of large volumes of CO_2 into chemicals and fuels.

Chapter 9 makes the analysis of bioprocesses (enzyme catalysed reactions) in which CO_2 is converted. Carboxylation processes and reduction of CO_2 to other C1-molecules such as HCOOH, H_2CO, CH_3OH and CH_4 are discussed here. A comparison to chemical processes discussed in previous chapters is straightforward.

Chapter 10 presents quite a new subject: the properties of hydrate-CO_2. This topic is a frontier research area and has great potential in the recovery of methane from hydrates. Only a limited amount of information exists in the scientific and technical literature, but the interest for future applications is very high.

In its entirety, this book presents the essential knowledge of the CO_2 conversion reaction mechanisms. We believe that it will be of great help to those who are newcomers in the fascinating field of carbon-cycling (CO_2 conversion) mimicking Nature and be for all readers a guide to discovering aspects of the recent history of CO_2 chemistry, a topic of great interest today for its potential impact on avoiding CO_2 emissions and reducing fossil carbon extraction.

The authors wish to thank Prof. Sibudjing Kawi and Dr. Yasotha Kathiraser of the Department of Chemical and Biomolecular Engineering of the National University of Singapore, Singapore, for their great help, having co-authored Chap. 7, and Prof. Praven Linga, Dr. Junjie Zheng and Ponnivalavan Babu of the same Department for having co-authored Chap. 10.

Thanks are due to Dr. Antonella Angelini, Department of Chemistry-University of Bari, Bari-IT, and Dr. Daniele Cornacchia, CIRCC-IT, for their assistance.

Michele Aresta is grateful to NUS, Chemical and Biomolecular Engineering Department, for their very kind hospitality during the writing of parts of this book and to IC^2R for some support.

Singapore-SG, June 2015
Bari-IT, June 2015

Michele Aresta
Angela Dibenedetto
Eugenio Quaranta

Contents

1 **The Carbon Dioxide Molecule** 1
1.1 Introduction 1
1.2 Electronic Properties of CO_2 2
1.2.1 Ground State of Carbon Dioxide 2
1.2.2 Lowest Excited States of Carbon Dioxide 6
1.3 Main Features of Carbon Dioxide Reactivity 8
1.3.1 Carbon Dioxide as O-Nucleophile 9
1.3.2 Carbon Dioxide as C-Electrophile 10
1.3.3 Amphoteric Reactivity of Carbon Dioxide 11
1.4 Carbon Dioxide Radical Anion, CO_2^- 12
1.5 Carbon Dioxide Radical Cation, CO_2^+ 19
1.6 Spectroscopic Techniques Applied to the CO_2 States 20
1.6.1 IR Spectroscopy 20
1.6.2 UV Spectrum of Carbon Dioxide 22
1.6.3 Nuclear Magnetic Resonance (NMR) Spectroscopy . . . 26
References 27

2 **CO_2 Coordination to Metal Centres: Modes of Bonding and Reactivity** 35
2.1 Mode of Bonding of Carbon Dioxide to Transition-Metal Centres 35
2.2 XRD Structure of η^2-C,O Mononuclear Metal Complexes 37
2.3 XRD Structural Data for η^1-C Metal Complexes 41
2.4 XRD Structural Characterization of O-End-On Complexes 43
2.5 Multinuclear Complexes 43
2.5.1 Side-On Bonded Complexes 43
2.5.2 O-End-On Bonded Complexes 44
2.6 Spectroscopic (IR and ^{13}C-NMR) Data for the Complexes Correlated to the Various Modes of Bonding of CO_2 47
2.6.1 Infrared Data Relevant to Transition Metal Complexes 47

2.6.2 NMR Data for Transition Metal Complexes 49
2.7 Fluxionality of the CO_2 Molecule Coordinated to Transition Metal Systems 51
2.8 Interaction of CO_2 with Metal Atoms in Low-Temperature Solid-Inert-Gas Matrices 55
2.9 Interaction of Metal Cations with CO_2 in the Gas Phase 57
2.10 Reactions of Coordinated CO_2 58
2.10.1 Reaction of Coordinated CO_2 with Electrophiles and O-Transfer from CO_2 to Produce CO 59
2.10.2 Reactions of Coordinated CO_2 with Nucleophiles 63
2.11 Conclusions 64
References 64

3 Interaction of CO_2 with Electron-Rich Moieties 71
3.1 Reaction with the Hydride Ion 71
3.2 Reaction with Hydroxide and Alkoxide Species 72
3.3 Reaction with Carbanions 73
3.4 Reaction with Amines 74
References 82

4 Insertion of CO_2 into E–X Bonds 85
4.1 Carbon Dioxide Insertion into M–H Bonds 85
4.2 CO_2 Insertion into M–OH Bonds 91
4.3 CO_2 Insertion into M–C Bonds 95
4.4 CO_2 Insertion into M–OR Bonds 97
4.5 Insertion into M–O_2 Bonds 98
4.6 CO_2 Insertion into M–N Bonds 102
4.6.1 Insertion into Transition Metal Amides 102
4.6.2 Insertion into Main Group and Post-Transition Metal Amides 108
4.6.3 Insertion into Amides of Non-Metallic Elements 112
4.7 CO_2 Insertion into M–P Bonds 115
4.8 CO_2 Insertion into C–C Bonds 117
4.9 CO_2 Insertion into C–N Bonds 120
4.10 Insertion into Other E–X Bonds 127
4.10.1 Insertion into M–M Bonds 127
4.10.2 Insertion into Si–H Bonds 128
4.10.3 Insertion into C–H Bonds 129
4.11 Conclusions 131
References 131

5 Interaction of CO_2 with C–C Multiple Bonds 143
5.1 Introduction 143
5.2 Oxidative Coupling with CO_2 144
5.3 Carboxylation of Olefins 147
5.4 Carboxylation of Alkynes 156

5.5 Carboxylation of Allenes 164
5.6 Carboxylation of Conjugated Dienes 169
References 176

6 Reaction Mechanisms in the Direct Carboxylation of Alcohols, Polyols, Cyclic Ethers, and Cyclic Amines to Afford Monomeric Compounds and Polymeric Materials 183
6.1 Utilization of Organic Carbonates and Conventional Synthetic Routes 183
6.2 Direct Carboxylation of Alcohols 185
6.2.1 Thermodynamic and Kinetic Issues 185
6.2.2 Reaction Mechanism 187
6.3 Direct Carboxylation of Diols and Polyols 210
6.4 Oxidative Carboxylation of Olefins to Afford Cyclic Carbonates 213
6.5 Carboxylation of Cyclic Ethers 216
6.5.1 Synthesis of Monomeric Cyclic Carbonates 217
6.5.2 Synthesis of Polycarbonates 221
6.6 Formation of Polyurethanes: Carboxylation of Cyclic Amines 223
6.7 Conclusions 225
References 226

7 Carbon Dioxide Conversion in High Temperature Reactions 237
7.1 Introduction 237
7.2 CO_2 as Oxidant 238
7.2.1 OCM Promoted by CO_2 238
7.2.2 Oxidative Dehydrogenation of Alkanes 243
7.2.3 Oxidative Dehydrogenation of Ethylbenzene to Styrene 254
7.2.4 CO_2 (Dry) Reforming of Methane 265
7.3 Hydrogenation of CO_2 277
7.3.1 Reverse Water Gas Shift Reaction (RWGS) 278
7.3.2 CO_2 Hydrogenation to Methanol and DME 278
7.3.3 Catalytic Systems for CO_2 Hydrogenation 280
7.3.4 Reaction Mechanism for CO_2 Hydrogenation 285
7.4 Conclusions 295
References 296

8 One- and Multi-electron Pathways for the Reduction of CO_2 into C1 and C1+ Energy-Richer Molecules: Some Thermodynamic and Kinetic Facts 311
8.1 Introduction 311
8.2 Key Steps and Aspects in CO_2 Reduction 312
8.3 One-Electron Transfer to CO_2 vs Multi-electron Transfer 321

8.4 Competitive Coordination of CO_2 and H^+ to a Catalytic Centre and Their Reduction . . . 322
8.5 Sequential "*One-Electron Plus One-Proton*" Pathways in Multi-electron Reduction of Bound CO_2 . . . 326
8.6 Photochemical and Photoelectrochemical Reduction of CO_2 . . . 332
8.7 Perspective Electrochemical, Photochemical and Photoelectrochemical Reduction of CO_2 . . . 339
References . . . 340

9 Enzymatic Conversion of CO_2 (Carboxylation Reactions and Reduction to Energy-Rich C1 Molecules) . . . 347
9.1 Introduction . . . 347
9.2 CO_2 Fixation in Biosynthesis . . . 348
9.2.1 Calvin–Benson–Bassham-Cycle . . . 349
9.2.2 Reductive TCA (Arnon–Buchanan) Cycle . . . 350
9.2.3 Reductive Acetyl-CoA (Wood–Ljungdahl) Pathway . . . 350
9.2.4 Acyl-CoA Carboxylation Pathways . . . 351
9.3 Carboxylation Reactions . . . 353
9.3.1 Bio-Carboxylation of Aromatic and Hetero-Aromatic Compounds . . . 353
9.3.2 Bio-Carboxylation of Epoxides . . . 357
9.4 Reduction Reactions . . . 358
9.4.1 Carbon Monoxide Dehydrogenases . . . 358
9.4.2 Formate Dehydrogenases . . . 360
9.4.3 Formaldehyde Dehydrogenase . . . 360
9.4.4 Alcohol Dehydrogenases . . . 361
9.4.5 Production of Acetic Acid . . . 361
9.4.6 Reduction of CO_2 to Carbon Monoxide or Formate . . . 362
9.4.7 Bioconversion of Carbon Dioxide into Methanol . . . 364
References . . . 366

10 Thermodynamics and Applications of CO_2 Hydrates . . . 373
10.1 Introduction . . . 373
10.2 Structure of Gas Hydrates . . . 376
10.2.1 Formation of Gas Hydrates from a Microscopic Perspective . . . 376
10.2.2 Crystal Structures of Gas Hydrates . . . 378
10.2.3 Characteristics of CO_2 Hydrates . . . 381
10.3 Physical Properties of CO_2 Hydrates . . . 382
10.3.1 Mechanical Properties . . . 384
10.3.2 Thermal Properties . . . 384
10.3.3 Other Physical Properties . . . 386
10.4 Phase Equilibrium of CO_2 Hydrate . . . 386

10.4.1 Experimental Methods to Study Hydrate Phase Equilibria . . . 386
10.4.2 Pressure–Temperature Phase Diagram of $CO_2 + H_2O$ System . . . 389
10.5 Applications of CO_2 Hydrates . . . 390
10.5.1 Formation of CO_2 Hydrate . . . 390
10.5.2 Dissociation of CO_2 Hydrate . . . 391
10.5.3 CO_2 Capture and Sequestration . . . 392
10.5.4 Replacement of CH_4 by CO_2 in Naturally Occurring Hydrates . . . 395
10.5.5 CO_2 Hydrates in Refrigeration Processes . . . 396
References . . . 399

Index . . . 403

About the Authors

Michele Aresta MRSC, Doctor Industrial Chemistry, University of Milan, Milan-IT; President of the Scientific Council of the Interuniversity Consortium on Chemical Reactivity and Catalysis, Italy; IMM chair at NUS Department of Chemical and Biomolecular Engineering, Singapore; Honorary Chair at the Chemical Engineering Faculty, University of Tianjin, Tianjin, China.

His scientific interests are in the fields of: Chemistry of small molecules (N_2, H_2, O_2, CO, CO_2, ethene, allene, propene, butadiene); Metals in low oxidation state (also negative); Co-ordination chemistry; Reactivity of co-ordinated molecules; Reaction mechanisms; Catalysis; CO_2 chemistry; CO_2 as building blocks for chemicals and carbon source for fuels; Photocatalysis; Utilization of biomass; Conversion of polyols; Conversion of long chain acids and esters; Polyunsaturated compounds; Valorization of polyols; New bio-sourced monomers for polymeric materials. Author of over 250 papers in international journals and of 10 books on CO_2 and biomass valorization; owner of several patents.

Founder and Honorary Chair of the International Conference on Carbon Dioxide Utilization-ICCDU.

Award of the Italian Chemical Society for "Pioneering work on carbon dioxide activation (and utilization)" 1990.

Renoir Award for the Dissemination of Science, 1990.

Award of the Société Française de Chimie, 1995 for Inorganic Chemistry.

Award of the Tianjin University, Tianjin China, for Green Chemistry 2002.

Several recognitions from the American Chemical Society for dissemination of "CO_2 Utilization".

Angela Dibenedetto Associate Professor at the University of Bari-UNIBA (IT)—Department of Chemistry. Her scientific interests are focused on carbon dioxide utilization in synthetic chemistry, catalysis, coordination chemistry and organometallic chemistry, green chemistry, marine biomass (algae) production by enhanced carbon dioxide fixation, marine biomass as source of fuels and chemicals applying the biorefinery concept.

She is director of the Interuniversity Consortium on Chemical Reactivity and Catalysis-CIRCC. In 2001 she was the winner of the RUCADI Prize for "Better Carbon Management—An Intelligent Chemical Use of CO_2" delivered by ACP-Belgium, Carburos Metalicos-Spain and ENIChem-Italy. Author of over 90 scientific papers on carbon dioxide utilization published in international journals since 1995 and several book chapters. She was invited speaker at several international conferences.

Eugenio Quaranta Graduated in Chemistry in 1983, received his Ph.D. in Chemistry in 1990 presenting a final dissertation entitled "Utilization of carbon dioxide in carboxylation reactions: the role of metal systems". Researcher at the University of Bari from 1990 to 1998, at present he is Associate Professor of General and Inorganic Chemistry at the Dipartimento di Chimica of the University of Bari. His main fields of research are coordination chemistry, organometallic chemistry, catalysis; activation of small molecules (CO_2, CO, CH_2O, C_2H_4, H_2, dimethyl carbonate, etc.); utilization of carbon dioxide in chemical synthesis; eco-sustainable synthetic methodologies.

Chapter 1
The Carbon Dioxide Molecule

Abstract The basic aspects of the reactivity of carbon dioxide (CO_2) are featured in this chapter and related to the electronic structure of the molecule. The electronic properties of neutral CO_2 are compared with those of the radical ions $CO_2^{\cdot-}$ and $CO_2^{\cdot+}$. The potential of a few spectroscopic techniques (infrared, ultraviolet–visible, nuclear magnetic resonance) in the characterization of CO_2 states is also highlighted.

1.1 Introduction

The carbon dioxide (CO_2) molecule is a relatively simple molecular system. Nevertheless, much of the life on Earth, as well as the fate of our planet itself, depends heavily on the properties of this small triatomic C_1 molecule.

Large amounts of CO_2 (~700 gigatonnes/year), are involved in the natural C-cycle, which includes the production of biomass by photosynthesis, biological respiration, carbonization processes, and so on. CO_2 is the waste end product of several human activities (e.g., fossil combustion, cement manufacture, fermentation, industrial processes). Since the beginning of the industrial revolution, in the second half of the nineteenth century, anthropogenic emissions of CO_2 have been growing continuously, reaching the level of more than 35 gigatonnes/year. Together with the indiscriminate deforestation of large areas of land, this has brought about a continuous increase of CO_2 concentration in the atmosphere, which is known to contribute severely to climate change. Much effort is currently being devoted to setting up efficient strategies to reduce emission/accumulation of CO_2 in the atmosphere. CO_2 chemical utilization (CCU), which is aimed at converting the heterocumulene into fuels or other valuable products, is currently under assessment as a sustainable technology for reducing waste and making better use of both carbon and energy.

There is, however, a net dichotomy between the sophisticated facilities exploited by Nature to convert CO_2 into valuable chemicals and the difficulties experienced by chemists in converting the heterocumulene through efficient catalytic processes. CO_2 reduction, or, more generally, CCU, is a challenging task and not only mere

M. Aresta et al., *Reaction Mechanisms in Carbon Dioxide Conversion*,
DOI 10.1007/978-3-662-46831-9_1

chemical routes but also electrochemical, photochemical, and photo-electrochemical approaches are currently under investigation. A great deal of effort is also addressed to converting this molecule by trying to mimic Nature.

In tackling the problem of CCU, it is of fundamental importance to have a clear view of the basic features of the reactivity of this molecule, which is strictly related to its electronic configuration. This chapter confronts this issue.

1.2 Electronic Properties of CO_2

1.2.1 Ground State of Carbon Dioxide

In its electronic ground state, the CO_2 molecule is linear and belongs to the point group $D_{\infty h}$. Both carbon–oxygen bonds are equivalent with the equilibrium C–O distance of 1.1600 Å, as obtained from an analysis of infrared (IR) spectra and confirmed by electron diffraction [1, 2]. Both carbon–oxygen bonds are polar as a result of the presence of net partial charges on the carbon and oxygen atoms [3, 4]. However, the molecule has no permanent electric dipole. In fact, the dipole moments associated with each of the C–O bonds cancel each other because they have equal magnitude but opposite directions because of the linear molecular geometry (Fig. 1.1).

The molecule has an electric quadrupole [5], whose moment is -4.3×10^{-26} esu/cm^2 [6] and thus exhibits significant intermolecular interactions, which may account for the formation of neutral aggregates. Neutral clusters of $(CO_2)_n$ $(2 \le n \le 5)$ have been produced in a molecular beam expansion system and observed via mass spectrometry [7, 8]. The intermolecular interactions can become even stronger when a cationic or anionic CO_2 molecule interacts with other CO_2 molecules, affording positively or negatively charged intermolecular aggregates, which may be regarded, respectively, as models for solvated CO_2^{+} radical cations or solvated CO_2^{-} radical anions [7–11]. Positively charged aggregates of $(CO_2)_n^{+}$ $(2 \le n \le 10)$ have been generated by nucleation of neutral CO_2 molecules to CO_2^{+} radical cation in a superionic expansion [9]. The stability of these clusters is significantly higher than that of neutral aggregates. The anionic clusters are also rather stable [10]. In contrast to the CO_2^{-} radical anion, which is unstable with respect to electron detachment (see Sect. 1.4), the dimeric species $(CO_2)_2^{-}$ is stable by 0.9 eV, and stability increases with increasing cluster size.

The electronic configuration of the linear ground state, ${}^1\Sigma_g{}^+$, of CO_2 is as shown below.

$$
\begin{aligned}
&{}^1\Sigma_g^+\ (\text{ground state}):\ 1\sigma_u^2\ 1\sigma_g^2(-541.1\ \text{eV})\ 2\sigma_g^2(-297.5\ \text{eV})\ 3\sigma_g^2(-37.6\ \text{eV})\\
&2\sigma_u^2(-37.6\ \text{eV})\quad 4\sigma_g^2(-19.4\ \text{eV})\ 3\sigma_u^2(-18.1\ \text{eV})\ 1\pi_u^4(-17.6\ \text{eV})\ 1\pi_g^4(-13.8\ \text{eV})
\end{aligned}
$$

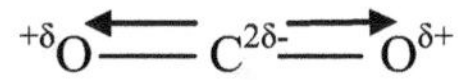

Fig. 1.1 Dipole moments in the carbon dioxide molecule

The values in parentheses provide an approximate estimate of CO_2 molecular orbital (MO) energies evaluated by measuring the ESCA (Electron Spectroscopy for Chemical Analysis) ionization energies for the molecule [12–14]. The ESCA spectrum does not allow one to measure any energy splitting for the molecular orbitals $3\sigma_g$ and $2\sigma_u$ because of the inherent broadening of these levels. Nevertheless, calculations clearly indicate that $3\sigma_g$ has a lower energy than $2\sigma_u$ [3, 4, 15–18]. Both $3\sigma_u$ and $1\pi_u$ MOs are very close in energy, with the $3\sigma_u$ MO slightly more stable than the $1\pi_u$. This conclusion is also supported by several computational studies. In general, the calculated values of orbital energies are usually affected by the computational approach used. Using more extended basis sets in calculations allows one to reproduce the correct energy ordering of the $3\sigma_u$ and $1\pi_u$ levels [3, 4, 15–18].

Figure 1.2 shows a qualitative energy diagram for the valence shell MOs of the CO_2 molecule. The atomic orbitals (AOs) on oxygens and carbon, which are considered in bond formation, are $2s_O$ (-32.4 eV) and $2p_O$ (-15.9 eV), and $2s_C$ (-19.4 eV) and $2p_C$ (-10.7 eV), respectively. The K-shell orbitals of oxygens and carbon are not taken into account because they lie very low on the energy scale: they remain basically unchanged in the molecule and become $1\sigma_u$, $1\sigma_g$, and $2\sigma_g$. In the linear molecule six σ molecular orbitals are generated by combining six atomic orbitals, consisting of 2s and $2p_z$ (directed along the internuclear axis) of all three atoms. These MOs are marked in Fig. 1.2 as $3\sigma_g$, $4\sigma_g$, $5\sigma_g$, $2\sigma_u$, $3\sigma_u$, and $4\sigma_u$. The remaining $2p_x$ and $2p_y$ orbitals form three doubly degenerate π-orbitals denoted as $1\pi_u$, $1\pi_g$, and $2\pi_u$.

Schematic and simplified representations of the CO_2 molecular orbitals are shown in Fig. 1.3. Figure 1.4 illustrates qualitatively the variation of the MO energies with the change of OCO bond angle (diagram of Walsh [19]) and the correlation of symmetry representations of the molecular orbitals of CO_2 among the point groups associated with the linear ($D_{\infty h}$) and bent (C_{2v}) configuration.

The main contributors to the energy lowest valence molecular orbital, $3\sigma_g(3a_1)$, are the $2s_O$ and $2s_C$ atomic orbitals, with a minor contribution from the $2p_{zO}$ AOs. The $2\sigma_u(2b_2)$ MO is mainly a bonding combination of $2s_O$ AOs and the $2p_{zC}$ AO. This molecular orbital presents a nodal plane passing through the central carbon and perpendicular to the molecular axis. The $4\sigma_g(4a_1)$ and $3\sigma_u(3b_2)$ MOs contain C–O antibonding combinations of $2s_O$ with $2s_C$ and $2p_{zC}$, respectively, but they are compensated by strong admixture of bonding $2p_{zO}$ states (see Fig. 1.3).

The doubly degenerate $1\pi_{ux}(1b_1)$ and $1\pi_{uy}(5a_1)$ orbitals are, respectively, bonding combinations of $2p_x$ and $2p_y$ states on all the three atoms. The $1\pi_{ux}(1b_1)$ MO is equivalent to the $1\pi_{uy}(5a_1)$ orbital in the linear configuration, but it is rotated by 90° about the internuclear axis. In the linear configuration the doubly degenerate

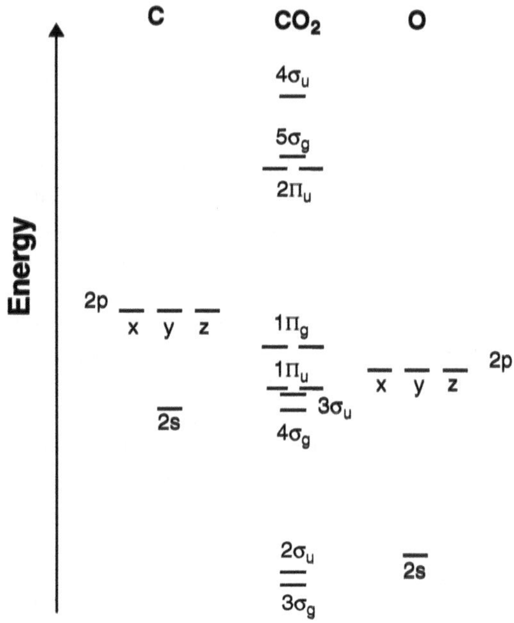

Fig. 1.2 Molecular orbital diagram of carbon dioxide

$1\pi_{gx}(1a_2)$ and $1\pi_{gy}(4b_2)$ molecular orbitals are equivalent and reciprocally rotated by 90° about the molecular axis. They have no weight on the carbon atom, and play the role of lone pairs. The $1\pi_{gx}(1a_2)$ MO consists of an antibonding combination of $2p_{xO}$ AOs, whereas the $1\pi_{gy}(4b_2)$ MO is an antibonding combination of $2p_{yO}$ AOs.

The first unoccupied molecular orbitals are the two $2\pi_u$ MOs. Both of them are degenerate and equivalent in the linear configuration. The $2\pi_{ux}(2b_1)$ MO has mainly a $2p_{xC}$ character with a little contribution from the $2p_{xO}$ AOs. Besides a nodal plane passing through the internuclear axis, this molecular orbital has nodal planes intersecting each CO bond, because the starting atomic orbitals are combined in an antibonding manner. The $2\pi_{uy}(6a_1)$ MO is equivalent to the $1\pi_u(1b_1)$ MO in the linear configuration, but it involves the $2p_y$ AOs of all the three atoms.

Both the $5\sigma_g(7a_1)$ and $4\sigma_u(5b_2)$ MOs are strongly antibonding. The $5\sigma_g(7a_1)$ MO is made up of $2s_C$ AO and $2s_O$ and $2p_{zO}$ AOs combined in an antibonding manner and shows four nodal planes perpendicular to the internuclear axis, which intersect the two oxygen atoms and the two C–O bonds. The $4\sigma_u$ $(5b_2)$ MO consists of $2p_z$ AOs from all three atoms with a $2s_O$ AO contribution.

The 16 valence electrons of CO_2 are distributed over the four σ orbitals (8 electrons) $3\sigma_g$, $2\sigma_u$, $4\sigma_g$, and $3\sigma_u$ and the four π-orbitals (8 electrons) $1\pi_u$ and $1\pi_g$. Of the eight occupied molecular orbitals, both $3\sigma_g$ and $2\sigma_u$ are responsible for the σ-bonding skeleton of the molecule, whereas the $1\pi_u$ orbitals account for the C–O π-bonds. By contrast, the electrons $4\sigma_g$, $3\sigma_u$, and $1\pi_g$ are, respectively, σ and π lone pairs. This interpretation agrees with a few features of the ESCA spectrum of CO_2. In fact, the fairly sharp nature of the bands at 13.8 ($1\pi_g$), 18.1 ($3\sigma_u$), and 19.4 eV

Fig. 1.3 Molecular orbitals of CO_2 (only $1\pi_{uy}$, $1\pi_{gy}$, and $2\pi_{uy}$ are shown)

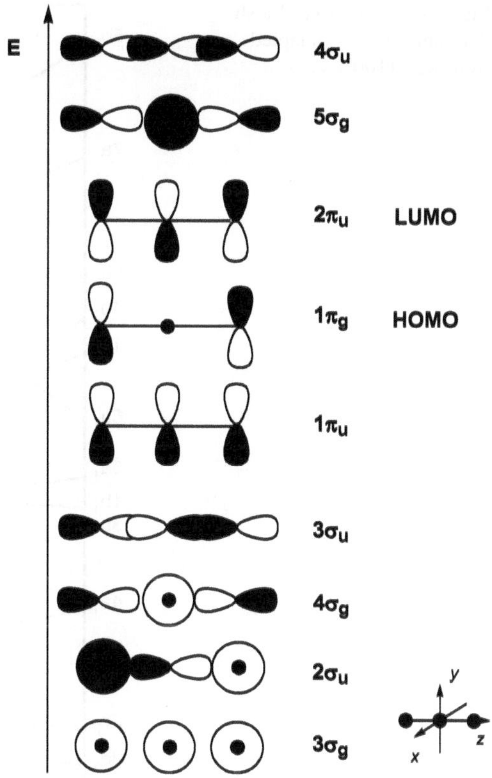

($4\sigma_g$) in the ESCA spectrum [12–14] is consistent with their assignment to electrons largely localized on the end oxygen atoms. Conversely, the level associated with the band at 17.6 eV ($1\pi_u$), which is broad and exhibits a fine vibrational structure, has apparently a bonding character.

Figure 1.4 shows that the energies of both occupied and unoccupied molecular orbitals may undergo significant alterations upon bending the molecule from linear geometry. In general, the energies of the σ-orbitals increase or are almost unaffected upon the change of geometry from linear to bent. However, the behavior of the π-orbitals is somewhat more diversified and their energies exhibit more significant changes. The degeneracy of the π-orbitals is lost in the bent geometry and each couple of formerly equivalent and degenerate π-orbitals splits in two non-equivalent orbitals with different energies. The splitting is relatively modest for both $1\pi_u$ and $1\pi_g$. In the case of $1\pi_u$ orbitals, the destabilization of $1\pi_u(5a_1)$ is roughly compensated by the stabilization of $1\pi_u(1b_1)$. By contrast, in the case of $1\pi_g$ orbitals, the energies of both components increase with diminution of the bond angle. The energy increase upon bending is much faster for $1\pi_g(4b_2)$ than for $1\pi_g(1a_2)$. The rapid increase of the energy of $1\pi_g(4b_2)$ upon molecule bending accounts for the linearity of the CO_2 molecule in the ground state.

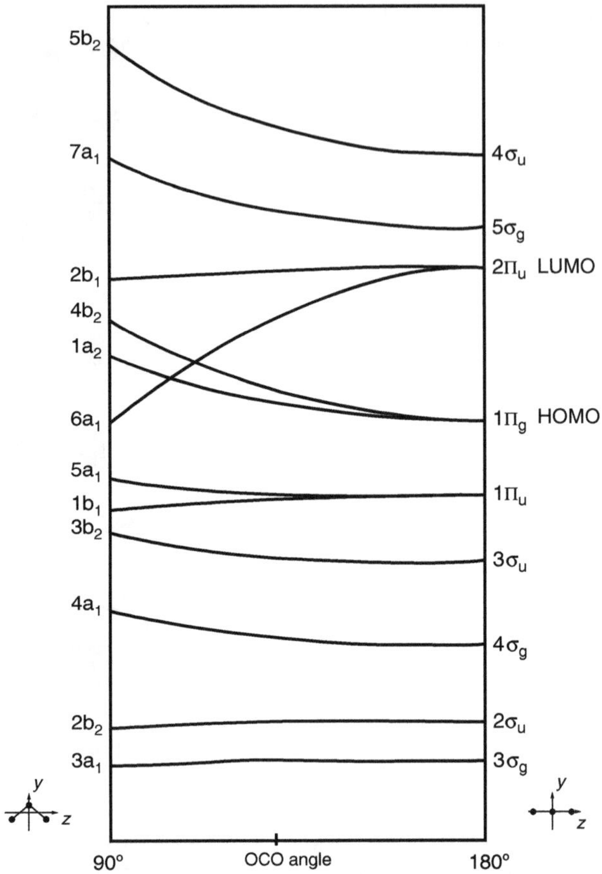

Fig. 1.4 Qualitative Walsh diagram for CO_2 (adapted from Ref. [100])

As for the $2\pi_u$ MOs, the energy of the component $2\pi_u(2b_1)$ is only negligibly affected by the extent of bending. However, bending causes a net decrease in the energy of the $2\pi_u(6a_1)$ MO, which may become even more stable than $1\pi_g(4b_2)$. The behavior of the $2\pi_u(6a_1)$ MO upon molecule bending is particularly intriguing as it suggests that *any event which enable the $2\pi_u(6a_1)$ level to be populated is also expected to produce a distortion of CO_2 molecule from linearity.*

1.2.2 Lowest Excited States of Carbon Dioxide

The above statement finds a first validation when considering the geometry of the lowest excited states of the CO_2 molecule [19–27]. Inspection of Fig. 1.2 shows that the molecular orbital excitation of lowest energy for the CO_2 molecule involves the promotion of one $1\pi_g$ electron into the $2\pi_u$ MO. The resulting excited electronic configuration, $1\sigma_u^2\ 1\sigma_g^2\ 2\sigma_g^2\ 3\sigma_g^2\ 2\sigma_u^2\ 4\sigma_g^2\ 3\sigma_u^2\ 1\pi_u^4\ 1\pi_g^3\ 2\pi_u^1$, gives rise to six

Table 1.1 Correlation of electronic state representations between $D_{\infty h}$ and C_{2v} point groups

Point group	Ground state	$(1\pi_g \rightarrow 2\pi_u)$ transition states			
$D_{\infty h}$	$^1\Sigma_g^+$	$^{1,3}\Sigma_u^-$	$^{1,3}\Delta_u$ [a]		$^{1,3}\Sigma_u^+$
C_{2v}	$^1A_1$	$^{1,3}A_2$	$^{1,3}A_2$	$^{1,3}B_2$	$^{1,3}B_2$

[a]The double degenerate Δ_u state splits into two components (A_2 and B_2) in the point group of lower symmetry (C_{2v})

Table 1.2 Structural parameters of bent valence excited states of CO_2

Structural parameter	Calculated values[a]				Experimental values		
	$^3A_2$	$^1A_2$	$^3B_2$	$^1B_2$	$^1A_2$ [b]	$^3B_2$	$^1B_2$ [c]
R_e(CO)/Å	1.261	1.262	1.251	1.260	1.262 ± 0.010	1.249(2)[d] 1.237(2)[e]	1.246 ± 0.008
α_e(OCO)/deg	127.4	127.0	118.5	117.8	129 ± 1	148.2(2)[d] 152.2(2)[e]	122 ± 2

[a][21]
[b][22]. Values for the longest wavelength band analyzed (see main text)
[c][25]
[d][27]. Band at 185.7 nm
[e][27]. Band at 175.6 nm

excited states in the linear configuration, $^3\Sigma_u^-$, $^3\Delta_u$, $^3\Sigma_u^+$, $^1\Sigma_u^-$, $^1\Delta_u$, and $^1\Sigma_u^+$, which correlate with A_2 and/or B_2 states in the point group of lower symmetry, C_{2v}, resulting from bending the molecule (Table 1.1).

According to the Walsh diagram of the molecule (Fig. 1.4), a few of the above excited states (A_2, B_2) are expected to be strongly bent, whereas the others are expected to be slightly bent or linear [19]. Spielfieldel et al. [21] studied the lowest lying valence excited states of CO_2 by ab initio calculations, and found that the electronic energy depended critically on the geometric parameters of the nuclear configuration, such as the C–O bond distances and OCO bond angle. Using large scale multireference configuration interaction, the ordering of the lowest valence excited states of CO_2 was calculated to be $^3B_2$, $^3A_2$, $^1A_2$, followed by $^1B_2$. These states have bent equilibrium structures. The calculated structural parameters of the bent nuclear configurations characterizing the lowest excited states of CO_2 differed markedly from those of the electronic ground state, $^1\Sigma_g^+$, and are summarized in Table 1.2 together with the available experimental data. According to the calculations, the lowest $^{1,3}A_2$ and $^{1,3}B_2$ states have much longer C–O equilibrium distances (around 1.26 Å) than the electronic ground state (1.16 Å). Moreover, regardless of the spin multiplicity, the B_2 states (OCO bond angle ≈ 118°) were found to be markedly more bent than the A_2 states (OCO bond angle ≈ 127°).

Dixon [25], in 1963, identified the chemiluminescence observed in the CO flame to be fluorescence from a bent $^1B_2$ state correlated to the $^1\Delta_u$ state. The analysis of the emission bands allowed one to measure the experimental excitation energy (5.7 eV) and to determine the structural parameters of the bent configuration (R (CO) = 1.25 Å; OCO angle = 122°). The values computed by Spielfieldel for the

1^1B_2 state ($E = 5.74$ eV, R(CO) = 1.26 Å, OCO angle = 118°; Table 1.2) compare well with the experimental results obtained by Dixon, fully supporting the Dixon assignment.

Cossart-Magos et al. [22] analyzed the high resolution absorption spectrum of gaseous CO_2 between 6.2 and 7.1 eV (200–175 nm) and were able to show that this part of spectrum is rich in very weak bands with a remarkable rotational structure. They achieved a coherent analysis of nine perpendicular-type bands defining an excited state bending progression and identified a bent excited valence state, different from that characterized by Dixon (1^1B_2), which was characterized as a $^1A_2$ state, rendered vibronically allowed by excitation of one quantum of the asymmetric stretching $\nu_3(b_2)$ vibration (Sect. 1.6.1). This state, which correlates with the $^1\Sigma_u^-$ state of the linear molecule, was the first with this symmetry. The structural parameters for the latter excited state were deduced from spectrum: the C–O bond length decreased from about 1.26 Å to 1.22 Å, whereas the OCO bond angle increased from about 129° to 138° depending on the wavelength of the analyzed band. These results are consistent with the R(CO) and the O$\hat{C}$O values of 1.262 Å and 127° calculated by Spielfieldel for the equilibrium geometry of the 1^1A_2 state. The excitation energy for the $^1A_2$ state (5.61 eV) calculated by Spielfieldel was somewhat lower than the energy region where the $^1A_2$ state was detected, probably because the absorption spectrum analyzed by Cossart-Magos et al. in the region of 6.2–7.2 eV involved transitions into highly excited vibrational states.

Mohammed et al. [26], studying the continuum emitted in the spectral region between 350 and 550 nm by CO_2 molecules isolated in an Ar matrix at 4.7 K and excited at 147 or 123.6 nm, assigned the continuum to the emission from the lowest excited triplet state $^3B_2$ of the bent CO_2 molecule to the linear ground state. They located the lowest triplet state of CO_2 at 3.5 ± 0.2 eV and measured its lifetime, which was around 570 ms. The experimental value of 3.5 ± 0.2 eV was close to that (3.7 eV) calculated by Winter et al. [24] for the adiabatic excitation energy to the lowest triplet state $^3\Sigma_u^+(^3B_2)$ of CO_2. In another of their studies, Cossart-Magos et al. [27] reinvestigated the weak absorption spectrum of CO_2 gas between 175 and 200 nm (see above) and focused on the parallel-type bands observed at 185.7 and 175.6 nm. The rotational structure of both bands was found to be consistent with the assignment of both absorptions to the lowest triplet–singlet transition, $^3B_2 \leftarrow {}^1\Sigma_g^+$, excited with different quanta of the bending $\nu_2(a_1)$ vibration. Table 1.2 gives, for the 1^3B_2 state, the values obtained for the R(CO) bond length and OCO bond angle, as deduced by the values of the rotational constants.

1.3 Main Features of Carbon Dioxide Reactivity

The molecular orbitals most relevant to the chemical reactivity of carbon dioxide are the $1\pi_g$ and $2\pi_u$ orbitals, which play the role of HOMOs and LUMOs, respectively. Although the doubly occupied non-bonding $1\pi_g$ MOs are mainly localized on the end oxygen atoms, the empty antibonding $2\pi_u$ orbitals are mostly centered on the central carbon. CO_2, therefore, is an amphoteric oxide, as, in principle, the oxygen atoms can exhibit a Lewis base character, whereas the carbon atom can play the role of a Lewis acid center. However, carbon dioxide is a better acceptor than donor of electron density and, consequently, the reactivity of the molecule is dominated by the electrophilic character of carbon rather than the weak nucleophilic properties of the oxygen atoms.

1.3.1 Carbon Dioxide as O-Nucleophile

A measure of the weak basic character of the oxygen atoms of CO_2 is provided by the value of proton affinity of this molecule. The determination of this quantity for CO_2 has been the subject of both experimental measurements and theoretical calculations [28–37]. The measured values range around the recommended value of 540.5 ± 2 kJ/mol [36], is appreciably lower than those found for other O-containing molecules, such as, for instance, H_2O, MeOH, Me_2O, H_2CO, MeCHO, Me_2CO, HCO_2H, $MeCO_2H$ [31, 35, 36].

The hydroxycarbonyl cation $HOCO^+$ (1.1) is certainly the simplest adduct in which CO_2 acts as an O-nucleophile. This "nonterrestrial" molecule is thought to be an important intermediate species in

$$CO_2 + H^+ \longrightarrow HOCO^+ \qquad (1.1)$$

gas phase reactions in interstellar clouds and space [38]. Submillimeter wave spectroscopy [39, 40] and infrared spectroscopy studies [41, 42] have definitively shown that the interstellar lines observed in 1981 by Thaddeus et al. [43] in the 85 GHz region, and tentatively assigned to either $HOCO^+$ or HOCN, belong to $HOCO^+$. By using different spectroscopic techniques, this molecule was shown to be formed in ion-molecule reactions by proton transfer between CO_2 and H_3^+, NH_2^+, or HBr^+ [44, 45].

The rotational spectra of $HOCO^+$ support the hypothesis of a planar structure but do not allow one to determine, experimentally, a precise and reliable structure for this molecule and, more specifically, to distinguish between a structure **I** (H–$O_1 = 0.9715$ Å, O_1–C = 1.2057 Å, C–O_2 = 1.1409 Å, $O_1\hat{C}O_2 = 178.83°$, $H\hat{O}_1C = 122.25°$) with a *cis* conformation and a quasi-linear heavy-atom chain, and a structure **II** (H–$O_1 = 0.9766$ Å, O_1–C = 1.2085 Å, C–O_2 = 1.1400 Å, $O_1\hat{C}O_2 = 174.39°$, $H\hat{O}_1C = 119.38°$) with a *trans* conformation and a small bend at the C atom [40]. Theoretical investigations agree on the fact that the OCO backbone of this molecule is slightly bent, with a *trans* configuration and the C–O and C–OH bond distances being, respectively, shorter and longer than what was observed in the non-protonated molecule [30–32, 46]. The equilibrium values for the OCO and COH angles, as recently computed by Hammani et al. [46] using the coupled electron pair approximation (CEPA) method with a correlation consistent basis set, are respectively equal to 174.3° and 117.6°, whereas the O–C, C–OH, and O–H bond lengths are, respectively, 1.124, 1.229, and 0.984 Å. The positive charge is mainly localized on the carbon and hydrogen atoms with the following repartition: C, +0.584; H, +0.350; O, +0.012; (H)O, +0.054. Another recent computational study [32], focusing on the protonation reaction of allenes and some heteroallenes, has shown that end-on protonation of CO_2 is by far more energetically favored than proton attack at the central carbon. Notably, carbon protonation of the heterocumulene does not result in the formation of open forms, but affords a cyclic cation with a very weak O–O (1.733 Å) bond and a very high positive charge (+0.89) at the central CH unit.

In Chap. 2 several examples of O-coordinated CO_2 to metal centers are discussed. A peculiar feature relates the adducts of CO_2 to other moieties (either metal centers or other molecules or ions) and deserves attention. In fact, the available structural data show unambiguously that, whenever CO_2 behaves exclusively as an O-nucleophile, the coordinated CO_2 molecule essentially retains its original linear geometry or undergoes only slight distortion from linearity. This suggests that the interaction of CO_2 with an electrophilic center through one of the $1\pi_g$ lone pairs is not accompanied by any significant back-donation of electron density into the $2\pi_u$ orbitals of the heterocumulene, which, therefore, are left empty.

End-on (η^1-OCO) coordination mode has remained elusive for a long time. Nevertheless, this coordination mode may play an important role in biological systems. In photosynthetic CO_2 fixation, an oxygen-bound CO_2 ligand has been proposed to be enzymatically reduced by ribulose-1,5-biphosphate carboxylase-oxygenase (RuBisCO) [47]. A crystallographic study on a deacetoxycephalosporin C synthase (DAOCS) mutant showed the presence of electron density close to the iron center of the active site, which was found to be consistent with the presence of a monodentate O-coordinated CO_2 molecule [48].

1.3.2 Carbon Dioxide as C-Electrophile

η^1-C coordination of CO_2 to metal centers has been clearly shown and is discussed in Sect. 2.3. The structural data of the adducts indicate an important pouring of electron density from metal filled d-orbitals into the empty antibonding $2\pi_u(6a_1)$ orbital, which weakens the C–O bonds and, as expected from the Walsh diagram (Fig. 1.4), causes a remarkable bending of the coordinated CO_2 molecule. Analogous structural changes always mark the geometry of the CO_2 molecule whenever the electrophilic carbon atom of the heterocumulene is involved in an interaction with electron-rich species (amines, alcohols, carbanions, etc.).

This interaction also affects the nucleophilicity of the O atoms of CO_2. For instance, in CO_2-adducts with amines or alcohols, the resulting carbamate or

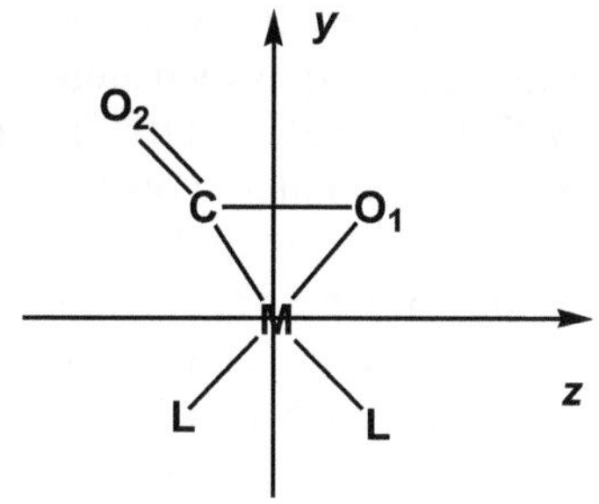

Fig. 1.5 Side-on coordination of CO_2 to a metal center

carbonate anions, under suitable conditions, can easily react with C-electrophiles to give organic carbamates [49, 50] or carbonates [51], respectively (see Chap. 4).

1.3.3 Amphoteric Reactivity of Carbon Dioxide

Most frequently, however, the reactivity of CO_2 fully reveals the intrinsically amphoteric nature of the molecule. Side-on (η^2-C,O) coordination of CO_2 to metal complexes (Fig. 1.5) clearly exemplifies this behavior. Different schemes can be proposed for describing the metal–CO_2 ligand bond in these systems. These adducts may be considered as three-membered oxametallacycles resulting from a formal oxidative addition of one of the π-bonds of the cumulene to the metal center: in this case, both the metal center, which increases by two units its formal oxidation state, and CO_2 ligand act simultaneously as both electron acceptor and electron donor. Otherwise, (η^2-C,O)-adducts can be regarded, more realistically, as π-complexes, wherein the metal–CO_2 interaction is reminiscent of that characterizing the metal–olefin bond and involves both electron donation from the O-centered $1\pi_{uy}(5a_1)$ and $1\pi_{gy}(4b_2)$ orbitals of CO_2 to empty d-orbitals of the metal center and, to a greater extent, electron back-donation from a filled d orbital of metal to the empty carbon centered $2\pi_{uy}(6a_1)$ orbital of the ligand [52]. Whatever model may be followed, in these systems the CO_2 ligand always acts simultaneously as electron donor through an end-oxygen and electron acceptor at the central carbon. Several examples are discussed in Chap. 2.

It may be worth mentioning the behavior of solid CO_2 under very high pressures [53–59]. CO_2 is a very stable compound, which, in the solid state, can exist in different molecular phases (CO_2-I (dry-ice) to CO_2-IV, and CO_2-VII) depending on pressure and temperature [53]. At very high pressure, the π molecular bond is replaced by an extended network of C–O single bonds. In 1999, a few experiments showed that CO_2-III, confined inside a diamond anvil cell under 40 GPa pressure and subjected to laser heating (ca. 1800 K), was transformed into a new crystalline "super-hard" phase, called CO_2-V [55], which can be quenched at ambient temperature. The crystal structure of CO_2-V was found to be orthorhombic ($P2_12_12_1$), analogous to that of SiO_2 tridymite (a distorted high temperature phase of β-quartz)

[56]. The new phase is composed of CO_4 tetrahedra, where each carbon atom is bonded to four oxygen atoms at a carbon–oxygen distance of 1.36 Å at 40 GPa and an O–C–O angle of 110°. The C–O–C angle of 130° is markedly smaller than those of SiO_2 tridymites (174–180°) or quartz (145°). CO_2-V is a "super-hard" polymeric form of CO_2. Accordingly, the bulk modulus for CO_2-V (362 GPa) is higher than for any of the SiO_2 polymorphs and very close to that of cubic-BN (369 GPa).

More recently, the synthesis of an amorphous, silica-like form of carbon dioxide (a-CO_2), which is called "a-carbonia" [57], has been reported. This material is homologous to other Group 14 dioxide glasses (a-SiO_2; a-GeO_2). Both CO_2-V and a-CO_2 convert back to the molecular state when pressure and temperature are brought back to ambient values.

Recent results suggest that carbon dioxide polymerization does not occur via intermediate states where molecules gradually distort as pressure increases, but is most likely caused by solid-state chemical reactions between CO_2 molecules [59]. Such studies are important to gain better insight into transformations of light element-based molecules under high pressure and temperature and may open up new horizons in solid state chemistry under extreme conditions.

1.4 Carbon Dioxide Radical Anion, CO_2^-

Coordination of carbon dioxide to a metal center results generally in marked modification of the bond properties (CO bond lengths, OCO angle, etc.) of CO_2 ligand and provides the heterocumulene in an "activated form" (Chap. 2). Conversion of CO_2 to CO_2^- anion is another important mode of activating carbon dioxide molecule, and represents the first of the reactive steps leading to the partial or full reduction of the heterocumulene (Chap. 8).

Carbon dioxide anion is a radical species, isoelectronic with NO_2. Qualitative arguments, such as those based on the Walsh rules [19] (see Sect. 1.2.1), predict a bent equilibrium geometry (C_{2v} symmetry) with elongated CO bonds for CO_2^- in the $^2A_1$ ground state, which may be expressed as $(1a_1)^2$ $(1b_2)^2$ $(2a_1)^2$ $(3a_1)^2$ $(2b_2)^2$ $(4a_1)^2$ $(3b_2)^2$ $(1b_1)^2$ $(5a_1)^2$ $(1a_2)^2$ $(4b_2)^2$ $(6a_1)^1$ (see also Fig. 1.4). This is supported also by quantum-mechanical studies [60–67]. According to a recent theoretical analysis at B3LYP6-31-G**//B3LYP/6-31G* level [65], the charge density distribution of CO_2^-, from a natural population analysis (NPA), shows negative character for the two O atoms ($-0.76e$) and a positive charge on the C atom ($+0.52e$). Spin (densities) populations indicate a 68 % electron delocalization on the C atom and 16 % on each oxygen. CO_2^- is, therefore, a radical-like species at carbon and basic at oxygen. Accordingly, calculations also show that O-protonation of the radical anion, which affords the hydroxycarbonyl radical OCOH, is favored over C-protonation, which should result in the formation of the less stable formiloxy radical HCOO [60, 66]. These studies also show that *trans*-OCOH is more stable than the *cis*-isomer and barely bound with respect to the H + CO_2 asymptote [60, 66, 67].

In the gas phase, CO_2^- is metastable against electron autodetachment (1.2). The existence of

$$CO_2^- \longrightarrow CO_2 + e^- \quad (1.2)$$

metastable autodetaching CO_2^- anion in the gas phase was first noted by Paulson [68] by studying the reactions of the O^- ion with CO_2. Afterwards, Cooper and Compton [69, 70] observed the formation of metastable, autodetaching CO_2^- ions in collisions of electrons or Cs atoms with organic molecules, such as cyclic anhydrides (maleic anhydride, succinic anhydride), which contained nonlinear OCO units. The lifetimes of the generated species, measured with a time-of-flight spectrometer, were in the range 30–60 μs. A subsequent study on CO_2^- ions formed in collisions of alkali metal atoms with linear CO_2 molecules resulted in a mean lifetime of 90 ± 20 μs and a value of -0.6 ± 0.2 eV for the adiabatic electron affinity (EA_{ad}) of the CO_2 ground state [71]. The measured value for EA_{ad} agrees well with the value of −0.67 eV computed by Gutsev et al. [4] and should be compared with the markedly higher value of the vertical electron affinity (−3.6 eV, as determined by means of electron scattering measurements) [72]. The large difference between the two values can be understood from the large stabilization of CO_2 upon bending. The reason for the relatively high stability of the anion is ascribed to the barrier caused by the change in molecular geometry on going from the $^2A_1$ state of CO_2^- to the ground state of CO_2, which has been calculated by Pacanski et al. to be 0.4 eV [61]. Therefore the CO_2^- molecule in its equilibrium geometry is metastable, being stabilized kinetically.

Ovenall and Whiffen [73] first produced CO_2^- in solid matrix by irradiation of sodium formate crystals with γ-rays at room temperature. Their ESR experiments allowed them to ascertain that the CO_2^- anion is bent with a bond angle of 134° and has a $^2A_1$ ground state. From the experimental values of ^{13}C hyperfine coupling coefficients and *g* values they derived the coefficients for the mixing of atomic orbitals in the half-filled $6a_1$ molecular orbital, occupied by the odd electron. The $6a_1$ molecular orbital proved to be 14 % carbon 2s, 66 % carbon $2p_z$, and 10 % for each oxygen $2p_z$. Chantry and Whiffen studied the UV–Vis absorption spectrum of CO_2^- trapped in sodium formate [74]. The radical anion did not absorb in the visible region, but it did in the accessible UV region not obscured by matrix absorption ($\lambda > 240.0$ nm). Three maxima were found at 340.0, 280.0, and 255.0 nm, the first being assigned to the $2b_1(^2B_1) \leftarrow 6a_1(^2A_1)$ transition. The assignments of the absorptions at 280 nm ($6a_1(^2B_2) \leftarrow 4b_2(^2A_1)$ or $2b_1(^2B_2) \leftarrow 4b_2(^2A_1)$ or $5b_2(^2B_2) \leftarrow 6a_1(^2A_1)$) and at 255.0 nm ($7a_1 \leftarrow 6a_1$ or $6a_1 \leftarrow 5a_1$) were less certain. Hartman and Hisatsune [75] observed the infrared spectrum of CO_2^- which they produced by γ-irradiation of sodium formate in alkali halide pressed disks. From the asymmetric stretching frequencies observed for ^{12}C and ^{13}C isotopomers they computed a bond angle of $127 \pm 8°$ and obtained a value of 8.40 mdyn/Å for the force constant of the asymmetric stretching frequency. The experimental data also allowed an estimation of the C–O bond length (1.25 Å), which is markedly longer than that measured for CO_2. At room temperature, the decay half-life of the radical anion in a KBr disk was estimated to be over 1 year.

The CO_2^- radical anion reacted easily with water, giving hydrogen carbonate and formate anions (1.3), but, under anhydrous conditions, it decayed by a bimolecular process to afford CO_3^{2-} and CO [76].

$$2CO_2^- + H_2O \longrightarrow HCO_3^- + HCO_2^- \tag{1.3}$$

The rate constants, determined by IR spectroscopy in the range 370–488 K, were $10^{9.4 \pm 0.1}\exp[(-18.9 \pm 0.2)/RT]$ and $10^{6.6 \pm 0.1}\exp[(-11.9 \pm 0.2)/RT]$ M^{-1} s^{-1} for KCl and KBr matrices, respectively [76]. Only in KCl was oxalate ion observed as a minor reaction product.

The second-order kinetics for the decay of the CO_2^- radical, under anhydrous conditions, is consistent with the mechanism summarized by the reactions (1.4) and (1.5). These reactions have also been proposed to explain the thermal decomposition pathway of oxalate ion in a KBr matrix. In the oxalate pyrolysis reaction, the rate-determining step was the reverse of the reaction (1.5), which was first-order in oxalate and had an activation energy of 60 kcal/mol [77]. The subsequent step was the rapid reaction (1.4) which was assumed to proceed through carbonylcarbonate $OCOCO_2^{2-}$ dianion, a species isoelectronic with a known isomer of dinitrogen tetroxide. Reaction (1.4) was proposed to be the dominant process for CO_2^- decay in KBr, where no oxalate was detected among the decay products. This matrix is not effective in deactivating the oxalate ion formed by the exothermic forward reaction (1.5). On the other hand, the presence of some oxalate among the decay products, when the decay process of CO_2^- was investigated in KCl, showed that the KCl matrix was partially effective in promoting CO_2^- decay also through reaction (1.5).

$$2CO_2^- \longrightarrow CO_3^{2-} + CO \tag{1.4}$$

$$2CO_2^- \rightleftharpoons C_2O_4^{2-} \tag{1.5}$$

Callens et al. [78] detected CO_2^- by ESR in KCl single crystals doped with $Na_2{}^{13}CO_3$ after X-ray irradiation at room temperature. The radical anion, which occupied an anion vacancy, had a bent geometry with an OCO angle of 130.5° and rotated rapidly about its O–O axis. From the spectroscopic parameters they calculated the weights of $2s_C$ and $2p_{zC}$ orbitals contributing to the $6a_1$ molecular orbital of the radical anion, which were found to be 11.8 % and 43.6 %, respectively. The spin density on the oxygen atoms was found to be 0.45, not significantly different from that found for the oxygen atoms of NO_2 in the same host lattice.

CO_2^- radicals have also been detected by ESR in biological apatites (tooth enamel, bone) and their synthetic analogues, when exposed to γ-rays, UV light, or thermal treatment [79]. A general mechanism for the formation of the CO_2^- radical in apatites under different working conditions (γ-rays, UV, thermal treatment) has been proposed and is summarized in Scheme 1.1. The precursors for CO_2^- radical formation are CO_3^{2-} anions. By capturing a free electron, which originates from the ionization of impurities by radiation/temperature, the CO_3^{2-} ion is transformed into a metastable short-lived CO_3^{3-} radical, which decays to CO_2^-.

$$CO_3^{2-} + e^- \longrightarrow CO_3^{3-} \longrightarrow CO_2^-$$

Scheme 1.1 Mechanism of formation of CO_2^- in apatites

Single crystals and polycrystalline samples of lithium formate monohydrate ($HCO_2Li{\cdot}H_2O$) were recently X-irradiated at 295 K and studied by using X-band EPR, ENDOR, and ENDOR-induced EPR (EIE) spectroscopy at 200 or 295 K [80]. The *g*- and ^{13}C hyperfine coupling tensors of the CO_2^- radical were determined. The 2s and 2p spin densities on the C-atom were estimated to be 0.15 and 0.48, respectively. A mechanism for the formation of the CO_2^- radical was proposed, which implies the ejection of an electron as the primary step, followed by deprotonation, (1.6) and (1.7). Electron loss is most probably from the anion HCO_2^- rather than Li^+, as the ionization potential of

$$HCOO^-Li^+ \xrightarrow{h\nu} HCOOLi^+ + e^- \quad (1.6)$$

$$HCOOLi^+ \longrightarrow COO^-Li^+ + H^+ \quad (1.7)$$

Li^+ far exceeds that of HCO_2^-. The above mechanism is different from that previously suggested by Symons and coworkers [81] for the formation of CO_2^- upon γ-irradiation of thallium formate, (1.8)–(1.11). In the latter case, the formiloxy radical is believed to form a hydrogen atom, which may attack a second formate ion with generation of CO_2^-. According to this mechanism, electron capture by formate anion should give radical anion HCO_2^{2-}. The latter species is extremely unstable and yields CO_2^- on annealing.

$$HCOO^- \xrightarrow{h\nu} HCOO + e^- \quad (1.8)$$

$$HCOO \longrightarrow H + CO_2 \quad (1.9)$$

$$H + HCOO^- \longrightarrow H_2 + CO_2^- \quad (1.10)$$

$$HCOO^- + e^- \longrightarrow HCOO_2^{2-} \quad (1.11)$$

Matrix isolation has been shown to be a very useful technique to stabilize CO_2^- generated by reduction of CO_2 with alkali metal atoms during their codeposition at low temperature in a neat or inert matrix [82–86]. An early matrix isolation spectroscopic (IR) study on alkali atoms and CO_2, codeposited at 14 K in solid Ar, was conducted by Jacox and Milligan [82]. Further studies showed a rich reactivity, except for sodium which was the least reactive among the alkali metals (Li, Na, K, Cs) investigated.

The reaction of carbon dioxide with Li, Na, K, and Cs in rare gas and nitrogen matrices leads to the reduction of the heterocumulene with formation of $M^+CO_2^-$ species. Two geometric isomers of $Li^+CO_2^-$ were isolated in solid argon or krypton

[83, 84]. One has a C_{2v} symmetry, consistent with a rhombus structure in which the metal interacts symmetrically with both the oxygen atoms of CO_2^-, whereas in the second isomer, showing C_s symmetry, the lithium center is bonded to only one of the two oxygens. Upon photolysis with a Nernst glower IR source, $Li^+CO_2^-$ (C_s) is rearranged to the symmetric $Li^+CO_2^-$ (C_{2v}) isomer [83]. The other alkali atoms (Na, K, Cs) formed only the $M^+CO_2^-$ (C_{2v}) isomer [85, 86]. These findings agree with the results of a few theoretical studies by Jordan [87], who predicted two geometric isomers for $Li^+CO_2^-$, which are close in energy, with the C_{2v} structure being more stable than that having a C_s symmetry (energy stabilization with respect to neutral CO_2 and metal atom: 0.85 eV (C_{2v}) vs 0.83 eV (C_s)). In contrast, for $Na^+CO_2^-$, the C_{2v} geometry was computed to be significantly more stable than the C_s structure (energy stabilization with respect to neutral CO_2 and metal atom: 0.34 eV (C_{2v}) vs 0.14 eV (C_s)). The OCO bond angle for $M^+CO_2^-$ (C_{2v}) (M = Li, K, Cs) was calculated from the experimental IR spectra using the triatomic molecule approximation [85]. The computed values of the OCO bond angle are close to 130° and depend on the matrix used and the nature of the alkali metal.

Under high metal concentrations, M_2CO_2 species can also form [83]. In the case of Li, only the most stable Li_2CO_2 (C_s) isomer was produced. With other alkali atoms (K, Cs), M_2CO_2 (C_{2v}) molecules, having, most likely, a W-shape, also formed [85, 86]. The CO_2^{2-} bond angles calculated for K_2CO_2 (C_{2v}) and Cs_2CO_2 (C_{2v}) were, respectively, 125.9° and 123.2°. M_2CO_2 (C_{2v}) species (M = Na, K, Cs) can be photolytically converted to the corresponding M_2CO_2 (C_s) isomers by IR radiation [85, 86]. It has been suggested that M_2CO_2 (C_{2v}) forms via reaction of M_2 molecules with CO_2, as opposed to M_2CO_2 (C_s), which is produced by sequential addition of M to MCO_2 (C_{2v}) [85]. In highly concentrated matrices, the formation of oxalates $M_2C_2O_4$ was also observed, which may decompose into M_2CO_3 and CO upon annealing above 200 K.

The above discussion emphasizes that CO_2^-, which has a limited lifetime in the gas phase, can be stabilized by interaction with a metal countercation. The nature of the environment wherein CO_2^- is trapped also significantly affects the spectroscopic features of the radical anion [88–99]. For instance, ESR spectra of CO_2^- measured in different environments show that the ^{13}C isotropic hyperfine coupling constant (a_{iso}) varies over a wide range depending on the environment. Moreover, spin density is frequently found on a neighboring metal cation, and there are large variations in the magnitude of $a_{iso}(M^+)$ with the environment. With the purpose of characterizing as closely as possible the unperturbed electronic structure of CO_2^-, the radical anion was also isolated in gas rare matrices (Ne, Ar), free from any interaction with alkali metal cations, and studied by ESR [96] and IR spectroscopy [97–99]. The results of these experiments have emphasized substantial differences with respect to those acquired under different conditions. FT-IR measurements [97–99] on CO_2^- samples generated as specified above have allowed the metal independent absorptions of the radical anion to be identified (see also Sect. 1.6.1). Notably, removal of the interaction with alkali metal cation, characteristic of $M^+CO_2^-$ ion pairs obtained through co-condensation experiments [82–86] increased the antisymmetric stretching frequency of CO_2^- to 1658,3 cm^{-1} [97], relatively close to the values (1665–1676 cm^{-1}) reported in pressed alkali halide pellets [75], and also lowered the energy threshold for CO_2^- electron photodetachment into the visible spectral region.

Of particular relevance to heterogeneous catalysis are both formation and characterization of CO_2^- radical anion at solid surfaces [100, 101]. Surface-stabilized CO_2^- radical is considered to play a key role in many photoreduction processes of CO_2 in the presence of semiconductor powders or TiO_2 [102–107]. Recently Chiesa et al. [108, 109] studied the activation of CO_2 by excess electrons trapped at the surface of an insulating oxide, such as polycrystalline MgO. A few morphological features on the surface of this oxide, such as steps, corners, or reverse corners, are able to promote, at temperatures as low as 77 K, the spontaneous ionization of H atoms (for instance, generated from H_2, under UV light) by stabilizing the two fragments H^+ and e^-, respectively, in the form of a surface hydroxyl unit and an excess electron, which can be described as an ion pair $(H^+)(e^-)$. The surface of MgO decorated by electron–proton pairs acquires an extraordinary chemical reactivity leading to the reduction, by electron transfer of gaseous molecules to the corresponding radical anions. Accordingly, CO_2^- radical anion was generated from CO_2 at the surface of MgO by direct electron transfer from surface trapped excess electrons and were characterized by CW-ESR spectroscopy [108]. The radical anion showed bent structure with an angle of about 126°. Both ^{13}C and ^{17}O hyperfine structures were resolved for the first time. This allowed one to obtain a detailed mapping of the unpaired electron spin density distribution (mainly located on the carbon atom) within the SOMO (Single Occupied Molecular Orbital) and to ascertain the magnetic equivalence of the oxygen atoms. On the basis of the experimental results it has been proposed that the CO_2^- radical ion displays a side-on structure with the two oxygen atoms symmetrically disposed with respect to four- and three-coordinated Mg^{2+} ions at MgO edges, reverse corners, and corners. A later pulsed ESR study showed that CO_2^- radical ion can be stabilized by the interaction with a nearby adsorbed proton (OH group) [109]. This was demonstrated unambiguously by the small hyperfine interaction with the proton, detected via matched three pulse and HYSCORE experiments. Based on the comparison of measured and computed hyperfine (^{13}C and ^{17}O) and superhyperfine (^1H) coupling constants, the dominant species observed in the experiments should correspond to CO_2^- formed at three-coordinated Mg corner sites.

Single electron reduction of CO_2 to CO_2^- (1.12) is the first, rate determining, step in multielectron electrochemical reduction of CO_2 to other valuable species, such as formic acid,

$$CO_2 + e^- \longrightarrow CO_2^- \qquad (1.12)$$

alcohols, hydrocarbons, CO, and oxalate [110–112]. Reaction (1.12) occurs at high voltage ($E(CO_2/CO_2^-) = -1.90$ V) as a result of the large reorganizational energy between the linear molecule and the bent radical anion [113]. Accordingly, CO_2 electrochemical reduction does not occur easily and the really applied electrolysis potentials for CO_2 reduction are more negative than the thermodynamic values. The fate of CO_2^-, once formed, depends on several factors, such as concentration of reactants, electrode potential, temperature, electrocatalyst material, and nature of electrolyte solution (i.e., aqueous vs non-aqueous electrolyte). These aspects are discussed in greater detail elsewhere in this book (see Chap. 8).

CO_2^- can also be generated by ionizing radiation in aqueous solutions [114]. For instance, the radical anion can be produced by reaction of hydroxyl radicals or H atoms with formic acid/formate ion, hydroxyl radicals with carbon monoxide, or solvated electrons with carbon dioxide [115]. In these experiments, the hydroxycarbonyl radical OCOH can also form, (1.13)–(1.16). The latter radical is the protonated form of the CO_2^- radical anion (1.16). The pK_a of OCOH has been a matter of several experimental determinations resulting in significantly different values found for this quantity. A pK_a ($OCOH/CO_2^-$) value of 2.3 has been quite recently determined by von Sonntag et al. [115] by using pulse radiolysis with conductometric detection.

$$OH + HCO_2H/HCO_2^- \longrightarrow H_2O + OCOH/CO_2^- \quad (1.13)$$

$$H + HCO_2H/HCO_2^- \longrightarrow H_2 + OCOH/CO_2^- \quad (1.14)$$

$$CO + OH \longrightarrow OCOH \quad (1.15)$$

$$OCOH \rightleftarrows CO_2^- + H^+ \quad (1.16)$$

In neutral to basic solutions the main product of the bimolecular decay of CO_2^- was found to be oxalate. In acidic solutions, in contrast, CO_2 was formed. It was generally believed that oxalate formation in neutral and alkaline solutions can occur by head-to-head recombination at the carbon of CO_2^- radical anions (1.5), whereas disproportionation between two OCOH radicals (1.17) was assumed to be responsible for the formation of CO_2 in acidic solutions of formic acid. CO_2 and formate may also form by reaction between OCOH and CO_2^-, as shown in (1.18).

$$2OCOH \longrightarrow CO_2 + HCO_2H \quad (1.17)$$

$$OCOH + CO_2^- \longrightarrow HCO_2^- + CO_2 \quad (1.18)$$

Von Sonntag and coworkers [115] have reinvestigated the mechanism of CO_2^- bimolecular decay. The bimolecular decay rate constant of CO_2^- ($\approx 1.4 \times 10^9$ dm^3 mol^{-1} s^{-1}) was found to be independent of pH in the range 3–8, at constant ionic strength. The yields of the products of the bimolecular decay of the carboxyl radicals, CO_2 and oxalate anion, were found to depend strongly on the pH of solution with an inflection point at pH 3.8. The observed pH dependence was explained by considering a new mechanism, which involves a common carbanion intermediate and is summarized in Scheme 1.2. A major feature of this mechanism is that CO_2^- radicals react mainly (>90 %) by head-to-tail recombination to give the carbanion intermediate **A**, which may rearrange to oxalate or undergo a competing proton-catalyzed disproportionation, which accounts for the formation of CO_2. According to the proposed mechanism, protonation of the intermediate **A** at the oxygen atom (step (d)) should be faster than step (e). However, once **B** was formed, it can undergo subsequent protonation at carbon (step (f)), possibly assisted by a molecule of water, to give the mixed anhydride **C**, which decomposes to CO_2 and formate.

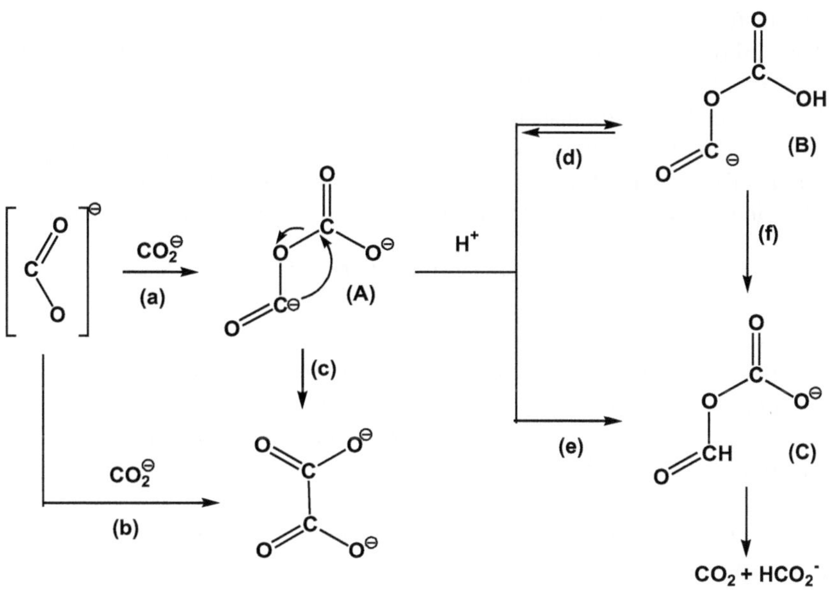

Scheme 1.2 Bimolecular decay of CO_2^- radical anion

1.5 Carbon Dioxide Radical Cation, CO_2^+

Removal of a $1\pi_g$ electron from the CO_2 molecule results in the formation of the CO_2^+ radical cation in the ground state $^2\Pi_g$. The energy required for this process is 13.79 eV [12–14, 116], which is much larger than that required for the formation of the radical anion CO_2^- (Sect. 1.4) or dissociation of neutral CO_2 to CO and atomic oxygen (Sect. 1.6.2). Spectroscopic measurements [117–120] show that CO_2^+ in its $^2\Pi_g$ ground state is a linear molecule with $D_{\infty h}$ point group symmetry: the two C–O bonds are equivalent with a length which is only slightly longer (1.1769 Å) than that measured in the neutral molecule.

Removal of one electron from the inner orbitals $1\pi_u$, $3\sigma_u$, $4\sigma_g$ of CO_2 requires a higher amount of energy [12–14, 116] (Sect. 1.2.1) and affords CO_2^+ radical cation in the excited states $^2\Pi_u$, $^2\Sigma_u^+$, and $^2\Sigma_g^+$, respectively. The above excited states are also linear ($D_{\infty h}$) [117–125]. The experimental C–O bond distances for the $^2\Pi_u$ and

$^2\Sigma_u^+$ excited states are, respectively, 1.228 and 1.180 Å. The available structural data, as well as theoretical calculations [121–125], show that a significant increase in the carbon–oxygen bond length takes place, upon ionization, when the electron is removed from the CO_2 $1\pi_u$ orbital, in accordance with the strongly bonding character of this orbital (Sect. 1.2.1), whereas a very modest change is found when electron removal occurs from the orbitals $1\pi_g$, $3\sigma_u$, or $4\sigma_g$, which exhibit, mainly or fully, non-bonding character (Sect. 1.2.1).

The carbon dioxide cation plays an important role in the dynamics of plasma discharges and chemistry of planetary atmospheres [126]. In general, two possible pathways can be distinguished in single ionization of CO_2. The first pathway (non-dissociative ionization; see also above), affords CO_2^+ ion, possibly in an electronically and/or rovibrationally excited state (1.19). The vibronic structures of CO_2^+ in the various states have been studied by means of different experimental

$$CO_2 \longrightarrow CO_2^+ \left[\text{or } (CO_2^+)^*\right] + e^- \quad (1.19)$$

techniques [127–129], such as emission, laser absorption, He(I) photoelectron, threshold photoelectron (TPE), photoionization efficiency, photoelectron–photon coincidence, photoion–photon coincidence, photoelectron–photoion coincidence, pulsed field ionization (PFI)–photoelectron (PE) (PFI–PE), PFI–PE–photoion (PI) (PFI–PE–PI) coincidence measurements, and resonance-enhanced multiphoton ionization.

However, CO_2 molecule may also dissociate into a charged and a neutral fragment (dissociative ionization; (1.20) and (1.21)). The dissociative ionization of CO_2 has also been the subject of

$$CO_2 \longrightarrow CO^+ + O + e^- \quad (1.20)$$

$$CO_2 \longrightarrow O^+ + CO + e^- \quad (1.21)$$

several studies employing a variety of techniques [127–129].

Multiple ionization of the molecule might also occur in collisional processes and in some cases it is followed by Coulomb explosion (1.22), which is fragmentation of the molecular ion into two or three fragments [130].

$$CO_2^{n+} \longrightarrow C^{p+} + O^{q+} + O^{r+} \quad (1.22)$$

1.6 Spectroscopic Techniques Applied to the CO_2 States

1.6.1 IR Spectroscopy

The vibrational (and rotational) spectra of carbon dioxide have been investigated extensively for their relevance to the study of planetary atmospheres [131–135]. Moreover, IR spectroscopy is a useful technique for continuous recording of carbon dioxide levels in the air and for analysis of remote sensing data.

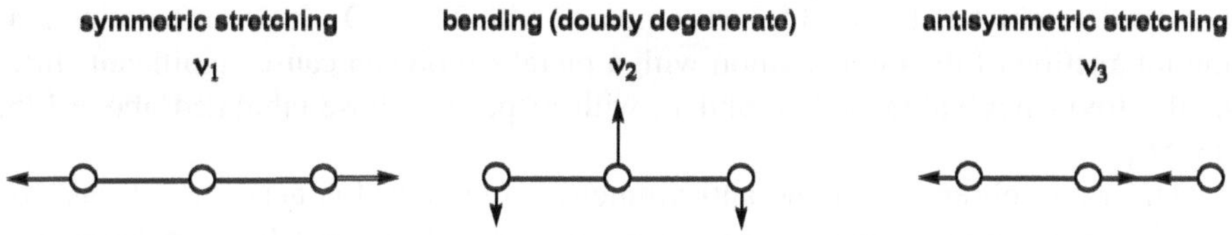

Fig. 1.6 Normal vibration modes of carbon dioxide molecule in its ground state

Table 1.3 Vibrational normal modes and related frequencies (cm^{-1}) for neutral, cationic, and anionic CO_2

	ν_1 [ν_s(OCO)]	ν_2 [δ(OCO)]	ν_3 [ν_a(OCO)]	Notes	Refs.
CO_2	1333	667.38	2349.16	Gas state	[136]
CO_2^+	1244.3	511.4	1423.08	Gas state	[139]
CO_2^-	1253.8	714.2	1658.3	Ne matrix	[96, 98]

The linear triatomic molecule exhibits four normal vibration modes ($3N-5$, where N is the number of atoms in the molecule), which are shown in Fig. 1.6. While the antisymmetric stretching (ν_3) and the doubly degenerate bending (ν_2) modes are IR-active, the symmetric stretching mode (ν_1) is IR-inactive, as this normal mode does not generate any change of the electric dipole moment of the molecule (Sect. 1.2.1). In the IR spectrum of free carbon dioxide the antisymmetric stretching frequency (ν_3) is found at 2349.16 cm^{-1} in the gaseous state [136] and 2344.0 cm^{-1} in the solid state (15 K) [137]. In aqueous solution this band is shifted by only 6 cm^{-1} from the vapor phase value (down to 2343 cm^{-1}), indicating the absence of hydrogen bonding between water molecules and dissolved CO_2 [138]. The infrared absorption produced by the bending vibration of the molecule is observed at 667.38 cm^{-1} for gaseous CO_2 [136]. In a pure CO_2-ice, the ν_2 spectrum at 15 K has two components, peaking at 654.7 and 659.8 cm^{-1} with a full width at half maximum of 1.8 and 3.1 cm^{-1}, respectively, showing Davidov splitting caused by a highly ordered ice structure [137]. Because a carbon dioxide molecule has a center of symmetry, the antisymmetric stretching (v_3) and the bending (ν_2) modes are Raman inactive. A Fermi dyad is observed in the Raman spectrum of gaseous CO_2 at 1285.40 and 1388.15 cm^{-1} [136], which originates from the resonance between the unperturbed energy levels associated with the fundamental transition of the ν_1 totally symmetric stretching mode ($\nu_1 = 1333$ cm^{-1}) and the harmonic transition $2\nu_2$ of the ν_2 bending mode.

Table 1.3 compares the IR fundamental frequencies of gaseous CO_2 [136] with those measured for gaseous CO_2^+ [139] and for CO_2^- isolated in solid neon [97, 99].

The IR spectrum of CO_2^- in a neon matrix, where the radical anion is free from interactions with metal cations (Sect. 1.4), exhibits marked differences with respect to that of the neutral parent molecule. Because of the bent geometry (C_{2v}) of the anion, all the three normal vibration modes of CO_2^- are IR-active. The changes of the stretching frequencies are pronounced. Their lower values reflect a reduced CO bond order, which is 1.5 in CO_2^- as compared with 2 for CO_2. It is worth noting that the interaction of the radical anion with a metal cation can cause significant shifts of the frequency values ν_1, ν_2, and ν_3 with respect to those tabulated above [75, 83–86].

The value observed for the antisymmetric stretching frequency ν_3 of gaseous CO_2^+ is very close to that measured for the radical cation when the latter was isolated in solid neon ($\nu_3 = 1421.7$ cm^{-1}) [97, 99]. It has long been recognized that ν_3 for CO_2^+ is anomalously low [118, 140, 141]. From the analysis of high resolution spectra of CO_2^+, Kawaguchi et al. [140] argued that the origin of the anomalously low ν_3 frequency could be explained in terms of a vibronic interaction between the $^2\Pi_g$ ground state and the $^2\Pi_u$ excited electronic state through the ν_3 vibration normal mode.

The formation of CO_2-adducts, which may involve the population of one of the LUMOs of CO_2, can also originate large modifications in the IR spectrum of the CO_2 moiety. The antisymmetric stretching mode, ν_a(OCO), is lowered in the range 2250–1400 cm^{-1}, the symmetric stretching mode, ν_s(OCO), becomes IR-active and can absorb in the region 1400–1100 cm^{-1}, the bending mode δ(OCO) is shifted from 667 cm^{-1}, and additional vibrational modes, such as metal-carbon and/or metal-oxygen stretching modes, and C=O out-of-plane deformation, may be observed in the low-frequency region (down to 300 cm^{-1}).

The spectroscopic features of these absorptions can provide useful information on the bonding mode of the heterocumulene in the adduct [142–145] (see Sect. 2.6).

1.6.2 UV Spectrum of Carbon Dioxide

The UV absorption spectrum of carbon dioxide has been widely studied both experimentally [20, 22, 27, 116, 146] and theoretically [18, 21, 24, 147–150] (see also Sect. 1.2.2). A huge effort has been lavished on identifying the excited states involved in the electronic transitions. To this end, electron impact spectroscopy has proved to be an additional useful diagnostic tool [151–153].

Carbon dioxide is transparent in the visible and near-to-middle ultraviolet regions, at least down to 210 nm ($h\nu$ <5.91 eV), both in the liquid phase [154] and in the gas phase [117].

Only weak absorptions can be observed below 11 eV. CO_2 shows three maxima in the vacuum ultraviolet region at 147.5 (8.41 eV), 133.2 (9.31 eV), and 112.1 nm (11.08 eV), respectively. The bands corresponding to the maxima at 147.5 nm and 133.2 nm exhibit a vibrational structure which is poor, irregular, and apparently complex for the maximum at 147.5 nm, while it is sharper and more regular for that at 133.2 nm. The assignment of these absorption maxima has been controversial and a much debated question. Several authors have assigned the maxima at 147.5 nm and 133.2 nm to transitions to $^1\Delta_u$ and $^1\Pi_g$ states, respectively. The maximum at 112.1 nm exhibits much higher intensity. This is the strongest absorption band of carbon dioxide. The intensity of the absorption indicates that this transition, identified as $^1\Sigma_u^+ \leftarrow {}^1\Sigma_g^+$, is optically allowed [148]. Above 11 eV there are several Rydberg series converging to the various states of CO_2^+.

The spectroscopic properties of CO_2 in the UV region are of fundamental importance for understanding the photochemistry of this molecule. In fact, upon UV irradiation, CO_2 can undergo photodissociation into CO and O (1.23) [155–160]. Studying the reaction (1.23) is of great help in

$$CO_2 \xrightarrow{h\nu} CO + O \tag{1.23}$$

elucidating the role and chemical behavior of CO_2, not only in the Earth's atmosphere but also in the atmospheres of comets or other planets such as Mars and Venus [161–165], where any build-up of oxygen is believed to occur only through this process.

In general, the energy states of O and CO generated by CO_2 photodissociation depend on the wavelength of the absorbed UV light. Table 1.4 summarizes the energy thresholds for the production of CO and O in their different states by photodissociation of CO_2. The electronically excited oxygen atoms generated during the photodissociation process, for instance $O(^1S)$ and $O(^1D)$, may undergo subsequent quenching to the $O(^3P)$ ground state by collisions. Recombination of O (^3P) atoms can afford molecular oxygen, O_2, through reaction (1.24), where M is some third body or, also, the vessel walls.

$$O(^3P) + O(^3P) + M \longrightarrow O_2 + M \tag{1.24}$$

When using UV photons of energy in the range 7.42–10.45 eV (120.0 nm $< \lambda <$ 167.2 nm), CO_2 can dissociate via two channels, affording $O(^3P)$ or $O(^1D)$, respectively (1.23a) and (1.23b).

Table 1.4 Energy (eV) and wavelength (nm) thresholds for the photodissociation of $CO_2(^1\Sigma_g^+)$ to $CO(X^1\Sigma^+, a^3\Pi, A^1\Pi)$ and $O(^3P, ^1D, ^1S)$[a]

CO_2	CO	O	E	λ
$^1\Sigma_g^+$	$X^1\Sigma^+$	^3P	5.45	227.5
$^1\Sigma_g^+$	$X^1\Sigma^+$	^1D	7.42	167.2
$^1\Sigma_g^+$	$X^1\Sigma^+$	^1S	9.65	128.6
$^1\Sigma_g^+$	$a^3\Pi$	^3P	11.47	108.2
$^1\Sigma_g^+$	$a^3\Pi$	^1D	13.45	92.3
$^1\Sigma_g^+$	$a^3\Pi$	^1S	15.68	79.2
$^1\Sigma_g^+$	$A^1\Pi$	^3P	13.50	92.0
$^1\Sigma_g^+$	$A^1\Pi$	^1D	15.46	80.3
$^1\Sigma_g^+$	$A^1\Pi$	^1S	17.69	70.2

[a]From [155]

$$CO_2 + h\nu \longrightarrow CO(^1\Sigma) + O(^3P) \tag{1.23a}$$

$$CO_2 + h\nu \longrightarrow CO(^1\Sigma) + O(^1D) \tag{1.23b}$$

Slanger and Black [156] revisited the dissociation of CO_2 at 147.0 and 130.2–130.6 nm and ascertained that the quantum yield for oxygen atom production was unity, thus closing a long-lasting dispute on this matter. Zhu and Gordon [157] determined, by a chemical scavenging method, the branching ratio between reactions (1.23a) and (1.23b), which was 94 % $O(^1D)$ and 6 % $O(^3P)$ for $h\nu$ corresponding to light at 157 nm.

The production of $O(^3P)$ under the conditions used (157 nm) has been a matter of further studies [158, 159] and requires a comment about the mechanism of formation. In fact, the 157-nm (7.91 eV) excitation is to the dissociative continuum of the lowest $^1B_2$ state (Sect. 1.2.1), corresponding to the dissociation products $CO(^1\Sigma)$ and $O(^1D)$ (7.42 eV) (Fig. 1.7). Moreover, the photodissociation of CO_2 to $CO(^1\Sigma)$ + $O(^3P)$ violates the spin conservation rule, as the spin of CO_2 in the ground state ($^1\Sigma_g{}^+$) is zero whereas the total spin of $CO(^1\Sigma) + O(^3P)$ is unity, and, therefore, reaction (1.23a) is a spin-forbidden process.

The mechanism proposed to explain the production of $O(^3P)$ implies a transition from the surface of the $^1B_2$ state to the surface of the lower $^3B_2$ state. Figure 1.7 shows that the two states intersect approximately 0.27 eV above the $CO(^1\Sigma) + O(^3P)$

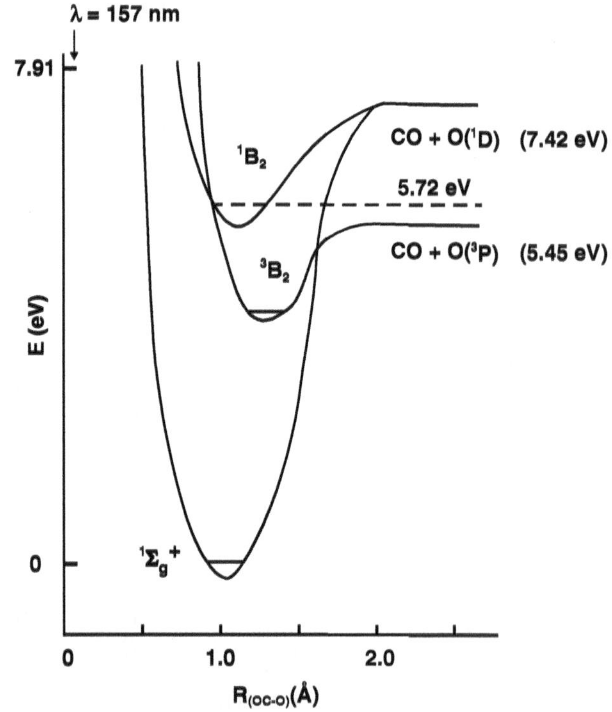

Fig. 1.7 Photodissociation of CO_2 at 157 nm (adapted from [157])

asymptote. In principle, the efficiency of this singlet–triplet transition depends on the lifetime of the excited state. According to Zhu and Gordon [157], although the molecule contains energy excess (11 kcal/mol; 0.48 eV) over the $CO(^1\Sigma) + O(^1D)$ dissociation threshold, it does not dissociate readily because the bent upper state ($^1B_2$) makes it difficult to concentrate enough energy along the reaction coordinate. This increases the lifetime of the excited state and the probability of curve crossing to the triplet state. This mechanism has been recently criticized by Cossart-Magos et al. [27], who have proposed the $^3A_2$ state (Sect. 1.2.2), instead of the $^3B_2$, as the CO_2 triplet state involved in the photodissociation process.

A new phenomenon has been observed by Bhattacharya et al. [160] who studied the formation of O_2 and CO produced by CO_2 photolysis at 185 ± 2 nm (≈6.7 eV; Hg lamp). Under the above conditions the only energetically possible primary process is the production of $O(^3P)$ ((1.23a); see also Table 1.4). They focused on oxygen and carbon isotope fractionation associated with the products, which was compared to the estimate based on a kinetic model. Both CO and O_2 were enriched in ^{17}O by about 81 %, but not in ^{18}O. Moreover, CO was enriched in ^{13}C by about 37 % with respect to the model composition.

Figure 1.8 helps to explain the generally accepted mechanism for CO_2 dissociation by ≈6–7 eV energy photons. It implies an electronic transition of the CO_2 molecule from the singlet ground state ($^1\Sigma_g^+$) to the upper (bound) singlet state $^1B_2$. In this state, CO_2 cannot dissociate, as this transition takes the molecule into a

Fig. 1.8 Photodissociation of CO_2 by ≈6–7 eV energy photons (adapted from [160])

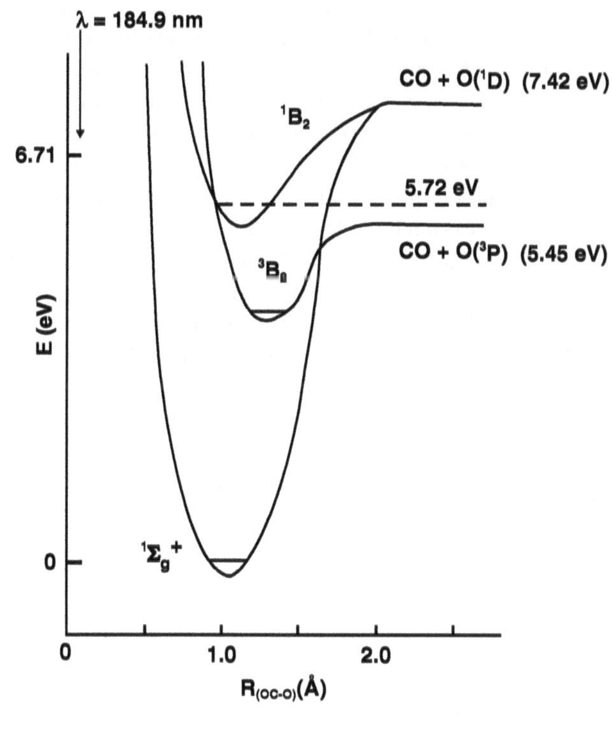

vibrational level which is below the asymptote of the state $^1B_2$ but above the crossover zone with the state $^3B_2$. Dissociation can occur if the molecule can reach the crossover region (by collisions, for instance) and undergo a transition to the triplet state $^3B_2$.

As already emphasized, the latter transition is spin forbidden, but can take place in the presence of a suitable perturbation, such as spin-orbit coupling involving the valence electrons. Because this perturbation is of same magnitude for the different isotopes, it cannot explain, per se, the anomalous ^{13}C and ^{17}O isotopic enrichment observed in CO and O_2 products. Therefore, it has been proposed that the nuclear spin of ^{13}C ($I = 1/2$) and ^{17}O ($I = 5/2$) may introduce additional coupling by hyperfine interaction, which may increase the dissociation rate of the isotopologues containing ^{13}C and ^{17}O. Interestingly, the ^{17}O enrichment is 2.2 ± 0.2 times higher than that of ^{13}C, a value very close to the value of 2.7, which gives the $\mu(^{13}C)/\mu(^{17}O)$ magnetic moment ratio for the two nuclei.

The influence of nuclear spin in photodissociation of molecules is a novel effect which may have important implications and potential applications in the study of terrestrial and planetary atmospheres [160].

1.6.3 Nuclear Magnetic Resonance (NMR) Spectroscopy

CO_2 dissolved in a nonpolar solvent, such as benzene or toluene, exhibits only one ^{13}C resonance around 124 ppm. In aqueous solutions the resonance is close to 125 ppm [166] and can be used for the quantification of free CO_2.

^{13}C-NMR has proved to be a very useful spectroscopic technique for identifying the COO moiety when the carbon dioxide molecule, as a whole, is incorporated into a new compound. The ^{13}C resonance of CO_2 shifts downfield upon fixation of the heterocumulene into organic products, such as carbamic acids or carbamates (150–160 ppm), carbonates (145–165 ppm), acids (160–180 ppm), esters (160–170 ppm), or their metallorganic analogues (metal carbamates $L_nM(O_2CNR_2)$, metal carbonates $L_nM(O_2COR)$, metallacarboxylates $L_nM(O_2CR)$, metallacarboxylic acids $L_nM(CO_2H)$, and metallaesters $L_nM(CO_2R)$.

The above remarks emphasize the usefulness of NMR spectroscopy in demonstrating the formation of a CO_2-complex: these aspects are discussed in Sect. 2.6. As noted in Sect. 1.6.1, IR spectroscopy may also be of great help in this context and several CO_2 complexes have been characterized adequately by the combined use of both the spectroscopic techniques. This is particularly useful in those cases wherein X-ray characterization is not possible.

Variable temperature NMR spectroscopy has been used to study the fluxional behavior of CO_2 adducts in solution (see Sect. 2.7).

References

1. Hirota E, Iijima T, Kuchitsu K, Lafferty WJ, Ramsay DA, Vogt J (1992) Structure data of free polyatomic molecules. In: Kuchitsu K (ed), Landolt-Börnstein, vol II/21. Springer, Berlin, p 151
2. Graner G, Hirota E, Iijima T, Kuchitsu K, Ramsay AD, Vogt J, Vogt N (1995) Structure data of free polyatomic molecules. In: Kuchitsu K (ed) Landolt-Börnstein, vol II/23. Springer, Berlin, p 146
3. Vučelić M, Ohrn Y, Sabin JR (1973) Ab initio calculation of the vibrational and electronic properties of carbon dioxide. J Chem Phys 59:3003–3007
4. Gutsev GL, Bartlett RJ, Compton RN (1998) Electron affinities of CO_2, OCS, and CS_2. J Chem Phys 108:6756–6762
5. Maroulis G, Thakkar AJ (1990) Polarizabilities and hyperpolarizabilities of carbon dioxide. J Chem Phys 93:4164–4171
6. Buckingham AD, Disch RL, Dunmur DA (1968) Quadrupole moments of some simple molecules. J Am Chem Soc 90:3104–3107
7. Lobue JM, Rice JK, Novick SE (1984) Qualitative structure of $(CO_2)_2$ and $(OCS)_2$. Chem Phys Lett 112:376–380
8. Brigot N, Odiot S, Walmsley SH, Whitten JL (1977) The structure of the carbon dioxide dimer. Chem Phys Lett 49:157–159
9. Johnson MA, Alexander ML, Lineberger WC (1984) Photodestruction cross sections for mass-selected ion clusters: $(CO_2)_n^+$. Chem Phys Lett 112:285–290
10. Bowen KH, Liesegang GW, Sanders RA, Herschbach DR (1983) Electron attachment to molecular clusters by collisional charge transfer. J Phys Chem 87:557–565
11. Rossi AR, Jordan KD (1979) Comment on the structure and stability of $(CO_2)_2^-$. J Chem Phys 70:4442–4444
12. Allian CJ, Gelius U, Allison DA, Johansson G, Siegbahn H, Siegbahn K (1972) ESCA studies of CO_2, CS_2 and COS. J Elect Spectrosc Relat Phenom 1:131–151
13. Turner DW (1968) Molecular photoelectron spectroscopy. In: Hill HAO, Day P (eds) Physical methods in advanced inorganic chemistry. Interscience, London
14. Turner DW, May DP (1967) Frank–Condon factors in ionization: experimental measurements using molecular photoelectron spectroscopy. II. J Chem Phys 46:1156–1160
15. Cremaschi P, Simonetta M (1974) A theoretical study of electrophilic aromatic substitution. I. The electronic structure of NO_2^+. Theoret Chim Acta 34:175–182
16. Müller JE, Jones RO, Harris J (1983) Density functional calculations for H_2O, NH_3, and CO_2 using localized muffin-tin orbitals. J Chem Phys 79:1874–1884
17. Moncrieff D, Wilson S (1995) On the accuracy of the algebraic approximation in molecular electronic structure calculations: IV. An application to a polyatomic molecule: the CO_2 molecule in the Hartree–Fock approximation. J Phys B (At Mol Opt Phys) 28:4007–4013
18. Nakatsuji H (1983) Cluster expansion of the wavefunction. Valence and Rydberg excitations, ionizations, and inner-valence ionization of CO_2 and N_2O studied by the SAC and SAC CI theories. Chem Phys 75:425–441
19. Walsh AD (1953) The electronic orbitals, shapes, and spectra of polyatomic molecules. Part II. Non-hydride AB_2 and BAC molecules. J Chem Soc 2266–2288
20. Rabalais JW, McDonald JM, Scherr V, McGlynn SP (1971) Electron spectroscopy of isoelectronic molecueles. II. Linear triatomic groupings containing sixteen valence electrons. Chem Rev 71:73–108
21. Spielfieldel A, Feautrier N, Cossart-Magos C, Werner H-J, Botschwina P (1992) Bent valence states of CO_2. J Chem Phys 97:8382–8388
22. Cossart-Magos C, Launay F, Parkin JE (1992) High resolution absorption spectrum of CO_2 between 1750 and 2000 Å. 1. Rotational analysis of nine perpendicular-type bands assigned to a new bent-linear electronic transition. Mol Chem Phys 75:835–856

23. Wang Y-G, Wiberg KB, Werstiuk NH (2007) Correlation effects in EOM-CCSD for the excited states: evaluated by AIM localization index (LI) and delocalization index (DI). J Phys Chem 111:3592–3601
24. Winter NW, Bender CF, Goddard WA III (1973) Theoretical assignments of the low-lying electronic states of carbon dioxide. Chem Phys Lett 20:489–492
25. Dixon RN (1963) The carbon monoxide flame bands. Proc R Soc A 275:431–446
26. Mohammed HH, Fournier J, Deson J, Vermeil C (1980) Matrix isolation study of the CO_2 lowest triplet state. Chem Phys Lett 73:315–318
27. Cossart-Magos C, Launay F, Parkin JE (2005) High resolution absorption spectrum of CO_2 between 1750 and 2000 Å. 2. Rotational analysis of two parallel-type bands assigned to the lowest electronic transition $1^3B_2 \leftarrow X^1\Sigma_g^+$. Mol Phys 103:629–641
28. Green S, Schor H, Siegbahn P, Thaddeus P (1976) Theoretical investigation of protonated carbon dioxide. Chem Phys 17:479–485
29. Seeger U, Seeger R, Pople JA, Schleyer Pvon R (1978) Isomeric structures of protonated carbon dioxide. Chem Phys Lett 55:399–403
30. Scarlett M, Taylor PR (1986) Protonation of CO_2, COS, CS_2. Proton affinities and the structure of protonated species. Chem Phys 101:17–26
31. Hartz N, Rasul G, Olah GA (1993) Role of oxonium, sulfonium, and carboxonium dications in superacid-catalyzed reactions. J Am Chem Soc 115:1277–1285
32. Gronert S, Keeffe JR (2007) The protonation of allene and some heteroallenes, a computational study. J Org Chem 72:6343–6352
33. Traeger JC, Kompe BM (1991) Determination of the proton affinity of carbon dioxide by photoionization mass spectrometry. J Mass Spectrom Org Mass Spectrom 26:209–214
34. Bohme DK, Mackay GI, Schiff HI (1980) Determination of proton affinities from the kinetics of proton transfer reactions. The proton affinities of O_2, H_2, Kr, O, N_2, Xe, CO_2, CH_4, N_2O, and CO. J Chem Phys 73:4976–4986
35. Lias SG, Liebman JF, Levin RD (1984) Evaluated gas phase basicities and proton affinities of molecules. J Phys Chem Ref Data 13:695–808
36. Hunter EP, Lias SG (1998) Evaluated gas phase basicities and proton affinities of molecules: an update. J Phys Chem Ref Data 27:413–656
37. Hayhurst AN, Taylor SG (2001) The proton affinities of CO and CO_2 and the first hydration energy of gasous H_3O^+ from mass spectrometric investigations of ions in rich flames of C_2H_2. Phys Chem Chem Phys 3:4359–4370
38. Fock W, McAllister T (1982) Probable abundance ratios for interstellar HCS_2^+, HCOS, HCO_2^+. Astrophys J 257:L99–L101
39. Bogey M, Demuynek C, Destombes JL (1986) The submillimeter wave spectrum of the protonated and deuterated carbon dioxide. J Chem Phys 84:10–15
40. Bogey M, Demuynek C, Destombes JL, Krupnov A (1988) Molecular structure of $HOCO^+$. J Mol Struct 190:465–474
41. Amano T, Tanaka K (1985) Difference frequency laser spectroscopy of the ν_1 band of $HOCO^+$. J Chem Phys 82:1045–1046
42. Amano T, Tanaka K (1985) Difference frequency laser spectroscopy of the ν_1 fundamental band of $HOCO^+$. J Chem Phys 83:3721–3728
43. Taddeus P, Guélin M, Linke RA (1981) Three new "nonterrestrial molecules". Astrophys J 246:L41–L45
44. Burt JA, Dunn JL, Mc Ewan MJ, Sutton MM, Roche AE, Schiff HI (1970) Some ion–molecule reactions of H_3^+ and the proton affinity of H_2. J Chem Phys 52:6062–6075
45. Adams NG, Smith D, Tichy M, Javahery J, Twiddy ND, Ferguson EE (1989) An absolute proton affinity scale in the 130–140 kcal mol^{-1} range. J Chem Phys 91:4037–4042
46. Hammami K, Jaidane N, Lakhdar ZB, Spielfeldel A, Feautrier N (2004) New ab initio potential energy surface for the ($HOCO^+$-He) van der Waals complex. J Chem Phys 121:1325–1330

47. Mauser H, King WA, Gready JE, Andrews TJ (2001) CO_2 fixation by Rubisco: computational dissection of the key steps of carboxylation, hydration, and C–C bond cleavage. J Am Chem Soc 123:10821–10829
48. Lee HJ, Lloyd MD, Harlos K, Clifton IJ, Baldwin JE, Schofield CJ (2001) Kinetic and crystallographic studies on deacetoxycephalosporin C synthase (DAOCS). J Mol Biol 308:937–948
49. Aresta M, Quaranta E (1997) Carbon dioxide: a substitute for phosgene. ChemTech 27:32–40
50. Quaranta E, Aresta M (2010) The chemistry of N-CO_2 bonds: synthesis of carbamic acids and their derivatives, isocyanates, and ureas. In: Aresta M (ed) Carbon dioxide as chemical feedstock. Wiley-VCH, Weinheim
51. Ballivet-Tkatchenko D, Dibenedetto A (2010) Synthesis of linear and cyclic carbonates. In: Aresta M (ed) Carbon dioxide as chemical feedstock. Wiley-VCH, Weinheim
52. Sakaki S (1990) Transition-metal complexes of nitrogen, carbon dioxide, and similar small molecules. Ab-initio MO studies of their stereochemistry and coordinate bonding nature. Stereochem Organomet Inorg Compd 4:95–177
53. Santoro M (2010) Non-molecular carbon dioxide at high pressure. In: Boldyreva E, Dera P (eds) High-pressure crystallography: from fundamental phenomena to technological applications. Springer, Dordrecht
54. Schettino V, Bini R, Ceppatelli M, Ciabini L, Citroni M (2005) Chemical reactions at very high pressure. Adv Chem Phys 11:105–242
55. Iota V, Yoo CS, Cynn H (1999) Quartzlike carbon dioxide: an optically nonlinear extended solid at high pressures and temperatures. Science 283:1510–1513
56. Yoo CS, Cynn H, Gygi F, Galli G, Iota V, Nicol M, Carlson S, Häusermann D, Mailhiot C (1999) Crystal structure of carbon dioxide at high pressure: "superhard" polymeric carbon dioxide. Phys Rev Lett 83:5527–5530
57. Santoro M, Gorelli FA, Bini R, Ruocco G, Scandolo S, Crichton WA (2006) Amorphous silica-like carbon dioxide. Nature 441:857–860
58. Yota V, Yoo CS, Klepeis JH, Jenei Z, Evans W, Cynn H (2007) Six-fold coordinated carbon dioxide VI. Nat Mat 6:34–38
59. Datchi F, Giordano VM, Munsch P, Saitta AM (2009) Structure of carbon dioxide phase IV: breakdown of the intermediate bonding state scenario. Phys Rev Lett 103:185701
60. Matoušek I, Fojtík A, Zahradník R (1975) A semiempirical molecular orbital study of radicals and radical ions derived from carbon oxides. Coll Czech Chem Commun 40:1679–1685
61. Pacansky J, Wahlgren U, Bagus PS (1975) SCF ab intio ground state energy surface for CO_2 and CO_2^-. J Chem Phys 62:2740–2744
62. England WB, Rosemberg BJ, Fortune PJ, Wahl AC (1976) Ab initio vertical spectra and linear bent correlation diagrams for the valence states of CO_2 and its singly charged ions. J Chem Phys 65:684–691
63. England WB (1981) Accurate ab initio SCF energy curves for the lowest electronic states of CO_2/CO_2^-. Chem Phys Lett 78:607–613
64. Sommerfeld T, Meyer H-D, Cederbaum LS (2004) Potential energy surface of CO_2^- anion. Phys Chem Chem Phys 6:42–45
65. Villamena FA, Locigno EJ, Rockenbauer A, Hadad CM, Zweier JL (2006) Theoretical and experimental studies of the spin trapping of inorganic radicals by 5,5-dimethyl-1-pyrroline *N*-oxide (DMPO). 1. Carbon dioxide radical anion. J Phys Chem 110:13253–13258
66. Feller D, Dixon DA, Francisco JS (2003) Coupled cluster theory determination of the heats of formation of combustion-related compounds: CO, HCO, CO_2, HCO_2, HOCO, HC(O)OH, and HC(O)OOH. J Phys Chem 107:1604–1617
67. Dixon DA, Feller D, Francisco JS (2003) Molecular structure, vibrational frequencies, and energetics of the HCO, HOCO and HCO_2 anions. J Phys Chem A 107:186–190
68. Paulson JF (1970) Some negative-ion reactions with CO_2. J Chem Phys 52:963–964
69. Cooper CD, Compton RN (1972) Metastable anions of CO_2. Chem Phys Lett 14:29–32

70. Cooper CD, Compton RN (1973) Electron attachment to cyclic anhydrides and related compounds. J Chem Phys 59:3550–3565
71. Compton RN, Reinhardt PW, Cooper CD (1975) Collisional ionization of Na, K, and Cs by CO_2, COS, and CS_2: molecular electron affinities. J Chem Phys 63:3821–3827
72. Boness MJW, Schulz GJ (1974) Vibrational excitation in CO_2 via the 3.8-eV resonance. Phys Rev A 9:1969–1979
73. Ovenall DW, Whiffen DH (1961) Electron spin resonance and structure of the CO_2^- radical anion. Mol Phys 4:135–144
74. Chantry GW, Whiffen DH (1962) Electronic absorption spectra of CO_2^- trapped in γ-irradiated crystalline sodium formate. Mol Phys 5:189–194
75. Hartman KO, Hisatsune IC (1966) Infrared spectrum of carbon dioxide anion radical. J Chem Phys 44:1913–1918
76. Hisatsune IC, Adl T, Beahm EC, Kempf RJ (1970) Matrix isolation and decay kinetics of carbon dioxide and carbonate anion free radicals. J Phys Chem 74:3225–3231
77. Hartman KO, Hisatsune IC (1967) Kinetics of oxalate ion pyrolysis in a potassium bromide matrix. J Phys Chem 71:392–396
78. Callens F, Matthys P, Boesman E (1989) Paramagnetic resonance spectrum of CO_2^- trapped in KCl. J Phys Chem Solids 50:377–381
79. Rudko VV, Vorona JP, Baran NP, Ishchenko SS, Zatovsky IV, Chumakova LS (2010) The mechanism of CO_2^- radical formation in biological and synthetic apatites. Health Phys 98:322–326
80. Vestad TA, Gustafsson H, Lund A, Hole EO, Sagstuen E (2004) Radiation-induced radicals in lithium formate monohydrate ($LiHCO_2{\cdot}H_2O$). EPR and ENDOR studies of X-irradiated crystal and polycrystalline samples. Phys Chem Chem Phys 6:3017–3022
81. Symons MCR, West DX, Wilkinson JG (1976) Radiation damage in thallous formate and acetate: charge transfer from thallous ions. Int J Radiat Phys Chem 8:375–379
82. Jacox ME, Milligan DE (1974) Vibrational spectrum of CO_2^- in an argon matrix. Chem Phys Lett 28:163–168
83. Kafafi ZH, Hauge RH, Billups WE, Margrave JL (1983) Carbon dioxide activation by lithium metal. 1. Infrared spectra of $Li^+CO_2^-$, $Li^+C_2O_4^-$ and $Li_2^{2+}CO_2^{2-}$ in inert gas matrices. J Am Chem Soc 105:3886–3893
84. Manceron L, Loutellier A, Perchard JP (1985) Reduction of carbon dioxide to oxalate by lithium atoms: a matrix isolation study of the intermediate steps. J Mol Struct 129:115–124
85. Kafafi ZH, Hauge RH, Billups WE, Margrave JL (1984) Carbon dioxide activation by alkali metals. 2. Infrared spectra of $M^+CO_2^-$ and $M_2^{2+}CO_2^{2-}$ in argon and nitrogen matrices. Inorg Chem 23:177–183
86. Bencivenni L, D'Alesssio L, Raimondo F, Pelino M (1986) Vibrational spectra and structure of $M(CO_2)$ and $M_2(CO_2)_2$ molecules. Inorg Chim Acta 121:161–166
87. Jordan KD (1984) Theoretical investigation of lithium and sodium complexes with CO_2. J Phys Chem 88:2459–2465
88. Bennett JE, Mile B, Thomas A (1965) Electron spin resonance of the CO_2^- radical ion at 77 K. Trans Faraday Soc 61:2357–2364
89. Borel JP, Faes F, Pittel A (1981) Electron paramagnetic resonance of Li-CO_2 complexes in a CO_2 matrix at 77 K. J Chem Phys 74:2120–2123
90. Cook RJ, Whiffen DH (1967) Endor measurements in X-irradiated sodium formate. J Phys Chem 71:93–97
91. Atkins PW, Keen N, Symons MCR (1962) Oxides and oxyions of the non-metals. Part II. CO_2^- and NO_2. J Chem Soc 2873–2880
92. Sharp JH, Symons MCR (1970) Unstable intermediates. Part LXXIX. Electron spin resonance studies of the effect of the environment upon the hyperfine parameters for CO_2^- and NO_2. J Chem Soc A 3075–3080
93. Dalal NS, McDowell CA, Park JM (1975) EPR and ENDOR studies of CO_2^- centers in X- and UV-irradiated single crystals of sodium formate. J Chem Phys 63:1856–1862

94. Bentley J, Carmichael I (1985) Electron spin properties of complexes formed by Li or Na with CO_2. J Phys Chem 89:4040–4042
95. Koppe R, Kasai PH (1994) $Li^+CO_2^-$ and $Na^+CO_2^-$ generated in argon matrices: an ESR study. J Phys Chem 98:11331–11336
96. Knight LB Jr, Hill D, Berry K, Babb R, Feller D (1996) Electron spin resonance rare gas matrix studies of $^{12}CO_2^-$, $^{13}CO_2^-$, and $C^{17}O_2^-$: comparison with ab initio calculations. J Chem Phys 105:5672–5686
97. Jacox ME, Thompson WE (1989) The vibrational spectra of molecular ions in solid neon. I. CO_2^+ and CO_2^-. J Chem Phys 91:1410–1416
98. Jacox ME, Thompson WE (1999) The vibrational spectra of CO_2^+, $(CO_2)_2^+$, CO_2^- and $(CO_2)_2^-$ trapped in solid neon. J Chem Phys 110:4487–4496
99. Zhou M, Andrews L (1999) Infrared spectra of the CO_2^- and $C_2O_4^-$ anions in solid argon. J Chem Phys 110:2414–2422
100. Freund HJ, Roberts MW (1996) Surface chemistry of carbon dioxide. Surf Sci Rep 25:225–273
101. Farkas AP, Solymosi F (2009) Activation and reaction of CO_2 on a K-promoted Au(111) surface. J Phys Chem C 113:19930–19936
102. Inoue T, Fujishima A, Konishi S, Honda K (1979) Photoelectrocatalytic reduction of carbon dioxide in aqueous suspensions of semiconductor powders. Nature 277:637–638
103. Thampi KR, Kiwi J, Gratzel M (1987) Methanation and photo-methanation of carbon dioxide at room temperature and atmospheric pressure. Nature 327:506–508
104. Ikeue K, Yamashita H, Anpo M, Takewaki T (2001) Photocatalytic reduction of CO_2 with H_2O on Ti−β zeolite photocatalysts: effect of the hydrophobic and hydrophilic properties. J Phys Chem B 105:8350–8355
105. Hwang JS, Chang JS, Psrk SE, Ikeue K, Anpo M (2005) Photoreduction of carbon dioxide on surface functionalized nanoporous catalysts. Top Catal 35:311–319
106. Saladin F, Alxneit I (1997) Temperature dependence of the photochemical reduction of CO_2 in the presence of H_2O at the solid/gas interface of TiO_2. J Chem Soc Faraday Trans 93:4159–4163
107. He H, Zapol P, Curtiss LA (2010) A theoretical study of CO_2 anions on anatase (101) surface. J Phys Chem C 114:21474–21481
108. Chiesa M, Giamello E (2007) Carbon dioxide activation by surface excess electrons: an EPR study of the CO_2^- radical ion adsorbed on the surface of MgO. Chem Eur J 13:1261–1267
109. Preda G, Pacchioni G, Chiesa M, Giamello E (2008) Formation of CO_2^- radical anion from CO_2 adsorption on an electron-rich MgO surface: a combined ab initio and pulse EPR study. J Phys Chem C 112:19568–19576
110. Barton Cole E, Bocarsly AB (2010) Photochemical, electrochemical, photoelectrochemical reduction of carbon dioxide. In: Aresta M (ed) Carbon dioxide as chemical feedstock. Wiley-VCH, Weinheim
111. Li W (2010) Electrocatalytic reduction of CO_2 to small organic molecule fuels on metal catalysts. In: Hu Y (ed) Advances in CO_2 conversion and utilization, ACS Symposium Series. American Chemical Society, Washington, DC, pp 55–76
112. Gennaro A, Isse AA, Severin M-G, Vianello E, Bhugun I, Savéant J-M (1996) Mechanism of the electrochemical reduction of carbon dioxide at inert electrodes in media of low proton availability. J Chem Soc, Faraday Trans 92:3963–3968
113. Wardman P (1989) Reduction potentials of one-electron couples involving free radicals in aqueous solutions. J Chem Ref Data 18:1637–1756
114. Von Sonntag C (1987) The chemical basis of radiation biology. Taylor and Francis, London
115. Flyunt R, Schuchmann MN, von Sonntag C (2001) A common carbanion intermediate in the recombination and proton-catalysed disproportionation of the carboxyl radical anion CO_2^-, in aqueous solution. Chem Eur J 7:796–799

116. Cossart-Magos C, Jungen M, Launay F (1987) High resolution absorption spectrum of CO_2 between 10 and 14 eV. Assignment of nf Rydberg series leading to a new value of the first ionization potential. Mol Phys 61:1077–1117
117. Herzberg G (1966) Molecular spectra and molecular structure. III. Electronic spectra and electronic structure of polyatomic molecules. Van Nostrand-Reinhold, New York, NY
118. Johnson MA, Rostas J (1995) Vibronic structure of the CO_2^+ ion: reinvestigation of the antisymmetric stretch vibration in the X, $\tilde{A}$, and B states. Mol Phys 85:839–868
119. Gauyacq D, Larcher C, Rostas J (1979) The emission spectrum of the CO_2^+ ion: rovibronic analysis of the $\tilde{A}^2\Pi_u$ - $X^2\Pi_g$ band system. Can J Phys 57:1634–1649
120. Gauyacq D, Horani M, Leach S, Rostas J (1975) The emission spectrum of the CO_2^+ ion: $B^2\Sigma_u^+$ - $X^2\Pi_g$ band system. Can J Phys 53:2040–2059
121. Horsley JA, Fink WH (1969) Study of the electronic structure of the ions CO_2^+ and N_2O^+ by the LCAO-MO-SCF method. J Phys B (Atom Mol Phys) 2(2):1261–1270
122. Carsky P, Kuhn J, Zahradnik R (1975) Semiempirical all-valence-electron MO calculations on the electronic spectra of linear radicals with degenerate ground states. J Mol Spectrosc 55:120–130
123. Grimm FA, Larsson M (1984) A theoretical investigation on the low lying electronic states of CO_2^+ in both linear and bent configurations. Phys Scr 29:337–343
124. Chambaud G, Gabriel W, Rosmus P, Rostas J (1992) Ro-vibronic states in the electronic ground state of CO_2^+ ($X^2\Pi_g$). J Phys Chem 96:3285–3293
125. Gellene GI (1998) CO_2^+: a difficult molecule for electron correlation. Chem Phys Lett 287:315–319
126. Dalgarno A, Fox JL (1994) Ion chemistry in atmospheric and astrophysical plasmas. In: Ng CY, Baer T, Powis I (eds) Unimolecular and bimolecular ion-molecule reaction dynamics, Wiley series in ion chemistry and physics. Wiley, Chichester, Chapter 1
127. Yang M, Zhang L, Zhuang X, Lai L, Yu S (2008) The [1 + 1] two-photon dissociation spectra of CO_2^+ via $\tilde{A}^2\Pi_{u,1/2}$ ($\nu_1\nu_2 0$) $X^2\Pi_{g,1/2}$ (000) transitions. J Chem Phys 128:164308 (1–7)
128. King SJ, Price SD (2008) Electron ionization of CO_2. Int J Mass Spectrom 272:154–164
129. Liu J, Chen W, Hochlaf N, Qian X, Chang C, Ng CY (2003) Unimolecular decay pathways of state-selected CO_2^+ in the internal energy range of 5.2–6.2 eV: an experimental and theoretical study. J Chem Phys 118:149–163
130. Siegmann B, Werner U, Lutz HO, Mann R (2002) Complete coulomb fragmentation of CO_2 in collisions with 5.9 MeV u^{-1} Xe^{18+} and Xe^{43+}. J Phys B (At Mol Opt Phys) 35:3755–3766
131. Guelachvili G, Rao KN (1997) Molecular constants. Guelachvili G (ed), Landolt-Börnstein, vol. II/20, subvol B2α. Springer, Berlin
132. Rothman LS, Hawkins RL, Wattson RB, Gamache RR (1992) Energy levels, intensities, and linewidths of atmospheric carbon dioxide bands. J Quant Spectrosc Radiat Transf 48:537–566
133. Tashkun SA, Perevalov VI, Teffo J-L, Rothman LS, Tyuterev VG (1998) Global fitting of $^{12}C^{16}O_2$ vibrational–rotational line positions using the effective Hamiltonian approach. J Quant Spectrosc Radiat Transf 60:785–801
134. Fox K (1972) High resolution infrared spectroscopy of planetary atmospheres. In: Rao KN, Mathews CW (eds) Molecular spectroscopy: modern research. Academic, New York, NY
135. White DW, Gerakines PA, Cook AM, Whittet DCB (2009) Laboratory spectra of the CO_2 bending-mode feature in interstellar ice analogues subject to thermal processing. Astrophys J Suppl S 180:182–191
136. Shimanouchi T (1972) Tables of molecular vibrational frequencies, consolidated Volume I. NSRDS-NBS (US) 39:1–164
137. van Broekhuizen FA, Groot IMN, Fraser HJ, van Dishoeck EF, Schlemmer S (2006) Infrared spectroscopy of solid CO-CO_2 mixtures and layers. A&A 451:723–731
138. Falk M, Miller AG (1992) Infrared spectrum of carbon dioxide in aqueous solution. Vibr Spectrosc 4:105–108

139. Jacox ME (1990) Vibrational and electronic energy levels of polyatomic transient molecules. Supplement 1. J Phys Chem Ref Data 19:1388–1546
140. Kawaguchi K, Yamada C, Hirota E (1985) Diode laser spectroscopy of the CO_2^+ ν_3 band using magnetic field modulation of the discharge plasma. J Chem Phys 82:1174–1177
141. Carter S, Handy NC, Rosmus P, Chambaud G (1990) A variational method for the calculation of spin-rovibronic levels of Renner-Teller triatomic molecules. Mol Phys 71:605–622
142. Gibson DH (1996) The organometallic chemistry of carbon dioxide. Chem Rev 96:2063–2095
143. Jegat C, Fouassier M, Mascetti J (1991) Carbon dioxide coordination chemistry. 1. Vibrational study of *trans*-$Mo(CO_2)_2(PMe_3)_4$ and $Fe(CO_2)$ $(PMe_3)_4$. Inorg Chem 30:1521–1529
144. Jegat C, Fouassier M, Tranquille M, Mascetti J (1991) Carbon dioxide coordination chemistry. 2. Synthesis and FTIR study of $Cp_2Ti(CO_2)$ (PMe_3). Inorg Chem 30:1529–1536
145. Jegat C, Fouassier M, Tranquille M, Mascetti J, Tommasi I, Aresta M, Ingold F, Dedieu A (1993) Carbon dioxide coordination chemistry. 3. Vibrational, NMR, and theoretical studies of $Ni(CO_2)(PCy_3)_2$. Inorg Chem 32:1279–1289
146. Ogawa M (1971) Absorption cross sections of O_2 and CO_2 continua in the Schumann and far-UV region. J Chem Phys 54:2550–2556
147. England WB, Ermler WC (1979) Theoretical studies of atmospheric triatomic molecules. New ab initio results for the $^1\Pi_g$-$^1\Delta_u$ vertical state ordering in CO_2. J Chem Phys 70:1711–1719
148. Spielfeldel A, Feautrier N, Chambaud G, Rosmus P, Werner H-J (1993) The first dipole-allowed electronic transition of $1^1\Sigma_u^+$ - X $^1\Sigma_g^+$ of CO_2. Chem Phys Lett 216:162–166
149. Buenker RJ, Honigmann M, Liebermann H-P, Kimura M (2000) Theoretical study of the electronic structure of carbon dioxide: bending potential curves and generalized oscillator strengths. J Chem Phys 113:1046–1054
150. Wiberg KB, Wang Y-G, de Oliveira AE, Perera SA, Vaccaro PH (2005) Comparison of CIS and EOM-CCSD-calculated adiabatic excited states structures. Change in charge density on going to adiabatic excited states. J Phys Chem 109:466–477
151. Lasettre EN, Skerbele A, Dillon MA, Ross KJ (1968) High-resolution study of electron-impact spectra at kinetic energies between 33 and 100 eV and scattering angles to 16°. J Chem Phys 48:5066–5097
152. McDiarmid R, Doering JP (1984) Electronic excited states of CO_2: an electron impact investigation. J Chem Phys 80:648–656
153. Chan WF, Cooper G, Brion CE (1993) The electronic spectrum of carbon dioxide. Discrete and continuum photoabsorption oscillator strengths (6–203 eV). Chem Phys 178:401–413
154. Eiseman BJ Jr, Harris L (1932) The transmission of liquid carbon dioxide. J Am Chem Soc 54:1782–1784
155. Okabe H (1978) Photochemistry of small molecules. Wiley, New York
156. Slanger TG, Black G (1978) CO_2 photolysis revised. J Chem Phys 68:1844–1849
157. Zhu Y-F, Gordon RJ (1990) The production of $O(^3P)$ in the 157 nm photodissociation of CO_2. J Chem Phys 92:2897–2901
158. Matsumi Y, Shafer N, Tonukura K, Kawasaki M, Huang Y-L, Gordon RJ (1991) Doppler profiles and fine structure branching ratios of $O(^3P_J)$ from photodissociation of carbon dioxide at 157 nm. J Chem Phys 95:7311–7316
159. Miller RL, Kable SH, Houston PL, Burak I (1992) Product distributions in the 157 nm photodissociation of CO_2. J Chem Phys 96:332–338
160. Mahata S, Bhattacharya SK (2009) Anomalous enrichment of ^{17}O and ^{13}C in photodissociation products of CO_2: possible role of nuclear spin. J Chem Phys 130:234312 (1–17)
161. Delsemme AH, Combi MR (1976) The production rate and possible origin of $O(^1D)$ in comet Bennett 1970 II. Astrophys J 209:L149–L151
162. Farquhar I, Thiemens MH, Jackson T (1998) Atmosphere-surface interactions on Mars: $\delta^{17}O$ measurements of carbonate from ALH 84001. Science 280:1580–1582

163. McElroy MB, Mc Connell JC (1971) Dissociation of CO_2 in the Martian atmosphere. J Atmos Sci 28:879–884
164. Rockmann T, Brenninkmeijer CAM, Saueressig G, Bergamaschi P, Crowley JN, Fischer H, Crutzen PJ (1998) Mass-independent oxygen isotope fractionation in atmospheric CO as a result of the reaction CO + OH. Science 281:544–546
165. Chakraborty S, Bhattacharya SK (2003) Experimental investigation of oxygen isotope exchange between CO_2 and $O(^1D)$ and its relevance to the stratosphere. J Geophys Res 108 (D23):4724–4738
166. Liger-Belair G, Prost R, Parmentier M, Jeandet P, Nuzillard J-M (2003) Diffusion coefficient of CO_2 molecules as determined by ^{13}C NMR in various carbonated beverages. J Agric Food Chem 51:7560–7563

Chapter 2
CO_2 Coordination to Metal Centres: Modes of Bonding and Reactivity

Abstract The modes of coordination of carbon dioxide (CO_2) to metal centres are presented in this chapter. The coordination at both room temperature and low temperature in gas matrices is discussed with the reactivity of the coordinated cumulene. X-ray diffraction structural data are presented and discussed together with spectroscopic properties of the complexes.

2.1 Mode of Bonding of Carbon Dioxide to Transition-Metal Centres

The Walsh diagram (Fig. 2.1), already discussed in Chap. 1 and recalled here for the reader's benefit, suggests that carbon dioxide (CO_2) can in principle act as an acceptor of electrons (through the lowest unoccupied molecular orbital [LUMO] centred at the C atom) and a donor, through the highest occupied molecular orbital (HOMO) centred at the O atoms.

However, electrons and electron-rich species (nucleophiles such as metals in a low oxidation state, bases, hydride ions, etc.) most likely interact with CO_2 by binding to the C atom, whereas electron-poor centres (electrophiles such as the proton, metal centres in high oxidation state, electron deficient molecules, etc.) attack one of the O atoms. The modes of bonding with a single metal centre of different natures and properties are schematized in Scheme 2.1.

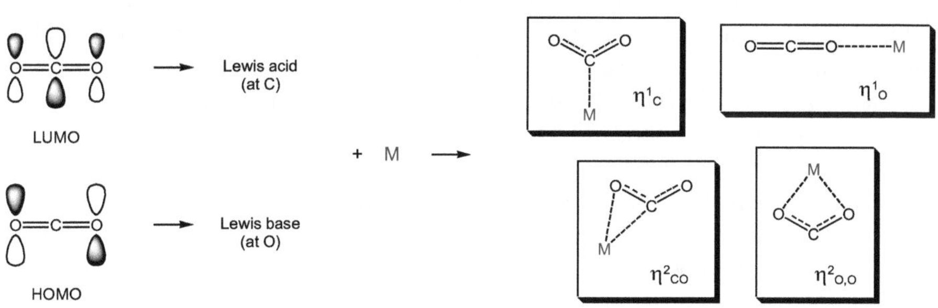

Scheme 2.1 Modes of bonding of CO_2 to a single metal centre of different natures. Reprinted with permission from [2]. Copyright (2010) WILEY-VCH Verlag GmbH & Co. KGaA, Weinheim

M. Aresta et al., *Reaction Mechanisms in Carbon Dioxide Conversion*,
DOI 10.1007/978-3-662-46831-9_2

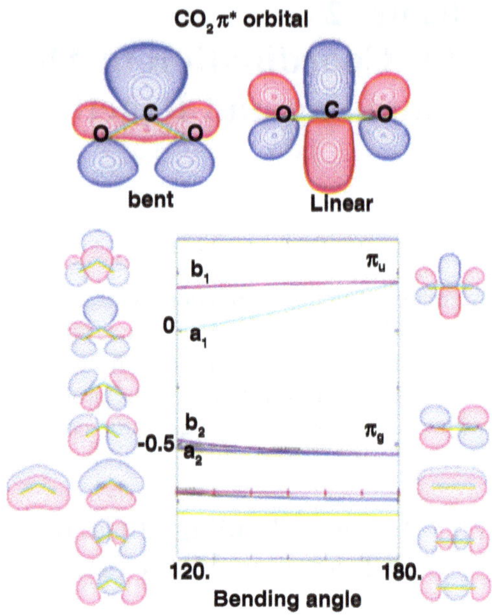

Fig. 2.1 The Walsh diagram of frontier molecular orbital energies for CO_2 in its linear and bent state. Reprinted with permission from [1]. Copyright (2007) American Chemical Society

The Walsh diagram also shows that if electron density is transferred to the LUMO, the cumulene molecule minimizes the energy by bending and the optimization of the energy state is reached with an OCO angle around 125°–133°. Conversely, if electron density is transferred from the O atoms to an acceptor, the cumulene may still maintain a quasilinear geometry. Such a forecast is demonstrated by experimental data (X-ray diffraction [XRD] structure of solid complexes) being correct, as discussed in the next paragraphs. Unfortunately, not all isolated complexes afforded good monocrystals for XRD structure determination. In unfortunate cases, if the XRD structure was unavailable, spectroscopic studies were used for getting information about the solid-state structure and the mode of bonding of the cumulene. Good information was obtained using isotope-labelled CO_2 (^{13}C, ^{17}O, ^{18}O) in infrared (IR) or even nuclear magnetic resonance (NMR) multinuclear studies. Although not so accurate as XRD, valuable information was collected with Fourier transform IR (FTIR) studies, which were able to connect the different coordination modes of bonded CO_2.

Theoretical calculations have more recently been used to collect information about bonding, as discussed below. The coordination of CO_2 to metal centres is presented in the next sections, following the history of discovery of the different classes of bonding modes.

2.2 XRD Structure of η^2-C,O Mononuclear Metal Complexes

The first transition metal complex to be structurally characterized was $(PCy_3)_2Ni(CO_2)$ (**1**) described by Aresta et al. [3] (Structure 2.1). The Ni atom shows a planar geometry, and coordinated CO_2 has two non-equivalent CO bonds (1.17 and 1.22 Å, to be compared with 1.16 of the free molecule) and an OCO angle of 133°, with a strong deviation from $D_{\infty h}$ symmetry.

The Ni–C and Ni–O_1 bond distances are 1.84 and 1.99 Å, respectively, showing that both C and one of the O atoms are within bonding distance from the Ni atom. The strong difference of the four ligands causes an irregular angle distribution at the Ni centre.

The deviation of the coordinated CO_2 molecule from linearity is well in accordance with the forecast done in Sect. 2.1 following the features of the Walsh diagram. Coordinated CO_2 has some features of $CO_2^{\cdot-}$ but is not identical to it. This point is relevant to the definition of the oxidation state of Ni. Because CO_2 is a neutral ligand and **1** can be made by substituting PCy_3 in $Ni(PCy_3)_3$, in which Ni has a "zero" oxidation state, the question is whether the oxidation state of Ni in **1** is still zero or +1 if CO_2 resembles $CO_2^{\cdot-}$, or +2 if Ni–O and Ni–C bonds are considered?

Most likely, no one of the three limit states in Scheme 2.2a–c represents the real situation.

A recent DFT study [5] described the electronic structure of $(PH_3)_2Ni(CO_2)$ (**2**), which is analogous to **1** but not identical because of the different nature and size of the phosphane ligands, which may vary both the electron donor–acceptor properties and the steric hindrance. As a matter of fact, the calculated OCO angle is 144°, much different from that in **1**, and the C–O bond lengths are 1.204 (C=O) and 1.256 A, also quite different from **1**. It has been suggested that in **2** the C–O

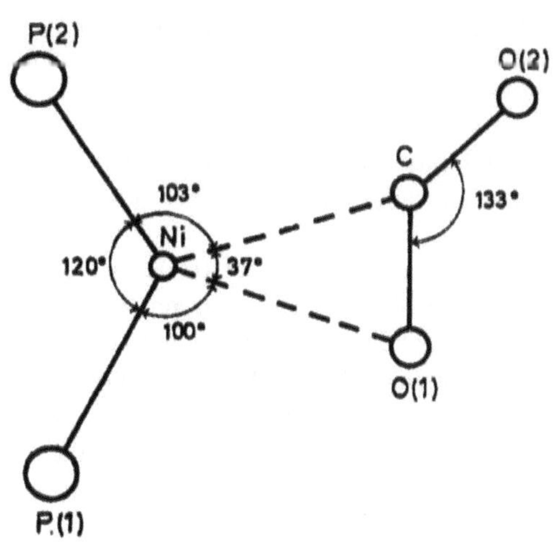

Structure 2.1 [4] Reproduced by permission of The Royal Society of Chemistry

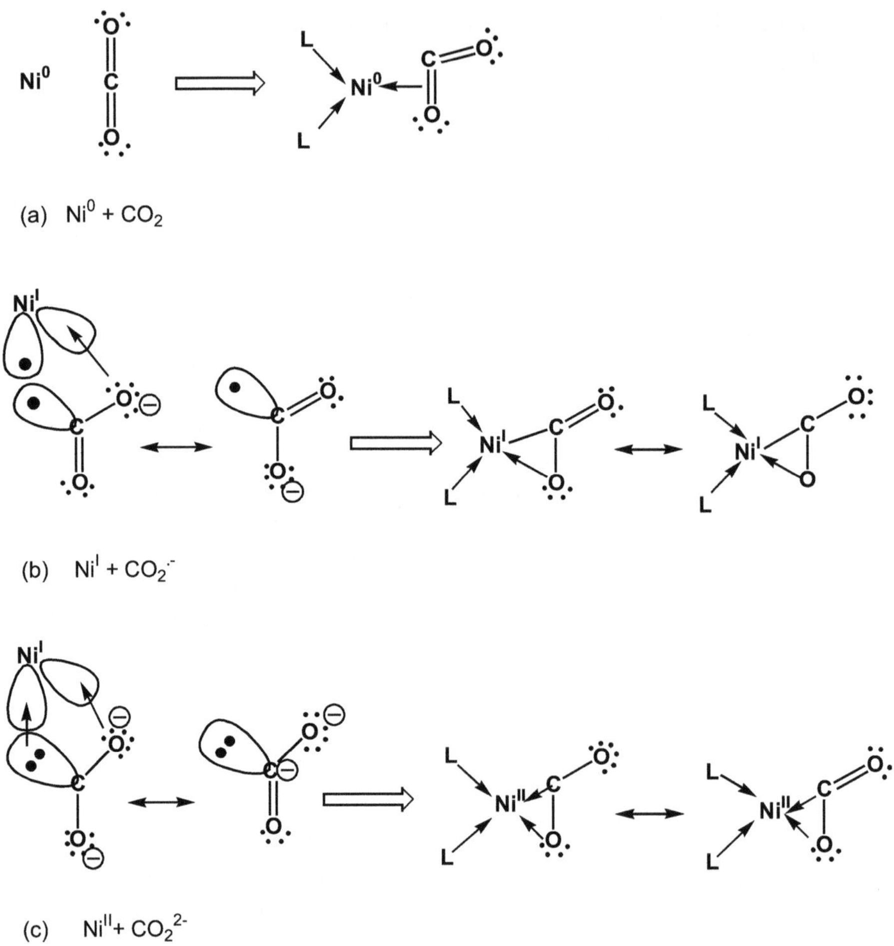

Scheme 2.2 Possible forms of the "Ni(CO_2)" moiety showing formal oxidation states (0), (+1), and (+2) for the Ni atom. Reprinted with permission from [6]. Copyright 1992 American Chemical Society

σ-bond may take part in electron donation to the metal, and that the P atom of the phosphane adjacent to the C=O moiety of CO_2 may transfer electron density to the C atom.

An example of bis(CO_2)-η^2 complex, namely, *trans*-$Mo(CO_2)_2HN(CH_2CH_2PMe_2)_2(PMe_3)$, was reported by Carmona et al. [5]. The complex has a distorted octahedral geometry with two staggered CO_2 ligands η^2-C,O-coordinated to molybdenum. The CO_2 ligands show C–O lengths ranging from 1.199 to 1.284 Å and OCO angles of 131°. The Mo–C distance (2.08 Å), as well as the distances between the molybdenum atom and the CO_2 oxygen atoms bound to the metal (2.111(6) and 2.138(6) Å), demonstrate strong binding of the coordinated C–O moieties to the metal center.

It must be emphasized that the η^2-C,O coordination represents by far the most popular mode of bonding of CO_2 and several metal complexes having such structure [7–10] have been isolated and structurally characterized.

The O,C-interaction is also found in other metal systems which are considered as primary η^1-C coordination to a metal centre stabilized by an external O–E interaction (where E is a different centre from that which binds the C atom of the cumulene).

The first structurally characterized complex of this type was Co(*n*-Pr-salen)(CO_2)K(THF) (salen = *N*,*N*′-ethylenebis(salicylideneaminato)), studied by Floriani et al. [11]. The crystals of this compound were dimorphous. The structural data, obtained for both crystalline forms (A and B), showed that the CO_2 ligand is markedly bent (OCO angle: 134.9(26)° for the A form, 132.0(15)° for the B form), with C–O bond lengths ranging from 1.20 to 1.24 Å. The CO_2 ligand is bound to cobalt through carbon (Co–C bond length: 1.99–2.00 Å), and to potassium ions through the oxygen atoms (K–O bond lengths: from 2.66 to 2.74 Å). One of the oxygen atoms of the CO_2 ligand is also bound to a second potassium ion (2.60–2.68 Å). The K^+---O interactions play a crucial role in the formation and stabilization of the Co–CO_2 adduct, as also documented by the behaviour of the strictly related compounds Co(salen)CO_2M (M = Li, Na, K). It was observed that the nature of the alkali metal exerts some control over the reversibility of CO_2 binding in Co(salen)CO_2M salts: the sodium and potassium salts lose CO_2 under vacuum, unlike the lithium salt which does not. Moreover, Co(salen)CO_2Na loses CO_2 partially upon addition of a sodium-complexing agent such as dicyclohexano-18-crown-6 (DCHC) and Co(salen)Na(THF) reacted only to a limited extent with CO_2 in a THF solution containing the crown-ether, where the complex [Co(salen) Na]$_2$(DCHC) is formed.

A spectroscopic study on prim,*rac*-CoL(CO_2)$^+$ (L = 5,7,7,12,14,14-hexamethyl-1,4,8,11-tetraazacyclotetradeca-4,11-diene), a cobalt-macrocycle CO_2 complex in which the CO_2 ligand occupies the axial metal coordination site on the primary macrocycle face, has

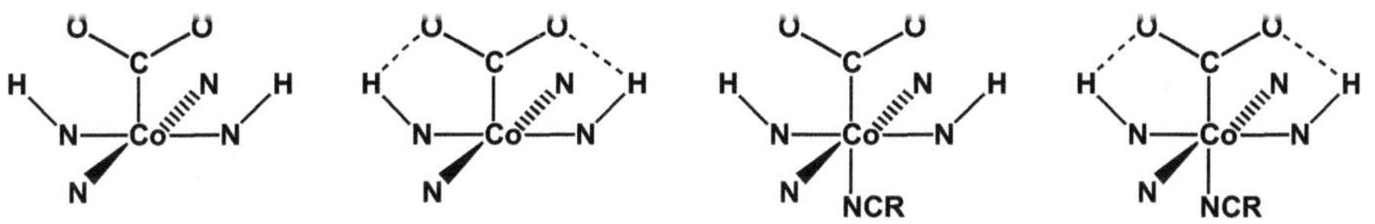

Scheme 2.3 Schematic representation of the penta- and hexa-coordinated CoLCO$_2^+$ adducts (R = Me, Pr; L = 5,7,7,12,14,14-hexamethyl-1,4,8,11-tetraazacyclotetradeca-4,11-diene; see main text). Reprinted with permission from [12]. Copyright (1993) American Chemical Society

provided evidence, in solution, of a CO_2 ligand interacting simultaneously with the nucleophilic cobalt(I) ion, through the electrophilic carbon, and the macrocyclic ancillary ligand through the oxygen atoms via hydrogen bonds [12]. The FT-IR

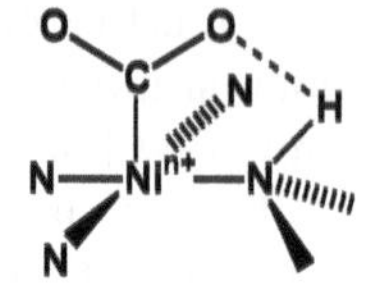

Structure 2.2 Adapted with permission from [13]. Copyright (1986) American Chemical Society

spectra of prim,*rac*-$CoL(CO_2)^+$, measured over the range 298–198 K in CD_3CN and in a CD_3CN/THF mixture, indicated the existence, in solution, of four CO_2 adducts (Scheme 2.3), two of which, respectively penta-(B) and hexa-coordinated (D), showed intramolecular hydrogen bonds between bound CO_2 and the amine hydrogens of the macrocyclic ligand.

An analogous edifice of interactions could also stabilize the putative adduct represented in Structure 2.2, which has been proposed as the key intermediate in the highly selective electrocatalytic reduction of CO_2 to CO by Ni(cyclam)Cl_2 (cyclam = 1,4,8,11-tetraazacyclotetradecane) in a purely aqueous medium [13, 14].

Sauvage et al. [13, 14] suggested that the high selectivity of Ni(cyclam)Cl_2 for electroreduction of CO_2 to CO over that of H_2O to H_2, in the presence of water as solvent, was related to the size of the macrocyclic ligand and to the presence of NH groups, which may stabilize, through hydrogen bonds, binding of the heterocumulene to the metal center and the intermediate formation of a CO_2 adduct as shown in Structure 2.2. Accordingly, the replacement of cyclam with 1,4,8,11-tetraazacyclotetra deca-1,4,8,11-tetraene or tetramethylated-cyclam (1,4,8,11-tetramethyl-1,4,8,11-tetraazacyclo tetradecane), which are unable to establish an analogous interaction, caused a marked decrease of CO/H_2 selectivity.

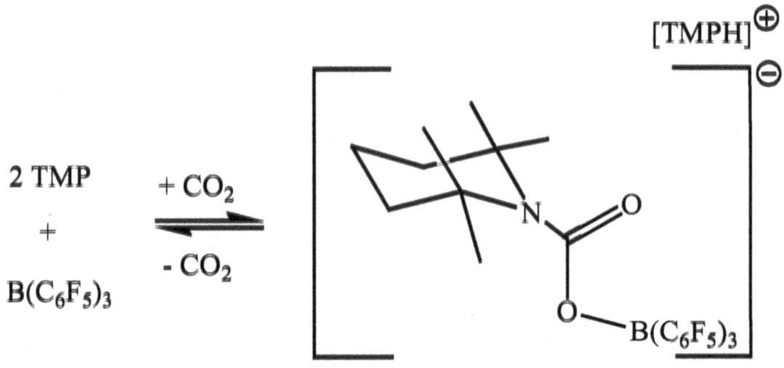

Scheme 2.4 Reactions of frustrated Lewis pairs with CO_2. Adapted with permission from [15]. Copyright (2010) American Chemical Society

Many simple reactions of carbon dioxide require the presence of acid–base systems (bifunctional systems). Recently, CO_2 was found to react easily with frustrated Lewis pairs [15, 16]. For instance, the components of the couple 2,2,6,6-tetramethylpiperidine (TMP)/tris(pentafluorophenyl)borane [$B(C_6F_5)_3$] activate CO_2, yielding boratocarbamate – TMPH ion pair (Scheme 2.4). CO_2

addition is reversible [15]. Analogous behavior was also observed with the couple $t Bu_3P/B(C_6F_5)_3$ and the intramolecular frustrated Lewis pair $(C_6H_2Me_3)_2PCH_2CH_2B(C_6F_5)_2$ [16, 17]. Other important examples of bifunctional activation of carbon dioxide are provided by carboxylation of sodium phenoxide, the key step of Kolbe–Schmitt synthesis of salicylic acid, [18, 19], carboxylation of Grignard reagents [20], and carbonic anhydrase promoted hydration of carbon dioxide [21]. It is now recognized that this enzymatic process implies the nucleophilic attack of a Zn^{2+}-bound hydroxyl group on free CO_2, which is held in place by hydrogen bonds within the pocket of the active site of the enzyme. The insertion reactions of CO_2 into M–E (E = H, C, N, O, etc.) (see Chap. 4) bonds provide other interesting examples of such bifunctional activation. The latter reactions may be relevant to catalytic processes implying CO_2 fixation into organic products with the generation of a new E–C bond. These reactions do not necessarily require the coordination of the heterocumulene to a metal centre as observed in isolated complexes. More often, they are initiated by nucleophilic attack of E at the electrophilic carbon of CO_2 and assisted by weak interactions between the metal centre and the lone pairs of one of the oxygen atoms of CO_2.

2.3 XRD Structural Data for η^1-C Metal Complexes

η^1-C transition metal complexes are much rarer than the η^2-C,O described above. The reaction of $Rh(diars)_2Cl$ (diars = *o*-phenylene-bis(dimethyl)arsine) with slightly pressurized CO_2 in CH_3CN affords $Rh(diars)_2(Cl)(CO_2)$ **3**, an example of the behaviour of CO_2 acting as an only-C-electrophile [22]. XRD of the isolated solid shows that, upon bonding to CO_2, the Rh-As framework remains almost unchanged. The coordination around Rh is essentially octahedral and CO_2 sits *trans* to chlorine and acts as a monodentate ligand, interacting with the nucleophilic metal centre through the central carbon (η^1-C coordination). In such an adduct, the C–O bond distances, 1.20(2) and 1.25(2) Å, respectively, are both elongated, although asymmetrically, with respect to the free ligand, while the OCO angle is 126° (Structure 2.3). Whether such asymmetric CO_2 structure is stabilized by a "*non conventional*" weak intramolecular C–O···H–C bonding, involving one of the methyl groups of the arsane ligand, or not has been a matter of discussion, but such interaction seems to be excluded by XRD.

η^1-C coordination has also been claimed in a few other Rh and Ir complexes [23–25], albeit only on the basis of spectroscopic (IR) studies.

$Ir(CO_2)(Cl)(dmpe)_2$ (dmpe = bis(dimethylphosphino)ethane), isostructural with **3**, reacts with FSO_3Me to afford the methoxycarbonyl–Ir complex $Ir(CO_2CH_3)(Cl)(dmpe)_2$ [24]. Such reactivity provides evidence that the oxygen atoms of CO_2 η^1-C-coordinated to a metal center are characterized by a significant nucleophilic character and can interact with electrophiles.

Structure 2.3 Reprinted with permission from [13]. Copyright (1983) American Chemical Society

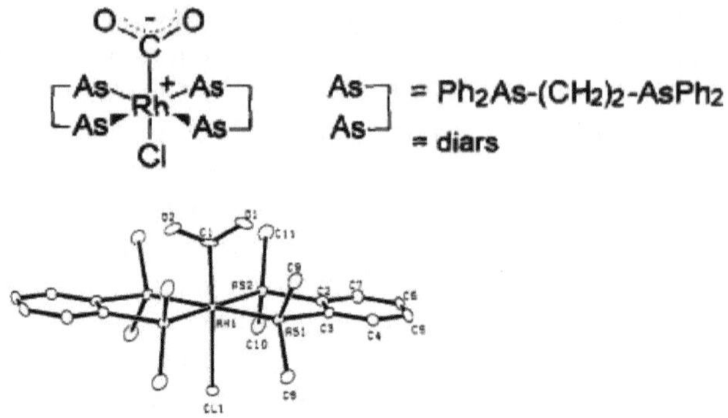

The iron complex $(PMe_3)_4Fe(CO_2)$ **4** described by Karsch in 1977 [26] is an example of how the metal centre and the ligands may influence the coordination of CO_2. **4** shows a trigonal bipyramidal geometry with CO_2 lying in the equatorial plane, with two relatively long asymmetric C–O bonds (1.25 and 1.28 Å) and a small OCO angle of 124°, smaller than that of the CO_2^- moiety. The structural data demonstrate an analogy of the nature of CO_2 in **4** and **3**, implying both C–O bonds are much elongated with respect to the free cumulene and the OCO angle close to 125°, smaller than that in the radical anion CO_2^-.

The η^1-*C* mode of bonding has been demonstrated to exist in an Ru complex described by Tanaka [27].

A comment is needed about the relative stability of η^1-*C* with respect to η^2-*C,O* complexes. In general, the latter are more stable than the former, which more easily lose CO_2 and can be considered as reversible carriers of the cumulene. The solvent can play a key role in the stabilization of the "M–CO_2" moiety which can be isolated if CO_2 is not strongly "solvated". The donor properties of the metal also determine the stability of the M–C bond and the η^1-*C* complex can often be stabilized by the interaction of the O atom with an external electrophile, as discussed in Sect. 2.2 for the Co–salen complexes.

A comparison is required at this point of such η^1-*C* metal–CO_2 complexes with the adduct of CO_2 with the ionic-liquid 1,3-dimethylimidazolium moiety [28]. The latter, upon reaction with dimethylcarbonate, affords 1,3-dimethylimidazolium-2-carboxylate. In the latter, the generated imidazolium C-carbene binds CO_2 at C by using the free couple of electrons. The C–COO bond length is 1.523(3) Å corresponding to a single C–C bond (two electrons bond) without delocalization from the ring. The two C–O bond lengths are equivalent (1.2398 Å) implying a single charge localized on each oxygen. This causes strong O···H bonding with pairing of two units in the solid state. The resulting $C \rightarrow CO_2$ moiety is formally similar to the $M \rightarrow CO_2$ discussed above: in the solid adduct 1,3-dimethylimidazolium-2-carboxylate, the carboxylate moiety is stabilized through the interaction of the nucleophilic O atoms with the imidazolium-ring hydrogens.

2.4 XRD Structural Characterization of O-End-On Complexes

The first evidence of a linear η[1]-O-coordinated CO_2 was reported in 2004 [29] by Castro-Rodriguez et al. who described the reaction of CO_2 with $(^{Ad}ArO)_3tacnU^{III}$ ($(^{Ad}ArO)_3tacn$ = 1,4,7-tris(adamantyl-5-*tert*-butyl-2-hydroxybenzyl)1,4,7-triazacyclononane). The sterically hindered ligands prevented the η^2-C,O bonding. The product obtained was the complex $(^{Ad}ArO)_3tacnU(\eta^1\text{-OCO})$ **5** in which the end-on CO_2 is η^1-O coordinated to uranium [30]. XRD of the solid has shown that the OCO angle is close to linear (178°), the U–O bond length is 2.351 Å, with a short bond and a longer terminal C–O bond: the terminal C–O bond length is 1.277(4) Å, which is longer than the internal C–O bond distance (1.122(4) Å).

The bonding model foresees an oxidized original U(III) to U(IV) with resonance structures represented by $U^{IV}{=}O{=}C{-}O^- \leftrightarrow U^{IV}{-}O{\equiv}C{-}O^-$ which justify the short and long C–O bonds. Magnetic moment (μ_{eff}) measurements have clearly established that U^{III} was oxidized to U^{IV} with the transfer of virtually one electron to CO_2. This transfer does not bring the electron in the empty LUMO centred on C and, therefore, the molecule is not bent as in η^1-C or η^2-C,O complexes.

2.5 Multinuclear Complexes

The coordination of CO_2 to more than one metal centre is possible. It has been clearly demonstrated through the XRD structure of isolated systems. Complexes bearing the CO_2 molecule coordinated to more than one metal centre in a μ-η^2, μ-η^3, μ-η^4 and μ-η^5 mode have been reported, and the geometries of bonding are shown in Scheme 2.5. Such complexes are in general quite stable.

2.5.1 Side-On Bonded Complexes

An interesting example of a μ-η^4 complex in which both C=O bonds of a single cumulene molecule are bonded to a metal centre was reported by Sadighi et al. in 2007 [31]. The μ-η^2, η^2-CO_2 coordination has been demonstrated to exist in a dinickel complex having the core $[(IPr)Ni]_2(\mu\text{-CO})(\mu\text{-}\eta^2,\eta^2\text{-}CO_2)$ **6**, where IPr is 1,3-bis(2,6-diisopropylphenyl)imidazol-2-ylidene. The CO originates from CO_2 and the two CO_2 ligands bridge two nickel atoms via two Ni–CO bonds in a side-on way (Structure 2.4).

The bridged CO_2 is bent at 133.4° and the CO bond lengths are 1.255 and 1.257 Å, much longer than for free CO_2 (1.16 Å) but comparable to the length found in similar side-on complexes. Similar Ni–biscarbene complexes, $(IMe)_2Ni(0)$, form analogous complexes (IMe = 1,3-dimesitylimidazol-2-ylidene).

Structure 2.4 Reprinted with permission from [31]. Copyright (2007) American Chemical Society

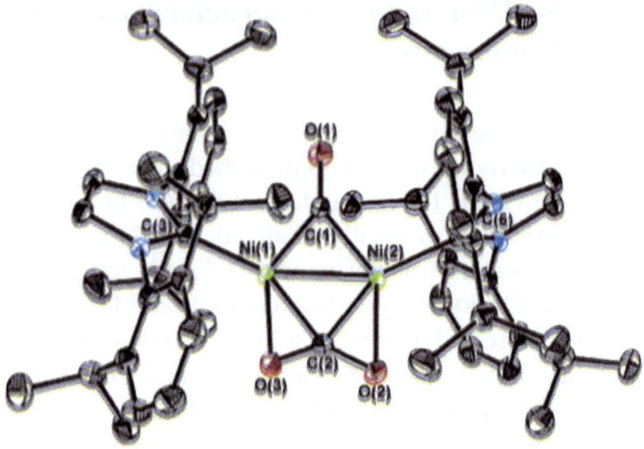

DFT studies have been carried out to demonstrate the bonding of CO_2 to Metal Organic Frameworks (MOF). A bridged coordination has been proposed by Hou et al. [22] in the interaction of CO_2 with M-MOF-74 (M = Mg and Zn), which shows a high CO_2 adsorption capacity. Metal ions are supposed to interact with the O atoms of CO_2, whereas the carbon atom of CO_2 gives strong interactions with the O atoms of the organic linkers of the MOF.

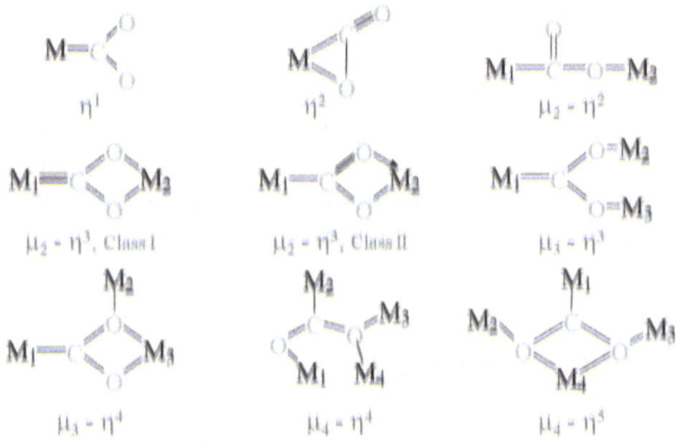

Scheme 2.5 Modes of bonding of CO_2 in multinuclear complexes with the metal centres bonded to O and C. Adapted with permission from [8]. Copyright (1999) Elsevier Science S.A

2.5.2 *O-End-On Bonded Complexes*

An interesting example of a multinuclear complex in which CO_2 is end-on bonded was provided by Chang et al. in 2005 [32] who reported trimeric aluminium–magnesium

Structure 2.5 Reprinted with permission from [32]. Copyright (2005) John Wiley and Sons

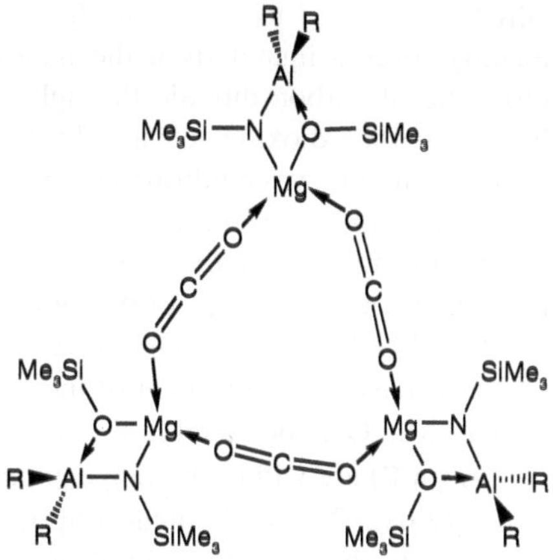

complexes of CO_2 coordinated in the linear μ-O,O′ mode (Structure 2.5). The CO_2 adducts were obtained by adding AlR_3 (R = Me, Et) to a THF solution of $Mg[N(SiMe_3)_2]_2$ at ambient temperature, and bubbling an excess of carbon dioxide through the stirred, ice-cooled reaction mixture (2.1).

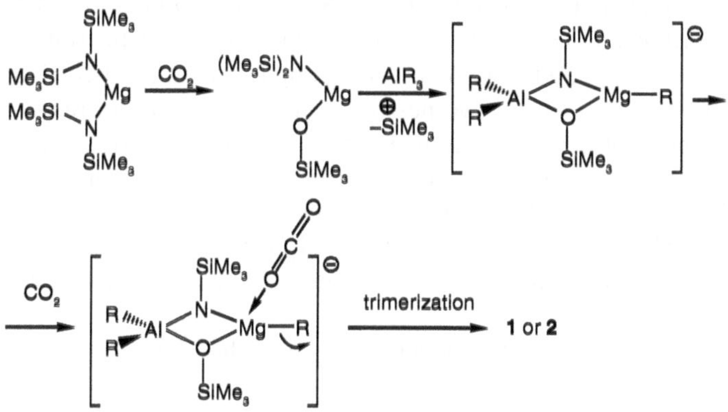

Scheme 2.6 Proposed reaction pathway for the formation of **7** (Me) and **8** (Et). Reprinted with permission from [32]. Copyright (2005) John Wiley and Sons

$$\mathrm{Mg}\left[\mathrm{N(SiMe_3)_2}\right]_2 + \mathrm{AlR_3} + \mathrm{CO_2} \rightarrow \left[\left\{\mathrm{R_2Al}(\mu\text{-}\mathrm{NSiMe_3})(\mu\text{-}\mathrm{OSiMe_3})\mathrm{Mg(THF)_2}(\mu\text{-}\mathrm{O_2C})\right\}_3\right]\ \mathbf{7}: \mathrm{R} = \mathrm{Me};\ \mathbf{8}: \mathrm{R} = \mathrm{Et} \tag{2.1}$$

According to the proposed mechanism (Scheme 2.6), carbon dioxide reacts with $Mg[N(SiMe_3)_2]_2$ to afford the oxo-transfer product $Mg[N(SiMe_3)_2](OSiMe_3)$, which is believed to form a bridged Al–Mg intermediate with AlR_3; this subsequently loses a ligand from the magnesium centre and is attacked by a second molecule of carbon dioxide through the oxygen atom as a weak electron donor. Finally, carbon dioxide acts as a bridging ligand to form a trimer.

The structure was confirmed by X-ray diffraction as having a C_3-symmetry axis, and three almost linear carbon dioxide bridges which form a 12-membered ring. The OCO angles are in the range 169.6°–175.3°, quite close to 180° and larger than the angles found in the η^1- or η^2-complexes. The C–O bond lengths are in the range 1.149–1.233 Å, with a longer and a shorter bond with respect to the free molecule, whose values are reminiscent of the U-complex reported above.

The Mg–O bond lengths (1.944(10)–2.232(12) Å for **7** and 2.004(16)–2.184 (11) Å for **8**) are within the range expected for Mg–O (monodentate donor ligand) bonds (2.012–2.236 Å), suggesting that each oxygen atom donates a lone pair of electrons to the vacant *p*-orbital of magnesium in a dative bond. The vibrational spectra showed that, in both molecules **7** (R = Me) and **8** (R = Et), only a small amount of electron density was transferred from coordinated CO_2 to the more electropositive Mg atoms.

Polyoxometalate anions such as $(H_2PMo_{11}CoO_{40})^{5-}$ were shown to be able to interact with CO_2 affording stable adducts of $(PMo_{11}CoO_{38})_n$ polymeric chains bridged by CO_2 in an axial direction. In the CO_2 adduct $(C_3H_5N_2)_3(C_3H_4N_2)$ $[PMo_{11}CoO_{38}(CO_2)]{\cdot}4H_2O$, single crystal X-ray diffraction showed that CO_2 is (μ-η^1,η^1)-coordinated [33]. The CO_2 moieties show slightly bent arrangements with OCO angles of 158.7° and C–O bond lengths of 1.287 Å. The two O atoms of the CO_2 link Mo^{VI} and Co^{II} from two neighbouring polyoxoanions, respectively. The coordinative bond of the two O atoms from CO_2 with the electropositive Mo^{VI} and Co^{II} ions reduces the electron density of the end O atoms, resulting in the increased OCO bond lengths. Similar results were also obtained for the strictly related compound $(C_3H_5N_2)_4(SiMo_{11}CoO_{38}(CO_2)){\cdot}4H_2O$. Such behaviour is of great interest as it opens the way to the use of polyoxometalates as CO_2 sequestering agents in aqueous medium.

X-ray diffraction [34] on the adduct of CO_2 with the MOF Ni_2(dhtp), where dhtp is 2,5-dihydroxyterephthalic acid, has shown that CO_2 is bound to Ni(II) in a η^1-O-end-on fashion and is slightly bent.

From the above data it is possible to conclude that CO_2 can act as an electron acceptor or an electron donor, or the two features can be present in the same adduct.

Low oxidation state metals and electron-rich moieties interact with the C atom, whereas electron-poor species (metal ions, protons) interact at the O terminal. As a consequence of such ability, CO_2 can give rise to divers bonds with metal centres depending on the nature of the latter.

2.6 Spectroscopic (IR and ^{13}C-NMR) Data for the Complexes Correlated to the Various Modes of Bonding of CO_2

Spectroscopic data of CO_2 metal complexes have been obtained and an attempt has been made to use them as an auxiliary technique for structure work-out in case XRD characterization of metal complexes was not available. Two spectroscopic techniques have mostly been used to this end, namely Infrared (IR, or FTIR, or DRIFT) and ^{13}C-Nuclear Magnetic Resonance (^{13}C-NMR), the former being most informative. Theoretical calculations have also been coupled with spectroscopic studies to get a better insight into the structural properties of the CO_2 complexes. Good quality information can be gathered by carrying out a detailed analysis of multiple data, although individual pieces of information may be quite deceptive.

2.6.1 Infrared Data Relevant to Transition Metal Complexes

Two IR modes have a relevance to the bonding mode of CO_2 to a metal centre and are tentatively used for roughly figuring out the interaction of CO_2 with a metal centre: the ν(C=O) and the δ(C=O).

It is now clear that the structural features of the metal complexes reflect on the position of the IR-bands characteristic of cumulene and a deviation from the original values is found; conversely, it is quite unlikely that the IR-features may be informative about the structure of the bonded CO_2. The infrared properties of CO_2 are summarized in Table 2.1, which collects data relevant to a number of T_M complexes.

As a general principle, as discussed above, any time an interaction occurs at the C atom the molecule deviates from its linearity, being bent from 180° to ca. 130° with a consequent low-energy deplacement of ν(C=O) from 2 340 cm^{-1} to ca. 1 750–1 650 cm^{-1}. When the interaction occurs at the oxygen atom without direct involvement of the C atom, the cumulene molecule almost maintains its linear geometry and the lowering of the energy of the above band is much lower (shift to ca. 2 200 cm^{-1}). The assignment of the bands is more precisely done and more information is gained if isotopic labelling is used, with ^{13}C- and ^{18}O-labelled CO_2. IR data, coupled to variable temperature NMR (VT-^{13}C-NMR, see Sect. 2.6.2) have been used to determine whether the coordinated side-on (O,C) cumulene is a rigid moiety or is fluxional (Sect. 2.7).

The use of isotopic labelled CO_2 has served to interpret correctly the intermediate steps in the geometry modification in solution [35].

Jegat et al. [36] have used labelled CO_2 to discover a general relationship that may better correlate FTIR data to the structure of T_M–CO_2 complexes. They have based their conclusion on the value of $\Sigma\Delta\nu$ that represents the sum of the frequency-

Table 2.1 IR ν_{OCO} bands (cm^{-1}) for several types of CO_2 complex. Adapted with permission from [9]. Copyright (1999) Elsevier

Compound	Type	ν_{asym}	ν_{sym}
$Rh(diars)_2(Cl)(CO_2)$	η^1	1 610	1 210
$Ni(PCy_3)_2(CO_2)$	η^2	1 740	1 140, 1 094
$[Pt(PEt_3)_2(Ph)]_2(CO_2)$	μ_2-η^2	1 495	1 290, 1 190
$[Co(en)_2(CO_2)](ClO_4)\cdot H_2O$		1 512	–
$CpRe(NO)(PPh_3)(CO_2)GePh_3$		1 545	1 048
$CpFe(CO)(PPh_3)(CO_2)Re(CO)_4(PPh_3)$		1 505	1 135
$Cp^*Re(CO)(NO)(CO_2)W(CO)_3Cp$		1 541	1 100
$CpFe(CO)_2(CO_2)SnPh_3$		1 499	1 159
cis,cis-$Ru(bpy)_2(CO)(CO_2)$ $-Ru(bpy)_2(CO)^+{}_2\ 2PF_6^-$		1 507	1 176
$[(PPh_3)_2(Cl)(t\text{Bupy})Ir$ $-(\mu\text{-}O)(\mu\text{-}CO_2)Os(O)_2(t\text{Bupy})_2]^+ClO_4^-$		1 593	1 022
$Cp^*Ir(\mu\text{-}t\text{BuN})(\mu\text{-}CO_2)ZrCp_2$		1 569	1 015
$Ru_2(\mu\text{-}CO_2)(CO)_4[(\mu\text{-}OPr)_2PNEtP(OPr)_2]_2$		1 710	–
$Cp^*Re(CO)(NO)(CO_2)Re(CO)_3(PPh_3)$	μ_2-η^3, Class I	1 437	1 282
$CpFe(CO)(PPh_3)(CO_2)Re(CO)_3[P(OEt)_3]$		1 435	1 252
$CpRu(CO)_2(CO_2)Zr(Cl)Cp_2$		1 348	1 290
$Cp^*Re(CO)(NO)(CO_2)Zr(Cl)Cp_2$		1 348	1 288
$Cp^*Re(CO)(NO)(CO_2)Mo(CO)_2Cp$		1 319	1 285
$Cp^*Re(CO)(NO)(CO_2)W(CO)_2Cp$		1 321	1 287
$CpRe(NO)(PPh_3)(CO_2)SnPh_3$	μ_2-η^3, Class II	1 395	1 188
$Cp^*Fe(CO)_2(CO_2)SnPh_3$		1 450	1 152
$Cp^*Re(CO)(NO)(CO_2)SnPh_3$		1 429	1 188 or 1 175
$[Cp^*Re(CO)(NO)(CO_2)]_2SnMe_2$		1 469	1 186
$(AdArO)_3tacnU(\eta^1\text{-}OCO)$	η^1	2 188	
$\{R_2Al(\mu\text{-}NSiMe_3)(\mu\text{-}OSiMe_3)Mg(THF)_2(\mu\text{-}O_2C)\}_3$		2 267(R = Me)	
		2 275(R = Et)	
$(C_3H_5N_2)_3(C_3H_4N_2)(PMo_{11}CoO_{38}(CO_2))\cdot 4H_2O$		2 169	

shifts observed for the two $\nu(CO)$ in ^{13}C and ^{18}O complexes. The relationships they have elaborated can be expressed as categorized below:

$$\text{side-on coordination}: \Sigma\Delta\nu(^{13}C) > \Sigma\Delta\nu(^{18}O) \text{ and } \Sigma\Delta\nu(^{18}O) < 60\,cm^{-1};$$
$$\eta^1\text{-C coordination}: \Sigma\Delta\nu(^{13}C) > \Sigma\Delta\nu(^{18}O) \text{ with } 60\,cm^{-1} < \Sigma\Delta\nu(^{18}O) > 70\,cm^{-1};$$
$$\text{end-on coordination}: \Sigma\Delta\nu(^{13}C) < \Sigma\Delta\nu(^{18}O) \text{ with } \Sigma\Delta\nu(^{18}O) > 70\,cm^{-1}.$$

Such a criterion requires use of expensive double-labelled CO_2 and a quite detailed analysis of the IR spectrum. Additional observations are reported by the same authors which may help in inferring the structure of T_M–CO_2 complexes in the absence of XRD data. For example, in η^1-C-coordinated complexes, the frequency splitting between the two ν(CO) stretching modes is observed to be less than 400 cm^{-1}.

O-end-on complexes show quite unique IR properties. The U^{IV} f^2 ion [29] exhibits a ν(CO) band at 2 188 cm^{-1}, shifted by 60 cm^{-1} using $^{13}CO_2$ and the out-of-plane bending mode γ(C=O), located in the 500–650 cm^{-1} region, exhibits preponderant ^{18}O effect, in contrast to other coordination modes, for which a larger ^{13}C effect is observed (namely $\Delta\gamma(^{13}C) = 10$–$20\ cm^{-1}$, vs $\Delta\gamma(^{18}O) = 5\ cm^{-1}$).

Other IR data relevant to O-bonded complexes are: $(C_3H_5N_2)_3(C_3H_4N_2)(PMo_{11}CoO_{38}(CO_2))\cdot 4H_2O$: 2 169 cm^{-1}, $^{13}CO_2$ red-shift: 61 cm^{-1}; $\{R_2Al(\mu\text{-}NSiMe_3)(\mu\text{-}OSiMe_3)Mg(THF)_2(\mu\text{-}O_2C)\}_3$: 2 267 (R = Me) and 2 275 cm^{-1} (R = Et), $^{13}CO_2$ red-shift: 73 and 65 cm^{-1}, respectively. In the above CO_2 complexes the shift of ν_a(OCO) absorption relative to ν_3 of free CO_2 is modest if compared with that observed when the heterocumulene is η^1-C or side-on (η^2-C, O) coordinated to a metal center. For instance, $Rh(diars)_2(Cl)(CO_2)$ [23] shows absorptions assigned to η^1-C-coordinated CO_2 at 1 660 (s) and 1 210 (s) cm^{-1}, whereas $Ni(CO_2)(PCy_3)_2$ [3] exhibits intense bands because of side-on coordination of the heterocumulene at 1 740 (vs) and 1 150 (s) cm^{-1}.

It is interesting to note that in η^1-C-coordinated CO_2 complexes the frequency splitting between the two ν(OCO) stretching modes is often less than 400 cm^{-1}, and, in end-on (η^1-OCO) complexes, the out-of-plane bending mode γ(C=O), absorbing in the 650–500 cm^{-1} range, shows a pronounced ^{18}O effect, different from the other coordination modes, where a larger ^{13}C effect is observed ($\Delta\gamma(^{13}C) = 10$–$20\ cm^{-1}$ vs $\Delta\gamma(^{18}O) = 5\ cm^{-1}$) [35, 36].

The IR spectra of the O-end-on complexes $Mg \leftarrow O{=}C{=}O \rightarrow Mg$ [32] exhibit strong absorptions at 2 267 and 2 275 cm^{-1}, slightly red-shifted by 73 and 65 cm^{-1} relative to free CO_2, whereas side-on complexes usually exhibit shifts greater than 300 cm^{-1}.

The adduct of CO_2 with $(H_2PMo_{11}CoO_{40})^{5-}$ [33] shows in the IR spectrum an absorption band at 2 169 cm^{-1}, shifted to 2 108 cm^{-1} upon ^{13}C labelling.

An extensive analysis of IR spectra of CO_2-T_M complexes can be found in [8, 9].

2.6.2 NMR Data for Transition Metal Complexes

As anticipated in Sect. 1.6.3, ^{13}C NMR spectra (or else ^{17}O-spectra) can contribute to structure investigation in solution and to gathering information about the behaviour of the CO_2-T_M complexes in the liquid phase. The multiplicity of peaks when P-ligands are present on active metal nuclei can help one to understand the nature of bonding. The shift from free CO_2 (124 ppm) may be an indication of how the C=O is affected by coordination to a metal. So, for example, O-end-on complexes show ^{13}C-resonance slightly up-field shifted (around ca. 121–110 ppm) with respect to the free cumulene, whereas side-on complexes usually show ^{13}C resonances downfield shifted around 200 ppm. The shift is obviously correlated to the transfer of electron density to/from the coordinated CO_2 and to the molecular geometry.

Table 2.2 reports a selection of data which can be helpful in illustrating the relation structure-NMR properties of the CO_2-T_M complexes.

Coordination of the heterocumulene to a metal center induces the shift of the ^{13}C resonance of coordinated CO_2. A modest up-field shift has been measured for end-on complexes, such as $\{R_2Al(\mu\text{-}NSiMe_3)(\mu\text{-}OSiMe_3)Mg(THF)_2(\mu\text{-}O_2C)\}_3$ (120.87 ppm, R = Me; 120.76 ppm, R = Et) [32], or $(C_3H_5N_2)_3(C_3H_4N_2)(PMo_{11}CoO_{38}(CO_2))\cdot 4H_2O$ (113.86 ppm) and $(C_3H_5N_2)_4(SiMo_{11}CoO_{38}(CO_2))\cdot 4H_2O$ (112.50 ppm) [33].

However, other known coordination modes, such as the side-on mode and the variety of bridging modes [8], shift to low fields the ^{13}C resonance of the CO_2 ligand relative to the value found for free CO_2.

Values in the range 150–270 ppm have been reported for different CO_2 complexes, and the chemical shift of coordinated CO_2 may differ widely even in complexes showing the same coordination mode.

Thus, in the side-on complexes $Ni(\eta^2\text{-}CO_2)(PCy_3)_2$ (159.28 ppm, at 298 K) and $Pd(\eta^2\text{-}CO_2)(PMePh_2)_2$ (166.2 ppm) coordinated CO_2 resonates at higher fields than in the metallocene derivatives $(Cp^*)_2Nb(\eta^2\text{-}CO_2)(CH_2SiMe_3)$ (200.5 ppm) and

Table 2.2 NMR data for carbon dioxide transition metal complexes. Adapted with permission from [37]. Copyright (1996) Published by Elsevier B.V

Complex	$\delta M\text{-}^{13}CO_2$	*d* M-C (Å)
$(PCy_3)_2Ni(\eta^2\text{-}C,O)$	159.28	1.84
$[(dppp)(CO)_3Re(\mu_2\text{-}\eta^3\text{-}CO_2)\ Re(CO)_3(dppp)]$	191.7	2.191(13)
$[(cp^*)(CO)(NO)Re(\mu_2\text{-}\eta^3\text{-}CO_2)W(Cp)_2][BF_4]$	247.4	2.04(4)
$[(cp^*)(CO)(NO)Re(\mu_2\text{-}\eta^3\text{-}CO_2)Ti(tmtaa)][BF_4]^a$	226.2	2.081(9)
$[(cp^*)(PPh_3)(NO)Re(\mu_2\text{-}\eta^3\text{-}CO_2)SnPh_3]$	207.6	2.058(9)
$[(cp^*)(CO)(NO)Re(\mu_2\text{-}\eta^3\text{-}CO_2)SnPh_3]$	206.9	2.100(9)
$[(cp^*)(CO)(NO)Re(\mu_2\text{-}\eta^2\text{-}CO_2)SnMe_3]$	196.1	2.103(5)
$[(cp^*)(CO)(NO)Re(\mu_2\text{-}\eta^3\text{-}CO_2)Re(CO)_3(PPh_3)]$	206.4	2.089(6)
$[(cp^*)(CO)(NO)Re(\mu_2\text{-}\eta^3\text{-}CO_2)Zr(cp)_2(Cl)]$	217.9	2.097(5)
$[(cp)(CO)(PPh_3)Fe(\mu_2\text{-}\eta^3\text{-}CO_2)Re(CO)_3\{P(OEt)_3\}]$	217.8	2.08(1)
$[(cp)(CO)(PPh_3)Fe(\mu_2\text{-}\eta^3\text{-}CO_2)SnBu_3]$	211.9	1.994(3)
$[(cp)(CO)(PPh_3)Fe(\mu_2\text{-}\eta^3\text{-}CO_2)SnMe3]$	245.9	1.932(7)
$[(cp^*)(CO)_2Fe(\mu_2\text{-}\eta^3\text{-}CO_2)SnPh_3]$	228.2	1.931(5)
$[(\eta^5\text{-}indenyl)(CO)(PPh_3)Fe(\mu_2\text{-}\eta^3\text{-}CO_2)SnPh_3]$	220.3	1.936(4)
$[(cp^*)(CO)(NO)Re(\mu_2\text{-}\eta^3\text{-}CO_2)Mo(CO)_2(cp)]$	220.4	1.934(6)
$[(CO)_5Re(\mu_2\text{-}\eta^3\text{-}CO_2)Re(CO)_4]_2$	218.4	1.956(3)
$[(cp^*)(CO)_2Ru(\mu_2\text{-}\eta^3\text{-}CO_2)Zr(cp)_2(Cl)]$	220.4	2.050(4)
$[(PMe_2Ph)_3Os(\mu\text{-}H)_2(\mu_3\ \text{-}\eta^3\text{-}CO_2)Rh_2(cod)_2]$	193	2.062(19)
$[(cp^*)Ir(\mu\text{-}tBuN)(\mu_2\text{-}\eta^2\text{-}CO_2)Zr(cp)_2]$	164.0	2.098(10)
$[(Et_3P)_2Pt(\mu_2\text{-}\eta^2\text{-}CO_2)Ge\{N(TMS)_2\}]$	171.4	2.086(9)

$Cp_2Ti(\eta^2\text{-}CO_2)(PMe_3)$ (212.3 ppm) or a few bis-(η^2-CO_2) complexes of molybdenum (201–217 ppm) [5]. It is unclear whether these differences in the chemical shift of side-on coordinated CO_2 are because of the difference in the coordination geometry of the various adducts or are related to the different nature of coordinating metal center.

Very high values of chemical shift have been found in a few complexes wherein CO_2 bridges two or more metal centers. For instance, in the bridged complexes $[(\eta^5\text{-}Cp)(CO)_2Fe{=}CO_2W(\eta^5\text{-}Cp)_2]PF_6$ and $[(\eta^5\text{-}Cp^*)(CO)_2Ru{=}CO_2W(\eta^5\text{-}Cp)_2]BF_4$ the coordinated CO_2 molecules resonate at 259.5 and 251.5 ppm, respectively, in agreement with their alternative formulation as dimetallated dioxocarbene ligands [38].

The coupling constant of the ^{13}C nucleus of coordinated CO_2 with the nuclei of the metal center or other ligands is a useful spectral parameter which may help to distinguish a CO_2-adduct from a metal–carbonate complex. For instance, the CO_2 ligand in Ni(dcpp)(η^2-CO_2) absorbs at 164.2 ppm, very close to 166.2 ppm, which is the resonance of carbonate ligand in Ni(dcpe)(CO_3). However, the carbonate ligand in Ni(dcpe)(CO_3) resonates as singlet, whereas coordinated CO_2 in Ni(dcpp)(η^2-CO_2) gives a doublet with a $^2J(C\text{–}P_{trans})$ equal to 45 Hz [39].

2.7 Fluxionality of the CO_2 Molecule Coordinated to Transition Metal Systems

VT-multinuclear NMR studies have allowed one to establish unequivocally whether the M-coordinated-CO_2 framework is a rigid structure or not. To this end, it is necessary that other nuclei (such as the P-donor atoms of phosphanes) are bonded to the metal centre so that, when the metal favours the coupling, it is possible to observe whether the solid state structure is maintained in solution or the framework exhibits some fluxionality. Complex **1** has been extensively investigated and the fluxional behaviour has been well explained by coupling IR and NMR studies [6, 40].

The dynamics of $Ni(\eta^2\text{-}CO_2)(PCy_3)_2$ in solution has been investigated by Aresta's group [6]. Both in the solid state and in solution, at temperatures below 200 K, $Ni(\eta^2\text{-}CO_2)(PCy_3)_2$ presents identical structures, as ascertained by CP-MAS-NMR and VT-NMR experiments. The CP-MAS ^{31}P-NMR spectrum of solid $Ni(\eta^2\text{-}CO_2)(PCy_3)_2$ showed two main peaks at 48.9 and 20.9 ppm, respectively, as expected for a planar arrangement of ligands around the Ni atom. Moreover, the CP-MAS ^{13}C-NMR spectrum of a $^{13}CO_2$-enriched solid sample of $Ni(\eta^2\text{-}^{13}CO_2)(PCy_3)_2$ exhibited the peak of CO_2 ligand at 159.28 ppm. In solution, below 200 K, two distinct signals were found in the ^{31}P spectrum for the phosphane ligands (51.7 and 20.9 ppm; $J(P\text{–}P) = 39.6$ Hz), showing that, at these temperatures, CO_2 is rigidly anchored at the planar nickel center in the (η^2-C,O)-coordination mode (Fig. 2.2). In accordance with the above findings, in the low-temperature

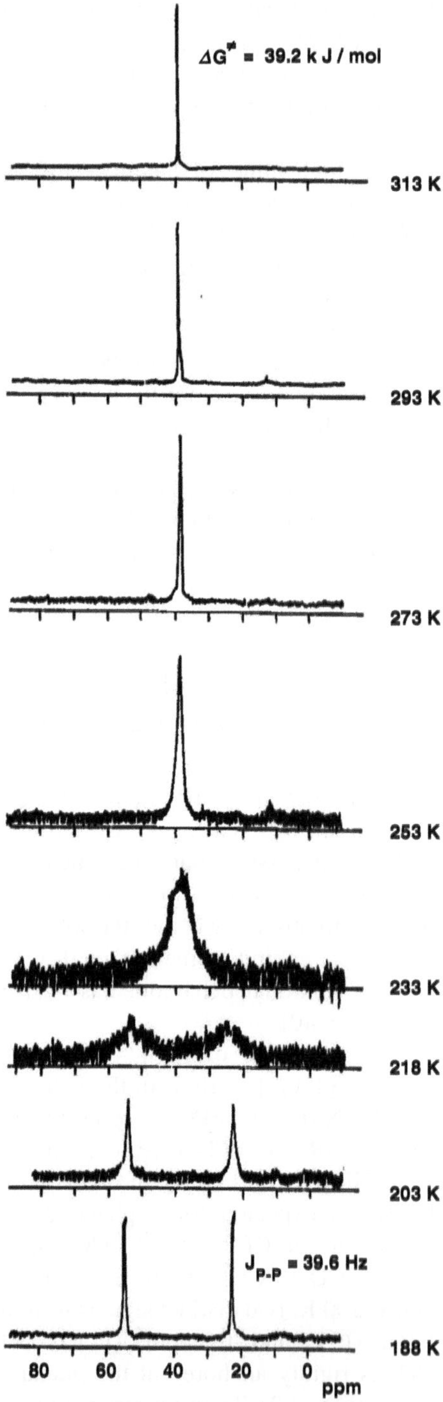

Fig. 2.2 VT $^{31}P\{^1H\}$ spectra of $Ni(CO_2)(PCy_3)_2$. Reprinted with permission from [6]. Copyright (1992) American Chemical Society

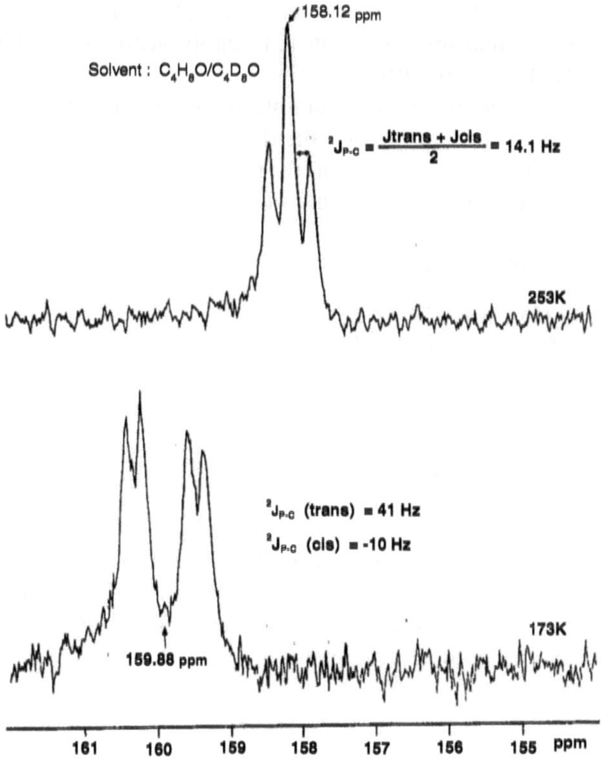

Fig. 2.3 $^{13}C\{^1H\}$ spectra of $Ni(^{13}CO_2)(PCy_3)_2$ at 253 K (*top*) and 173 K (*bottom*). Reprinted with permission from [6]. Copyright (1992) American Chemical Society

limiting ^{13}C spectrum (173 K) of $Ni(\eta^2\text{-}^{13}CO_2)(PCy_3)_2$ the resonance of the CO_2 ligand appeared as a doublet of doublets at 159.88 ppm with J(C–P_{trans}) and J(C–P_{cis}) of 41 and 10 Hz, respectively (Fig 2.3).

Upon increasing the temperature above 233 K, the two ^{31}P signals coalesced into a singlet, showing that the phosphane ligands became equivalent. $\Delta G^{\ddagger}$ for the exchange process was calculated to be 39.3 kJ/mol (9.4 kcal/mol). In the fast-exchange regime, at 253 K, a triplet at 158.12 ppm was found for the coordinated CO_2, with an averaged value of J(P–C) equal to 14.1 Hz.

The persistence of P–C coupling, even in the fast-exchange regime, ruled out the hypothesis that equilibration of PCy_3 ligands might involve a fast CO_2 decordination-recoordination mechanism. Averaging of the phosphane ligands ^{31}P resonances must involve an intramolecular process. The following mechanisms can be envisaged (Scheme 2.7):

- Fast rotation around the Ni η^2-CO bond (*i*)
- Equilibration as a result of the formation of an η^1-C intermediate (*ii*)
- Equilibration through the formation of an η^1-O intermediate (*iii*)

The available NMR data did not allow differentiating between mechanisms (*i*)–(*iii*) in Scheme 2.3. However, additional FT-IR studies and CAS-SCF calculations of $Ni(CO_2)L_2$ isomers showed that formation of the end-on coordinated intermediate (mechanism (*iii*)) may reasonably account for the fluxional behavior $Ni(\eta^2$-$CO_2)(PCy_3)_2$ [6, 40].

Another interesting example of fluxionality has been provided by Carmona's group [41]. They studied the dynamic behavior of *trans*-$L_4Mo(\eta^2$-$CO_2)_2$ adducts, showing that it consists of a synchronous motion of the two CO_2 ligands in which both molecules rotate in the same direction (Structure 2.6).

Structure 2.6 Reprinted with permission from [41]. Copyright (1991) American Chemical Society

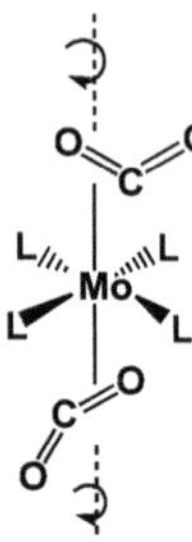

The above data emphasize the usefulness of NMR spectroscopy in demonstrating the state of a CO_2 complex in solution and comparing the solution state with the solid state. The solution mode of being of a T_M–CO_2 complex may be relevant to its reactivity with several agents.

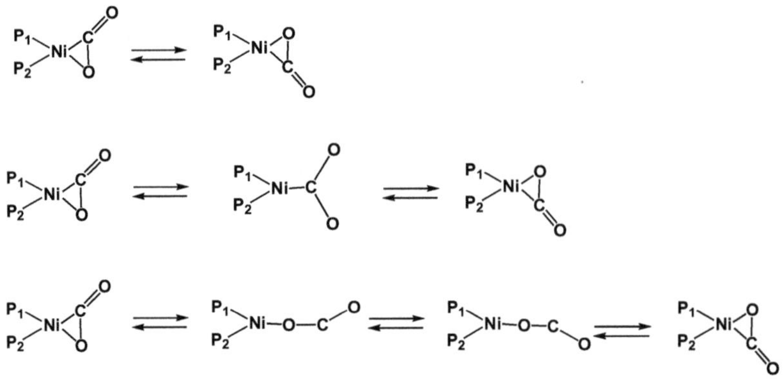

(*i*)

(*ii*)

(*iii*)

Scheme 2.7 Fluxional behavior of $Ni(CO_2)(PCy_3)_2$. Reprinted with permission from [6]. Copyright (1992) American Chemical Society

The combination of IR and NMR spectroscopies may be of great help, not only in inferring the structure of a CO_2-metal complex when XRD data are not available, but also for understanding their reactivity in solution.

2.8 Interaction of CO_2 with Metal Atoms in Low-Temperature Solid-Inert-Gas Matrices

Matrix isolation is a technique utilized to immobilize a "reactive species", such as naked metal atoms, at very low temperature (a few K) in an inert matrix such as a noble gas [42]. Atoms can be thermally produced and the combination of the matrix "cage effect" and low temperature makes that dynamic processes almost hindered. As mentioned above, the temperature of a few K, often below 10 K, makes the thermal factor kT as low as <0.083 kJ mol^{-1}, preventing most reactions from occurring. Under such conditions weakly bonded adducts can be frozen and studied. By raising the temperature by a few K it is possible to observe reactions. The most used techniques for generating atoms are vapour deposition and laser ablation. The latter should not produce atoms with excess kinetic energy which might overpass energy barriers and prevent the observation of adducts and reaction intermediates. Photochemical effects must be controlled, as they may occur, promoted by the light emitted during the ablation process.

Infrared spectroscopy is the key technique for the characterization of adducts and is often coupled to theoretical calculations (mainly DFT) for a good standard identification of adducts. The classes of adducts which have been investigated can be separated into two, namely (1) the adducts with Group 1, 2, 13 elements and (2) the adducts with T_M atoms.

Group 1 metals form MCO_2 complexes (IR around 1 600 cm^{-1}) having various structures (C_{2v}, C_s), and may lead to the formation of oxalate-like species MC_2O_4 by concentration effects or annealing. The latter may decompose into carbonates M_2CO_3 and CO above 200 K [43–46]. The decreasing order of reactivity of alkaline metals with CO_2 is Li > Cs > K > Na, the latter being active in neat CO_2-matrices.

Group 2 metals Be [47], Mg [48–50] and Ca [51] have been studied, both experimentally and/or theoretically and shown to promote the reaction $M + CO_2 \rightarrow MO + CO$, endothermic by 26, 66 and 35 kcal mol^{-1} for Be, Mg, and Ca atoms, respectively. The reaction steps through the cyclic (η^2-O,O)-MOCO structures having barriers of 23, 20 and 14 kcal mol^{-1}, respectively. Thermal Mg atoms [48] led to the formation of the MgOCO C_{2v} cyclic structure (calculated OCO angle of 128°) and IR bands located at 1 580, 1 385 and 866 cm^{-1} with only a small number of Mg atoms being active (even in pure CO_2). The product of condensation of Mg atoms with a $CO_2/C_2H_4/Ar$ mixture exhibits IR bands at 1 768, 1 284 and 1 256 cm^{-1}, assigned to the formation of a five-member cycle $MgC_2H_4CO_2$ through the formation of a C–C bond between ethylene and carbon dioxide (binding energy 18 kcal mol^{-1}).

B [52, 53] promotes the CO_2 to CO conversion and the reaction $B + CO_2 \rightarrow BO +$ CO is exothermic by 64 kcal mol^{-1}. The recombination of BO and CO leads to OBCO ($\nu_{CO} = 1\ 863$ cm^{-1}) and is barrier-free [53].

Al atoms react with CO_2 in Ar matrices [54] leading to the formation of $AlCO_2$ moieties from which two isomers originate, according to the temperature. At low temperature, the adduct presents a C_s symmetry, with two non-equivalent CO bonds, whereas at higher temperature (25 K) a ring structure is formed in which Al interacts with both oxygen atoms. Above 30 K, the cluster structure yields Al_2O and CO.

The first attempts to isolate T_M–CO_2 systems were made by Ozin et al. [55] who used Ag and Au atoms. Fragmentary information remained available on T_M-systems (a π-adduct reported for Cu [56] and the formation of OM(CO) moieties with Cr and W [57]) until a systematic study on metals belonging to the first row was made [58, 59]. It was shown that early metals (Ti, V, Cr) cause splitting of one of the C=O bonds and the formation of the O=M(CO) moiety, whereas late metals form adducts of formula M–CO_2 in which the cumulene shows divers modes of bonding dependent on the metal. A side-on bonding was found for Ni, an O-end-on for Cu, and a η^1-C for Fe and Co. Interestingly, the same coordination modes were found for atoms and metal complexes (see above).

DFT calculations have been used for modelling the conversion of M–CO_2 into O=M(CO). Papai et al. [60] have shown that no energy barrier exists for the conversion of Ti(CO_2) into O=TiCO with ground state Ti (Fig. 2.4). The comparison of the calculated data with those Ti(CO_2) complexes in various coordination modes were located on the triplet and quintet potential surfaces, with the triplet state (O,O) coordination being the most stable, and lying above the OTiCO molecule by about 30 kcal mol^{-1}. Hwang et al. [61] have shown that the most energetically favourable reaction mechanism is the insertion of the Ti atom into the C=O bond of the η^2(C,O) Ti(CO_2) complex, to produce the triplet OTi(CO) molecule.

It should be noted that the early transition metal atoms (Sc, Ti, V) are more active than Group 3 metal atoms in promoting the M–CO_2 into O=M(CO) conversion, most probably because of a better stabilization of the C=O fragment formed from CO_2.

Ni has been intensively investigated and shown that the reaction is initiated with an electron transfer from the Ni atom to the CO_2 molecule, followed by the C=O

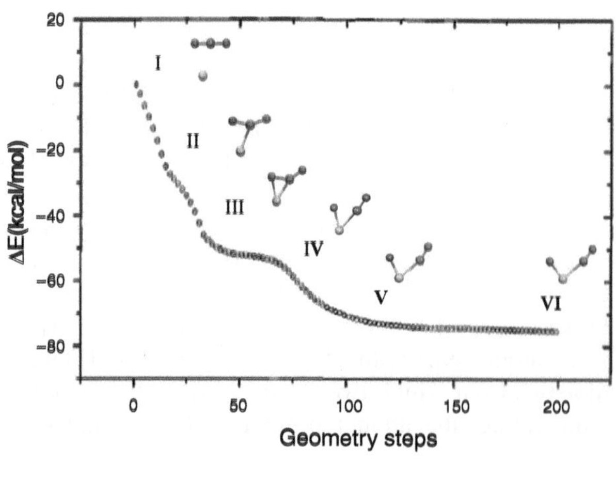

Fig. 2.4 Mechanism of conversion of Ti–CO_2 into O=Ti(CO). Adapted with permission from [60]. Copyright (1997) American Chemical Society

bond cleavage: such steps occur in a concerted mode together with the charge transfer processes, without a preliminary formation of a η^2-CO_2 complex.

Metals belonging to the second and third rows, which are used in several catalytic processes, have been intensively studied. Zr [62], Nb [63] and Ta [64] atoms show the formation of OM(CO) and $O_2M(CO)_2$ products. The process is barrierless for Zr, whereas Nb and Ta transfer electrons to CO_2 and promote the formation of adducts as intermediates. Laser ablated Co and Rh atoms [65] show similar reactions with formation of neutral species such as OM(CO), $O_2M(CO)$ and $OCo_2(CO)$. Laser ablated Re, Ru and Os atoms in Ar and Ne matrices [66, 67] afford divers species of the type OM(CO), $O_2M(CO)$ and $O_2Os(CO)_2$ (formed by reaction of OOs(CO) with a second CO_2 molecule), and $OCRu(O_2)(CO)$ obtained by addition of a CO_2 molecule to ORu(CO). Osmium is more reactive than ruthenium.

Heavy metals such as lanthanides or actinides have been investigated. La atoms [65] afford OLa(CO) as the primary product. Upon UV-visible photo-rearrangement, La(η^2-CO)O and OLa(η^2-CO) are formed. Noticeably, the comparison of the CO stretching frequency of OMCO species in the series Sc, Y and La indicates an increase of the metal d orbital → CO π^* back-donation from Sc to La. In the case of uranium [68], pulsed laser evaporated U atoms are sufficiently energetic to insert into CO_2 upon condensation in argon matrices, forming OUCO (1 799, 804 cm^{-1}), with a larger charge transfer from OU to CO than for OBCO [53].

On the basis of the above studies, it can be ruled out that metal systems interact with CO_2 in low temperature matrices with formation of metal complexes, which may or may not be the precursors of the O=M(CO) species, the product of a *formal insertion* of the M-atom into one of the two C=O bonds of CO_2. Although "naked" metal atoms cannot be compared to "LnM" systems, the ensemble of the two types of metal systems allow the identification of general trends in the periodic table for what concerns the description of the modes of bonding of CO_2 in $M(CO_2)$ moieties and can depict a reaction mechanism for their further conversion into O=M (CO) species.

2.9 Interaction of Metal Cations with CO_2 in the Gas Phase

CO_2 reacts with early transition metal cations (Sc, Ti, Y, Zr, Nb, La, Hf, Ta, W) [68, 69] with O atom transfer to afford "MO^+ + CO". The driving force is the high O-affinity of the cations which leads to stable MO^+ moieties.

Transition metals from Groups 8 to 12 bind CO_2 and form $M^+(CO_2)_n$ adducts with n varying from 1 to 4. Metal cations are electrostatically bound to the negative quadrupole moment of CO_2 in an end-on way. Obviously, the major stabilizing strength is the donation O-to-metal cation which does not contribute to the stabilization of the adduct as the π-back-donation is close to zero [70]. The M^+–CO_2 bond distances depend on both the size of the cation and its electronic configuration, and the binding energies follow the trend of the ion size and the repulsion between the metal cation d-orbitals and the CO_2 occupied orbitals, the order of repulsion being $3d\sigma > 3d\pi > 3d\delta$. The binding energies usually vary between 10 (Mn and Fe) and 25 kcal mol^{-1}.

CO_2-clusters are also formed with V, Ni and Fe adducts. The general formula is $M^+(CO_2)_n$, with n up to 11 for V [71], and 14 for Ni [72] and Fe [73]. Reactions of iron clusters Fe_n^+ ($n = 1$–18) with CO_2 give Fe_nO^+ as primary product. $Fe_nCO_2^+$ adducts are formed for large cluster ions ($n > 9$) at low kinetic energies [74]. In agreement with theoretical calculations [70–72], the dissociation energy of the linear end-on Fe^+–CO_2 adduct has been measured to be 14.3 kcal mol^{-1} [75]. Reaction of ground state copper Cu^+ [76] with CO_2 might form CuO^+ and CO, a reaction which is endothermic and spin limited. Pt^+ [77] forms an associative nonlinear end-on complex $Pt^{+}\cdot OCO$, with subsequent formation of the final product O–Pt^+–CO. Al^+ [78] produces AlO^+ + CO. Finally, U^+ can give UO^+ + CO, an exothermic reaction by 2.1 eV, which has a low cross section [79].

2.10 Reactions of Coordinated CO_2

It is uncontroversial that the coordination of CO_2 to a metal centre causes a perturbation of the cumulene molecular structure, which is made evident by the changes of the properties of CO_2. The shift from the free molecule may be quite important or less relevant, depending on the mode of bonding to the metal centre. However, the deviation of the molecular structure from linearity, the elongation or shortening of C–O bonds, and the shift of the ν(C=O) IR absorptions towards lower-energy and low field appearance of ^{13}C-NMR signals are evidence of the changes occurring in coordinated CO_2 with respect to the free molecule. One can now ask whether such changes give an indication of the reactivity of the bonded cumulene, or else whether the coordination of CO_2 is a strict pre-requisite for further conversion, or else if there is a preferred bonding mode for CO_2 conversion, which must be the electronic properties of the M–CO_2 system for this may represent a reactive form of CO_2. There is no general answer to such questions, as the behaviour of M–CO_2 systems is very dependent on the reaction conditions and the co-reagents. Whether a substrate reacts with coordinated CO_2 or free CO_2 reacts with a coordinated substrate, or even whether a three-centre reaction takes place implying the metal centre, CO_2 and the substrate, is not a priori definable in a general way and whether coordinated CO_2 is necessary or not is the leitmotif for several reactions in which CO_2 is implied. Let us consider a simple example: the reaction of **1** with ethene. It has been shown in previous paragraphs that **1** in solution may not be a rigid molecule: temperature and solvent determine for CO_2 the mode of being attached to the metal. If ethene is added to a solution of **1** in toluene at room temperature, the exchange reaction depicted in (2.2) takes place

$$P_2Ni(CO_2) + CH_2 = CH_2 \rightarrow P_2Ni(CH_2 = CH_2) + CO_2 \qquad (2.2)$$

and the ethene-Ni complex is isolated in quantitative yields. Increasing the pressure of CO_2 and ethene and lowering the temperature favours the coupling of both reagents (2.3) onto a putative "(L–L)Ni" moiety. At 240 K ethene and CO_2 may also couple at sub-atmospheric pressure of CO_2. The coupling is favoured when bidentate ligands (dppe, depe, bpy) are used:

$$(L - L)Ni + CO_2 + C_2H_4 \rightarrow (L - L)\underline{NiCH_2 - CH_2 - C(O)O} \qquad (2.3)$$

Such a reaction, as stated in the rest of this book, strongly depends on the metal as to what is the nature of the C–C coupling product.

Nevertheless, although several reactions of CO_2 can be categorized as insertion of CO_2 into active M–E bonds (E = H, O, C, N, P, S and others), there are reactions which clearly require coordinated CO_2 to occur. The following chapters make a detailed analysis of the reaction mechanism of CO_2 conversion; here we discuss a few examples of reactivity of coordinated CO_2 with nucleophiles or electrophiles.

2.10.1 Reaction of Coordinated CO_2 with Electrophiles and O-Transfer from CO_2 to Produce CO

The O-transfer reaction has great relevance to energy production using the couple CO_2–CO.

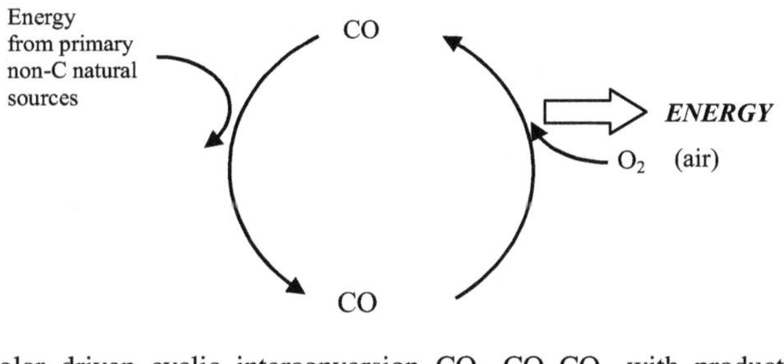

Scheme 2.8 Solar driven cyclic interconversion CO_2–CO–CO_2 with production of thermal energy. Reprinted with permission from [80]. Copyright (2014) American Chemical Society

When the conversion of CO_2 into CO upon coordination to a metal centre was first discovered [3, 81] it opened up great expectations, as running such a cycle as that represented in Scheme 2.8 would be of great practical interest.

It must be said that the non-catalyzed CO_2 conversion into CO and "O" requires high temperature (>1 200 K) or high radiation energy (<160 nm) (see Chap. 1). The effect of coordination to a metal is now quite evident as the "deoxygenation" of CO_2 can be performed at room temperature using the appropriate metal centre and "oxophile". Unfortunately, the best O-acceptors are phosphanes (PR_3) [82, 83] which are converted into their oxides ($O{=}PR_3$), products of very limited utility and difficult to reduce back to the parent compounds without the use of energy-intensive reducing agents (metals) which cancel out the benefit of having the CO_2 splitting running at room temperature. Similar reactions have been shown in previous paragraphs to occur with metal atoms, which also cannot be easily and advantageously converted back to their reduced forms. As a matter of fact, today the O-transfer process (or CO_2 reduction to CO) is still under examination and theoretical calculations are useful tools for the design of metal systems (MOFs are the preferred ones [84]) which might be used as promoters of cyclic processes. The theoretical approach is particularly interesting when one wishes to explore the role of particular metals. In fact, DFT calculations have shown that although the d^3 Mo $(NRAr)_3$ triamide complexes (R = aliphatic group, Ar = aryl moiety) are known not to be able to react with CO_2, by changing the central atom to a d^2 configuration a favourable interaction with CO_2 was postulated.

Discovering a cyclic process that runs on CO_2–CO and makes use of perennial energy sources (solar energy) as energy input would be of great interest. Metal oxides have been studied as catalysts for such reaction with use of Solar Power Concentrators allowing one to operate at high temperature. An interesting metal mediated deoxygenation of CO_2 is implemented based on reactions (2.4)–(2.7).

$$MO_x + \text{Energy} \rightarrow MO_{x-1} + 1/2O_2 \tag{2.4}$$

$$MO_{x-1} + CO_2 \rightarrow MOx + CO \tag{2.5}$$

$$CO + 1/2O_2 \rightarrow CO_2 + \text{Energy} \tag{2.6}$$

$$CO + 2H_2 \rightarrow \text{Liquid fuels} + H_2O \tag{2.7}$$

The produced CO can either be used as a fuel (2.6) or used in combination with H_2 obtained from water splitting for the production of liquid fuels (2.7).

An intriguing use of CO_2 as O-transfer agent driven by metal catalysts has been investigated by Park [85] who used CO_2 as oxidant in combination with dioxygen. The reaction investigated was the dehydrogenation of ethylbenzene to afford styrene (2.8). As H_2O and CO can be processed to afford H_2 and CO_2 (WGS) (2.9), the net reaction is represented in

$$C_6H_5 - CH_2CH_3 + CO_2 \rightarrow C_6H_5 - CH = CH_2 + CO + H_2O \tag{2.8}$$

$$CO + H_2O \rightarrow CO_2 + H_2 \tag{2.9}$$

$$C_6H_5 - CH_2CH_3 \rightarrow C_6H_5 - CH = CH_2 + H_2 \tag{2.10}$$

(2.10) in which CO_2 acts as "promoter" of the dehydrogenation, favouring milder conditions and higher selectivity with respect to pure dioxygen. Reactions (2.8)–(2.10) are discussed in detail in Chap. 7. Coordinated CO_2 undergoes attack by several electrophiles, as shown in Scheme 2.9. The proton "H^+" is a common reagent.

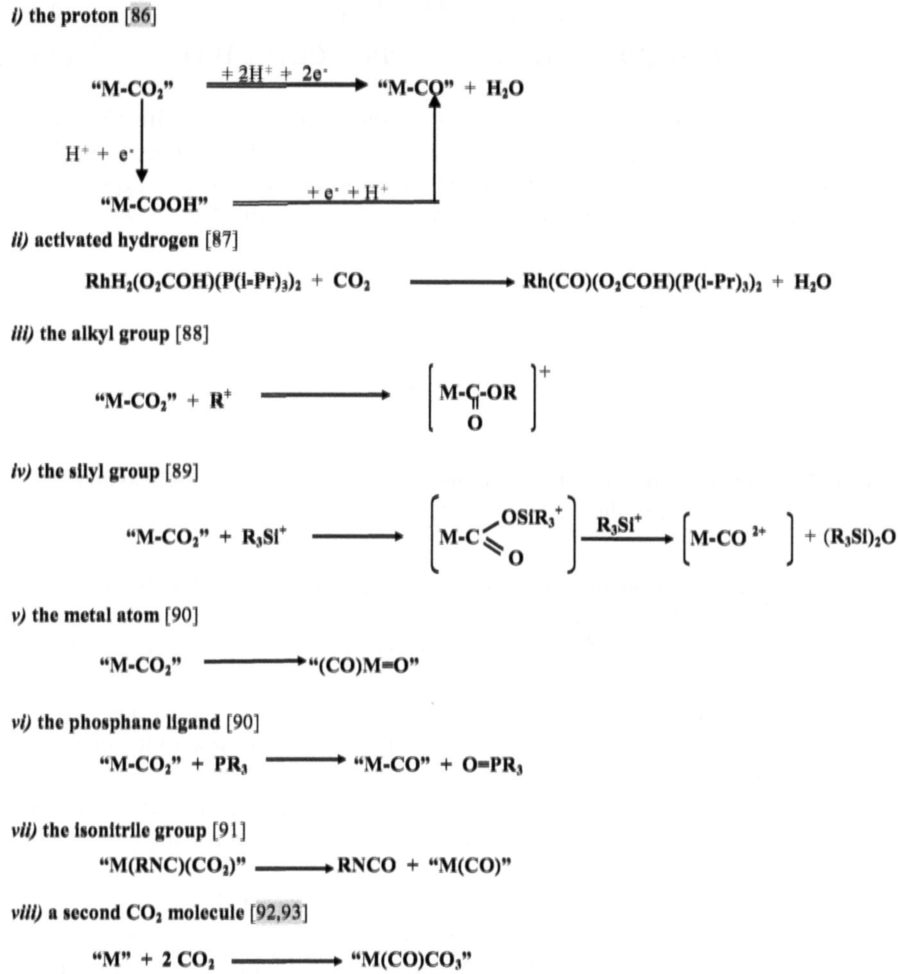

Scheme 2.9 Electrophilic attack on coordinated CO_2

The protonation of coordinated CO_2 occurs in very mild conditions (<273 K) and affords CO as product, mimicking natural systems (2.11).

$$\mathrm{LnM(CO_2) + 2H^+ \rightarrow LnM^{2+} + CO + H_2O} \tag{2.11}$$

The reduction of CO_2 to CO occurs at the expense of the metal, which is oxidized: in order to be able to repeat the cycle, the metal must be reduced to its original oxidation state. The utilization of PhSH as a source of protons and electrons has been demonstrated and is an interesting biomimetic system. Both diphenyldisulphide and the carbonyl species were detected [3, 86–93].

$$\mathrm{PhSH \rightarrow PhS^- + H^+ + e^-} \tag{2.12}$$

$$2\,\mathrm{PhS} \rightarrow \mathrm{PhS} - \mathrm{SPh} \tag{2.13}$$

$$\mathrm{P_2Ni(CO_2)} + 2\mathrm{H^+} + 2\mathrm{e^-} \rightarrow \text{“}\mathrm{P_2Ni(CO)}\text{”} + \mathrm{H_2O} \tag{2.14}$$

As shown in Scheme 2.4, the hydrido complex can interact with CO_2 and a carbonyl bicarbonate complex is formed. In recent years the chemistry of "M–H CO_2" systems has reached high levels of attention because of the T_M catalysed conversion of CO_2 and H_2 into formic acid (see Chaps. 3 and 4).

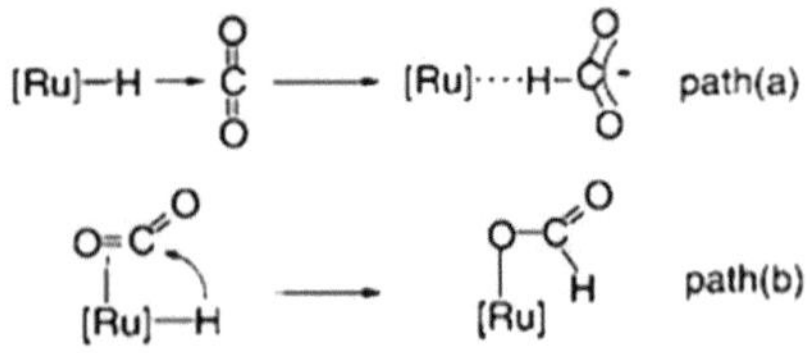

Scheme 2.10 Putative transition states for the insertion of CO_2 into a Ru–H bond. Reprinted with permission from [94]. Copyright (2001) American Chemical Society

The hydride migration from Fe, Ru and Os hydrides to CO_2 has been theoretically studied using DFT. The key issue was again to determine whether an insertion of free CO_2 did occur in the M–H bond, or the hydride migrated to CO_2 coordinated to the metal [94] (Scheme 2.10).

A phosphine bearing a pending amine was used as ligand, with the amine as potential support in H-transfer or as stabilizing agent. The reactivity order was found to be Fe > Ru > Os.

The proposed mechanism says that with Fe the hydride migration to coordinated CO_2 takes place, whereas Ru and Os follow the alternative path with attack of free CO_2 onto the M–H bond.

$Fe(CO_2)(depe)_2$ has been shown to react with MeI or MeOTf to give Me_2O and the cationic iron(II) carbonyl complexes $[FeX(CO)(depe)_2]^+X^-$, with X = I or OTf [95]. Conversely, R_3SiCl (R = Me or Ph) in Et_2O at 195 K converts $Fe(CO_2)(depe)_2$ into the iron carboxylate complex $FeCl(CO_2SiR_3)(depe)_2$ which clearly results from the attack of the electrophilic R_3Si^+ moiety on one oxygen of the coordinated CO_2. In the latter case, the moiety "Fe-C(O)OSiR_3" is less prone to further reaction than the analogous "Fe-C(O)OR" which is quite stable per se [96, 97], but undergoes reaction with the alkylating agent RX to form the carbonyl complex $[FeX(CO)(depe)_2]^+X^-$. The formation of the latter takes place via direct addition of the methylcation to the O atom of the -OR moiety of the carboxylate.

$$\left[\mathrm{Fe(CO_2)\,(depe)_2}\right] + 2\mathrm{RX} \rightarrow \mathrm{R_2O} + \left[\mathrm{FeX(CO)\,(depe)_2}\right]^+\mathrm{X}^- \tag{2.15}$$

Silylation reactions have long been studied [89] with the hope of finding a catalytic system for CO_2 to CO conversion. Stoichiometric reaction or TON of a very few units were reported. Recently, a combined experimental–theoretical study was carried out on the hydrosilylation of carbon dioxide by Me_2PhSiH with ruthenium nitrile complexes *mer*-($RuX_3(MeCN)_3$) and *cis*/*trans*-($RuX_2(MeCN)_4$) with X = Br and Cl [98]. The Me_3Si moiety was proposed to be transferred first to a coordinated halide and from such a position to the side-on coordinated CO_2 to Ru with elimination of $Me_3SiOCHO$. The order of reactivity for CO_2 hydrosilylation was reported to be $RuCl_mL_n > RuBr_mL_n$ and $Ru^{III} > Ru^{II}$. The silyl transfer from $R_3Si\cdots Cl–M$ to the oxygen atom of the η^2-bound CO_2 has been reported to be the highest in energy in the proposed pathway. More applications of the CO_2 insertion into the Si–H bond are discussed in Chap. 4.

2.10.2 Reactions of Coordinated CO_2 with Nucleophiles

Reactions with nucleophiles are much less documented in the literature. On the basis of what has been discussed above, one can observe that if the C atom of the cumulene is engaged with bonding to a metal in a side-on C,O-coordination mode, then it is less likely to become engaged in the reaction with a second nucleophile. Conversely, if O-end-on forms of coordinated CO_2 are considered, then the nucleophilic attack at C should be likely. Any electron-rich species is a potential reagent: H^-, OH^-, OR^-, NR_3, NR_2^-, $RR'R''C^-$ and so on. In this case too, the question is open as to whether coordinated CO_2 is necessary or whether free CO_2 may interact with the "catalyst activated" nucleophile to afford a carboxylate which is then bound to the metal centre.

Recently, in a theoretical study [99] of the hydrogenation of CO_2 to formic acid using Ru^{II} catalysts in presence of water, no coordination of CO_2 to the metal centre was identified, but low energy assemblages with the C and O atoms of CO_2 interacting with H^- bound to the metal or the H^+ of H_2O. In the absence of water, CO_2 directly coordinates to the Ru center to afford $Ru(H)_2(\eta^2\text{-}CO_2)(PMe_3)_3$. The Ru-($\eta^1$-formate) intermediate is produced via CO_2 interaction with the M–H bond more than attack by the H^- on coordinated CO_2.

Using kinetic measurements, Konno et al. [100] have shown that the nucleophilic attack of the hydride ligand of $Ru(tpy)(4,4'\text{-}X_2bpy)H^+$ (X = H, MeO) to the carbon atom of CO_2 was the rate-determining step for the formation of the formate complex $Ru(tpy)(4,4'\text{-}X_2bpy)(OCHO)^+$, whereas the coordination of CO_2 to the metal, either in η^1-O or η^2-C,O mode, does not play any important role in the transition state.

A nucleophilic attack has been invoked in the case of reaction of $(PCy_3)_2Ni(\eta^2\text{-}CO_2)$ with $R_3P{=}CH_2$ in a Wittig-like reaction [101]. In the Wittig reaction, a (2 + 2) cycloaddition, strong nucleophiles such as trialkyl phosphorus ylides $R_3P{=}CH_2$ and organic ketones or aldehydes react to form alkenes. Actually, CO_2 bubbled into THF containing $Me_3P{=}CH_2$ affords the free ketene $CH_2{=}C{=}O$ which immediately dimerizes to methylene-β-propiolactone. Ketenes can be stabilized by coordination to transition metal complexes in a η^2-C,O (for early transition metals) or η^2-C,C (for late transition metals) manner. In this study, complex **1** was reacted with $Me_3P{=}CH_2$ in toluene at 253 K for 2 h and a nickel ketene complex $(Cy_3P)_2Ni(\eta^2\text{-}C,O)–CH_2{=}CO$ was isolated and characterized (IR bands at 1 611 and

1 570 cm^{-1}). The reaction mechanism is not clearly stated, as it could well be a nucleophilic attack by the ylide onto the coordinated CO_2, but could also be explained in terms of CO_2 dissociation from Ni, reacting with the ylide in solution to form the ketene which is then coordinated to nickel. The unusual η^2-C,O ketene bonding to Ni seems to be more in favour of a nucleophilic attack on the coordinated CO_2.

2.11 Conclusions

The coordination of CO_2 to metal centres has been demonstrated to occur via the C or O atom, or else through the O,C or O,O atoms, or else through the three OCO atoms. The molecular geometry and structural parameters have been defined by XRD. Alternative tools for getting information and multinuclear NMR about the structure of T_M–CO_2 complexes have been developed which imply IR studies and use of theoretical calculations. The rigidity or fluxionality of the complexes has also been investigated. The effective involvement of coordinated CO_2 with metal centres in reactions involving the O atoms has been documented much more frequently than the reaction at the C atom of the coordinated molecule. The knowledge of the modes of bonding of CO_2 has considerably increased since the first stable CO_2 complex $Ni(CO_2)(PCy_3)_2$ was structurally characterized 40 years ago [3]. The relevance of such modes of bonding to the reactivity of the cumulene still leaves open questions which require an answer.

The huge advancement of theoretical calculations (parallel computers) and their application to systems that are more and more complex and close to real metal complexes are good support to foreseeing the properties of the T_M–CO_2 complexes and simulating their reactivity. Spectroscopic techniques, more and more sophisticated, allow the identification and characterization of stable complexes or of short-lived species so far difficult to identify, and, thus, give support to the definition of the reaction mechanisms. Still a number of issues remain open and need investigation. One of these is the co-operative effect of ancillary ligands. Several basic reactions still need a correct understanding for they can be exploited in synthetic applications.

References

1. Wang YG, Wiberg KB, Werstiuk NH (2007) Correlation effects in EOM-CCSD for the excited states: evaluated by AIM localization index (LI) and delocalization index (DI). J Phys Chem A 111:3592–3601
2. Mascetti J (2010) Carbon dioxide coordination chemistry and reactivity of coordinated CO_2. In: Aresta M (ed) Carbon dioxide as chemical feedstock. Wiley-VCH Verlag GmbH & Co. KGaA, Weinheim

3. Aresta M, Nobile CF, Albano VG, Forni E, Manassero M (1975) New nickel-carbon dioxide complex: synthesis, properties, and crystallographic characterization of (carbon dioxide)bis (tricyclohexylphosphine)nickel. J Chem Soc Chem Commun:636–637
4. Kégl T, Ponec R, Kollar L (2011) Theoretical insights into the nature of nickel−carbon dioxide interactions in $Ni(PH_3)_2(\eta^2\text{-}CO_2)$. J Phys Chem C 115:12463–12473
5. Contreras L, Paneque M, Sellin M, Carmona E, Perez PJ, Gutierrez-Puebla E, Monge A, Ruiz C (2005) Novel carbon dioxide and carbonyl carbonate complexes of molybdenum. The X-ray structures of *trans*-$[Mo(CO_2)_2\{HN(CH_2CH_2PMe_2)_2\}(PMe_3)]$ and $[Mo_3(\mu_2\text{-}CO_3)(\mu_2\text{-}O)_2(O)_2(CO)_2(H_2O)(PMe_3)_6]\cdot H_2O$. New J Chem 29:109–115
6. Aresta M, Gobetto R, Quaranta E, Tommasi I (1992) A bonding-reactivity relationship for Ni $(PCy_3)_2(CO_2)$: a comparative solid-state-solution nuclear magnetic resonance study (^{31}P, ^{13}C as a diagnostic tool to determine the mode of bonding of CO_2 to a metal center). Inorg Chem 31:4286–4290
7. Bristow GS, Hitchcock PB, Lappert DM (1981) A novel carbon dioxide complex: synthesis and crystal structure of $[Nb(\eta\text{-}C_5H_4Me)_2(CH_2SiMe_3)(\eta^2\text{-}CO_2)]$. J Chem Soc Chem Commun 21:1145–1146
8. Gibson DH (1996) The organometallic chemistry of carbon dioxide. Chem Rev 96:2063–2095
9. Gibson DH (1999) Carbon dioxide coordination chemistry: metal complexes and surface-bound species. What relationships? Coord Chem Rev 185–186:335–355
10. Yin X, Moss JR (1999) Recent developments in the activation of carbon dioxide by metal complexes. Coord Chem Rev 181:27–59
11. Gambarotta S, Arena F, Floriani C, Zanazzi PF (1982) Carbon dioxide fixation: bifunctional complexes containing acidic and basic sites working as reversible carriers. J Am Chem Soc 104:5082–5092
12. Fujita E, Creutz C, Sutin N, Brunschwig BS (1993) Carbon dioxide activation by cobalt macrocycles: evidence of hydrogen bonding between bound CO_2 and the macrocycle in solution. Inorg Chem 32:2657–2662
13. Beley M, Collin JP, Ruppert R, Sauvage JP (1986) Electrocatalytic reduction of carbon dioxide by nickel cyclam^{2+} in water: study of the factors affecting the efficiency and the selectivity of the process. J Am Chem Soc 108:7461–7467
14. Collin JP, Sauvage JP (1986) Electrochemical reduction of carbon dioxide mediated by molecular catalysts. Coord Chem Rev 1993:245–268
15. Berkefeld A, Piers WE, Parvez M (2010) Tandem frustrated Lewis pair/tris (pentafluorophenyl)borane-catalyzed deoxygenative hydrosilylation of carbon dioxide. J Am Chem Soc 132(31):10660–10661
16. Stephan DW, Erker G (2010) Frustrated Lewis pairs. Angew Chem Int Ed 49:46–76
17. Appelt C, Westenberg H, Bertini F, Ehlers AW, Slootweg JC, Lammertsma K, Uhl W (2011) Geminal phosphorous/aluminum-based frustrated Lewis pairs: C-H versus C≡C activation and CO_2 fixation. Angew Chem Int Ed 50: 3925–3928 and references therein
18. Zevaco T, Dinjus E (2010) Main group element- and transition metal promoted carboxylations of organic substrates (alkanes, alkenes, alkynes, aromatics, and others). In: Aresta M (ed) Carbon dioxide as chemical feedstock. Wiley-VCH Verlag GmbH & Co, KGaA, Weinheim
19. Haruki E (1982) Organic synthesis with carbon dioxide. In: Inoue S, Yamazaki N (eds) Organic and bioorganic chemistry of carbon dioxide. Halsted, New York
20. Takay I, Yamamoto A (1982) Organometallic reactions of carbon dioxide. In: Inoue S, Yamazaki N (eds) Organic and bioorganic chemistry of carbon dioxide. Halsted, New York
21. Bertini I, Luchinat C (1994) The reaction pathway of zinc enzymes and related biological catalysts. In: Bertini I, Gray HB, Lippard SJ, Valentine JS (eds) Bioinorganic chemistry. University Science, Mill Valley

22. Hou XJ, He P, Li H, Wang X (2013) Understanding the adsorption mechanism of C_2H_2, CO_2, and CH_4 in metal-organic frameworks with coordinatively unsaturated metal sites. J Phys Chem C 117:2824–2834
23. Calabrese JC, Herskovitz T, Kinney JB (1983) Carbon dioxide coordination chemistry. 5. Preparation and structure of Rh(η^1-CO_2)(Cl)(diars)$_2$. J Am Chem Soc 1983:5914–5915
24. Harlow RL, Kinney JB, Herskovitz T (1980) Carbon dioxide co-ordination chemistry: preparation and *X*-ray crystal structure of the methoxycarbonyl complex [IrCl(CO_2Me)-($Me_2PCH_2CH_2PMe_2$)$_2$]FSO_3 from a CO_2 adduct. J Chem Soc Chem Commun:813–814
25. Aresta M, Nobile CF (1977) Carbon dioxide-transition metal complexes. III. Rh(I)-CO_2 complexes. Inorg Chim Acta 24:L49–L50
26. Karsch HH (1977) Funktionelle Trimethylphosphinderivate, III. Ambivalentes Verhalten von Tetrakis (trimethylphosphin) eisen: Reaktion mit CO_2. Chem Ber 110:2213–2221
27. Tanaka K, Ooyama D (2002) Multi-electron reduction of CO_2 via Ru-CO_2, –C(O)OH, –CO, –CHO, and –CH_2OH species. Coord Chem Rev 226:211–218
28. Holbrey JD, Reichert WM, Tkatchenko I, Bouajila E, Walter O, Tommasi I, Rogers RD (2003) 1,3-Dimethylimidazolium-2-carboxylate: the unexpected synthesis of an ionic liquid precursor and carbene-CO_2 adduct. Chem Commun (Camb) 1:28–29
29. Castro-Rodriguez I, Nakai H, Zakharov LN, Rheingold AL, Meyer K (2004) A linear, O-coordinated η^1-CO_2 bound to uranium. Science 305:1757–1759
30. Lam OP, Anthon C, Meyer K (2009) Influence of steric pressure on the activation of carbon dioxide and related small molecules by uranium coordination complexes. Dalton Trans 44:9677–9691
31. Lee CH, Laitar DS, Mueller P, Sadighi JP (2007) Generation of a doubly bridging CO_2 ligand and deoxygenation of CO_2 by an (NHC)Ni(0) Complex. J Am Chem Soc 129:13802–13803
32. Chang CC, Liao MC, Chang TH, Peng SM, Lee GH (2005) Aluminum-magnesium complexes with linear bridging carbon dioxide. Angew Chem Int Ed 44:7418–7420
33. Gao G, Li F, Xu L, Liu X, Yang Y (2008) CO_2 coordination by inorganic polyoxoanion in water. J Am Chem Soc 130:10838–10839
34. Dietzel PDC, Johnsen RE, Fjellväg H, Bordiga S, Groppo E, Chavan S, Blom R (2008) Adsorption properties and structure of CO_2 adsorbed on open coordination sites of metal–organic framework Ni_2(dhtp) from gas adsorption, IR spectroscopy and X-ray diffraction. J Chem Soc Chem Commun:5125–5127
35. Jegat C, Fouassier M, Mascetti J (1991) Carbon-dioxide coordination chemistry. 1. Vibrational study of trans-(CO_2)$_2$Mo(PMe_3)$_4$ and Fe(CO_2)(PMe_3)$_4$. Inorg Chem 30:1521–1529
36. Jegat C, Fouassier M, Tranquille M, Mascetti J (1991) Carbon-dioxide coordination chemistry. 2. Synthesis and FTIR study of Cp_2Ti(CO_2)(PMe_3). Inorg Chem 30:1529–1536
37. Leitner W (1996) The coordination chemistry of carbon dioxide and its relevance for catalysis: a critical survey. Coord Chem Rev 153:257–284
38. Pilato RS, Housemekerides CE, Jernakoff P, Rubin D, Geoffroy G, Rheingold AR (1990) Net [2 + 2] cycloaddition reactions of the oxo complexes $Cp_2M=O$ (M = Mo, W) with electrophilic organic and organometallic substrates. Formation of bimetallic μ^2-η^3-CO_2 complexes. Organometallics 9:2333–2341
39. Mastrorilli P, Moro G, Nobile CF, Latronico M (1992) Carbon dioxide-transition metal complexes. IV. New Ni(0)-CO_2 complexes with chelating diphosphines: influence of P-Ni-P angle on complex stabilities. Inorg Chim Acta 192:189–193
40. Jegat C, Fouassier M, Tranquille M, Mascetti J, Tommasi I, Aresta M, Inglod F, Dedieu A (1993) Carbon dioxide co-ordination chemistry 3. Vibrational, NMR, and theoretical studies of Ni(CO_2)(PCy_3)$_2$. Inorg Chem 32:1279–1289
41. Carmona E, Hughes AK, Munoz MZ, O'Hare DM, Perez PJ, Poveda ML (1991) Rotational isomerism and fluxional behavior of bis(carbon dioxide) adducts of molybdenum. J Am Chem Soc 113:9210–9218
42. Almond MJ, Downs AJ (1989) Spectroscopy of matrix isolated species. Wiley, Chichester

43. Jacox ME, Milligan DE (1974) Vibrational spectrum of CO_2^- in an argon matrix. Chem Phys Lett 28:163–168
44. Hauge RH, Margrave JL, Kauffmann JW, Rao NA, Konarski MM, Bell JP, Billups WE (1981) Matrix isolation studies on the lithium-induced reductive coupling of carbon dioxide. J Chem Soc Chem Commun:1258–1260
45. Kafafi ZH, Hauge RH, Billups WE, Margrave JL (1984) Carbon dioxide activation by alkali metals. 2. Infrared spectra of $M^+CO_2^-$ and $M_2^{2+}CO_2^{2-}$ in argon and nitrogen matrice. Inorg Chem 23:177–183
46. Manceron L, Loutellier A, Perchard JP (1985) Reduction of carbon dioxide to oxalate by lithium atoms: a matrix isolation study of the intermediate steps. J Mol Struct 129:115–124
47. Hwang DY, Mebel AM (2000) Theoretical study on reforming of CO_2 catalyzed with Be. Chem Phys Lett 325:639–644
48. Solov'ev VN, Polikarpov EV, Nemukhin AV, Sergeev GB (1999) Matrix isolation and ab initio study of the reactions of magnesium atoms and clusters with CO_2, C_2H_4, and CO_2/C_2H_4 mixtures: formation of cyclic complexes. J Phys Chem A 103:6721–6725
49. Hwang DY, Mebel AM (2000) Theoretical study on the reaction mechanism of CO_2 with Mg. J Phys Chem A 104:7646–7650
50. Polikarpov EV, Granovsky AA, Nemukhin AV (2001) On the potential-energy surface of the $Mg + CO_2$ (C_{2v}) system. Mend Commun 11:150–151
51. Hwang DY, Mebel AM (2000) Reaction mechanism of CO_2 with Ca atom: a theoretical study. Chem Phys Lett 331:526–532
52. Burkholder TR, Andrews L, Bartlett RJ (1993) Reaction of boron atoms with carbon dioxide: matrix and ab initio calculated infrared spectra of OBCO. J Phys Chem 97:3500–3503
53. Chin CH, Mebel AM, Hwang DY (2003) Theoretical study of the reaction mechanism of boron atom with carbon dioxide. Chem Phys Lett 375:670–675
54. Lequere AM, Xu C, Manceron L (1991) Vibrational spectra, structures, and normal-coordinate analysis of aluminum-carbon dioxide complexes isolated in solid argon. J Phys Chem 95:3031–3037
55. Huber H, Mc Intosh D, Ozin GA (1978) Metal atom chemistry and surface chemistry: (carbon dioxide)silver, $Ag(CO_2)$. A localized bonding model for weakly chemisorbed carbon dioxide on bulk silver. Inorg Chem 17:1472–1476
56. Kurtikyan TS, Kazaryan SG (1981) IR spectral study of low-temperature interaction between copper vapours and carbon dioxide. Arm Khim Zh 34:375–379
57. Almond MJ, Downs AJ, Perutz RN (1985) Matrix photooxidation of the metal carbonyls M $(CO)_6$ (M = Cr, W) by the isoelectronic molecules carbon dioxide and nitrous oxide. Inorg Chem 24:275–281
58. Mascetti J, Tranquille M (1985) IR evidence for the formation of CO_2 transition-metal atom complexes in low-temperature matrices. Surf Sci 156:201–205
59. Mascetti J, Tranquille M (1988) Ab initio investigation of several low-lying states of all-trans octatetraene. J Phys Chem 92:2177–2184
60. Papai I, Mascetti J, Fournier R (1997) Theoretical study of the interaction of the Ti atom with CO_2: cleavage of the C – O bond. J Phys Chem A 101:4465–4471
61. Hwang DY, Mebel AM (2002) Theoretical study of TiO-catalyzed hydrogenation of carbon dioxide to formic acid. J Chem Phys 116:5633–5642
62. Zhang LN, Wang XF, Chen MH, Qin QZ (2000) Activation of CO_2 by Zr atom. Matrix-isolation FTIR spectroscopy and density functional studies. Chem Phys 254:231–238
63. Chen MH, Wang XF, Zhang LN, Qin QZ (2000) IR spectroscopic and DFT studies on the reactions of laser-ablated Nb atoms with carbon dioxide. J Phys Chem A 104:7010–7015
64. Wang XF, Chen MH, Zhang LN, Qin QZ (2000) Spectroscopic and theoretical studies on the reactions of laser-ablated tantalum with carbon dioxide. J Phys Chem A 104:758–764
65. Jiang L, Xu Q (2007) Infrared spectroscopic and density functional theory study on the reactions of lanthanum atoms with carbon dioxide in rare-gas matrices. J Phys Chem A 111:3519–3525

66. Tague TJ, Andrews L, Hunt RD (1993) Matrix infrared spectra of the products of uranium-atom reactions with carbon monoxide and carbon dioxide. J Phys Chem 97:10920–10924
67. Liang BY, Andrews L (2002) Reactions of laser-ablated rhenium atoms with carbon dioxide: matrix infrared spectra and density functional calculations on OReCO, O_2ReCO, $ORe(CO)_2$, $O_2Re(CO)_2$, $OReCO^-$, and $ORe(CO)_2^-$. J Phys Chem A 106:595–602
68. Herman J, Foutch JD, Davico GE (2007) Gas-phase reactivity of selected transition metal cations with CO and CO_2 and the formation of metal dications using a sputter ion source. J Phys Chem A 111:2461–2468
69. Koyanagi GK, Bohme DK (2006) Gas-phase reactions of carbon dioxide with atomic transition-metal and main-group cations: room-temperature kinetics and periodicities in reactivity. J Phys Chem A 110:1232–1241
70. Sodupe M, Branchadell V, Rosi M, Bauschlicher CW (1997) Theoretical study of M^+-CO_2 and OM^+CO systems for first transition row metal atoms. J Phys Chem 101:7854–7859
71. Walker NR, Walters RS, Duncan MA (2004) Infrared photodissociation spectroscopy of $V^+(CO_2)_n$ and $V^+(CO_2)_nAr$ complexes. J Chem Phys 120:10037–10045
72. Walker NR, Walters RS, Grieves GA, Duncan MA (2004) Growth dynamics and intracluster reactions in $Ni^+(CO_2)_n$ complexes via infrared spectroscopy. J Chem Phys 121:10498–10507
73. Gregoire G, Duncan MA (2002) Infrared spectroscopy to probe structure and growth dynamics in Fe+–(CO_2)n clusters. J Chem Phys 117:2120–2130
74. Griffin JB, Armentrout PB (1997) Guided ion beam studies of the reactions of Fen +(n = 1–18) with CO_2: iron cluster oxide bond energies. J Chem Phys 107:5345–5355
75. Tjelta BL, Walter D, Armentrout PB (2001) Determination of weak Fe^+–L bond energies (L = Ar, Kr, Xe, N_2, and CO_2) by ligand exchange reactions and collision-induced dissociation. Int J Mass Spectrom 204:7–21
76. Rodgers MT, Walker B, Armentrout PB (1999) Reactions of Cu^+ (1 S and 3 D) with O_2, CO, CO_2, N_2, NO, N_2O, and NO_2 studied by guided ion beam mass spectrometry. Int J Mass Spectrom 182(183):99–120
77. Zang XG, Armentrout PB (2003) Activation of O_2, CO, and CO_2 by Pt^+: the thermochemistry of PtO^+. J Phys Chem A 107:8904–8914
78. Clemmer DE, Weber ME, Armentrout PB (1992) Reactions of aluminum (1+)(1S) with nitrogen dioxide, nitrous oxide, and carbon dioxide: thermochemistry of aluminum monoxide and aluminum monoxide (1+). J Phys Chem 96:10888–10893
79. Armentrout PB, Beauchamp JL (1980) Reactions of U^+ and UO^+ with O_2, CO, CO_2, COS, CS_2 and D_2O. Chem Phys 50:27–36
80. Aresta M, Dibenedetto A, Angelini A (2014) Catalysis for the valorization of exhaust carbon: from CO_2 to chemicals, materials, and fuels. Technological use of CO_2. Chem Rev 114 (3):1709–1742
81. Albano P, Aresta M, Manassero M (1980) Interaction of carbon dioxide with coordinatively unsaturated rhodium(I) complexes with the ligand 1,2 bis(diphenylphosphino)ethane. Inorg Chem 19(4):1069–1072
82. Nicholas KM (1980) Catalytic oxidation of phosphines by transition metal-activated carbon dioxide. J Organomet Chem 188:C10–C12
83. Ohnishi T, Seino H, Hidai M, Mizobe Y (2005) The C = O and C = S bond cleavage in carbon dioxide and tolyl isothiocyanate by reactions with the Mo(0) tetraphosphine complex [Mo{*meso-o*-$C_6H_4(PPhCH_2CH_2PPh_2)_2$}($Ph_2PCH_2CH_2PPh_2$)]. J Organomet Chem 690:1140–1146
84. Brookes NJ, Ariafard A, Stranger R, Yates BF (2011) Tuning the laplaza-cummins 3-coordinate $M[N(R)Ph]_3$ catalyst to activate and cleave CO_2. Dalton Trans 40:5569–5578
85. Park SE, Koo HM, Park YK, Park SM, Park JC, Lee OK, Park YC, Seo JH (2011) Expression of aldehyde dehydrogenase 6 reduces inhibitory effect of furan derivatives on cell growth and ethanol production in Saccharomyces cerevisiae. Bioresour Technol 102(10):6033–6038
86. Aresta M, Quaranta E, Tommasi I (1988) Reduction of co-ordinated carbon dioxide to carbon monoxide via protonation by thiols and other Brønsted acids promoted by Ni-systems: a

contribution to the understanding of the mode of action of the enzyme carbon monoxide dehydrogenase. J Chem Soc Chem Commun:450–452
87. Yoshida T, Thorn DL, Okano T, Ibers JA, Otsuka S (1979) Hydration and reduction of carbon dioxide by rhodium hydride compounds. Preparation and reactions of rhodium bicarbonate and formate complexes, and the molecular structure of $RhH_2(O_2COH)(P(i\text{-}Pr)_3)_2$. J Am Chem Soc 101:4212–4221
88. Aresta M (1977) Carbon dioxide-transition metals complexes. III. Rhodium(I)-CO_2 complexes. Inorg Chim Acta 24:L49–L50
89. Tsai JC, Khan M, Nicholas KM (1989) Reactivity of coordinated carbon dioxide: reactions of $(C_5H_5)_2Mo(.eta.2\text{-}CO_2)$ with electrophiles. Organometallics 8:2967–2968
90. Demerseman B, Bouquet G, Bigorgne M (1978) Reduction de l'anhydride carbonique par le systeme chlorure de titanocene—metal reducteur (zinc, aluminium). Nouvelle methode de preparation de $Cp_2Ti(CO)_2$ et synthese d'un complexe carbonate de titane(III). J Organomet Chem 145:41–48
91. Tsuda T, Sanada S, Saegusa T (1976) Copper-promoted deoxygenation of carbon dioxide by isocyanide. J Organomet Chem 116:C10–C11
92. Chatt J, Kubota M, Jeffery Leigh G, March FC, Mason R, Yarrow DJ (1974) A possible carbon dioxide complex of molybdenum and its rearrangement product di-μ-carbonato-bis {carbonyltris(dimethylphenylphosphine) molybdenum}: X-ray crystal structure. J Chem Soc Chem Commun:1033–1034
93. Herskowitz T, Guggenberger L (1976) Carbon dioxide coordination chemistry. The structure and some chemistry of the novel carbon dioxide addition product chlorobis(carbon dioxide) tris(trimethylphosphine)iridium. J Am Chem Soc 98:1615–1616
94. Matsubara T, Hirao K (2001) Density functional study on the hydrido migration to CO_2 and CS_2 of the $(\eta^5\text{-}C_5H_4(CH_2)_3NH_3^+)MH(H_2PCH_2PH_2)$ (M = Fe, Ru, and Os) complexes promoted by the protonated amine arm. Which path does the reaction take, abstraction or insertion? Organometallics 20:5759–5768
95. Hirano M, Akita M, Tani K, Kumagai K, Kasuga N, Fukuoka A, Komiya S (1997) Activation of coordinated carbon dioxide in $Fe(CO_2)(depe)_2$ by group 14 electrophiles. Organometallics 16:4206–4213
96. Busetto L, Angelici R (1968) Reactions of cyclopentadienyliron carbonyl cations with amines. Inorg Chim Acta 2:386–390
97. Bryndza HE, Tam W (1988) Monomeric metal hydroxides, alkoxides, and amides of the late transition metals: synthesis, reactions, and thermochemistry. Chem Rev 88:1163–1185
98. Deglmann P, Ember E, Hofmann P, Pitter S, Walter O (2007) Experimental and theoretical investigations on the catalytic hydrosilylation of carbon dioxide with ruthenium nitrile complexes. Chem Eur J 13:2864–2879
99. Ohnishi YY, Nakao Y, Sato H, Sakaki S (2006) Ruthenium(II)-catalyzed hydrogenation of carbon dioxide to formic acid. Theoretical study of significant acceleration by water molecules. Organometallics 25:3352–3363
100. Konno H, Kobayashi A, Sakamoto K, Fagalde F, Katz N, Saitoh H, Ishitani O (2000) Synthesis and properties of $[Ru(tpy)(4,4'\text{-}X_2bpy)H]^+$(tpy = 2,2′:6′,2″-terpyridine, bpy = 2,2-′-bipyridine, X = H and MeO), and their reactions with CO_2. Inorg Chim Acta 299:155–163
101. Gong JK, Wright CA, Thorn M, McCauley K, McGill JW, Sutterer A, Hinze SM, Prince RB (1998) In: Inui T, Anpo M, Izui K, Yanagida S, Yamaguchi T (eds) Advances in chemical conversions for mitigating carbon dioxide, vol. 114. Elsevier, Amsterdam, pp 491–494

Chapter 3
Interaction of CO_2 with Electron-Rich Moieties

Abstract In this chapter the direct, non-metal-mediated interaction of carbon dioxide with electron-rich elemental or molecular species is discussed. Anionic species such as H^-, OH^-, and R_3C^- and covalent species such as amines have been taken into consideration, in view of their relevance to systems of potential or real industrial interest.

3.1 Reaction with the Hydride Ion

Hydrides can be categorized in various classes according to the nature of the bond that links the metal and hydrogen. The following categories can usually be found in the literature: ionic hydrides (hydrides with group 1 and 2 elements), covalent hydrides, polymeric hydrides, metallic hydrides, borderline hydrides, and transition metal complex hydrides. Here the reaction of ionic hydrides with carbon dioxide (CO_2) is considered. Transition metal hydrides with the formula L_nM–H are discussed in Chap. 4, where the interaction of CO_2 with analogous E–X systems (where E and X are different elements) is discussed.

Group 1 and group 2 hydrides are ionic, insoluble in aprotic solvents, and not usable in protic solvents such as water and alcohols because they promptly react with H_2 evolution:

$$M^+H^- + H_2O \rightarrow M^+OH^- + H_2 \tag{3.1}$$

The ionic hydrides are thermally very stable and have quite high melting temperatures. Depending on the element, they can melt without (CaH_2) or with (LiH, NaH, and others) decomposition. NaH decomposes above 573 K, whereas LiH and CaH_2 are stable up to 900 K.

Therefore, the reaction of ionic hydrides with CO_2 mostly implies a heterogeneous, biphasic system in which pressurized CO_2 reacts at a temperature of ~473 K with the MH (M = Li, H, K, Rb) or MH_2 (Ca, Ba, Sr) species to afford the relevant formate:

M. Aresta et al., *Reaction Mechanisms in Carbon Dioxide Conversion*,
DOI 10.1007/978-3-662-46831-9_3

$$M^+H^- + CO_2 \rightarrow HCO_2^- M^+ \tag{3.2}$$

Group 2 elements afford the $M(HCO_2)_2$ formates, too.

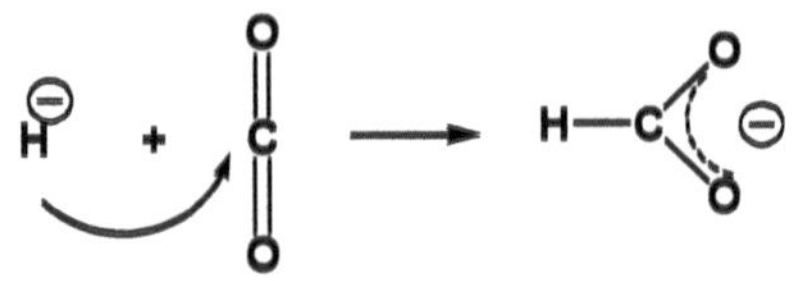

Scheme 3.1 Attack of the hydride anion at the electrophilic C of CO_2

Although such a reaction has been known for a long time, it has never found any practical application for the synthesis of formates, which are more conveniently prepared by routes involving CO and metal hydroxides:

$$CO + MOH \rightarrow HCOO^- M^+ \tag{3.3}$$

The reaction mechanism of (3.2) is quite simple and linear because it involves the attack of the ionic hydride at the electrophilic C of CO_2 (Scheme 3.1).

The resulting formate anion is stabilized by interaction with the metal cation. The literature does not offer much information on such reactive systems in terms of kinetic studies, thermodynamics of reaction, and so on, most likely because of the harsh reaction conditions and limited practical interest of the reactions. The available information points to a kinetic law of an overall second-order reaction, i.e., rate $= k[H^-][CO_2]$.

3.2 Reaction with Hydroxide and Alkoxide Species

The reaction of CO_2 with ionic metal hydroxides $[M(OH)_n]$ or alkoxides $[M(OR)_n]$ is of quite different interest than the reaction with hydride systems. The kinetics of the reaction was investigated as early as 1924 [1, 2] in systems involving an infinite dilution of hydroxide ions or the co-presence of both OH^- and RO^- anions that were competing for small amounts of CO_2. Such studies were repeated later [3, 4] in an attempt to determine the kinetic constants for the reactions better. However, such studies demonstrate the long-lasting interest of such topic. More recently, more sophisticated techniques have been used [5, 6]. Such studies have clearly shown that the reaction mechanism can be described as shown in Scheme 3.2, in which the hydroxide anion attacks the electrophilic carbon of CO_2.

The rate constants were dependent on the cation and the ionic strength [7]. Today this topic is of great interest, linked to the utilization of the reaction depicted in Scheme 3.2 for the separation of CO_2 from flue gases. However, new emphasis has been put on the topic for the correct determination of the rate constants in a wider range of temperatures and conditions for better reversibility of the reactive systems. The reaction is first-order regarding both hydroxide and CO_2:

$$r = k[OH^-][CO_2] \tag{3.5}$$

Kenig et al. [8, 9] have investigated such systems with dynamic experiments in the temperature range 293–343 K, identifying the differences between various cations (Na^+, K^+) and elaborating an equation

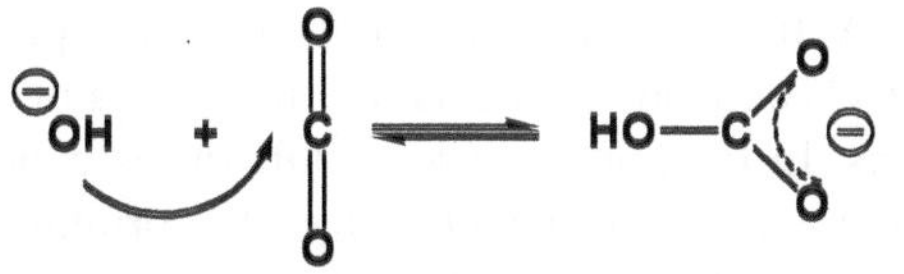

Scheme 3.2 Reaction of the hydroxide anion with CO_2

that takes into account the temperature and ionic strength. The influence of the nature of the cations may imply an effect of the latter in the stabilization of the hydrogencarbonate or else a role in the attack of the hydroxide moiety at CO_2.

Reaction (3.4) is also of great interest for the fixation of CO_2 into inorganic carbonates (3.6) which have been considered as potential compounds for the safe storage of CO_2 for a long time:

$$\begin{aligned} CaO + H_2O &\rightarrow Ca^{2+}2OH^- \xrightarrow{CO_2} Ca^{2+} + 2HCO_3^- \\ &\rightarrow CaCO_{3(s)} + H_2O + CO_2 \end{aligned} \tag{3.6}$$

3.3 Reaction with Carbanions

The reaction of carbanions with CO_2 has also been known for a long time. Such reaction has a key role in natural (de)carboxylation processes. The most classic example is given by the reaction of a Grignard reagent with CO_2:

$$R^{-+}MgBr + CO_2 \rightarrow R\text{-}CO_2^- + MgBr^+ \tag{3.7}$$

R = alkyl or aryl moiety

In this case the reaction mechanism is also quite straightforward: the carbanion attacks the electrophilic carbon of CO_2 producing the carboxylate. The latter can be converted into the relevant acid by reaction with an inorganic strong acid such as HCl or, even better, H_2SO_4. Such reaction, even if it addresses a synthetic requirement of great industrial relevance such as the synthesis of carboxylic acids, has no practical application as it consumes 1 mol of Mg per mol of carboxylate produced, resulting in an energy-intensive methodology (the preparation of the Grignard demands elemental Mg produced by electrolysis) and is not too friendly from the environmental point of view because of the use of halogens and production of large amounts of inorganic salts as waste. Therefore, its application is only tutorial or, at best, usable on a small laboratory scale for some specialty products.

3.4 Reaction with Amines

The reaction of CO_2 with organic amines (RR′R″N, $ArNH_2$; R = H or aliphatic moiety, Ar = aromatic moiety) is of great interest as the aliphatic members have been used for up-taking CO_2 from a gas stream or for synthetic purposes (3.8). The reactivity of amines with CO_2 depends on the nature of the organic moiety they bear and the reaction conditions. In fact, under strictly anhydrous conditions, although tertiary aliphatic amines do not absorb CO_2, aliphatic primary and secondary amines, as well as ammonia itself, react quickly with carbon dioxide to form the corresponding ammonium carbamates (3.8) which, in a few cases, have been fully characterized in the solid state by X-ray diffraction [10, 11].

$$2RR'NH + CO_2 \leftrightharpoons RR'N - CO_2^- RR'NH_2^+ \tag{3.8}$$

$$RR'N - CO_2^- RR'NH_2^+ \rightarrow 1/2\left[RR'N - COOH\right]_2 + RR'NH \tag{3.9}$$

Aromatic amines hardly react with carbon dioxide because of their poorer basicity/nucleophilicity [12]. However, in the presence of suitable auxiliary strong bases (B), such as amidines or pentaalkylguanidine superbases, carbamate salts $(BH)O_2CNRAr$ (R = H, alkyl) can be generated in solution, as revealed by spectroscopic and reactivity data [13]. It has been shown that even tributylamine may be effective if a suitable alkali metal salt is present in the reaction mixture: in the latter case, the *N*-arylcarbamate has been isolated as an alkali salt (3.10) [14, 15].

$$ArNH_2 + B + MBPh_4 + CO_2 \rightarrow M(O_2CNHAr) + [HB]BPh_4 \tag{3.10}$$

Ar = Aryl; B = NR_3 (R = alkyl); M = Li, Na, K

The reaction of ^{15}N-labeled primary amines with $^{13}CO_2$ in the presence of a strong base such as *N*-cyclohexyl-*N′*,*N′*,*N″*,*N″*-tetramethylguanidine (CyTMG) deserves attention. In CH_3CN, in the presence of 2 equiv. of CyTEG, $Et^{15}NH_2$ undergoes double carboxylation with formation of $Et^{15}N(^{13}CO_2)_2^{2-}(^+HCyTMG)_2$, as evidenced by means of ^{15}N and ^{13}C NMR spectroscopy. However, with $Ph^{15}NH_2$ only $Ph^{15}NH^{13}CO_2^-$ $^+$HCyTMG was formed, with $Cy^{15}NH_2$ both species, $Cy^{15}NH^{13}CO_2^-$ $^+$HCyTMG and $Cy^{15}N(^{13}CO_2)_2^{2-}(^+HCyTMG)_2$, were observable, and the ratio of the two species was highly dependent on the temperature at which the spectra were obtained [13].

In a very few cases, a different reactivity has been observed. The Co-aminophosphane complex $Co(NO)_2[PhP(OCH_2CH_2)_2NH]Cl$ reacted at 243 K with CO_2 (0.1 MPa) to afford $CoCl(NO)_2[PhP(OCH_2CH_2)_2N\text{–}CO_2H]$ [16, 17]. Dibenzylamine, Bn_2NH (neat, or in heptane solution), also reacted with carbon

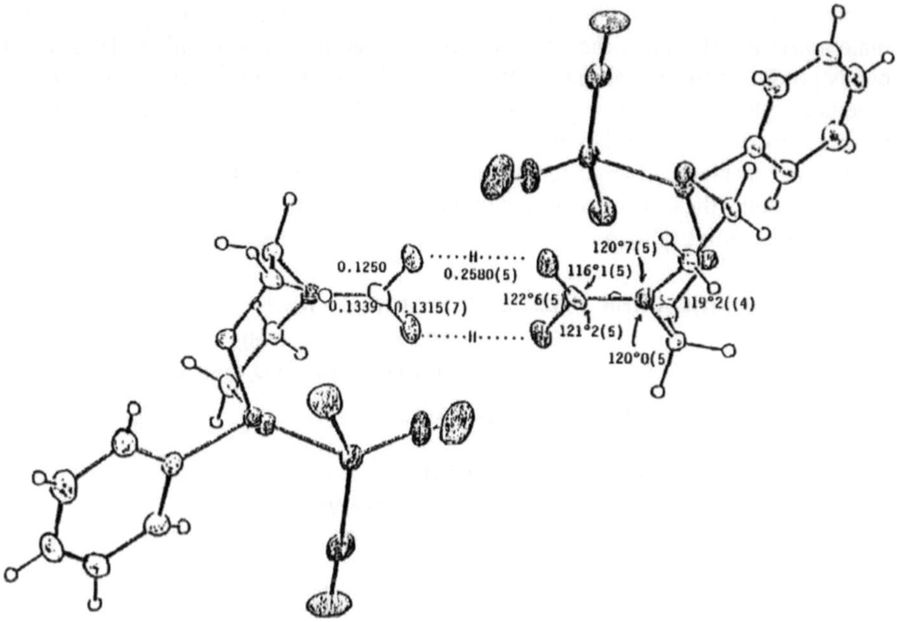

Fig. 3.1 Molecular structure of $CoCl(NO)_2[PhP(OCH_2CH_2)_2N–CO_2H]$ (bond distances in nm)

dioxide to give $Bn_2NC(O)OH$ [17]. Spectroscopic (IR) and gas volumetric measurements showed that $Bn_2NC(O)OH$ formed by carbonation of dibenzylamine through the intermediacy (3.9) of the dialkylammonium carbamate salt $(Bn_2NH_2)O_2CNBn_2$ [11, 17]. Compounds $CoCl(NO)_2[PhP(OCH_2CH_2)_2N–CO_2H]$ and $Bn_2NC(O)OH$ are unique examples of derivatives of the elusive carbamic acid H_2NCO_2H, which have been isolated and fully characterized. In both cases single-crystal X-ray analysis has shown that the relevant carbamic acid molecules are organized in an H-bonded dimeric structure (Fig. 3.1).

The isolation of both $CoCl(NO)_2[PhP(OCH_2CH_2)_2N–CO_2H]$ and $Bn_2NC(O)OH$ demonstrates that carbonation of secondary amine to the corresponding carbamic acid can be a thermodynamically favorable process, depending on the nature of the amine substituents and the selected reaction conditions. It should be noted that whether the formation of the carbamic acid is or is not a direct insertion of CO_2 into the N–H bond (3.11) has been argued over for a long time and is still a controversial issue (see also Chap. 4). A key factor in the stabilization of the acid is the establishment of the intermolecular hydrogen bond network in the solid state.

$$RR'N - H + CO_2 \rightarrow RR'N - CO_2H \tag{3.11}$$

Both mono- and dimeric *N*-alkyl substituted carbamic acids have also been detected in solution [12, 18, 19]. At 293 K, under N_2, the partial decomposition of the $(BnNH_3)O_2CNHBn$ salt, suspended in THF, was followed by IR spectroscopy [18]. In addition to the absorption at 1545 cm^{-1}, assigned to carbamate anion $BnNHCO_2^-$, the IR spectrum of an equilibrated solution also showed bands produced by free amine (3390 and 3320 cm^{-1}) and free CO_2 (2340 cm^{-1}), together with absorptions at 1725 and 1685 cm^{-1} (shifted to 1678 and 1643 cm^{-1}, respectively, upon $^{13}CO_2$ labeling) which were produced by the mono- and dimeric *N*-alkyl substituted carbamic acids $BnNHCO_2H$ and $[BnNHCO_2H]_2$ (Scheme 3.3). On the whole, these results also demonstrate that, in solution, alkylammonium carbamates exhibit quite a

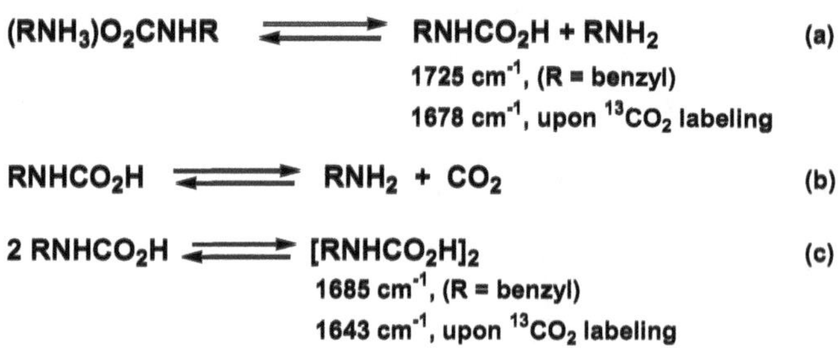

Scheme 3.3 Chemical stability of alkylammonium carbamates

complex behavior and can convert into the corresponding monomeric carbamic acid $RNHCO_2H$, which can either decarboxylate or undergo self-association with formation of dimers [18].

A somewhat different behavior has been described for unsubstituted carbamic acid H_2NCO_2H [10, 11]. The latter species has not yet been isolated as a pure compound, but it has been detected by several methods, including IR spectroscopy. Based on its low-temperature IR spectrum, it has been proposed that the stability of such a compound in the solid phase may be caused by intermolecular hydrogen bonding of the zwitterion form $H_3N^+COO^-$ [20]. It is worth noting that gaseous carbamic acid is unstable with respect to decomposition into $NH_4(O_2CNH_2)_{(s)} + CO_{2(g)}$ and into $NH_{3(g)} + CO_{2(g)}$ ($\Delta H° = -26$ kJ/mol) [21, 22]. According to a few theoretical studies, the zwitterion $H_3N^+CO_2^-$ has been calculated to be less stable than monomeric acid H_2NCO_2H and, therefore, the zwitterion form has been proposed as the most probable intermediate for the decomposition of gaseous carbamic acid to ammonia and CO_2 [22].

The reaction of amines with CO_2 to afford the corresponding carbamate has been the subject of extended kinetics studies [23]. In most cases, water was used as reaction medium, wherein hydrolytic processes forming hydrogencarbonate ((3.12)–(3.14)) can also occur and compete with carbamate

$$OH^- + CO_2 \rightarrow HCO_3^- \quad (3.12)$$

$$CO_2 + H_2O \leftrightarrows HCO_3^- + H^+ \quad (3.13)$$

$$R_2N - CO_2^- + H_2O \rightarrow HCO_3^- + R_2NH \qquad (3.14)$$

formation. The reaction of carbon dioxide with primary or secondary amines was found to be of first kinetic order with respect to CO_2 concentration, but can exhibit complicated reaction orders depending on the amine.

Scheme 3.4 illustrates the mechanism proposed by Danckwerts for the conversion of amines and CO_2 into alkylammonium carbamates [24]. This two-step pathway implies the direct reaction of the amine with CO_2 to give a zwitterion intermediate species [25], which undergoes deprotonation by a base B, thus converting into the carbamate product. The base B is the amine itself under anhydrous conditions, but in aqueous medium this role may also be played by H_2O or OH^-.

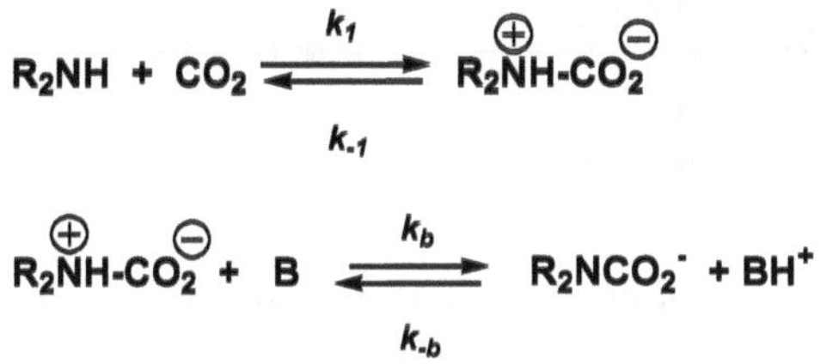

R_2NH = I or II amine; B = R_2NH, H_2O, OH^-

Scheme 3.4 Carbamate formation from amines and carbon dioxide: zwitterion mechanism

Applying the steady-state approximation to the zwitterion intermediate and assuming that the terms containing the rate constant k_{-b} are small by comparison with the other terms, the rate of carbamate formation can be expressed as

$$\text{rate} = k_1[CO_2][R_2NH]/(1 + k_{-1}/\Sigma k_b[B]) \qquad (3.15)$$

where $\Sigma k_b[B]$ is the contribution to the removal of the proton by all bases present in solution (H_2O, OH^-, R_2NH). For non-aqueous solvents, only the amine can be regarded as the deprotonating base.

According to (3.15), the reaction rate may show a fractional kinetic order, between one and two, with respect to amine concentration. When zwitterion deprotonation is much faster than zwitterion splitting ($\Sigma k_b[B] \gg k_{-1}$), zwitterion formation is the rate determining step and (3.15) becomes (3.16):

$$v = k_1[CO_2][R_2NH] \qquad (3.16)$$

In this case the reaction rate is of first order with respect to the amine and the overall reaction order is two. However, when zwitterion deprotonation is the rate determining step ($k_{-1} \gg \Sigma k_b[B]$), a more complicated expression is obtained for the reaction rate:

$$v = (k_1/k_{-1})[CO_2][R_2NH](\Sigma k_b[B]) \qquad (3.17)$$

According to (3.17), a fractional overall reaction order between two and three may be observed. In the asymptotic case wherein the amine contribution to zwitterion deprotonation is much more important than that of H_2O and OH^-, an overall kinetic order of three can be found, as, in this case, the reaction rate are of first order with respect to [CO_2] and of second order with respect to [R_2NH].

The case of sterically hindered amines deserves a comment. In general, steric hindrance at the nitrogen atom does not seem to prevent carbamate formation under anhydrous conditions [17]. However, sterically hindered amines are poorly reactive with CO_2 in water, where hydrogencarbonate rather than carbamate may form, not only according to reactions (3.12)–(3.14), but also through a

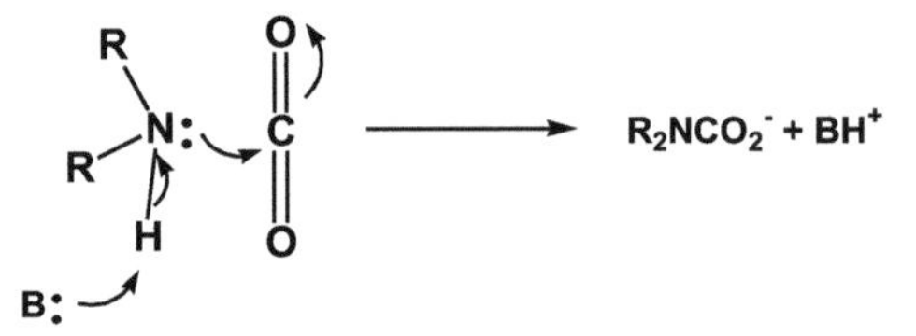

Scheme 3.5 Carbamate formation from amines and carbon dioxide: termolecular mechanism

zwitterion mechanism. In the latter case, the zwitterion intermediate is supposed to react more easily with H_2O according to (3.18). Equation (3.19) summarizes the overall reaction which accounts for the formation of hydrogencarbonate anion.

$$R_2NH^+ - CO_2^- + H_2O \leftrightarrows HCO_3^- + R_2NH_2^+ \tag{3.18}$$

$$R_2NH - CO_2 + H_2O \leftrightarrows HCO_3^- + R_2NH_2^+ \tag{3.19}$$

Crooks and Donnellan [26], studying the kinetics of formation of *N*,*N*-dialkylcarbamate from DEA (2,2′-iminodiethanol; diethanolamine) and carbon dioxide in anhydrous ethanol, emphasized that the kinetics results were consistent with both the zwitterion mechanism and a single-step termolecular reaction (Scheme 3.5). In subsequent work [27], they showed that the termolecular mechanism is also a good model for the carbamation of several primary and secondary amines in water, as supported by the large negative values of activation entropy. This mechanism may be regarded as the borderline case of the Danckwerts mechanism for $k_{-1} \gg \Sigma k_b[B]$. The initial product is not a zwitterion, but a loosely-bound encounter complex. Most of these complexes break up to give reagent molecules (CO_2 and amine) again; however, a few of them can react with a second molecule of amine or a water molecule to give ionic products. This mechanism has been revisited from a theoretical point of view by da Silva and Svendsen for the specific case of MEA (2-aminoethanol; monoethanolamine) and DEA [28]. The ab initio calculations suggested that a single-step, third-order reaction was the most likely. It would seem unlikely that a zwitterion intermediate with a significant lifetime may be present in the system.

Tertiary amines cannot form carbamates, but they react with CO_2 in aqueous solvents to form hydrogencarbonate (3.20). Yu and Astarita [29] studied the kinetics of the absorption of carbon dioxide in MDEA (bis(2-hydroxyethyl) methylamine; *N*-methyldiethanolamine) solutions and

$$R_3N + CO_2 + H_2O \leftrightarrows HCO_3^- + R_3NH^+ \tag{3.20}$$

concluded that MDEA acted as a homogeneous catalyst for CO_2 hydrolysis. As a result, the rate of CO_2 absorption in aqueous MDEA solutions was significantly larger than one would calculate by simply taking into account the alkalinity of the reaction. They proposed a zwitterion-type mechanism summarized by (3.21) and (3.22). Equation (3.22) describes the homogeneous hydrolysis reaction of an unstable amine–CO_2 adduct, which reacts with water to form hydrogencarbonate anion.

$$R_3N + CO_2 \leftrightarrows R_3N^+ - CO_2^- \tag{3.21}$$

$$R_3N^+ - CO_2^- + H_2O \leftrightarrows HCO_3^- + R_3NH^+ \tag{3.22}$$

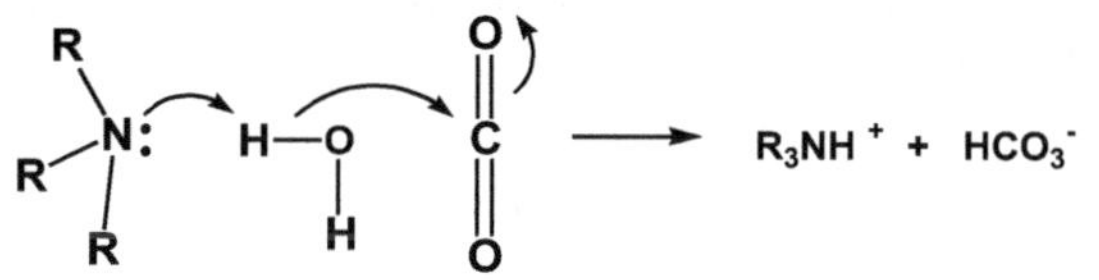

Scheme 3.6 Hydrogencarbonate formation from tertiary amines and carbon dioxide: termolecular mechanism

Crooks and Donnellan [30] investigated the kinetics of the reaction between CO_2 and tertiary amines, such as MDEA and TEA (2,2′,2″-nitrilotriethanol; triethanolamine) in water. A single step termolecular mechanism was proposed (Scheme 3.6) to be operative in this case, as suggested by the reaction order and the large negative activation entropy.

N-Substituted carbamic acids or their anions play a key role as intermediates in several biological processes (carboxylations by biotin enzymes; photosynthetic activation of CO_2 promoted by Rubisco (ribulose 1,5-biphosphate carboxylase); glucuronidation, etc.) [10, 11, 31, 32].

Nowadays, the chemistry of amines with CO_2 is drawing more and more attention, mainly because of its potential applications [10, 11]. In addition to the traditional uses in chemical industry (Solvay process, synthesis of urea), new applications have begun to emerge. Because of their modest thermal stability, alkylammonium carbamates can release CO_2 upon heating (Scheme 3.3). This property may be useful in several circumstances. For instance, amines (MEA, DEA, TEA, MDEA, etc.) and polymer-bound amines are being investigated as reusable "CO_2 scrubbers" for removing carbon dioxide from industrial exhaust streams [23, 33–36]. To this end, ionic liquids and MOF (Metal Organic Framework) incorporating –NH_2 groups in their structure are also under study [33–36]. This method has also been extended to the use of multiple amine-containing dendrimers [37, 38]. The dynamic CO_2–amine chemistry has been successfully exploited for the reversible formation of organogels by exposing solutions of some long-chain alkylamines to CO_2 [39, 40] or to build up supramolecular polymers [41, 42]. Thermally reversible carbamate chemistry has recently been employed for molecular imprinting of polymers [43] and for preparing switching-polarity solvents [44, 45].

Alkylammonium, as well as metal and p-block carbamates (see Chap. 4), are potential carriers or sources of carbamic group and, if the transfer of "RR′NCO_2" moiety to a suitable organic substrate is accomplished, can be used in the synthesis of organic carbamates, which are potential precursors of ureas and/or isocyanates [10]. To date, much effort has been made to develop new safe CO_2-based synthetic methodologies for these classes of chemicals, usually obtained through hazardous phosgenation methods [46, 47].

The transfer of the carbamic group from carbamate sources to a suitable electrophilic organic substrate is not a banal step. Carbamate anion, $RR'NCO_2^-$, is a remarkable case of an ambident nucleophile as it shows two sites for interacting with an electrophile, E (Scheme 3.7): the oxygen ends and the carbamic nitrogen atom. Formation of carbamate ester requires the electrophilic attack by E to be addressed to one of the O-atoms (Scheme 3.7a). This results in the incorporation of originally fixed CO_2 into the

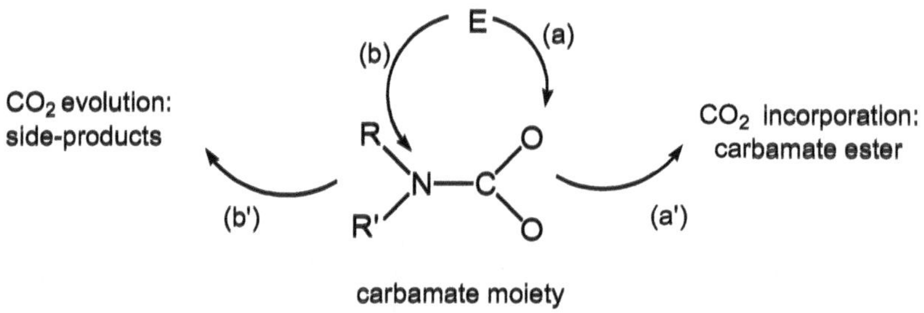

Scheme 3.7 Interaction of carbamate group with an electrophile E

"RR′NCO_2" group of a carbamate ester (Scheme 3.7a′). Conversely, electrophilic attack directed to the nitrogen atom (Scheme 3.7b) causes elimination of CO_2 from the carbamic moiety and formation of undesired side-products (Scheme 3.7b′).

The reactivity of ionic carbamates, such as alkylammonium or alkali carbamates, with alkyl halides can be directed towards *O*-carbamation, rather than *N*-alkylation, if a suitable complexing agent for the countercation is used [14, 48–50]. In the absence of any complexing agent, the strong interactions between the cation and carbamate anion depress the nucleophilicity of carbamic oxygens and alkyl halides react with the nitrogen atom, despite the delocalization of the nitrogen lone-pair on the carboxyl group. The complexing agent, by spreading and shielding the positive charge of the cation, originates much looser interactions between the cation and the O-ends of carbamate anion, whose O-nucleophilicity is thus increased. Thus, anion–cation interactions seem to play a key role in this chemistry.

Aresta and Quaranta [18] investigated the direct reaction of aliphatic primary amines with alkyl halides in the presence of carbon dioxide. Under the working conditions (293–353 K; 0.1 MPa CO_2; solvent = THF, MeOH, $PhCH_3/CH_2Cl_2$ mixtures) the formation of carbamate ester was not observed at all or occurred only to a very low extent. The reaction afforded mainly *N*-alkylation products, which may form by direct alkylation of free amine present at equilibrium or be the result of electrophilic attack by alkyl halide at the carbamic nitrogen atom of carbamate anion or other carbamic species ($RNHCO_2H$, $[RNHC(O)OH]_2$; see Scheme 3.3) present in the reaction mixture.

The synthesis of carbamate esters RNHC(O)OR′ from primary amines, CO_2, and R′X implies *O*-alkylation of the intermediate ionic carbamate $(RNH_3)O_2CNHR$ and requires that the reactivity of $(RNH_3)O_2CNHR$ towards R′X has to be modified. This result was achieved by working in the presence of a suitable macrocyclic polyether. In the presence of 18-crown-6, *O*-alkylation can occur in competition with *N*-alkylation and organic carbamates were obtained in satisfactory yields at ambient temperature [18]. The change of reactivity was ascribed to the formation of a "host–guest" adduct (Scheme 3.8) between the crown-ether and the alkylammonium cation RNH_3^+. The crown-ether can both increase the solubility of carbamate salt, usually poorly soluble in most organic solvents, and change the reactivity of carbamate anion as an ambident nucleophile. Ionic association phenomena, caused by hydrogen bonding between alkylammonium cation and carbamate anion, can strongly depress the O-nucleophilicity of these salts. Complexation of RNH_3^+ ions by crown-ether molecules weakens the

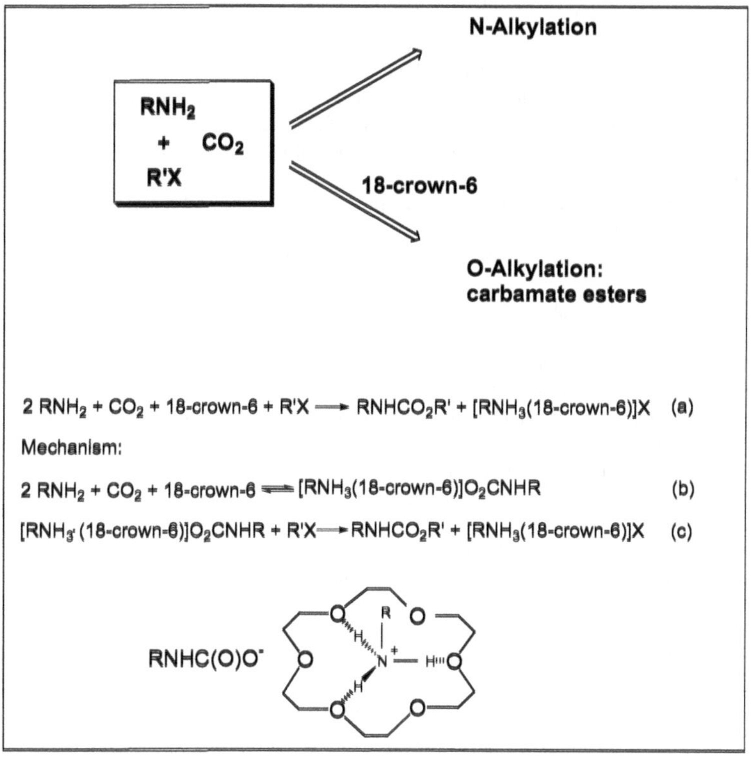

Scheme 3.8 Direct synthesis of carbamate esters from primary amines, carbon dioxide and alkyl halides in the presence of 18-crown-6-ether

anion–cation interactions and enhances O-nucleophilicity of carbamate anion. An analogous change of reactivity was observed by complexing the alkali cation of Group 1 metal carbamates with suitable crown-ethers or cryptands [48–50].

A remarkable enhancement of the O-nucleophilicity of carbamate anion can be achieved if the counterion (BH^+) is a highly polarizable cation, characterized by an extended delocalization of the positive charge. As already noted (see above), carbamate salts $(BH)O_2CNRR'$ can be generated by reacting amines with CO_2 in the presence of strong bases, such as amidines, pentaalkylguanidines, and phosphazenes [10, 13]. In the presence of the above superbases, amines and CO_2 react with alkyl halides, generating carbamate esters with high yield. The strong base B not only drives the formation of carbamate salt, favoring a higher salt concentration in solution, but also facilitates the formation of a "naked" $RR'NCO_2^-$ carbamate ion, more effectively separated from the countercation, thus making the O-ends of the anion more prone to electrophilic attack.

References

1. Faurholt C (1924) Studies on aqueous solutions of carbonic anhydride and carbonic acid. J Chim Phys 21:400–401
2. Faurholt C (1927) Studies on monoalkylcarbonates. Z Physik Chem 126:85–86
3. Heston BO, Dermer OCV, Woodside JA (1942) Acad Sci 67–68
4. Pinsent BRW, Pearson L, Roughton FJW (1956) The kinetics of combination of carbon dioxide with hydroxide ions. Trans Faraday Soc 52:1512–1514
5. Himmelblau DM, Babb AL (1958) Kinetic studies of carbonation of reactions using radioactive tracers. AIChE J 4:143–147
6. Sirs JA (1958) Electrometric stopped flow measurements of rapid reactions in solution. Trans Faraday Soc 54:201–205
7. Astarita G, Savage DW, Bisio A (1983) Gas treating with chemical solvents. Wiley, New York, NY
8. Kenig EI, Kucka L, Gorak A (2002) Rigorose Modellierung von Reactiveabsorptionprozesses. Chem Ing Tech 74:745–750
9. Kucka L, Kenig EY, Gorak A (2002) Kinetics of gas-liquid phase reaction between carbon dioxide and hydroxide ions. Ind Eng Chem Res 41:5962–5967
10. Quaranta E, Aresta M (2010) The chemistry of N-CO_2 bonds: synthesis of carbamic acids and their derivatives, isocyanates, and ureas. In: Aresta M (ed) Carbon dioxide as chemical feedstock. Wiley-VCH, Weinheim, pp 121–167
11. Belli Dell'Amico D, Calderazzo F, Labella L, Marchetti F, Pampaloni G (2003) Converting carbon dioxide into carbamato derivatives. Chem Rev 103:3857–3897
12. Masuda K, Ito Y, Horiguchi M, Fujita H (2005) Studies on the solvent dependence of the carbamic acid formation from ω-(1-naphthyl)alkylamines and carbon dioxide. Tetrahedron 61:213–229
13. McGhee WD, Riley D, Kevin C, Pan Y, Parnas B (1995) Carbon dioxide as a phosgene replacement: synthesis and mechanistic studies of urethanes from amines, CO_2, and alkyl chlorides. J Org Chem 60:2820–2830
14. Aresta M, Dibenedetto A, Quaranta E (1995) Reaction of alkali-metal tetraphenylborates with amines in the presence of CO_2: a new easy way to aliphatic and aromatic alkali-metal carbamates. J Chem Soc Dalton Trans 3359–3363

15. Aresta M, Quaranta E (1995) Novel, CO_2-promoted synthesis of anhydrous alkylammonium tetraphenylborates: a study of their reactivity as intra- and inter-molecular proton transfer agents. J Organomet Chem 488:211–222
16. Aresta M, Ballivet-Tkatchenko D, Bonnet MC, Faure R, Loiseleur H (1985) Synthesis and structural characterization of $Co(NO)_2[PhP(OCH_2CH_2)_2NH]Cl$: a novel carbon dioxide carrier. J Am Chem Soc 107:2994–2995
17. Aresta M, Ballivet-Tkatchenko D, Belli Dell'Amico D, Bonnet MC, Boschi D, Calderazzo F, Faure R, Labella L, Marchetti F (2000) Isolation and structural determination of two derivatives of the elusive carbamic acid. Chem Commun 1099–1100
18. Aresta M, Quaranta E (1992) Role of the macrocyclic polyether in the synthesis of *N*-alkylcarbamate esters from primary amines, CO_2 and alkyl halides in the presence of crown-ethers. Tetrahedron 48:1515–1530
19. Hampe EM, Rudkevich DM (2003) Exploring reversible reactions between CO_2 and amines. Tetrahedron 59:9619–9625
20. Khanna RK, Moore MH (1999) Carbamic acid: molecular structure and IR spectra. Spectrochim Acta Part A 55:961–967
21. Remko M, Rode BM (1995) Ab initio study of decomposition of carbamic acid and its thio and sila derivatives. J Mol Struct (THEOCHEM) 339:125–131
22. Wen N, Brooker MH (1995) Ammonium carbonate, bicarbonate, and carbamate equilibria: a Raman study. J Phys Chem 99:359–368
23. Vaydya PD, Kenig EY (2007) CO_2-alkanolamine reaction kinetics: a review of recent studies. Chem Eng Technol 30:1467–1474
24. Danckwerts PV (1979) The reaction of CO_2 with ethanolamines. Chem Eng Sci 34:443–446
25. Caplow M (1968) Kinetics of carbamate formation and breakdown. J Am Chem Soc 90:6795–6803
26. Crooks JE, Donnellan JP (1988) Kinetics of formation of *N,N*-dialkylcarbamate from diethanolamine and carbon dioxide in anhydrous ethanol. J Chem Soc Perkin Trans 2:191–194
27. Crooks JE, Donnellan JP (1989) Kinetics and mechanism of the reaction between carbon dioxide and amines in aqueous solution. J Chem Soc Perkin Trans 2:331–333
28. da Silva EF, Svendsen HF (2004) Ab initio study of the reaction of carbamate formation from CO_2 and alkanolamines. Ind Eng Chem Res 43:3413–3418
29. Yu W-C, Astarita G, Savage DW (1985) Kinetics of carbon dioxide absorption in solutions of methyldiethanolamine. Chem Eng Sci 40:1585–1590
30. Crooks JE, Donnellan JP (1990) Kinetics of the reaction between carbon dioxide and tertiary amines. J Org Chem 55:1372–1374
31. Schaefer WH (2006) Reaction of primary and secondary amines to form carbamic acid glucuronides. Curr Drug Metab 7:873–881
32. Walther D, Ruben M, Rau S (1999) Carbon dioxide and metal centres: from reactions inspired by nature to reactions in compressed carbon dioxide as solvent. Coord Chem Rev 182:67–100
33. Bara JE, Camper DE, Gin DL, Noble RD (2010) Room-temperature ionic liquids and composite materials: platform technologies for CO_2 capture. Acc Chem Res 43:152–159
34. Brennecke JF, Gurkan BE (2010) Ionic liquids for CO_2 capture and emission reduction. J Phys Chem Lett 1:3459–3464
35. Choi S, Watanabe T, Bae T-H, Sholl DS, Jones CW (2012) Modification of the Mg/DOBDC MOF with amines to enhance CO_2 adsorption from ultradilute gases. J Phys Chem Lett 3:1136–1141
36. Yang Z-H, He L-N, Gao J, Liu A-H, Yu B (2012) Carbon dioxide utilization with C-N bond formation: carbon dioxide capture and subsequent conversion. Energy Environ Sci 5:6602–6639
37. Kovvali AS, Sirkar KK (2001) Dendrimer liquid membranes: CO_2 separation from gas mixtures. Ind Eng Chem Res 40:2502–2511
38. Fadhel B, Hearn M, Chaffee A (2009) CO_2 adsorption by PAMAM dendrimers: significant effect of impregnation into SBA-15. Micropor Mesopor Mat 123:140–149

39. Carretti E, Dei L, Baglioni P, Weiss RG (2003) Synthesis and characterization of gels from polyallylamine and carbon dioxide as gellant. J Am Chem Soc 125:5121–5129
40. George M, Weiss RG (2006) Molecular organogels. Soft matter comprised of low-molecular-mass organic gelators and organic liquids. Acc Chem Res 39:489–497
41. Stastny V, Anderson A, Rudkevivh DM (2006) Supramolecular structures from lysine peptides and carbon dioxide. J Org Chem 71:8696–8705
42. Rudkevivh DM, Xu H (2005) Carbon dioxide and supramolecular chemistry. Chem Commun 2651–2659
43. Ki CD, Oh C, Oh S-G, Chang JY (2002) The use of a thermally reversible bond for molecular imprinting of silica spheres. J Am Chem Soc 124:14838–14839
44. Phan L, Andreatta JR, Horvey LK, Edie CF, Luco AL, Mirchandani A, Darensbourg DJ, Jessop PJ (2008) Switchable-polarity solvents prepared with a single liquid component. J Org Chem 73:127–132
45. Jessop PG, Mercer SM, Eldebrant DJ (2012) CO_2-triggered switchable solvents, surfactants, and other materials. Energy Environ Sci 5:7240–7253
46. Aresta M, Quaranta E (1997) Carbon dioxide, a potential substitute for phosgene. ChemTech 27:32–40
47. Carafa M, Quaranta E (2009) Synthesis of organic carbamates without using phosgene: carbonylation of amines with carbonic acid diesters. Mini-Rev Org Chem 6:168–183
48. Aresta M, Quaranta E (1988) Reactivity of phosphacarbamates: transfer of the carbamate group promoted by metal assisted electrophilic attack at the carbon dioxide moiety. J Org Chem 53:4153–4154
49. Aresta M, Quaranta E (1992) Alkali-metal-assisted transfer of carbamate group from phosphocarbamates to alkyl halides: a new easy way to alkali-metal carbamates and to carbamate esters. J Chem Soc Dalton Trans 1893–1898
50. Belforte A, Calderazzo F (1989) Formation of alkylurethanes from carbon dioxide by regioselective *O*-alkylation of alkali-metal *N*,*N*-diethylcarbamates in the presence of complexing agent. J Chem Soc Dalton Trans 1007–1009

Chapter 4
Insertion of CO_2 into E–X Bonds

Abstract This chapter deals with the "insertion" reactions of carbon dioxide (CO_2) into E–X bonds, where E and X represent several different (sets of) atoms, such as M–H, M–OH, M–C, M–OR, M–O_2, M–N, M–P, C–C, C–N, Si–H, and M–M (M = metal). Such reactions are relevant to catalysis for the formation of new bonds in which CO_2 may be implied (C–C bonds or C–E bonds) and thus to the conversion of CO_2 into added-value chemicals. The insertion product can be thermodynamically and kinetically stable or labile, offering in the latter case the opportunity of a catalytic path.

4.1 Carbon Dioxide Insertion into M–H Bonds

The study of the interaction of carbon dioxide (CO_2) with covalent-polar M–H bonds reveals quite a different complexity and intriguing reaction mechanisms with respect to the reaction with ionic metal hydrides, as discussed in Sect. 3.1. As a matter of fact, the interaction of M–H bonds with CO_2 has been debated for quite a long time [1], and the insertion of the cumulene into M–H bonds has been shown to follow two routes, as depicted in Scheme 4.1, affording either the M–OCHO group or the M–COOH moiety, both having an applicative interest.

The two methods of insertion are known as "normal" (**1**) and "abnormal" insertion (**2**). The M–H bond can be considered as a polar bond in which the metal center usually bears the fraction of positive charge and the negative charge is accumulated on H (**3** in Scheme 4.2). The charge density can be influenced by the nature of the metal itself and by the ancillary ligands L_n. The effect can be such that a sensible modification of the charge distribution may also take place, resulting in the inversion of polarity (**4** in Scheme 4.2). This is well known in transition metal hydrides chemistry; the H atom bonded to a metal can be transferred as an atom [2], a hydride [3], or a proton [4]. Moreover, the same metal hydride may show a dual behavior (proton or hydride transfer) according to the receiving substrate [5].

The two forms **3** and **4** in Scheme 4.2 interact with CO_2 in different ways and generate two different transition states, **5** and **6**, which eventually generate the formate and the metallacarboxylate carboxylate. The effect of such interaction is

M. Aresta et al., *Reaction Mechanisms in Carbon Dioxide Conversion*,
DOI 10.1007/978-3-662-46831-9_4

that the partial charge on the H atom is released to the O atom in **5** or to the C atom in **6**. The conversion of **5** into the relevant end product **1** (Scheme 4.1) is proposed to take place

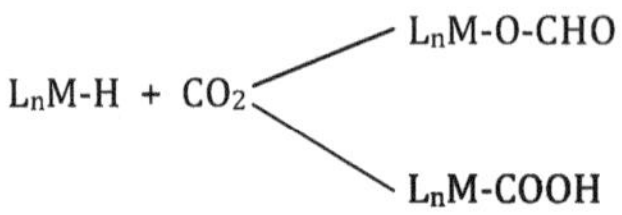

1 metalla formate

2 metalla carboxylic acid

Scheme 4.1 Insertion of CO_2 into a M–H bond

Scheme 4.2 Transition states for the "normal" and "abnormal" mode of interaction of CO_2 with M–H bonds

through the interaction of the O atom with the metal centre and concurrent cleavage of the M–H bond. As a matter of fact, a second transition state for such normal insertion **7** can be foreseen, which corresponds to a formal "2+2" interaction (Structure 4.1).

There are a number of early kinetic investigations on such reactions that reveal their reversibility [6–8] and irreversibility [9] and the reaction mechanism [6, 7, 9]. More recent studies of the reaction of CO_2 with monohydrido complexes such as $[Re(bpy)(CO)_3H]$ [10] and $[Os(bpy)_2(CO)H]$ [10] have clearly shown that the insertion is a second-order reaction (4.1) influenced by the solvent.

$$r = k[\mathrm{Re-H}][\mathrm{CO_2}] \tag{4.1}$$

The first order in both the metal hydride and CO_2 confirms that **5** and **7** are possible transition states (TSs). The reaction shows an inverse isotope dependence ($k_H/k_D = 0.55$) [8]. The activation parameters are reported to be $\Delta H^{\#} = 53.5$ kJ/mol and $\Delta S^{\#} = -138$ J/mol K. If the ancillary ligands are changed, one can observe an effect on the reaction rate. This effect can be explained by considering that ancillary ligands may

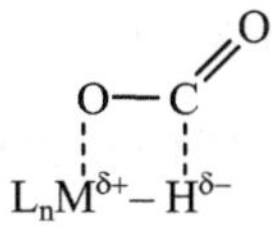

7

Structure 4.1 Possible transition state (**7**) for normal CO_2 insertion into a M–H bond

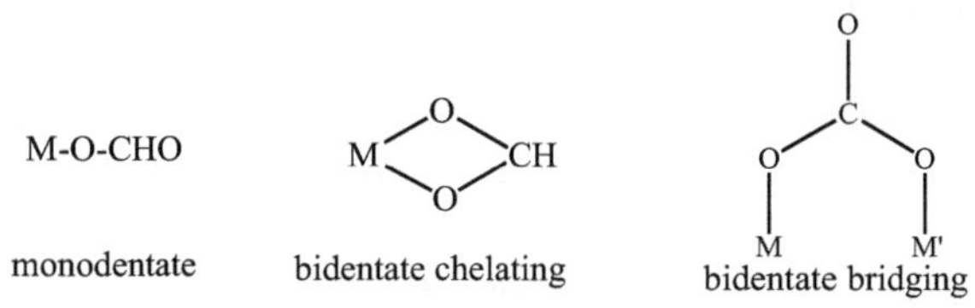

Scheme 4.3 Modes of bonding of the formate ion to a metal center

influence the charge separation on the M–H moiety. In particular, one can envisage that donor ligands would increase the electron density at the metal center, thus favoring the transfer of negative charge to H, increasing the polarity of the M–H bond. Such prediction is confirmed by experimental facts that show that increasing the donor capacity of the ancillary ligands facilitates the reaction. The variable polarity of the M–H bond also causes a dependence of the reaction rate on the nature of the solvent. Such dependence is also explained by taking into account the asymmetry of charge distribution in species **5** and **7**. It is worth emphasizing that the formate ion can be linked to the metal center in different ways, as shown in Scheme 4.3.

The mode depends on the nature of the metal center to which the formate is linked: A coordinative unsaturation of the metal favors the bidentate-chelating mode of bonding. The oxophilicity of the metal also favors the bonding of oxygen over other ligands. The linear mode of bonding has been shown to exist in *fac*-(bpy)$(CO)_3$Re(OCHO) [2–4], in which the formate is linked to a "pentacoordinated Re" moiety to afford the final esacoordinated Re complex. Conversely, the formate is bidentate chelating in $Ru(PPh_3)_3H(OCHO)$, in which the formate [11] is linked to the unsaturated "tetracoordinated Ru" moiety to afford the coordinatively saturated Ru(II) complex. The different modes can be distinguished by infrared (IR) spectroscopy.

The tendency of a transition metal hydride to transfer H^- to a substrate is called hydricity [12]. It is possible to determine the Gibbs free energy of the splitting of the covalent polar M–H bond to afford a metal cation and the hydride ion in solution. The hydricity is not parallel to the polarity of the M–H bond, nor can it be predicted on the basis of the electronic structure of the metal atom. It is a complex property that can be modeled for transition metal hydrides using multiparameter approaches. The hydricity concept applies to the interaction of M–H bonds with CO_2 as well [13]. The reactivity of M–H bonds toward CO_2 is linked to reactions that may have industrial interest, such as the hydrogenation of CO_2 to afford formic acid (4.2) and the electrochemical reduction of CO_2 to other C1 or C1+ molecules (4.3).

$$CO_{2(g)} + H_{2(g)} \rightarrow HCOOH_{(1)} \quad \Delta G^\circ = 33\ kJ\ mol^{-1} \tag{4.2}$$

$$CO_2 \xrightarrow{e^-,\ cat} HCOOH,\ CH_2O,\ CO,\ CH_3OH,\ C1 + \ldots \tag{4.3}$$

The direct synthesis of HCO_2H from CO_2 and H_2 catalyzed by transition metal systems has been much investigated in the past 20 years, and catalysts have been developed that have a high TON (turnover number) and TOF (turnover frequency) close to industrial utilization. The drawback in this field is the fact that the intermediate hydrido-formate species $L_nMH(OCHO)$ are kinetically stable and require a shuttle that helps to eliminate the hydride (amines that are converted into ammonium compounds) or the formate anion (group 1 metal

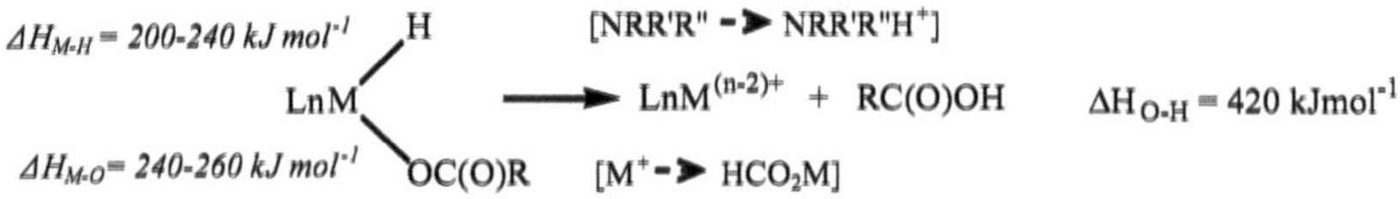

Scheme 4.4 ΔH for the elimination of free formic acid from a metal-hydrido-carboxylate. R represents an alkyl group that may substitute H; NRR′R″ is an amine; M^+ is a Group 1 cation)

cations) (Scheme 4.4). However, as well as free formic acid, formates (salts of group 1 elements or ammonium salts) are obtained [14]. The first example of the homogeneous hydrogenation of CO_2 was reported by Inoue et al. [15] in 1976. Since then, a conspicuous number of papers and reviews [16–19] have been published that have highlighted recent progress. The net reaction (4.2) between gaseous H_2 and CO_2 to afford liquid formic acid is endoergonic ($\Delta G^\circ = +33\ kJ\ mol^{-1}$) because of the strong entropic contribution (Table 4.1).

Conversely, if the reaction is carried out in a condensed medium such as water, despite the latter causing multiple equilibria with formation of CO_2 hydrated forms (4.4) in addition to the forms of dissociated formic acid (4.5), or if a base (ammonia or other bases) is used that produces formate salts instead of free formic acid, the thermodynamics of the process is favored ($\Delta G^\circ = -4\ kJ\ mol^{-1}$ for (4.6a) and (4.6b) and $\Delta G^\circ = -35\ kJ\ mol^{-1}$ for (4.7), respectively) [16–21].

Table 4.1 Thermodynamic data for the conversion of CO_2 and H_2 into HCO_2H under various conditions

Reagents and their status	ΔH° kJ mol^{-1}	ΔS° J mol^{-1} K^{-1}	ΔG° kJ mol^{-1}	Product and its status
$CO_{2(g)} + H_{2(g)}$	−31.2	−215	+32.9	$HCO_2H_{(l)}$
$CO_{2(aq)} + H_{2(aq)}$	−69	−230	−4	$HCO_2H_{(aq)}$
$CO_{2(aq)} + H_{2(aq)} + NH_{3(aq)}$	−89.3	−250	−9.5	$HCO_2{}^-{}_{(aq)}NH_4{}^+{}_{(aq)}$
$CO_{2(aq)} + H_{2(aq)} + Base_{(aq)}$			−35	$HCO_2{}^-{}_{(aq)} + BaseH^+{}_{(aq)}$
$MHCO_{3(aq)} + H_{2(aq)}$	−20.5	−66.4	−0.72	$MHCO_{2(aq)} + H_2O_{(l)}$

$$CO_{2(g)} + H_2O_{(l)} \rightarrow [CO_2 \cdot H_2O]_{(aq)}, \quad [H_2CO_3]_{(aq)}, HCO_3^-(aq), CO_3^{2-}(aq) \quad (4.4)$$

$$CO_{2(aq)} + H_{2(aq)} \rightarrow HCO_2^-(aq) + H^+_{(aq)} \quad (4.5)$$

$$CO_{2(aq)} + H_{2(g)} + NH_{3(aq)} \rightarrow HCO_2^-(aq)NH_4^+(aq) \quad (4.6a)$$

$$MHCO_{3(aq)} + H_{2(aq)} \rightarrow MO_2CH_{(aq)} + H_2O_{(l)} \quad (4.6b)$$

$$H_{2(g)} + CO_{2(g)} + B \rightarrow HCOO^-_{(aq)}BH^+(aq) \quad (4.7)$$

Interestingly, the hydrogenation of $HCO_3^-{}_{(aq)}$ to aqueous formate, $HCO_2^-{}_{(aq)}$ (4.6b) is exoergonic ($\Delta G° = -0.72$ kJ mol^{-1}and $\Delta H° = -20.5$ kJ mol^{-1}) [22]. The calculated thermodynamic properties for the formation of the formate anion are $\Delta H = -427$ kJ mol^{-1}, $\Delta S = -260$ J mol^{-1} K^{-1}, and $\Delta G = -350$ kJ mol^{-1}.

As mentioned above, the spontaneous release of formic acid from the catalyst (Scheme 4.4) does not take place for kinetic and thermodynamic reasons under neutral or basic conditions, as 20–80 kJ mol^{-1} indicates that the process is not thermodynamically favoured. Moreover, breaking of the M–O and M–H bonds can represent the real kinetic barrier to the process.

This has for a long time prevented the exploitation of such reaction as a catalytic direct route to free HCO_2H. In fact, either the ammonium- or the Group 1 metal formate were produced which need treatment with inorganic acids (such as H_2SO_4) for the production of the free acid (4.8).

$$R_3NH^+(M^+)HCO_2^- + HX \rightarrow HCO_2H + R_3NH^+(M^+)X^- \quad (X = HSO_4^-, Cl^-) \quad (4.8)$$

Such acid treatment produces equimolar (to HCO_2H) amounts of waste inorganic salts. Very recent reports are quite encouraging about the possible exploitation of the direct synthesis of free formic acid. In fact, Lawrenczy has shown that it is possible to produce free formic acid by carrying out the reaction in acidic media (in water or DMSO). Although limited to tests carried out directly in NMR tubes, 1.9 M formic acid has been detected in DMSO (0.2 M in water) using $RuCl_2$(1,3,5-triaz-7-phosphaadamantane) and other related ligands [23]. This finding represents a breakthrough in this field and can open the way to exploitation of the reaction. The conversion yield is influenced by the temperature over the range 303–363 K. At the latter temperature the yield is halved with respect to that at 303 K.

Much less is known about the "abnormal insertion" path bearing to the metallacarboxylic species (**2**) for which only a few claims exist. Nevertheless, the M–C(O)OH moiety can be stable and is easily formed upon reaction of a metal carbonyl species M(CO) with a hydroxo anion (OH^-) (4.9) or by protonation of co-ordinated CO_2 (4.10 and 4.11) or by insertion of CO_2 in the M–H bond (Scheme 4.1). The formation of Ni–C(O)OH via proton attack at the oxygen of a coordinated η^2(C,O)–CO_2 has been reported for the reaction of $Ni(PCy_3)_2(CO_2)$ with PhSH [24]. The resulting Ni–CO_2H moiety undergoes further protonation with water elimination and formation of a carbonyl species Ni–CO. Tanaka [25] has reported that the η^1(C)–CO_2 Ru(diphos)$_2$(CO_2) can be protonated to afford the stable Ru–COOH species characterized by XRD.

$$L_nM^{n+} - CO + OH^- \rightarrow M^{(n-1)+} - COOH \tag{4.9}$$

$$(PCy_3)_2Ni(CO_2) + H^+ + e^- \rightarrow \left[(PCy_3)_2Ni - COOH\right] \tag{4.10}$$

$$Ru(diphos)_2(CO_2) + H^+ \rightarrow \left[Ru(diphos)_2(COOH)\right]^+ \tag{4.11}$$

Dubois et al. [26] have proposed the formation of a Pd–CO_2H species as intermediate in the electrochemical reduction of CO_2 catalyzed by a Pd complex. However, despite the M–COOH species being proposed to be present and active in several processes implying H-species and CO_2, a clear-cut demonstration of CO_2 insertion into the M–H bond of a transition metal complex to afford M–CO_2H is not found in the literature. Consequently, it is difficult to find any demonstration of the reaction mechanism or the energies and geometries of TSs. Conversely, some thermodynamic and kinetic observations have been reported for the reverse reaction (4.12), which is for the conversion of the Ru(bpy)$_2$(CO)(COOH)$^+$ species, [27] generated through the attack of a hydroxo anion on a carbonyl moiety linked to Ru in the complex Ru(bpy)$_2$(CO)$_2^{2+}$, into the hydride Ru–complex L_nRu–H and CO_2. For such a kind of reaction it has been found that basic ancillary ligands favor the CO_2-elimination process.

$$L_nM - COOH \rightarrow L_nM - H + CO_2 \tag{4.12}$$

This implies that ligands with poor basicity or good Lewis acidity should favor the "abnormal" insertion process, which is in agreement with what was discussed above. The metallacarboxylic species M–CO_2H is

$(PCy_3)_2$Ni-[C=O(OH)] + e^- + H^+ = $(PCy_3)_2$Ni-[C=O(H_2O)] ⟶ $(PCy_3)_2$Ni(CO) + H_2O (a)

L_nPd-[C=O(OH)] + e^- + H^+ = L_nPd=C(OH)$_2$ ⟶ L_nPd(CO) + H_2O (b)

Scheme 4.5 Reaction of a M–CO_2H moiety with "e^- + H^+"

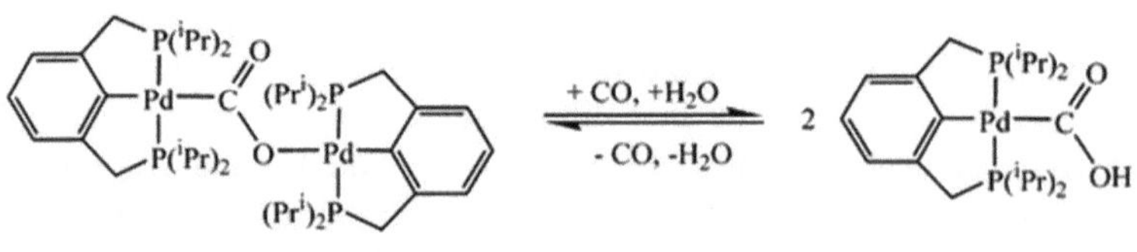

Scheme 4.6 Pd–CO_2H complex that reversibly undergoes intramolecular formal "HCO_2H" elimination (H_2O + CO) with formation of a μ^2-η^2-CO_2-complex

an intermediate in the reduction of CO_2 to CO via "$e^- + H^+$" transfer. Two paths have been proposed for such "$e^- + H^+$" transfer, as shown in Scheme 4.5: the protonation at the –OH group bound to C (Scheme 4.5a) [24, 25] and the protonation at the C=O of the carboxylic moiety [28] bearing to the formation of a metal-dihydroxycarbene analogous to similar species proposed as intermediates in Fisher–Tropsch processes (Scheme 4.5b). This point still requires clarification.

The conversion of M–CO_2H moieties does not always goes towards CO_2 elimination and M–H formation. Scheme 4.6 shows the case of a Pd–CO_2H complex which formally eliminates HCO_2H (better H_2O and CO) undergoing dimerization to afford a μ^2-η^2-complex (see Sect. 2.5) with the CO_2 molecule bridging two Pd atoms, i.e., being bound via C to one Pd and via O to the second one. The reaction is reported to be reversible [29].

More details about the reactivity of the M–H:CO_2 system can be found in Chap. 8, where the electrochemical, photochemical, and electrophotochemical reduction of CO_2 in water is discussed.

4.2 CO_2 Insertion into M–OH Bonds

The insertion of CO_2 into discrete covalent polar M–OH bonds has long been investigated because as the reaction is related to the conversion of CO_2 into its hydrated forms (anionic or coordinated HCO_3^- and the elusive acid H_2CO_3 which cannot be isolated as a pure compound), a reaction that plays a key role in CO_2 elimination in humans and animals in the respiratory process. Such a reaction is relevant to the enzyme carbonic anhydrase (CA) which accelerates the reaction of hydration–dehydration of CO_2, thus facilitating its uptake at the cellular level and its elimination in the lungs from where it is expelled [30]. Scheme 4.7 shows the mechanism of reaction of CO_2 with the Zn–OH moiety, active center of CA.

The insertion process is pH dependent: for CA at pH 7 it is a second-order process, the rate constant for step 3 in Scheme 4.7 being estimated to be $k_{cat}/K_M \simeq 1.1 \times 10^5$ to 1.6×10^8 M^{-1} s^{-1}. Such a value increases if the pH is increased. The above k_{cat}/K_M value must be compared with those commonly found for the interaction of transition metal systems L_nM–OH with CO_2 (45–590 M^{-1} s^{-1}) [31], with the value reported for OH^- in basic solutions (7.9×10^{-3} M^{-1} s^{-1}) [32] and with that reported for $[Ni(cyclen)(OH)]^+$ (3.3×10^3 M^{-1} s^{-1}) [33], a complex with unusual performance. The enzymatic system is, thus, several orders of

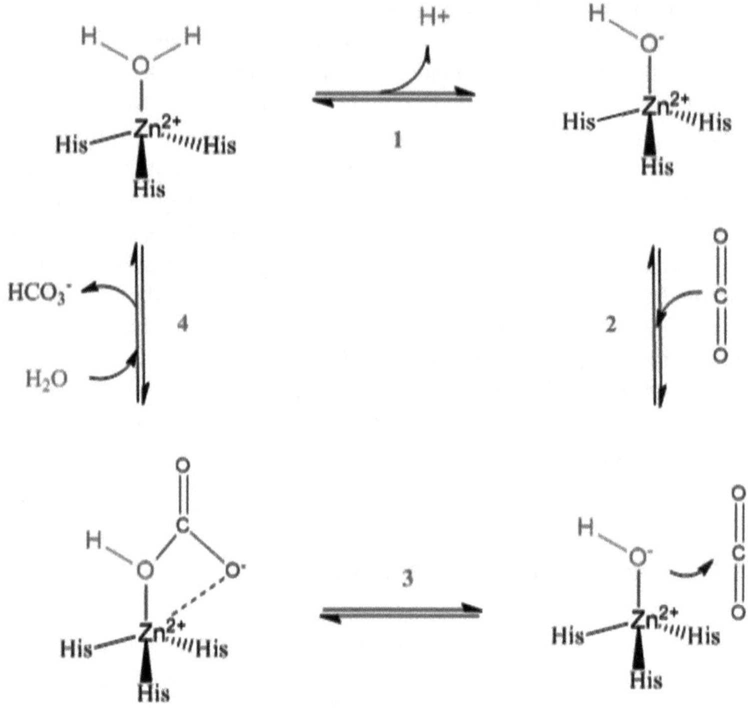

Scheme 4.7 Interaction of CO_2 with the Zn–OH active center of CA

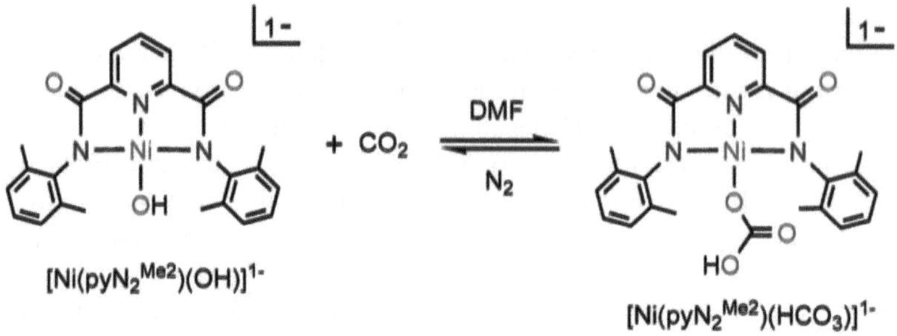

Scheme 4.8 Ni complex approaching CA

magnitude faster (2 to >5) than normal metal-systems at pH 7. A Ni-complex was recently characterized [34] that has properties much closer to those of CA. In fact, the planar nickel complex $[Ni^{II}(NNN)\ OH]^{1-}$ containing the tridentate 2,6-pyridinedicarboxamidate pincer ligand and a terminal hydroxide ligand is reported to have a rate constant equal to $9.5 \times 10^5\ M^{-1}\ s^{-1}$ in *N,N*-′-dimethylformamide at 298 K (Scheme 4.8).

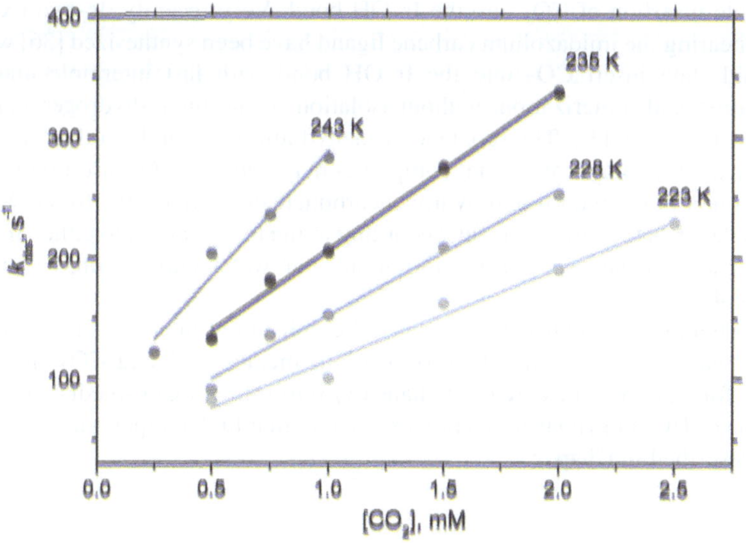

Fig. 4.1 Plots of the observed rate constants (k_{obs}) in DMF as a function of CO_2 concentration at 223–243 K and $[Ni(NNN)OH]^{-1}$ 0.1 mM

The complex is reported to be strictly monomeric as the steric hindrance of the ligand prevents any association in solution. The reaction is first order in $[CO_2]$ as shown in Fig. 4.1. It is also first order in [Ni] so that the overall second order is obtained as shown in (4.13):

$$\text{rate} = k_{cat}/K_M\left[Ni(NNN)OH^{1-}\right][CO_2] \quad (4.13)$$

Several other transition metal–hydroxo systems have been investigated. The insertion reaction affords in general a hydrogen carbonate species (4.14) which can be stable or can react further.

$$M - OH + CO_2 \rightarrow M - OC(O)OH \rightarrow \text{products} \quad (4.14)$$

One of the earliest reported studies was made on $Ir(OH)(CO)(PPh_3)_2$ [35] which was shown to undergo a

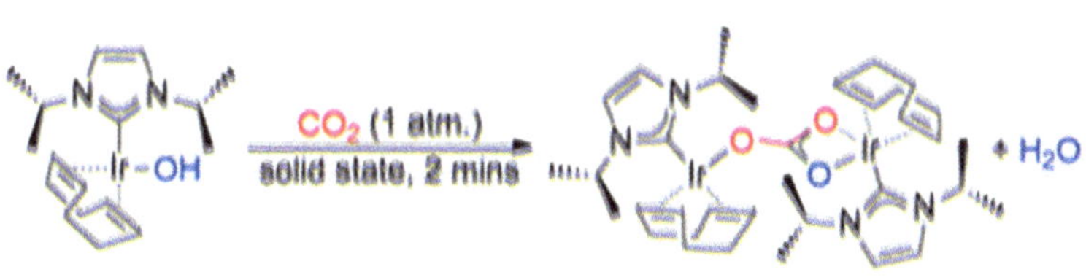

Scheme 4.9 Insertion of CO_2 into the Ir–OH bond with water elimination

reversible insertion of CO_2 into the Ir–OH bond. Very recently, Ir–hydroxo complexes bearing the imidazolium carbene ligand have been synthesized [36] which at the solid state insert CO_2 into the Ir–OH bond with fast intermolecular water elimination and dimerization, without isolation of the mono-hydrogen carbonate complex (Scheme 4.9). The spontaneous dehydration is found in other systems.

For example, dihydroxo metal complexes may undergo CO_2 insertion [37] with formation of either mono- or dihydrogencarbonate derivatives: the former have not been isolated as they further react, according to the temperature, to either afford the latter or a carbonato species via intramolecular water elimination, as shown in Scheme 4.10.

The chemistry of hydroxo species is quite rich and of interest as it is relevant to several reactive systems in which (transition) metals–H_2O and CO_2 are present. This is, for example, the case of methane dry reforming or CO_2-mediated dehydrogenation of HCs or syngas utilization or other similar high temperature processes as those described in Chap. 7.

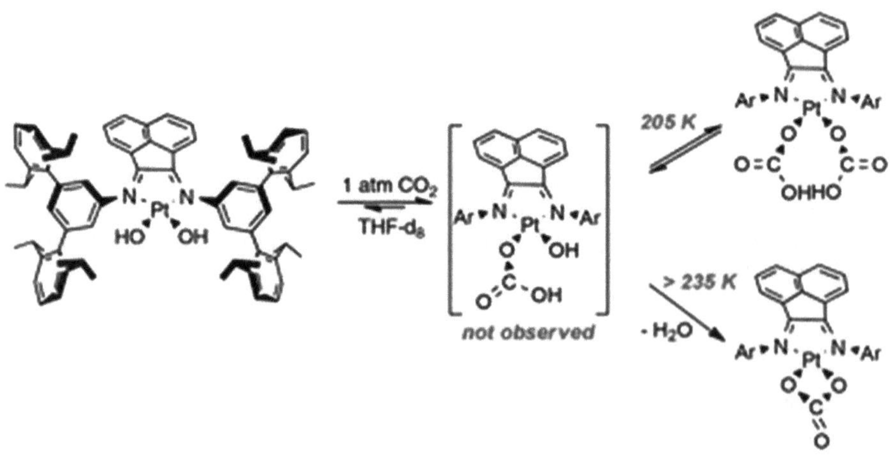

Scheme 4.10 Reaction of CO_2 with Pt-(dihydroxo) complexes and the spontaneous intramolecular water elimination

Scheme 4.11 Insertion of CO_2 into the metal–carbon bond

4.3 CO_2 Insertion into M–C Bonds

In this chapter the elementary reaction of CO_2 with a discrete M–C bond is discussed. The coupling of CO_2 with unsaturated hydrocarbons, which may also imply insertion of CO_2 into an M–C bond, is discussed in Chap. 5. The insertion–deinsertion of CO_2 into M–R bonds (R = alkyl or aryl moiety) is relevant to carboxylation–decarboxylation processes which are very important in biological systems [38]. In this case, as seen for the insertion into M–H bonds, two mechanisms are possible, as depicted in Scheme 4.11.

What was discussed for the M–H bond may hold here too. In fact, the M–C bond is polar with the same formal charge separation observed in the M–H bond. Therefore, the normal mode of insertion would be considered the one that bears to species (**8**) in Scheme 4.11. The nature of the metal and its electronic structure-oxidation state may influence the insertion path. Metals in a high oxidation state would favor the formation of **8** in Scheme 4.11, whereas lower oxidation state metal centers might favor the production of **9**, also depending on the ancillary ligands. Both metal–alkyl and metal–aryl bonds undergo such insertion reactions, although different reactivities are observed. Early studies on the reaction of M–R systems with CO_2 were reported by Volpin et al. [39]. Interestingly, they demonstrated that Ti(IV) inserts CO_2 into the Ti–CH_3 bond of $Cp_2Ti(CH_3)_2$ affording only the metalla ester $Cp_2Ti(O_2C–CH_3)_2$, whereas Co(I) was able to insert CO_2 into the Co–C_2H_5 bond of $(PPh_3)_2(CO)CoC_2H_5$, affording both the $L_nCo–CO_2C_2H_5$ and the $L_nCoO_2C–C_2H_5$ forms. The acidolysis of the Co-complexes afforded propionic acid, C_2H_5COOH, from the former complex and ethyl formate, $HCO_2C_2H_5$, from the latter, confirming the two different modes of insertion.

The complex (diphos)Rh-phenyl [40], generated from (diphos)Rh(η^6-BPh_4), was shown to undergo facile insertion of CO_2 into the Rh-phenyl bond to afford the benzoato complex (diphos)Rh($O_2C–C_6H_5$), following the normal insertion route (**8**, Scheme 4.11).

Of interest are systems in which both M–H and M–C bonds are present as they offer the possibility of comparing the relative rates and mechanisms. A study on one of such systems, namely on *trans*-Ru(dmpe)$_2$(H)CH_3, has been reported by Field and coworkers, who have shown that [41] the Ru-formate obtained by insertion of CO_2 into the Ru–H bond is the kinetic product. Such reaction also occurs instantaneously at 273 K. Conversely, the insertion of CO_2 into the Ru–CH_3 bond occurs only at 333 K, a temperature at which the formate is decomposed. However, either the formate or the acetate can be isolated. The easy reversibility of the CO_2 insertion into the Ru–H bond is in agreement with the catalytic properties of Ru (II) complexes which are quite good catalysts in the formation of HCO_2H from CO_2 and H_2 [19] in both basic [42] and acid media [43], affording the formate anion and free formic acid (see above), respectively. The kinetics and thermodynamics of the CO_2 reaction with M–H and M–CH_3 bonds in the complex *trans*-Ru(dmpe)$_2$(H) CH_3 has been recently investigated [43]. The authors have confirmed the order of reactivity Ru–H $\gg$ Ru–CH_3, and have shown that, although the insertion into the

Ru–H bond readily occurs at 273 K or even lower, the insertion into the Ru–CH_3 bond has well defined activation parameters, namely: $\Delta H^{\#} = 12.7 \pm 0.6$ kcal/mol and $\Delta S^{\#} = -31.9 \pm 2.0$ J mol^{-1} K^{-1}, inferring a well defined and ordered transition state. Such findings are important as they confirm that the reaction occurs at the metal center driven by co-ordination and exclude a dissociative pathway. The reactions were monitored by using in situ IR and ^{1}H-NMR techniques. A computational study was also carried out which provided precious information on the *trans*-effect of a ligand on the insertion rate. So if –CH_3 was substituted with –Cl in the *trans*-Ru(dmpe)$_2$$CH_3$(X) complex (X = H, CH_3), the insertion of CO_2 into the Ru–H bond, so facile in *trans*-Ru(dmpe)$_2$(H)CH_3, was not observed, also under forcing conditions (1.3 MPa CO_2): the trans-labilizing effect of the CH_3 moiety has an important role in fastening the insertion reaction into the *trans*-Ru–H bond. This trend was confirmed by studying the reactivity towards CO_2 of the dimethyl complex *trans*-Ru(dmpe)$_2$(CH_3)$_2$. A double consecutive insertion has been observed [38] bearing to the mono- and di-acetato complexes *trans*-Ru(dmpe)$_2$(O_2C–CH_3)CH_3 and *trans*-Ru(dmpe)$_2$(O_2C–CH_3)$_2$, respectively. Both insertions are exothermic (−28 and −21 kcal mol^{-1}, for the first and second steps, respectively), with the second insertion being two orders of magnitude slower than the first one at ca. 300 K [43]. The first insertion benefits from the strong *trans*-effect of the methyl moiety, whereas in the second insertion the formate moiety is *trans* to Ru–CH_3, making the insertion less favored. The reaction rate was shown to be second order overall (4.17) and dependent on the P_{CO2}. More precisely, as the [CO_2] was four times higher than that of [Ru] at $P_{CO2} = 0.1$ MPa, an apparent rate first order in [Ru] was determined (4.15). However, being that the concentration of CO_2 in solution is dependent on its gas phase pressure P_{CO2},

$$\text{rate} = k_{obs}[\text{Ru}] \tag{4.15}$$

$$k_{obs} = k[\text{CO}_2] \tag{4.16}$$

$$\text{rate} = k[\text{Ru}][\text{CO}_2] \tag{4.17}$$

and k_{obs} having been found to be directly proportional to the [CO_2] in solution (4.16), the overall second order (4.17) results.

Computational studies on the Ru-complexes have confirmed that ancillary ligands with different basicity play a key role in fastening-slowing down the reaction rate; the following order has been calculated: tmeda > dmpe > dppe. The role of the metal has also been highlighted, showing that the analogous Fe–H and Fe–CH_3 complexes are kinetically and thermodynamically more active that the Ru-complexes, confirming that first row transition metals are better catalysts than the analogous second row.

4.4 CO_2 Insertion into M–OR Bonds

The reaction of CO_2 with M–OR (R = alkyl, aryl) bonds is of industrial interest as it is related to the synthesis of molecular organic carbonates and polycarbonates based on the direct carboxylation of substrates such as alcohol, polyols, and epoxides. In this paragraph the kinetics and thermodynamics of the elementary step (4.18) are discussed, and the synthesis of organic carbonates (acyclic and cyclic) is discussed in Chap. 6.

Because of the well defined polarity of the M–O bond, there is no ambiguity about the mode of insertion of CO_2. The electrophilic carbon of the heterocumulene is attacked by the electron-rich terminal of the polar $M^{\delta+}$–$O^{\delta-}$ bond. Therefore, the insertion product is a metal-alkylcarbonate complex (4.18).

$$L_nM - OR + CO_2 \rightarrow L_nM - OC(O)OR \qquad (4.18)$$

The parameters which govern the insertion of CO_2 into the M–OR bond are influenced by all the factors which may determine the polarity and the energy of the M–O bond, namely, the nature of the metal, its oxidation state, the ancillary ligands L, and the nature of R. In general, early transition metals, high oxidation states, and basic ancillary ligands favor the reaction. M–O–alkyl moieties are more reactive than M–O–aryl groups, because of the σ-donor character of the alkyl groups. A variety of metal complexes have been investigated, which have also been tested as catalysts in synthetic chemistry and copolymerization processes [44–61]. Despite such a wide availability of chemical information, mechanistic studies centered on the reaction of CO_2 with the M–OR systems are not very abundant. Early studies [62] propose that the TS of such a reaction can be represented as in Structure 4.2 and implies the interaction of the weak electrophilic C of CO_2 with the oxygen of the OR group. Such representation has also been considered valid for other CO_2-like cumulenes such as CS_2 and COS [63], and suggests that preliminary coordination of CO_2 to a metal center is unnecessary in this class of reaction, as in other insertion reactions considered above.

A well studied case of insertion of CO_2 into the M–OR bond is represented by the insertion of CO_2 into the M–OCH_3 bond of the complex *fac*-$(CO)_3$(dppe)Mn (OCH_3) to afford *fac*-$(CO)_3$(dppe)Mn–OC(O)OCH_3, a complex with scarce application as a copolymerization catalyst. Kinetic and mechanistic studies have been aimed at the definition of the TS and kinetic parameters [64]. Such reaction was reported to be very fast already at 195 K [65, 66]. Darensbourg and coworkers have used low temperature in situ IR for their studies. Despite the difficulty linked to the experimental conditions, they were able to determine at 195 K the value of $2.0 \times 10^{-3}\ M^{-1}\ s^{-1}$ for the rate constant k_1 for the second-order reaction (4.19). The authors say that such a value is most likely underestimated.

Structure 4.2 Proposed TS for the insertion of CO_2 into a M–OR bond

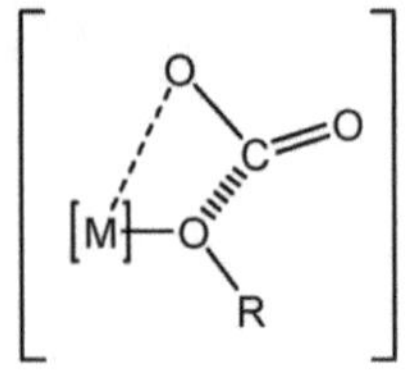

$$\text{fac-(CO)}_3\text{(dppe)Mn(OCH}_3\text{)} + CO_2 \underset{k_2}{\overset{k_1}{\rightleftharpoons}} \text{fac-(CO)}_3\text{(dppe)Mn} - \text{OC(O)OCH}_3 \quad (4.19)$$

The reaction of the carbonato complex with $^{13}CO_2$ has allowed the demonstration of a facile insertion–deinsertion equilibrium. The study of the deinsertion reaction has allowed estimation of the activation parameters as being $\Delta H^{\#} = 130 \pm 4.0$ kJ mol^{-1} and $\Delta S^{\#} = 121.6 \pm 11.9$ J mol^{-1} K^{-1}. From the above values the authors have calculated an approximate value of the equilibrium constant for the carboxylation reaction equal to 3×10^{13} M^{-1} at 195 K (or a ΔG value for the same reaction of < -50 kJ mol^{-1}), showing that the insertion of CO_2 into the M–O–alkyl bond is both kinetically and thermodynamically very favored. This trend has also been confirmed for the insertion of CO_2 into the Nb–OR bond in $[Nb(OR)_5]_2$ (R = methyl, ethyl, allyl) (see Sect. 6.2.2.1), a catalyst for the synthesis of dialkyl carbonates [67]. Very recently, the facile insertion of CO_2 into metal–phenoxide bonds has been reported [68] for cobalt and zinc complexes (Fig. 4.2). It should be noted that such metal systems are used as catalysts in the copolymerization of CO_2 with epoxides.

4.5 Insertion into M–O_2 Bonds

The reaction of a CO_2–metal system $L_nM(CO_2)$ with O_2 or of a dioxygen complex of a transition metal $L_nM(O_2)$ with CO_2 may lead to the formation of peroxocarbonates (Scheme 4.12).

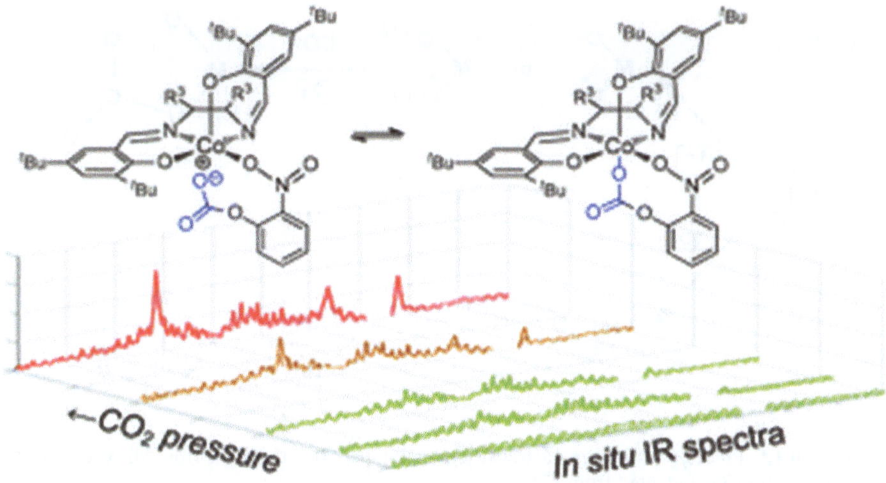

Fig. 4.2 Identified products of the insertion of CO_2 into the Co–phenoxide bond

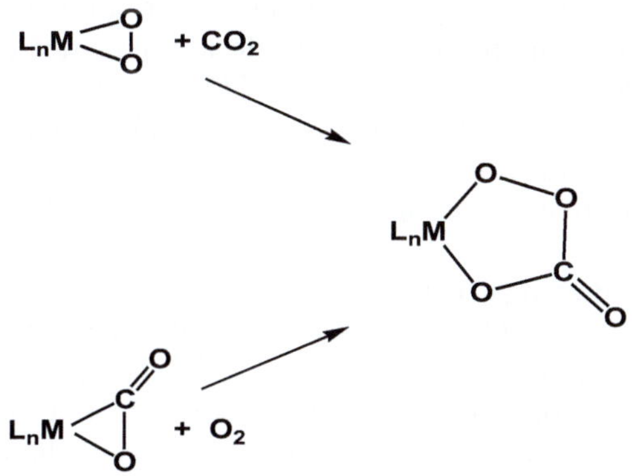

Scheme 4.12 Possible routes to transition metal peroxocarbonates from O_2 and CO_2 in the presence of a transition metal complex [L_nM] with the metal in a low oxidation state

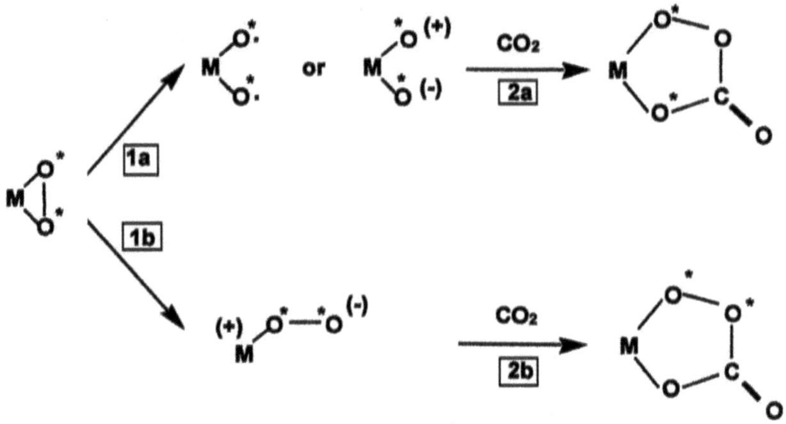

Scheme 4.13 Possible pathways to transition metal peroxocarbonates from dioxygen-metal-complexes and CO_2 (adapted from [73])

The reaction, known for a long time, is the one implying the addition of CO_2 to a dioxygen–metal-complex [69–72]. Nevertheless, very little was known about the reaction mechanism. Only recently have studies been carried out to shed light on the reaction of dioxygen complexes with CO_2 [73–75] or of carbon dioxide complexes with O_2 [76]. Two ways are possible, which imply the O–O (Scheme 4.13, routes 1a, 2a) or M–O bond opening (Scheme 4.13, routes 1b, 2b) with subsequent CO_2 insertion and ring closure to afford the peroxocarbonate structure.

Aresta's group investigated the formation of Rh(III)-peroxocarbonates $RhCl(CO_4)P_3$ ($P = PEt_2Ph$, $PEtPh_2$) from $RhCl(\eta^2\text{-}O_2)P_3$ and CO_2, making extensive use of labeled compounds ($^{16}O_2$, $^{18}O_2$, $C^{16}O_2$, $C^{18}O_2$) [73, 74].

Remarkably, reactions A and B in Scheme 4.14 afforded, respectively, isotopomers II and IV, instead of III and V. These results showed unambiguously that, in the specific case under study, the formation of the peroxocarbonate moiety did not proceed through the insertion of CO_2 into the metal–oxygen bond of the parent dioxygen complex, but implied O–O bond cleavage and CO_2 insertion. This conclusion was further supported by a modeling study, using the impulse oscillation model (IOM)

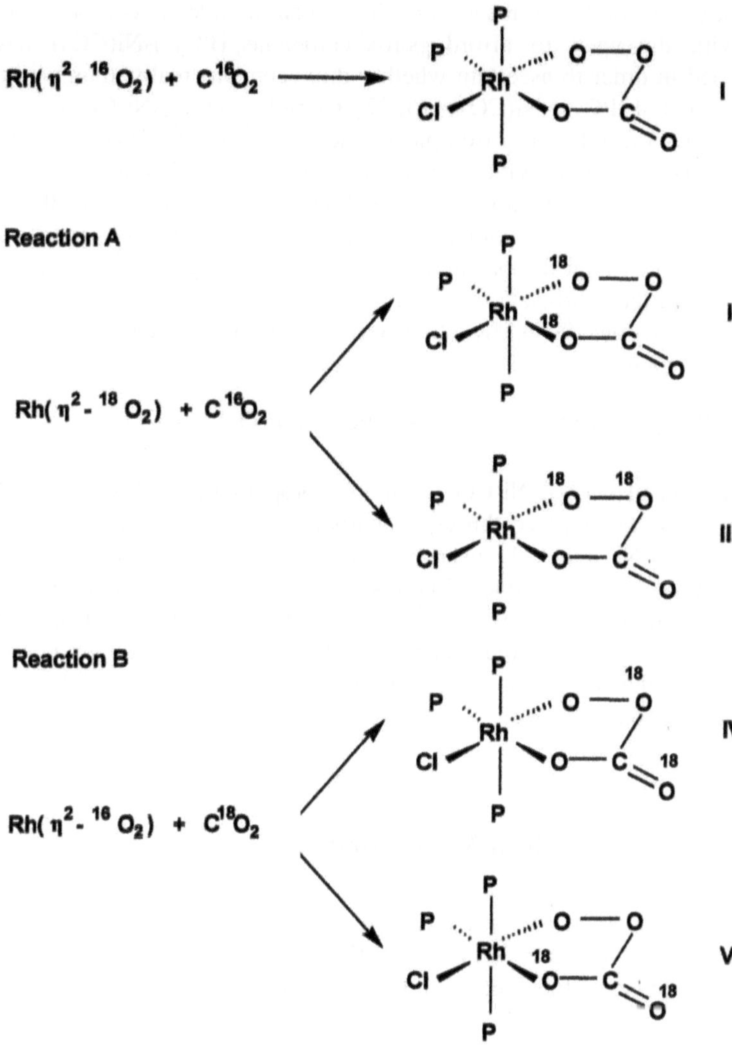

Scheme 4.14 Possible peroxocarbonate isotopomers from reactions A and B (adapted from [73])

developed by Borowiak and coworkers [75]. Such a mechanism is justified by the fact that the splitting of the O–O bond requires less energy (36 kcal mol^{-1}) than the cleavage of the M–O bond (ca. 65–90 kcal mol^{-1}) [76]. Most likely, the above reaction pathway is more general and widespread than can be imagined. The reactions of O_2 with CO_2-adducts of transition metals, both in the solid state and in solution, to give peroxocarbonate metal complexes, have long been regarded as examples of oxidative coupling reactions proceeding through the insertion of dioxygen into the M–C bond of the coordinated CO_2 ligand.

Recently, the solid state reaction of the carbon dioxide Ni–complex $(PCy_3)_2Ni(CO_2)$ with dioxygen to afford peroxocarbonate $(PCy_3)_2Ni(CO_4)$ has been re-examined in order to ascertain whether dioxygen was really able to insert into the Ni–C bond of $(PCy_3)_2Ni(CO_2)$ [75, 77]. Complex $(PCy_3)_2Ni(CO_2)$ was reacted with O_2 using the following isotopic species: $^{12}C^{16}O_2$, $^{13}C^{16}O_2$, $^{12}C^{18}O_2$, $^{16}O_2$, $^{18}O_2$. The first important evidence was that, when $(PCy_3)_2Ni(CO_2)$, or its labeled analogues $^{13}CO_2$ or $C^{18}O_2$, was reacted with O_2 ($^{16}O_2$ or $^{18}O_2$), the FTIR spectrum of the gas phase showed immediately the appearance of CO_2 (or its labeled analogues), which then disappeared with time. This suggested reaction (4.20) as the first step of the solid-state reaction.

The spectroscopic (IR, NMR) characterization of the Ni-peroxocarbonate product, as well as a normal

$$(PCy_3)_2Ni(CO_2)_{(s)} + O_{2(g)} \rightarrow (PCy_3)_2Ni(O_2)_{(s)} + CO_{2(g)} \qquad (4.20)$$

mode analysis of $(PCy_3)_2Ni(CO_4)$ and its isotopologues, substantiated that the overall reaction takes place through a carbon dioxide decoordination, dioxygen coordination to Ni, and CO_2 insertion into the O–O bond of the newly formed, reactive $(PCy_3)_2Ni(O_2)$ complex. The results of this study show that the formation of the peroxo-moiety requires an activated dioxygen molecule that interacts with carbon dioxide in an intermolecular mode. The incoming hetero-cumulene attacks one oxygen of the co-ordinated O_2 and this causes the O–O bond splitting, rather than the Ni–O bond cleavage.

4.6 CO_2 Insertion into M–N Bonds

Numerous examples of carbon dioxide insertion into the M–N bond of transition metal amides [78–84] are described in the literature. In comparison, however, mechanistic studies in this area are relatively few.

4.6.1 Insertion into Transition Metal Amides

Chisolm [85–87] has extensively studied the reaction of CO_2 with amides of early transition metals to give metal carbamates (4.21). Insertion of the heterocumulene into the metal–amido bond (4.21)

$$M(NMe_2)_n + mCO_2 \leftrightarrows M(NMe_2)_{n-m}(O_2CNMe_2)_m$$
$$(M:\ Ti,\ Zr,\ V\ (m = n = 4);\ Nb,\ Ta\ (m = n = 5);\ Mo,\ W\ (n = 6;\ m = 3)) \tag{4.21}$$

was believed, at first, to take place through a mechanism analogous to the insertion of unsaturated hydrocarbons into M–C(alkyl) bonds [85]. Later, it was ascertained [86], at least for the reactions of $W(NMe_2)_6$ and $W_2(NEt_2)_4Me_2$ with CO_2, to give, respectively, $W(NMe_2)_3(O_2CNMe_2)_3$ and $W_2(O_2CNEt_2)_4Me_2$, that the insertion reaction can proceed through an amine-catalyzed mechanism ((4.22a) and (4.23)). Trace amounts of amine can be fortuitously present in solutions of $M(NMe_2)_n$ compounds as a result of $M(NMe_2)_n$ hydrolysis produced by adventitious water. Under the working conditions (CO_2 excess relative to amine), free amine can interact with the heterocumulene to produce carbamic acid R_2NCO_2H, which reacts as shown in (4.23). According to this mechanism, the formation of the metal carbamate complex should be better regarded as a ligand substitution rather than a true insertion process.

$$CO_2 + R_2NH \leftrightarrows R_2NCO_2H \tag{4.22a}$$

$$M - NR_2 + R_2NCO_2H \leftrightarrows MO_2CNR_2 + R_2NH \tag{4.23}$$

As shown in Scheme 4.15, this reaction pathway has been largely exploited from the synthetic point of view for the preparation of a great number of metal carbamate adducts [78–84].

Energetics of (4.22a) seems to depend markedly on the substituents on the N atom [88–95]. Gaseous carbamic acid $H_2NC(O)OH$ has been reported to be unstable [82, 88] with respect to $NH_{3(g)}$ and $CO_{2(g)}$ ($\Delta H^\circ_r = -58$ kJ/mol; $\Delta G^\circ_r = -93$ kJ/mol [82]). However, theoretical calculations show that in other cases, wherein $R \neq H$, reaction (4.22a) may be close to thermoneutral or may exhibit negative reaction

$$L_nM\text{-}X + R_2NH + CO_2 \longrightarrow L_nMO_2CNR_2 + HX$$

X = H, alkyl, Cl, OR, ClO_4, etc

Scheme 4.15 Preparation of carbamate adducts by ligand substitution

heats [89–92]. Accordingly, carbamic acid formation (4.22a) is an equilibrium process. By using a number of spectroscopic techniques, carbamic acids have been detected in solution at concentrations that depend on the experimental conditions: nature of solvent, temperature, CO_2 pressure, CO_2/amine molar ratio [96–103]. High $[CO_2]/[amine]$ ratios favor acid with respect to carbamate salt, which, however, predominates in the presence of excess of amine (4.24). Monomeric acid, R_2NCO_2H, can be stabilized upon self-association to form a dimer, $(R_2NCO_2H)_2$ [104–108], which is expected to be a

$$R_2NCO_2H + R_2NH \leftrightarrows (R_2NH_2)O_2CNR_2 \quad (4.24)$$

less reactive species than the monomeric form. H-bonded dimeric carbamic acids $(R_2NCO_2H)_2$ have been detected in solution (R = benzyl) [108] and, depending on the nature of R substituents and the experimental conditions, have also been isolated and structurally characterized [104, 105]. The isolated dimer $[(PhCH_2)_2NCO_2H]_2$ forms from benzylamine and CO_2 through the intermediacy of $(R_2NH_2)(O_2CNR_2)$ (R = $PhCH_2$), after prolonged exposure of the carbamate salt to CO_2 [105]. Plausibly, under the working conditions, proton transfer between dialkylammonium cation and carbamate anion gives the amine $(PhCH_2)_2NH$ and the monomer acid $(PhCH_2)_2NCO_2H$ may be the initial step towards the formation of the dimer. However, when CO_2 is in excess relative to amine (see above), a different mechanism cannot be excluded. By assuming a 1:1 CO_2/RNH_2 reaction model, it was calculated that formation of the relevant monomer acid $RNHCO_2H$ is an asynchronous process involving a four-membered ring transition state where an N–C bond is formed and one proton is transferred [89–91]. This reaction pathway can be roughly assimilated to a CO_2 insertion into the NH bond of an amine. The calculated energy barriers, which are markedly affected by the used calculation method, were relatively high (45–70 kcal/mol, RNH_2 = methylamine, methylurea), showing that formation of carbamic acid by CO_2 insertion into the NH bond of the amine is not an easy process. Borowiak and coworkers [93] modeled, using the ab initio RHF/3-21G method, the reaction of CO_2 with dimethylamine in the presence of a second dimethylamine molecule. Starting from the most stable 1:2 CO_2/R_2NH complex [93, 94] they found that the reaction product is a cyclic H-bonded complex of dimethylcarbamic acid with dimethylamine. Remarkably, a significantly lower activation energy (9.54 kcal/mol) was calculated for this reaction pathway. Analogous calculations showed that the reaction of CO_2 with two ammonia molecules affords an H-bonded $H_2NCO_2H\cdots NH_3$ adduct analogous to that found for the 1:2 CO_2/R_2NH system [95]. However, the estimated energy barrier was, in this case, higher (29 kcal/mol).

In general, $M(NMe_2)_{n-m}(O_2CNMe_2)_m$ adducts are labile with respect to CO_2 exchange (4.25). The activation energy required for the exchange process was estimated by VT-^{1}HNMR

$$ML^*_n + x^{12}C\,O_2 \leftrightarrows ML_mL^*_{n-m} + m^{13}CO_2 + (x-m)^{12}CO_2\,(\text{where } L = O_2{}^{12}CNMe_2 \text{ and } L^*O_2{}^{13}CNMe_2) \quad (4.25)$$

measurements to be 54 ± 8 kJ/mol for M=Zr ($m = n = 4$) and significantly higher (≈96 kJ/mol) for $W(NMe_2)_3(O_2CNMe_2)_3$ [85]. The exchange process was found to be catalyzed by trace amounts of

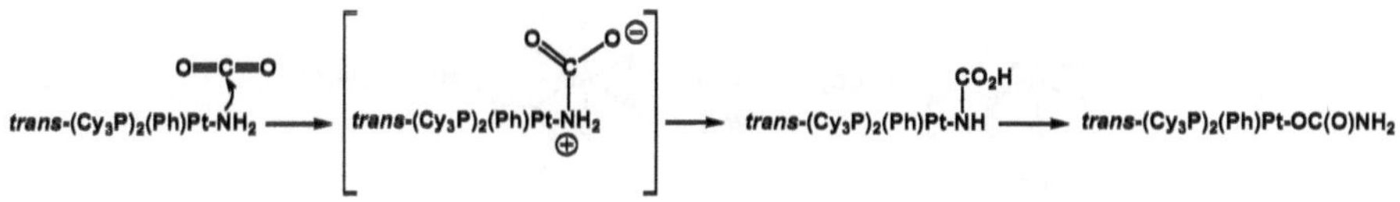

Scheme 4.16 Insertion CO_2 into the Pt–N bond of *trans*-Pt(Ph)(NH_2)(PCy_3)$_2$

amine, which, in solutions of ML_n compounds, can form as a result of ML_n trace hydrolysis. Chisolm and Extine [86] demonstrated that CO_2 exchange (4.25) proceeded through attack of carbamic acid formed by reaction of carbon dioxide and the free amine ((4.22b) and (4.26)), and did not imply any attack by carbamate anion that, under the working conditions (excess of CO_2 relative to free amine), was expected to be present in the reaction system at very low concentration.

$$^{12}CO_2 + Me_2NH \leftrightarrows Me_2N^{12}CO_2H \quad (4.22b)$$

$$MO_2{}^{13}CNMe_2 + Me_2N^{12}CO_2H \leftrightarrows MO_2{}^{12}CNMe_2 + Me_2N^{13}CO_2H \quad (4.26)$$

The Ti- and Zr-carbamato complexes can also undergo exchange according to (4.27) and (4.28) [87]. The latter exchange processes ((4.27) and (4.28)) were also found to be catalyzed by trace amount of fortuitous amine [86, 87].

$$MO_2{}^{13}CNMe_2 + MO_2{}^{12}CN(CD_3)_2 \leftrightarrows MO_2{}^{12}CNMe_2 + MO_2{}^{13}CN(CD_3)_2 \quad (4.27)$$

$$MO_2CNMe_2 + M'O_2CN(CD_3)_2 \leftrightarrows MO_2CN(CD_3)_2 + M'O_2CNMe_2 \quad (4.28)$$

The aptitude of metal–amides and metal–carbamates to hydrolyze in the presence of adventitious water, as well as the occurrence of (4.22a), are major caveats in studying the mechanistic aspects of this insertion reaction [109] and the mechanism summarized by (4.22a) and (4.23) may also be working in other cases. However, CO_2 attack on metal bonded nitrogen is also possible and has been documented experimentally.

Roundhill and coworkers [110] described the formation of N-bonded carbamate *trans*-Pt(Ph)($NHCO_2H$)(PCy_3)$_2$ from the reaction of *trans*-Pt(Ph)(NH_2)(PCy_3)$_2$ and carbon dioxide. The N-bonded carbamate complex can be easily isolated, as it is precipitated in benzene solvent. When this species was dissolved in dichloromethane, isomerization to the O-bonded complex *trans*-Pt(Ph)($OCONH_2$)(PCy_3)$_2$ took place. The O-bonded isomer can be obtained in a single step by reacting *trans*-Pt(Ph)(NH_2)(PCy_3)$_2$ with CO_2 in CH_2Cl_2. In the latter case, the N-isomer *trans*-Pt (Ph)($NHCO_2H$)(PCy_3)$_2$ can be observed as an intermediate in solution (Scheme 4.16).

Hartwig, Bergmann, and Andersen [111] carried out a very informative mechanistic study of the reaction of the *ortho*-metalated complex (Me_3P)$_4$Ru(η^2-NHC_6H_4) (**10**) with CO_2 to give the M–N insertion product (Me_3P)$_4$Ru(η^2-OC(O) NHC_6H_4) (**11**) (4.29). Scheme 4.17 summarizes the

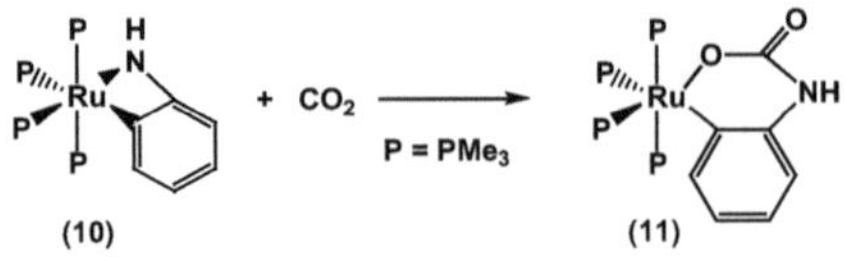

(4.29)

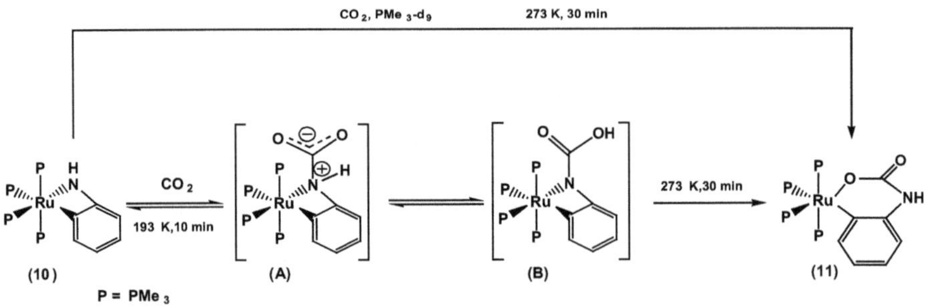

Scheme 4.17 Insertion of CO_2 into the Ru–N bond of $(Me_3P)_4Ru(\eta^2\text{-}NHC_6H_4)$ (adapted from [109])

mechanistic pathway for reaction (4.29), as supported by NMR investigations in which extended use of labeled derivatives ($^{13}CO_2$, $(Me_3P)_4Ru(\eta^2\text{-}^{15}NHC_6H_4)$ was made. The reaction proceeds via attack of CO_2 on the nitrogen atom of **10** to give intermediates **A** and **B** (Scheme 4.17), which were detected spectroscopically when the reaction was carried out at low temperature. Remarkably, when the reaction was carried out at 273 K in the presence of $PMe_3\text{-}d_9$, the formation of **A** and **B**, as well as the conversion of these intermediates into product **11**, occurred without any exchange of free deuterated phosphine with coordinated undeuterated phosphine, showing that the insertion of CO_2 must occur without dissociation of PMe_3 ligands. This result is consistent with the direct attack of CO_2 at the nitrogen atom of **10** to form **A** and **B** (Scheme 4.17) and demonstrates that a preliminary coordination of CO_2 to the metal center is not a prerequisite for the insertion reaction to occur. Therefore, the insertion reaction of CO_2 into the metal–nitrogen bond of the *ortho*-metalated anilide **10** proceeds through a pathway different from the classical coordination/migration mechanism operating for the metal–carbon bond insertion reactions. Further investigations have shown that the insertion into the M–N bond is the kinetically preferred pathway. In fact, upon heating at 393 K, **11** did not convert into the N,O-bonded anthranilate complex $(Me_3P)_4Ru(\eta^2\text{-}OC(O)C_6H_4NH)$ **12**, the product of formal insertion of CO_2 into the metal–carbon bond of **10**, but rearranged to carboxyamide complex $(Me_3P)_4Ru(\eta^2\text{-}N(Ph)C(O)O)$ **13**. Moreover, upon heating, $(Me_3P)_4Ru(\eta^2\text{-}OC(O)C_6H_4NH)$ **12**, prepared by an independent route, did not convert into **11**, but afforded the O,C-bound anthranilate complex $(Me_3P)_4Ru(\eta^2\text{-}OC(O)C_6H_3NH_2)$ **14** (Scheme 4.18).

An important issue is the regioselectivity of the insertion reaction in those cases where more than one reaction site is available for carbon dioxide insertion or, more generally, for reaction with the heterocumulene. Cowan and Trogler [112] found that *trans*-$Pt(H)(NHPh)(PCy_3)_2$ reacted reversibly with CO_2 to give a carbamate rather than a formate or metallo-acid complex. Other studies have established that metal–amide bonds were more reactive towards CO_2 insertion than metal–alkyl or metal–alkoxo bonds [110, 111, 113, 114]. Nevertheless, the η^5-cyclopentadienyl silyl-η-amide zirconium dimethyl complex $[\eta^5\text{-}C_5H_4SiMe_2(N^tBu)]ZrMe_2$ inserted CO_2 regioselectively into the Zr–Me bonds [115]. The metal–amide bond seems to be more reactive than the metal–imido bond [116, 117]. DFT studies on the reaction of $[W(NPh)(N_2N_{py})Me]^+$ (N_2N_{py}=MeC(2-C_5H_4N)(CH_2NSiMe_3)$_2$) with carbon dioxide showed that CO_2 insertion into the W–Me bond is thermodynamically preferred, but that insertion into W–N_{amide} is kinetically more facile [117]. Cycloaddition of CO_2 to the W=NPh imido bond was neither kinetically nor thermodynamically viable.

However, CO_2 can react with metal–imido bonds through a [2+2] cycloaddition reaction to give N,O-bound metal carbamates [118]. The latter species can react further and either extrude the corresponding isocyanate with formation of a metal–oxo complex [83, 119] or also insert a second CO_2

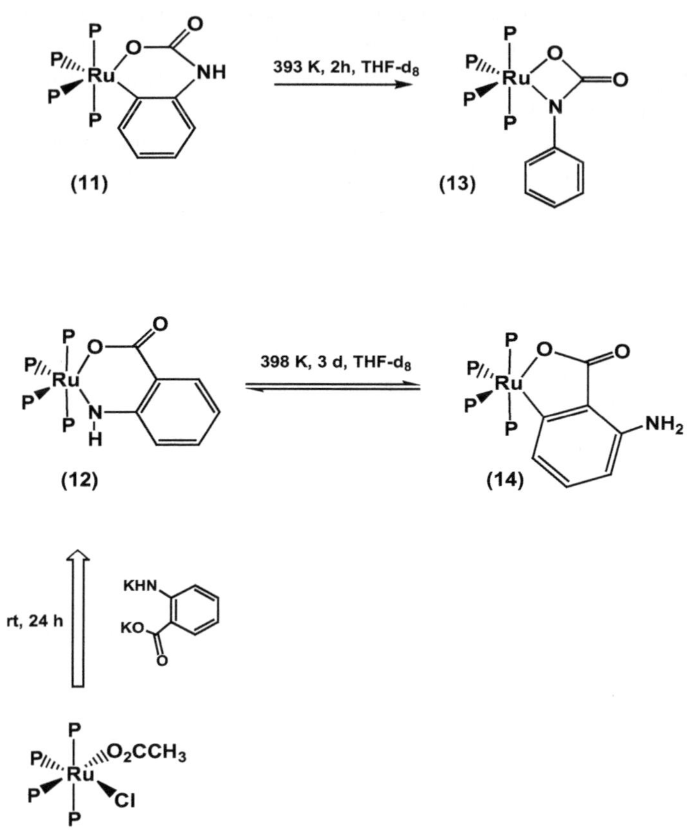

Scheme 4.18 Thermal behavior of complexes **11** and **12** (see main text) (adapted from [111])

molecule to give a dicarbamate adduct [119]. DFT(B3PW91) calculations on Ti (η-C_5H_5)(NR){MeC(NMe)$_2$} (R = Me, Ph, 4-C_6H_4Me, 4-$C_6H_4NMe_2$, 4-$C_6H_4CF_3$) reacting with CO_2 showed that the second CO_2 insertion is thermodynamically favored over isocyanate extrusion, and that the rates of the two processes are similar [119]. Calculations on $Ti(\eta\text{-}C_5R_5)(N^tBu)\{MeC(N^iPr)_2\}$ ($R = H$ or Me) showed that increasing the steric bulk increases the thermodynamic favorability of the isocyanate extrusion process and significantly raises the activation barrier for the second CO_2 insertion, making the latter process impossible.

4.6.2 Insertion into Main Group and Post-Transition Metal Amides

Carbon dioxide has been recognized for a long time to insert formally into the metal–nitrogen bond of amides of main group and post-transition metals [78–84, 120–130]. Nevertheless, experimental studies aimed at shedding light on the pathway of these carbamation reactions are scarce. For the reasons mentioned above, this is by no means an easy matter, which is made more complicated by the fact that the reactant amide is often involved in aggregation/disaggregation as well as solvation processes.

In early work, Inoue and Yokoo [121] found that carbon dioxide insertion into the Al–N bond of $Et_2Al–NEt_2$ was accelerated by tertiary amines such as, for instance, 1-methyl-imidazole, *N*-methylbenzimidazole, or pyridine. The role of the base was explained by taking into account that

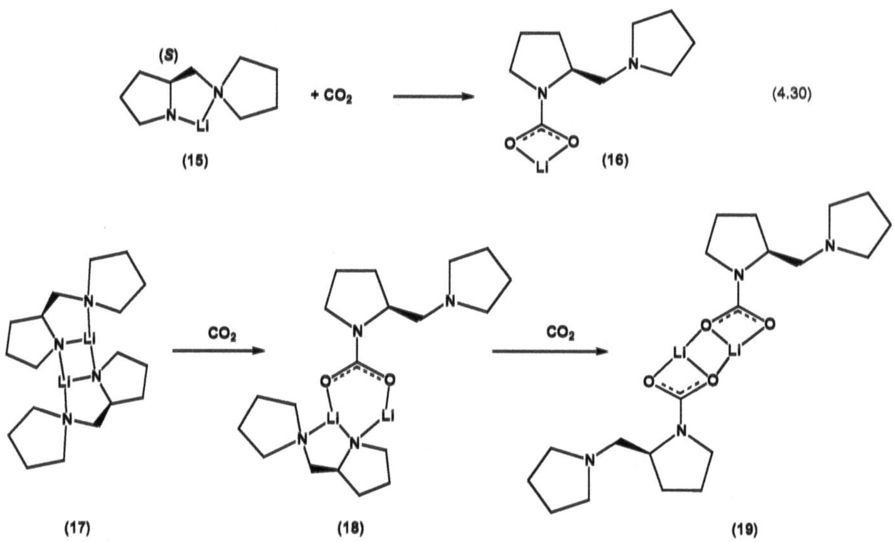

Scheme 4.19 Mechanism of formation of lithium carbamate **16**

diethyl aluminum diethylamide exists in a dimeric form with Al–N–Al bridges. In the presence of some complexing agent a monomeric complex can form, in which the nucleophilicity of the N atom, freed from the bridge, is enhanced. Based on this, the authors speculated that the metal bound amidic nitrogen should be more prone to attack from CO_2 with formation of carbamate moiety.

A picture of the mechanistic features characterizing the insertion reactions of carbon dioxide in a few of these systems has been provided by means of computational studies.

Nilsson Lill et al. [131] have modeled the mechanism of formation of lithium carbamate **16** (4.30) from chiral lithium amide (*S*)-2-(pyrrolidin-1-ylmethyl) pyrrolidide (**15**) and CO_2 by using detailed DFT calculations. The selected chiral lithium amide **15** can exist as a mixture of THF solvated and unsolvated dimers, in which the second pyrrolidine ring acts as an intramolecular ligand and, therefore, symmetrical (see **17** in Scheme 4.19) and unsymmetrical (where one Li atom is intramolecularly tetracoordinated and the other Li is dicoordinated) dimers of **15** were taken into account. On the energetically preferred reaction pathway amide **15** exists predominantly as a (solvated) symmetrical dimer **17**. Attack by CO_2 takes place in an "end-on" manner to give several heterodimers **18** which are composed of a lithium amide and a lithium carbamate. This exergonic step is rate-determining, and the activation barrier was found to be about 10 kcal/mol. Interestingly, it was found that the transition states without solvent coordination to lithium cation were more stable than those containing coordinated solvent molecules. Addition of a second carbon dioxide molecule to heterodimers **18** then leads, through a new encounter-complex, to several almost isoenergetic homodimers **19** of lithium carbamate **16**. This strongly exergonic insertion reaction has a barrier of 9.7 kcal/mol.

The mechanism of CO_2 fixation by alkylmagnesium amides and magnesium bisamides has been analyzed by Himmel [132] with the aid of quantum chemical calculations relying on pure DFT (BP86)

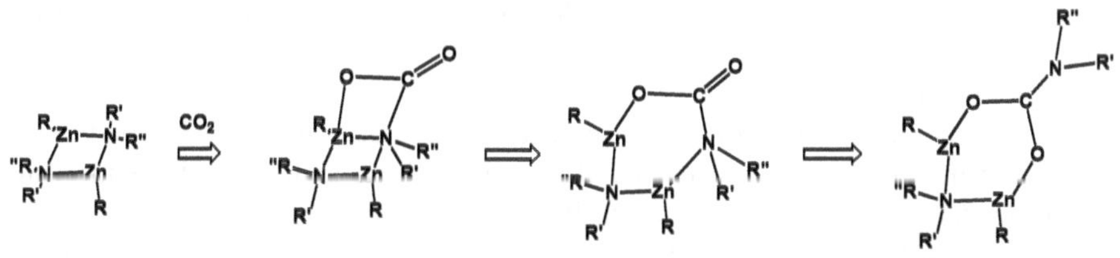

Scheme 4.20 Mechanism of formation of carbamates in dimeric alkylzinc amides

as well as hybrid-DFT(B3LYP) methods. In agreement with the body of experimental results, the calculations showed that magnesium amides prefer the dimeric over the monomeric form. In the dimeric form the two Mg atoms are bridged by two amido ligands. The presence of chelating ligands, such as ethylenediamine (EDA) which preferentially binds to monomer, shifts the equilibrium to the monomer side. According to the calculations, insertion of CO_2 into the Mg–N bond of monomeric alkylmagnesium amides and magnesium bisamides forms a [MgOCN] four-membered cyclic intermediate without any significant reaction barrier. Formation of the carbamate product from the latter species exhibits a relatively small barrier

($\Delta G^\circ = +33$ kJ/mol for $H_3CMgN(CH_3)_2$ and +48 kJ/mol for $Mg[N(CH_3)_2]_2$) to be overcome. Monomeric complexes with chelating donor molecules such as EDA react with CO_2, affording the carbamate product without any barrier and any intermediate. However, in the case of CO_2 insertion into one of the Mg–N bonds of dimeric alkylmagnesium amides, a significant barrier ($\Delta G^\circ = +78$ kJ/mol) has to be surmounted, which is associated with a transition state featuring a four-membered MgOCN ring.

Analogous studies [133] have also been carried out for the insertion of CO_2 in the Zn–N bond of zinc amides, which usually results in the formation of tetranuclear zinc–carbamate adducts [128–130]. Similar to what was found for monomeric alkylmagnesium amides, a four-membered cyclic [ZnNCO] intermediate is first obtained from the reaction of monomeric alkylzinc amides with CO_2. The activation energy ΔG of this first, rate-determining, step is around +60 kJ/mol. From this local minimum on the potential energy surface, the system has to overcome a much smaller barrier to convert into the monomeric carbamate product, which can then undergo dimerization and tetramerization (both showing negative ΔG values). The energetics of the CO_2 fixation reaction by different zinc amides of formula RZnNR′R″ were found to be very sensitive to changes in the electronic properties of the amido group, whereas modifications on the alkyl (R) side had almost no influence. Calculations indicated that the reaction of dimeric alkylzinc amides with CO_2 showed a larger barrier (ΔG value more than +100 kJ/mol) connected with a transition state featuring a four-membered [ZnNCO] cycle (Scheme 4.20), which evolves into a six-membered cyclic intermediate. The barrier separating the latter intermediate from the product is very small. The calculations also showed that the barrier for CO_2 fixation by dimeric alkylzinc amides was higher than the energy required for decomposition of the dimer into two monomers. It is also higher than the barrier for CO_2 fixation by monomeric alkylzinc amides ($\approx$60 kJ/mol). Therefore, it is plausible that, although dimeric alkylzinc amides are preferred over monomers, the reaction with CO_2 could mainly take place from the monomeric state.

As already emphasized for transition metal carbamates, the O_2C–N bond of the carbamate moiety of these systems is somewhat labile, and the carbamate group undergoes CO_2 exchange processes easily.

Kinetics and reaction mechanism for CO_2 exchange in 2-imidazolidinone-1-carboxylic acid lithium salt (Li **20**) were investigated by Lihs and Caudle [134]. *N*-Carboxyimidazolidone anion, **20**, was probed as an analogue for N^1-carboxybiotin and synthesized as the lithium salt by deprotonation of 2-imidazolidone **21** with phenyllithium and further reaction of the resulting lithium amide with carbon dioxide. The study was addressed to ascertain the viability of unimolecular CO_2 elimination from

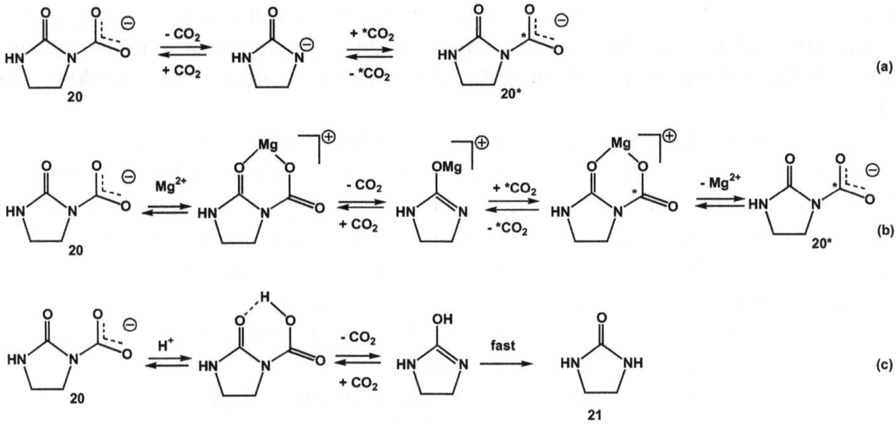

Scheme 4.21 CO_2 exchange and decarboxylation in **20** (adapted from [81])

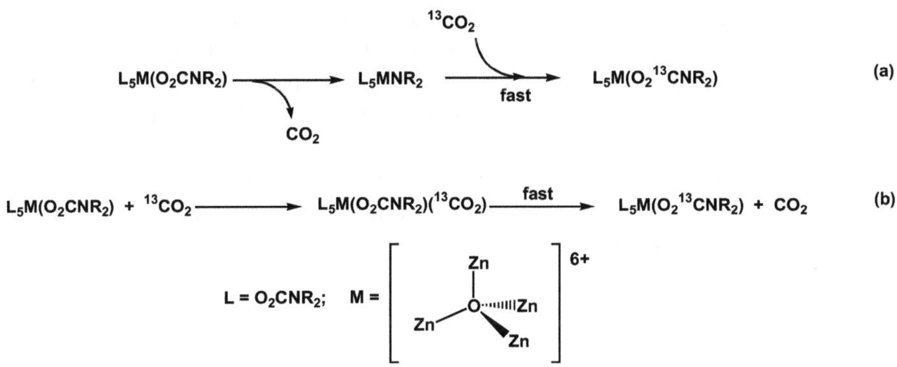

Scheme 4.22 CO_2-independent (**a**) and CO_2-dependent (**b**) exchange processes in tetranuclear zinc carbamates (adapted from [135])

N^1-carboxybiotin as a microscopic step in biotin-dependent carboxyl transfer enzymes. Time-dependent FTIR spectroscopy and NMR measurements showed that Li**20** undergoes carboxyl exchange with free carbon dioxide, with kinetics indicative of rate-limiting unimolecular dissociation of the $N{-}CO_2$ bond (Scheme 4.21a). The Lewis acid Mg^{2+} catalyzed the CO_2 exchange because of its ability to coordinate to **20** (Scheme 4.21b), whereas the reaction of Li**20** with carboxylic acids in DMSO resulted in acid-dependent decarboxylation of **20** (Scheme 4.21c), with a rate which depended on the concentration and strength (pK_a) of the acid. Both Lewis acid-catalyzed carboxyl exchange and acid-dependent decarboxylation exhibited a common mechanistic framework (Scheme 4.21) involving, as the initial step, the interaction of Mg^{2+} or H^+ with the carbonyl group of the ureido ring. These interactions act as a trigger for dissociation of CO_2 as they provoke polarization of the nitrogen lone pair toward the imidazolidone ring rather than the carboxyl group and weaken the $N{-}CO_2^-$ bond.

Tetranuclear zinc carbamates $Zn_4O(O_2CNR_2)_6$ ($NR_2 = N$-diethylamino, *N*-piperidyl, *N*-pyrrolidyl) (Scheme 4.22) were found, by transient FTIR spectroscopy to undergo C–N bond metathesis reactions, resulting in the exchange of the carboxyl group with CO_2 and exchange of the amino group Am with bulk secondary amine [135]. The net rate for CO_2 exchange was measured by monitoring the uptake of $^{13}CO_2$ and the concomitant release of $^{12}CO_2$. The CO_2 exchange kinetics revealed a CO_2-dependent (Scheme 4.22b) and a CO_2-independent pathway (Scheme 4.22a) for the CO_2 exchange process.

The CO_2-dependent process was rationalized in terms of pre-association of the incoming CO_2 with the complex prior to N–CO_2 bond cleavage, whereas the CO_2-independent process was explained as caused by unimolecular elimination of CO_2 from the complex. The transamination reaction, involving exchange of amino groups, showed rates that were independent from the concentration and nature of the incoming amine. Moreover, the average rates for transaminations were by a factor of four higher than the CO_2-independent CO_2 exchange rate for the corresponding complex. These data have been rationalized in terms of a common rate-limiting process for CO_2-independent CO_2 exchange and transamination which implies unimolecular elimination of CO_2. In these systems the ligand exchange dynamics was fast relative to the overall rate of C–N bond metathesis, thus showing that the unimolecular CO_2 exchange process cannot involve a rate-limiting dissociation of a carbamate ligand.

4.6.3 Insertion into Amides of Non-Metallic Elements

Carbon dioxide can also form carbamates by reaction with amides of non-metallic elements [78–84, 136–144]. Most examples reported in the literature deal with amides of silicon, germanium, and phosphorus. Among these, the insertion of the heterocumulene into the P–N bond of hexaalkylphosphorus triamides $P(NR_2)_3$, which affords phosphocarbamates species of formula $P(NR_2)_{3-x}(O_2CNR_2)_x$, has received particular attention for its potential in chemical synthesis, because this reaction opens a route to the phosgeneless synthesis of carbamate esters [138, 139] and ureas [140].

Mechanistic information on the pathway of these insertion reactions are sparse. A mechanism similar to that summarized by (4.22a) and (4.23), and involving the intermediacy of carbamic acid formed by the action of CO_2 on adventitious amine present in the reaction medium, seems to be operative for the "formal" insertion of

CO_2 into the Si–N bond of Me_3SiNEt_2 (4.31) [141] and in the P–N bond of the aminophosphorane $CH_3(CF_3)_3PN(CH_3)_2$ (4.32) [142]. Accordingly, traces of free

$$Me_3SiNEt_2 + CO_2 \rightarrow Me_3SiO_2CNEt_2 \quad (4.31)$$

$$CH_3(CF_3)_3PN(CH_3)_2 + CO_2 \rightarrow CH_3(CF_3)_3PO_2CN(CH_3)_2 \quad (4.32)$$

amine were found to be essential in order that reaction (4.31) may proceed. Moreover, Cavell and Griend [142] studied the reaction of $CH_3(CF_3)_3PN(CH_3)_2$ with CO_2 in the absence and in the presence of an amine trap. In the absence of any amine trap, labeled $^{13}CO_2$ was rapidly (15 min) incorporated in the carbamate ligand of $CH_3(CF_3)_3PO_2CN(CH_3)_2$. However, in the presence of an amine trap, such as Me_3SiCl, no detectable incorporation of CO_2 into the P–N bond of $CH_3(CF_3)_3PN(CH_3)_2$ occurred even over long times (5 days).

A bifunctional mechanism has been ascertained by means of IR and NMR spectroscopic measurements for the insertion of CO_2 into the P–N bond of bicyclophosphorane **22** (8-phenyl-1,7-dioxa-4-aza-8-phospha(V) bicyclo[3.3.0] octane) [143]. ^{1}H, ^{31}P, and ^{13}C NMR spectroscopies showed that, under strictly anhydrous conditions, CO_2 interacts reversibly with **22** to give the intermediate adduct **23**, which is stabilized by an intramolecular hydrogen bond. Intermediate **23** then undergoes irreversible insertion into the P–N bond to give the final product **24**, which has been isolated and characterized (Scheme 4.23).

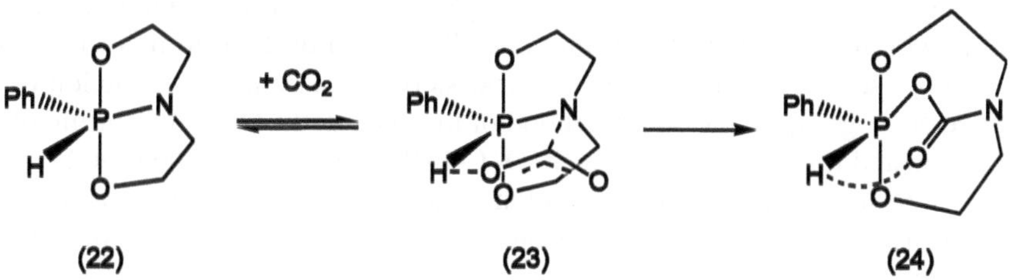

Scheme 4.23 Insertion of CO_2 into the P–N bond of bicyclophosphorane **22** (adapted from [143])

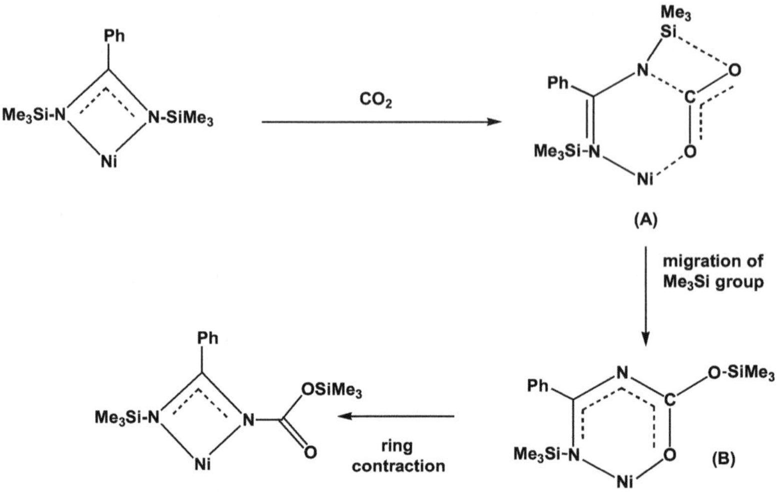

Scheme 4.24 CO_2 insertion into the N–Si bond of $(siam)_2Ni$ (adapted from [144])

Walther and coworkers [144], by studying the reactivity of CO_2 with $(siam)_2Ni$ (**25**) (*siam* = bistrimethylsilyl-benzamidinato, $[(Me_3SiN)_2CPh]^-$), observed, in the final product $(siamCO_2)_3(siam(CO_2)_2)Ni_2$ (characterized by single-crystal X-ray diffraction), the insertion of the heterocumulene into the N–Si bonds (Scheme 4.24). They suggested that CO_2 uptake by $(siam)_2Ni$ is the result of a stepwise reaction starting with the insertion of CO_2 into one Ni–N bond of **25** to form a carbamato-type group (see **A**, Scheme 4.24). The next step would be the fast reaction of the reactive N–Si group in the neighborhood of the activated CO_2 with splitting of the N–Si bond and formation a of a new Si–O–CO–N group (see **B**, Scheme 4.24). Ring contraction of **B** results in the formation of a modified amidinato ligand which, through an analogous pathway, is capable of inserting another molecule of CO_2 into the remaining N–Si bond.

CO_2 insertion into the N–Si bond of silylamido ligands can play a key role in a few metathetical reactions of carbon dioxide with bis-silyl-amido complexes $M[N(SiMe_3)_2]_n$ (Na, $n = 2$; Pr, Nd, $n = 3$; Ge, Sn, $n = 2$; Ti(IV), Zr(IV)) [82, 83]. These metathetical reactions result in the oxo-transfer from CO_2 to a silyl group of the bis-silylamido ligand and provide an entry into silyl isocyanates or 1,3-bis(silyl) carbodiimides (Scheme 4.25).

The mechanism through which the metathetical reaction is believed to occur with $M[N(SiMe_3)_2]_2$ (M = Ge, Sn) amides is shown in Scheme 4.26 [145]. The initial insertion step of the heterocumulene into the M–N bond is apparently followed by a facile molecular

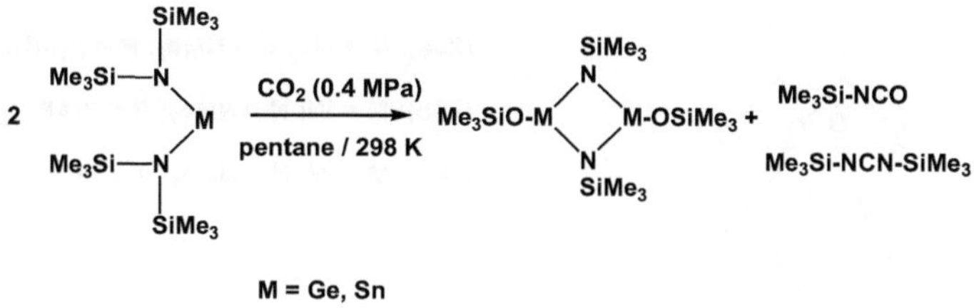

Scheme 4.25 Reaction of CO_2 with divalent Group 14 bis(silyl) amides (adapted from [145])

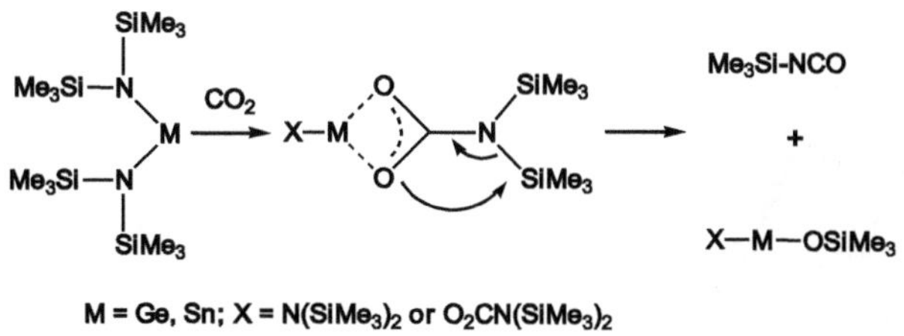

Scheme 4.26 Reaction of carbon dioxide with divalent Group 14 bis(silyl) amides: proposed mechanism (adapted from [145])

rearrangement which involves the elimination of an isocyanate fragment. Because of the subvalent nature of the Group 14 metal center, it has been proposed that carbamate ligand binds in a bidentate fashion. This mode of bonding might facilitate the 1,3-shift of a trimethylsilyl group by the enhancement of its electrophilic character because of that of the carbamic carbon.

4.7 CO_2 Insertion into M–P Bonds

Only a few examples of CO_2 insertion into M–P bonds of metal-phosphido complexes are known to date [146–149]. Carbon dioxide reacts selectively with the (amino)(phosphanyl)alumane $(tmp)_2AlPR_2$ (tmp = 2,2,6,6-tetramethyl-piperidino; R = $SiMe_3$) to give $(tmp)_2AlO_2CPR_2$, thus showing that insertion of the heterocumulene into the Al–P bond of this substrate is preferred over the insertion into the Al–N bond [146]. However, the reaction of 1,2-$M_2(NMe_2)_4(P(t\text{Bu})_2)_2$ complexes (M = Mo, W) with CO_2 (>4 equiv.) yielded $M_2(O_2CNMe_2)_2(O_2CP(t\text{Bu})_2)_2(NMe_2)_2$ (**26a**: M = Mo; **26b**: M = W; Structure 4.3) [148].

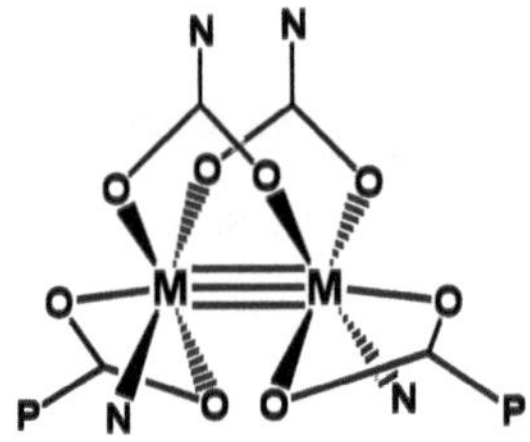

(26)

(26a): M = Mo; N = NMe_2; P = $P(t\text{-}Bu)_2$

(26b): M = W; N = NMe_2; P = $P(t\text{-}Bu)_2$

(26c): M = W; N = NMe_2; P = PCy_2

Structure 4.3 Insertion of CO_2 into M–N and M–P bonds (adapted from [148])

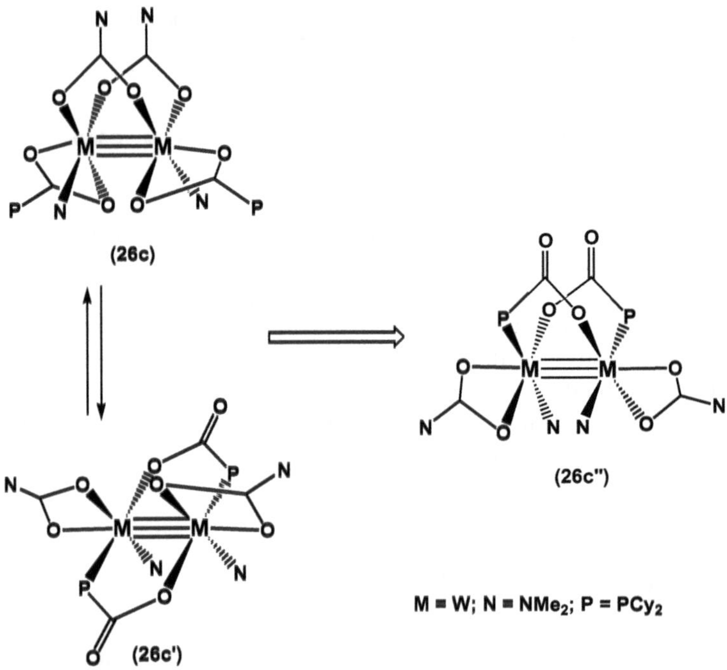

Scheme 4.27 Isomerization of **26c** (see main text) (adapted from [148])

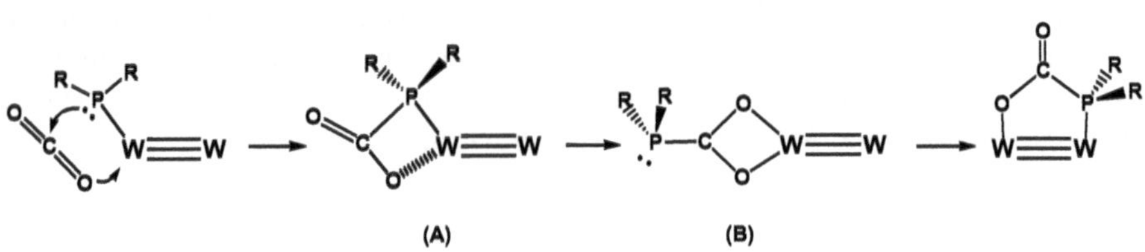

Scheme 4.28 CO_2 insertion into M-phosphido bond: proposed mechanism (adapted from [148])

The reaction occurred rapidly at room temperature, both in solution and in the solid state. The molybdenum complex **26a** was labile to ligand-exchange reactions in solution and attempts to grow good quality crystals afforded $Mo_2(O_2CP(tBu)_2)_4{\cdot}2C_6H_6$, a binuclear complex having a Mo–Mo quadruple bond supported by a paddle-wheel arrangement of four phosphinecarboxylate ligands which span the Mo–Mo bond [149]. The complex $W_2(O_2CNMe_2)_2(O_2CP(tBu)_2)_2(NMe_2)_2$, **26b**, was fully characterized by X-ray analysis, which confirmed the presence of a $(W{\equiv}W)^{6+}$ central unit in the molecule. A strictly related reaction occurred between 1,2-$W_2(NMe_2)_4(PCy_2)_2$ and CO_2 affording $W_2(O_2CNMe_2)_2(O_2CPCy_2)_2(NMe_2)_2$ (**26c**), which is structurally analogous to **26b**, as ascertained by NMR spectroscopy. In solution, **26c** rearranges with formation of the isomeric species **26c′** and **26c″**, having phosphinecarboxylate ligands bridging the $(W{\equiv}W)^{6+}$ unit through P- and O-to-W bonds (Scheme 4.27).

Low-temperature mechanistic studies have shown a kinetic preference for the insertion of CO_2 into metal–phosphido bonds as opposed to metal–amido bond and suggested, as the first step, the initial formation of a P-bound phosphine-carboxylate (**A**, Scheme 4.28). This intermediate plausibly

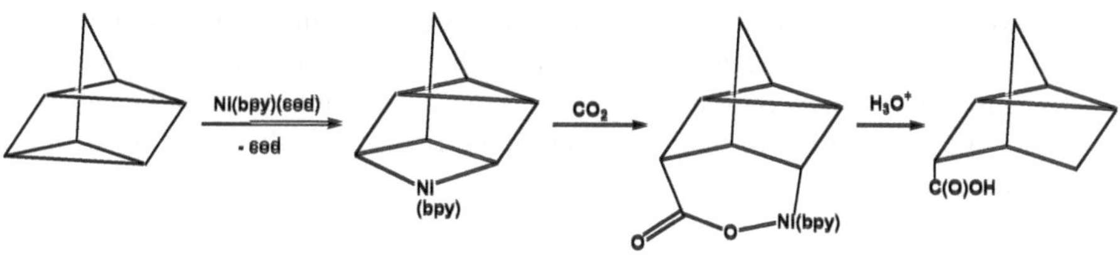

Scheme 4.29 Ni(0)-promoted carboxylation of quadricyclane (adapted from [151])

rearranges to a more stable O-bound phosphinecarboxylate (**B**, Scheme 4.28) because of the thermodynamic preference for W–O bonds. The subsequent rearrangement of **B** generates the isomers containing phosphinecarboxylate ligands O-, P-anchored to the $(W{\equiv}W)^{6+}$ center.

4.8 CO_2 Insertion into C–C Bonds

Formal insertion of CO_2 into the carbon–carbon bond has been observed only in a limited number of cases, dealing with strained cyclic hydrocarbons. Incorporation of the heterocumulene into a C–C bond of these substrates is a potential synthetic route to lactones, a class of products which find interesting applications in pharmaceutical, cosmetics, and foodstuffs industry.

In the presence of electron-rich low-valent transition metal systems, strained cyclic hydrocarbons can undergo ring opening and react with CO_2 to afford carboxylated products. The carboxylation reaction proceeds stoichiometrically with Ni(0) systems [150, 151]. For instance, Ni(bpy)(cod) (bpy = 2,2′-bipyridine; cod = 1,4-cyclooctadiene) promoted the stoichiometric carboxylation of quadricyclane, probably through the reaction sequence illustrated in Scheme 4.29

[151]. One of the three-membered strained rings of quadricyclane opens through an oxidative addition step to give a strained nickelacyclobutane containing Ni(II). Further insertion of CO_2 into a Ni–C σ bond yields a 2-oxa-3-nickelacyclohexanone complex. However, the latter species, which was isolated, is much too stable (as can also be inferred from the high decomposition point (517 K)) to release the corresponding lactone in a subsequent reductive elimination step, and crystalline *endo*-nortricyclane-3-carboxylic acid can be obtained only upon acid hydrolysis. The proposed reaction pathway is in agreement with the rhodium- and nickel-catalyzed ring-opening reactions of quadricyclane. The high strain energy of the cyclic hydrocarbon (423 kJ/mol) seems to be crucial for the feasibility of the carboxylation process. As a matter of fact, other cyclic three-membered hydrocarbons with a lower strain energy than quadricyclane, such as cyclopropane (118 kJ/mol) and norcarane (127 kJ/mol), did not afford carboxylated products in the presence of Ni(0). Most probably, because of their lower strain energy, the three-membered rings of the latter substrates do not open to give the corresponding nickelacyclobutane species.

Methylenecyclopropanes have been carboxylated with gaseous carbon dioxide in the presence of a stoichiometric amount of nickel(0) complexes [150]. Depending on the polarity of the reaction solvent and the nature of amine ligand (amidine or guanidine base), several reactions pathways have been observed, which may involve either oxidative coupling of CO_2 with the C–C double bond of the substrate (see Sect. 5.3) or opening of the three-membered ring and further reaction with the heterocumulene.

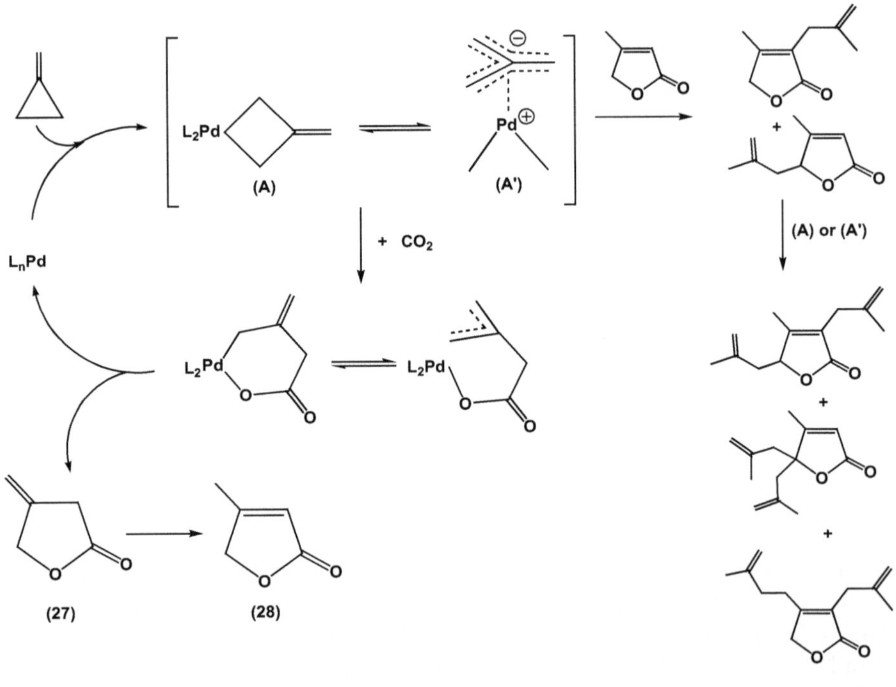

Scheme 4.30 Proposed mechanism for Pd-catalyzed carboxylation of methylenecyclopropane

Catalytic cycloaddition of CO_2 to methylenecyclopropanes with formation of γ-lactones and/or furanones has been achieved using palladium(0) catalysts (4.33) [152, 153].

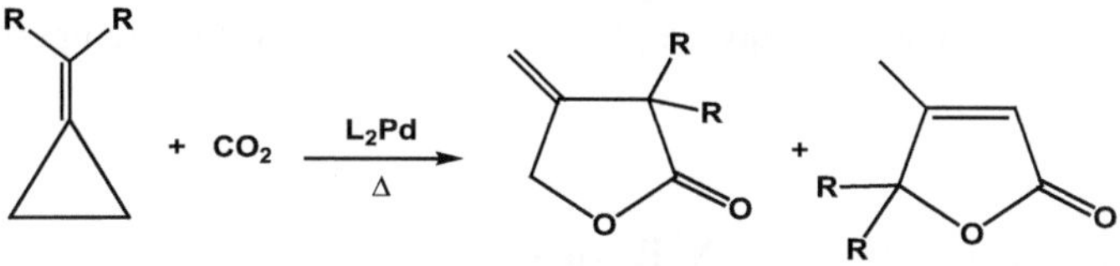

(4.33)

Scheme 4.30 summarizes the proposed reaction mechanism. The substrate would react with palladium(0) species to form a palladacyclobutane intermediate **A** in equilibrium with the tautomeric trimethylenemethane complex **A′**. Insertion of CO_2 into the palladium–carbon bond affords a palladium carboxylate, which can release γ-lactone **27** by reductive elimination. The latter species can isomerize to the thermodynamically more stable furanone **28**. α,β-Unsaturated lactone **28** contains acidic protons and, in principle, can react with intermediates **A** and/or **A′** (Scheme 4.30) to give cooligomers, which have been observed as by-products.

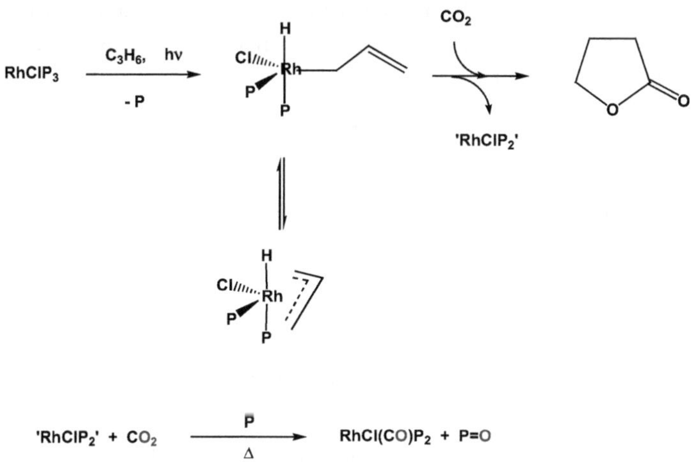

Scheme 4.31 Proposed mechanism for UV–vis driven rhodium-catalyzed carboxylation of cyclopropane (adapted from [154])

Catalytic incorporation of CO_2 into the cyclopropane ring can also be promoted by $RhCl(PMe_2Ph)_3$ in a light-driven reaction (4.34) [154]. At 263 K, UV–visible irradiation of a THF or toluene solution of

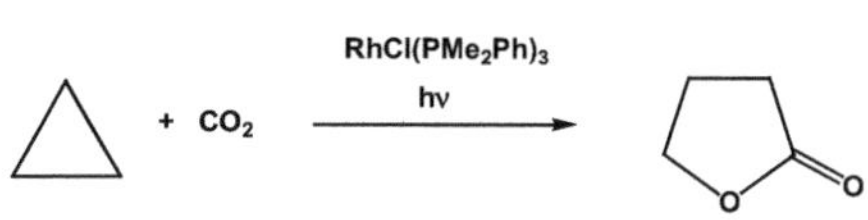

(4.34)

$RhCl(PMe_2Ph)_3$ in the presence of cyclopropane and pressurized carbon dioxide produced γ-butyrolactone (Scheme 4.31) in a variable molar ratio relative to rhodium, depending on the reaction conditions. The proposed mechanism for this reaction involves a hydrogen atom transfer to Rh to generate a hydrido-allyl intermediate that, after CO_2 uptake, evolves into butyrolactone. Formation of the carbonyl complex $RhCl(CO)(PMe_2Ph)_2$ corresponds to the deactivation of the catalyst. The lactone was not produced thermally, probably because the reduction of CO_2 to CO with phosphine oxidation is favored by polar solvents and high temperatures.

4.9 CO_2 Insertion into C–N Bonds

Incorporation of carbon dioxide into the aziridine ring gives oxazolidin-2-ones (4.35) [83, 155–174].

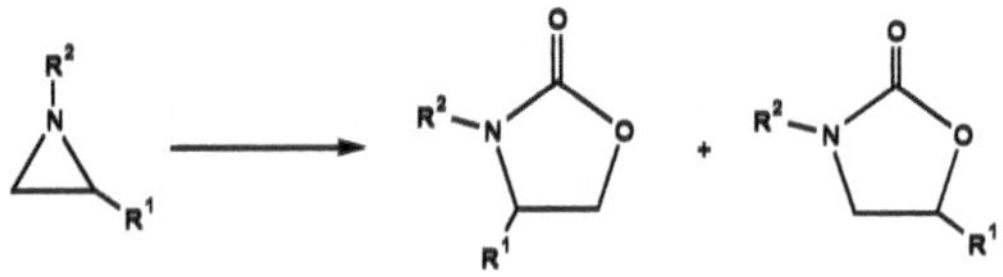

(4.35)

This reaction can proceed in the absence of any added catalyst, either under drastic conditions (393 K, 9 MPa, 24 h) [155] or, by using high speed ball milling (HBSM), at room temperature, under

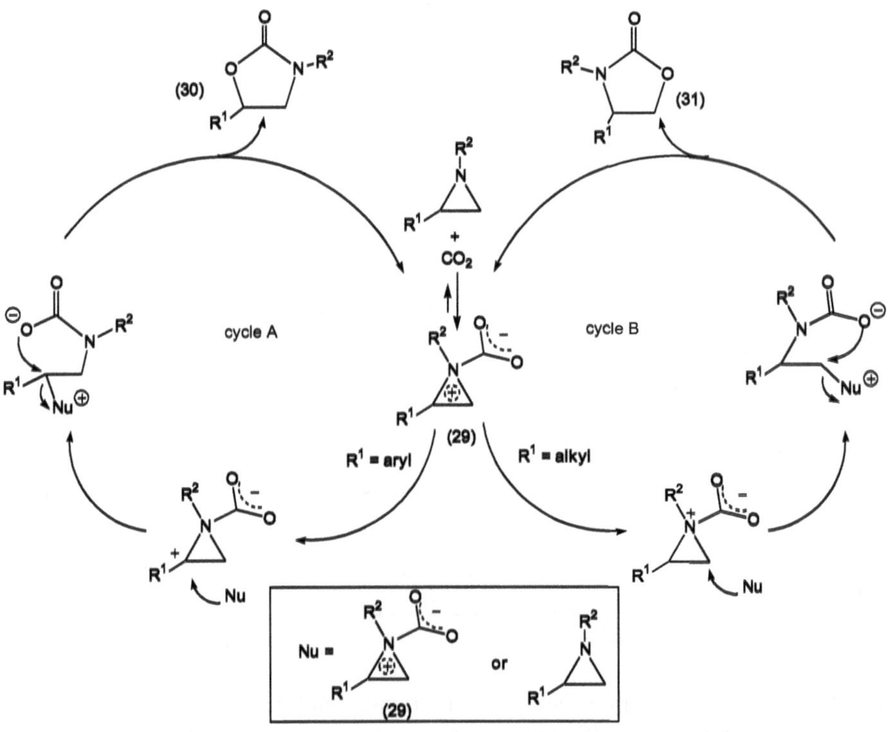

Scheme 4.32 Catalyst-free cycloaddition of CO_2 to aziridines: proposed mechanism (adapted from [155])

solventless conditions [156]. A key role in the catalyst-free process seems to be played by zwitterionic adduct **29** (Scheme 4.32) [155]. This species, which can form by coordination of CO_2 to the aziridine substrate, may undergo ring opening through nucleophilic attack by another similar molecule or aziridine itself [155, 156]. Final cyclization via intramolecular nucleophilic attack leads to the product as well as to regeneration of **29** or the substrate. It is worth noting that adduct **29** cannot undergo a unimolecular reaction and attack the 3-position because this would be a 5-*endo*-tet reaction, which is disfavored by Baldwin's rules; moreover, such a pathway would give net inversion, which is inconsistent with a few stereochemical findings observed in dedicated experiments. In accordance with the proposed mechanism, branched R^2 substituents at the N atom, which make the formation of adduct **29** more difficult, lowered activity, as well as aryl or electron-withdrawing groups (*N*-tosyl, for instance), which can reduce the electron density at the N atom and hinder CO_2 coordination. There exist two possible cycles (A or B) in this catalytic cycle, depending on the nature of R^1. When R^1 is an aryl group, cycle A is more favorable, leading to preferential formation of **30**, whereas, if R^1 is alkyl, cycle B is predominant, thus resulting mainly in the generation of **31**.

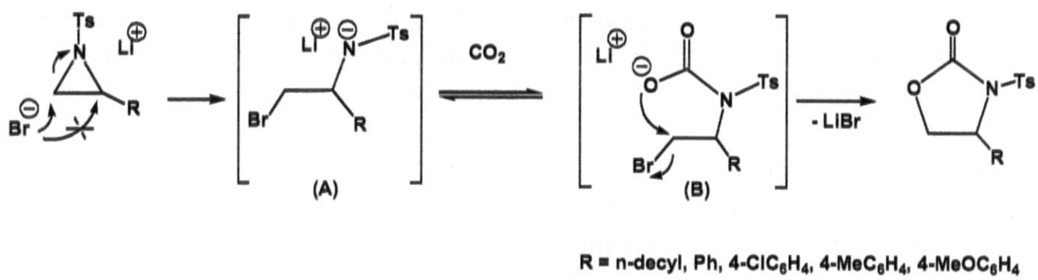

Scheme 4.33 LiBr-catalyzed carboxylation of *N*-tosylaziridines: proposed mechanism (adapted from [161])

The cycloaddition reaction can be promoted by a variety of catalysts. Depending on the used catalytic system, different typologies of mechanisms can be categorized.

Alkali or tetraalkylammonium halides have been shown to be effective catalysts for insertion of CO_2 into the C–N bond of aziridines [157–163]. The role of halide ion is crucial in these catalyzed processes as it acts not only as a nucleophile, thus promoting ring opening of adduct **29**, but also as a leaving group in the intramolecular cyclization step leading to cyclic carbamate. The mechanism of the LiI-catalyzed cycloaddition of CO_2 to aziridines has been modeled by means of DFT calculations, which have shown that a complex, incorporating one molecule of aziridine and a molecule of carbon dioxide together with Li^+ and I^-, is formed prior to the opening of the aziridine ring [160].

A different mechanism (Scheme 4.33) has been proposed by Endo and co-workers for the LiBr-catalyzed cycloaddition of CO_2 to 2-substituted *N*-tosylaziridines [161]. This reaction afforded selectively the 4-substituted regioisomer and was accelerated by electron-donating R substituents. In this case, the first step has been proposed to be a ring-opening reaction by nucleophilic attack of bromide anion from the catalyst to give intermediate **A** (Scheme 4.33). Regioselectivity of the opening reaction is responsible for regioselectivity observed in the final product. The fact that the reactivity of *N*-tosylaziridines was repressed by electron-withdrawing groups rules out that this step might be the rate-determining one. The next step is the reaction of anionic nitrogen with CO_2 to form intermediate **B** (Scheme 4.33), whose intramolecular cyclization gives the product. This step is believed to be the rate-determining one, as it is expected to be accelerated by electron-donating substituents through an inductive effect.

Activation of aziridine towards ring-opening can be promoted by Lewis acids other than CO_2 and capable of coordinating the substrate [164–166], or also by = (R)N···HX (X = N, O) hydrogen bond donors, such as polymer supported diol functionalised ionic liquids or amino acids [167–169], which have been used as catalysts. Scheme 4.34 illustrates the mechanism proposed by Miller and Nguyen [164] for the system (salen)chromium(III)/DMAP (salen = *N*,*N*′-ethylenebis (salicyliminate) dianion; DMAP = 4-dimethylaminopyridine), which proved to be an effective catalyst for the synthesis of 5-substituted oxazolidinones from carbon dioxide and aziridines. The step of aziridine activation by coordination to the Lewis acidic (salen)Cr metal center results in the formation of a partially cationic N atom and is followed by the nucleophilic ring-opening of aziridine by the Lewis base (LB) cocatalyst at the more substituted carbon to give an ionic intermediate. From this intermediate, after CO_2 uptake, the product was obtained. The presence of too much LB repressed the reaction because of competitive coordination of LB to the Lewis acidic Cr center. Moreover, the mechanism shown in Scheme 4.34 explains the catalytic activity observed in the absence of cocatalyst, as aziridine itself is a LB and can act as cocatalyst. An early report described the direct synthesis of oxazolidin-2-ones (5–80% yield) from

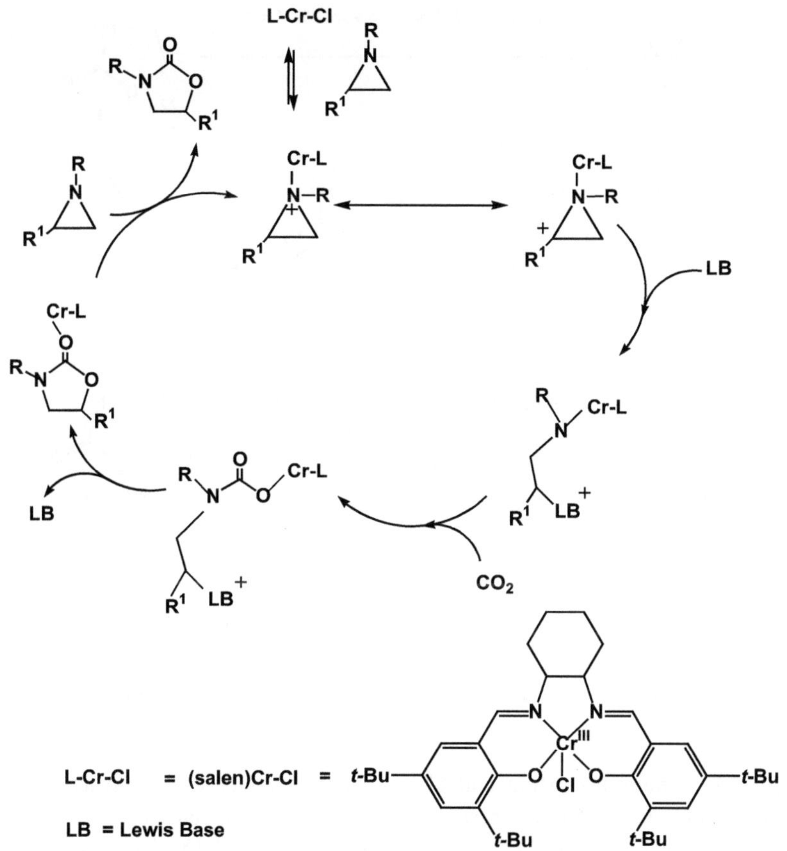

Scheme 4.34 Proposed mechanism for the cycloaddition of CO_2 with aziridines promoted by the system (salen)CrCl/DMAP (adapted from [164])

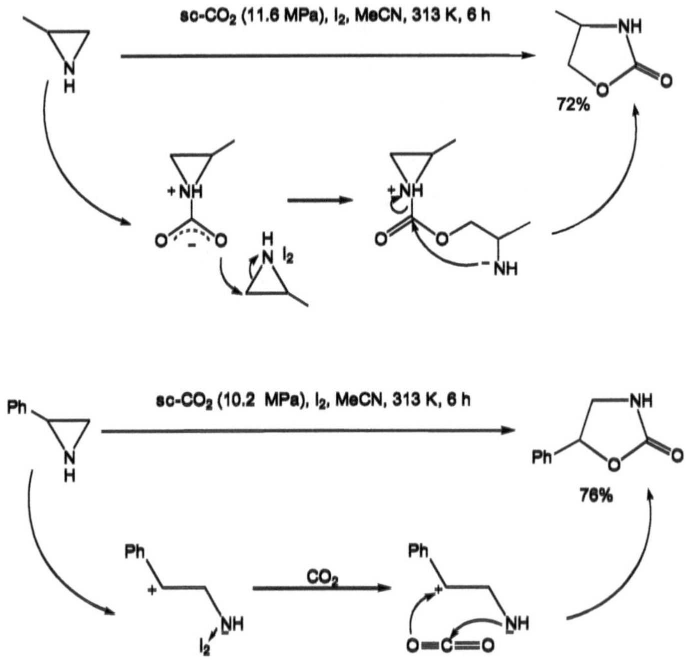

Scheme 4.35 Proposed mechanism for the cycloaddition of CO_2 with aziridines promoted by I_2 (adapted from [171])

aziridine or 2-methylaziridine and CO_2, in the presence of iodine as Lewis acid catalyst, at 333–353 K, under pressure of CO_2 [170]. This reaction was revisited by Kawanamy and Ikushima [171] using sc-CO_2 (Scheme 4.35). They found that the nature of substrate strongly affected the regioselectivity of the

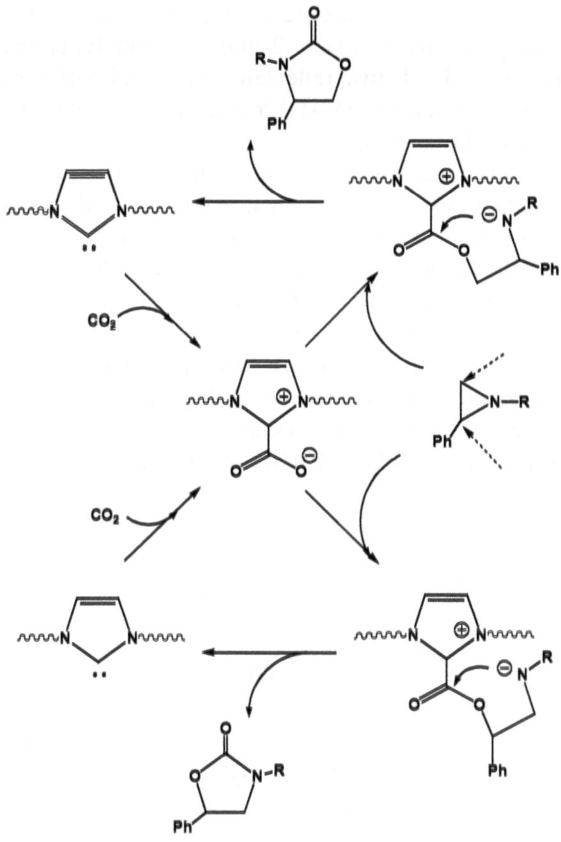

Scheme 4.36 Proposed mechanism for the cycloaddition of CO_2 with aziridines promoted by MCM-41-IPr

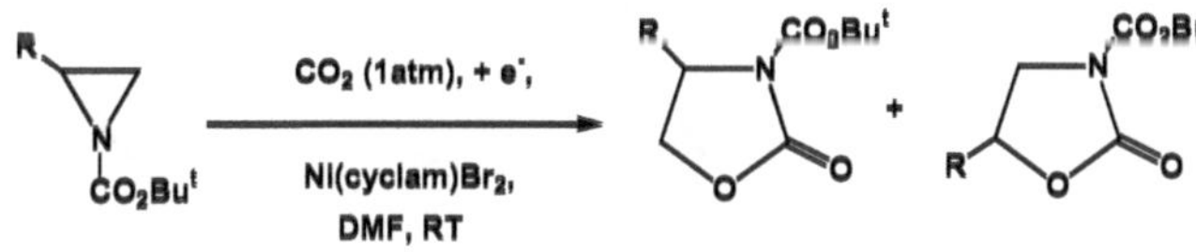

Scheme 4.37 Electrocarboxylation of *N*-Boc-aziridines

process. Using 2-methylaziridine as the substrate, only 4-methyloxazolidin-2-one was obtained, whereas 2-phenylaziridine converted into 5-phenyloxazolidin-2-one. The different regioselectivity was rationalized assuming that, in the first case, a carbamic acid formed, which then attacked a second aziridine molecule, whereas, in the other case, the formation of the 5-substituted regioisomer was ascribed to the formation of a 1,3-dipolar intermediate.

Cycloaddition of CO_2 with *N*-substituted phenylaziridines can be promoted by 1,3-bis-(2,6-diisopropylphenyl)imidazol-2-ylidene (IPr) functionalized MCM-41 (MCM-41-IPr) [172]. In situ diffuse reflectance infrared Fourier transform spectroscopy (DRIFTS) showed that MCM-41-IPr was able to capture CO_2 reversibly to give an MCM-41-IPr-CO_2 adduct.

Based on this result, it has been proposed that the zwitterionic moiety IPr-CO_2 on MCM-41 may attack one of the carbon atoms of the aziridine ring to generate a new zwitterion (Scheme 4.36). Nucleophilic attack of the formed nitrogen anion to the carbon atom of the carbonyl group produced oxazolidinones by intramolecular cyclic elimination. Finally, the MCM-41-IPr produced reacted quickly with CO_2 to regenerate the MCM-41-IPr-CO_2 adduct.

Duñach reported the insertion of the heterocumulene into *N*-Boc protected 2-substituted aziridines (Scheme 4.37) under mild conditions (ambient temperature, atmospheric CO_2 pressure), using electrochemical methods and dibromo(1,4,8,11-tetraazocyclotetradecane)-nickel(II), $Ni(Br)_2$(cyclam),

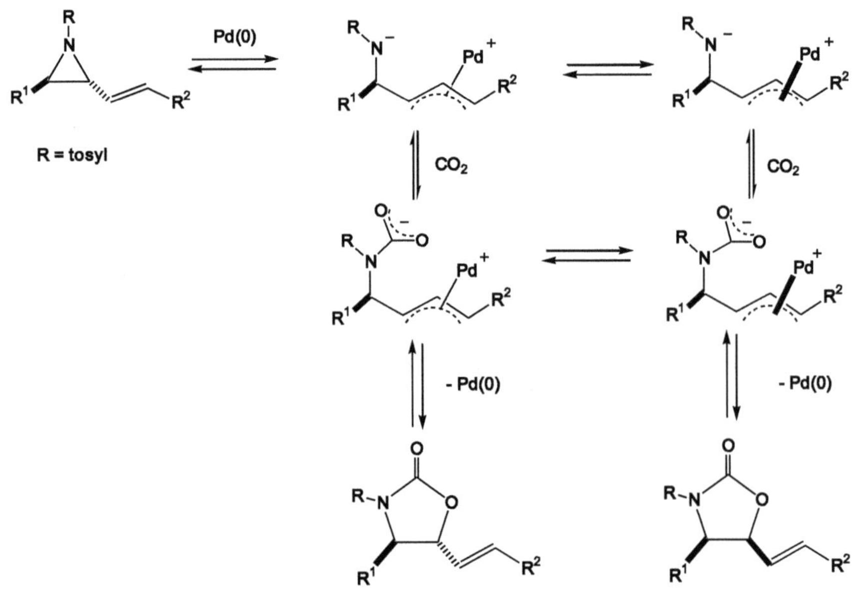

Scheme 4.38 Proposed mechanism for the cycloaddition of CO_2 with vinylaziridines promoted by Pd(0) (adapted from [174])

as the catalyst (10 mol%) [173]. However, the process was poorly selective as a mixture of the 4- and 5-regioisomers was obtained in all the cases investigated. The mechanism proposed for the electrocarboxylation process involves, as a first step, cathodic reduction of Ni(II) to Ni(0), which adds oxidatively the aziridine ring to give a four-membered azanickelacycle. Further CO_2 uptake and ring closure generate the cyclic carbamate. Under the working conditions, the latter process competes with oxalate formation resulting from cathodic reduction of CO_2 to CO_2^- radical anion, which undergoes head-to-head dimerization to oxalate.

Ring opening of vinylaziridines can be catalytically promoted by Pd(0). Under ambient atmosphere of carbon dioxide, 5-vinyloxazolidinones were obtained regio- and stereoselectively (Scheme 4.38) [174].

Recently, the reaction of insertion of CO_2 into the aziridine ring has been investigated in sc-CO_2 [175] and the formation of copolymers has been shown to occur under good control.

4.10 Insertion into Other E–X Bonds

CO_2 may react with several E–X systems to afford insertion products. Here we wish to discuss some cases that may be relevant either to the synthesis of metal complexes that may have an application as catalysts or maybe relevant to synthetic applications.

4.10.1 Insertion into M–M Bonds

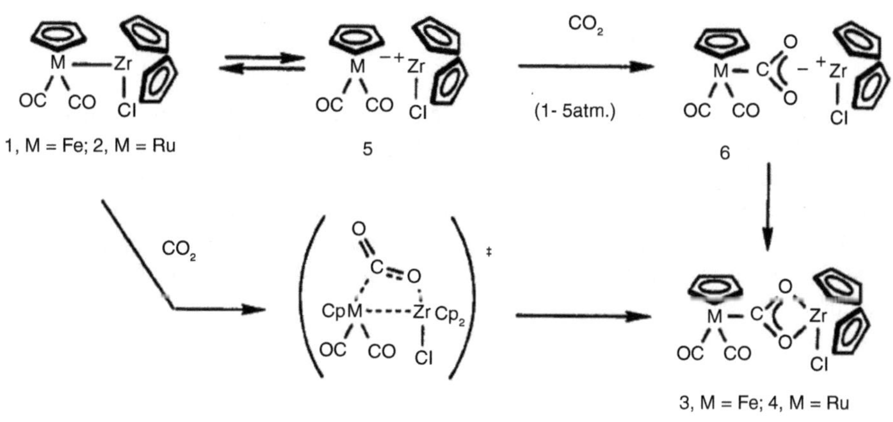

Scheme 4.39 Insertion of CO_2 into M–M bonds

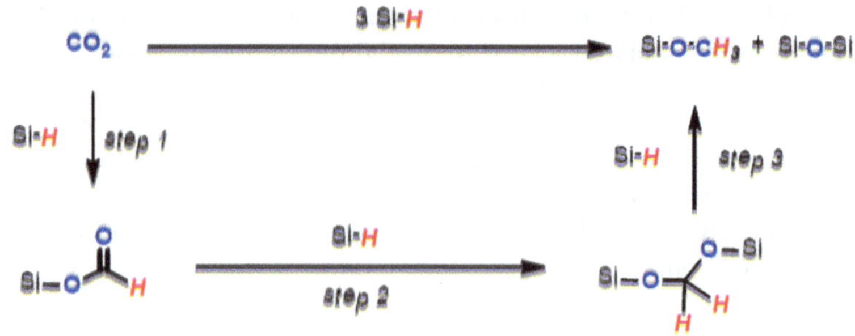

Scheme 4.40 Multiple attack of Si–H bonds on CO_2 with formation of reduced species of the cumulene. Si=R_3Si for sake of simplicity

The insertion of CO_2 into metal–metal bonds may lead to the formation of reduced species of the cumulene. So, it has been found that CO_2 may react with Fe–Zr and Ru–Zr bonds to give M–C(O)O–Zr species as product of interaction (where M = Fe, Ru) [176]. It is interesting to note that the insertion takes place in full respect of the oxophilicity and carbophilicity of the elements. In fact, in all cases the more oxophylic Zr is linked to the O atom, whereas the more carbophilic Fe or Ru is linked to the C atom. On the other hand, as discussed in Chap. 2, both Fe and Ru are able to afford M-η^1(C)–CO_2 complexes (Scheme 4.39).

4.10.2 *Insertion into Si–H Bonds*

The insertion of CO_2 into Si–H bonds leads, as a first step, to the formation of R_3SiOCH(O) moieties, formally considered as formate-silyl esters [177]. Such systems can be of interest as intermediates in catalysis in view of the easy hydrolysis of the Si–O–C bond or for their further reaction with Si–H bonds or any acid E–H bond. Scheme 4.40 shows that the silyl-formate can react with other R_3Si–H moieties to afford the R_3Si–O–CH_2O–Si species and, finally, the R_3Si–O–CH_3, precursor of methanol, CH_3OH.

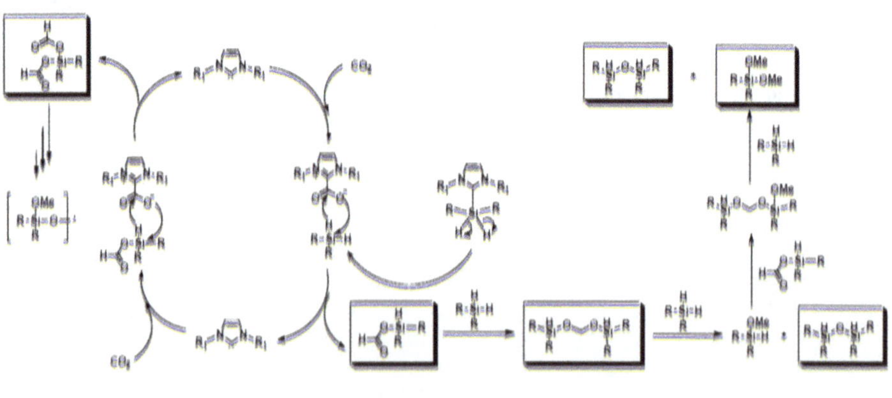

Scheme 4.41 Reaction of CO_2 with R_3Si–H under imidazolium salts catalysis [178–180]

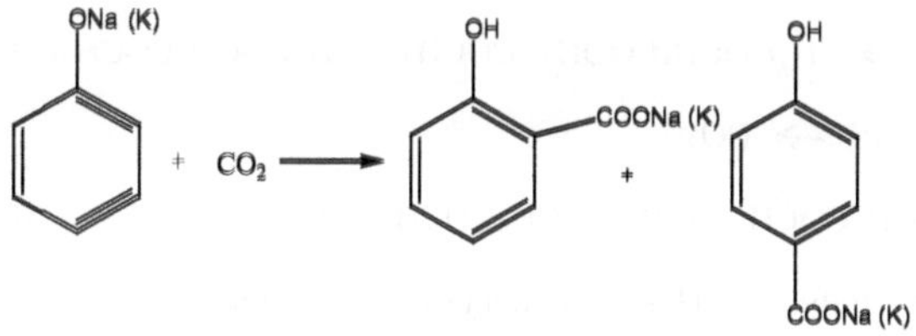

Scheme 4.42 Reaction of CO_2 with sodium(potassium)phenoxide to afford 2(4)-hydroxybenzoic acid

The series of reactions above gives an insight into the reaction mechanism of reduction of CO_2 to methanol, with three consecutive $2e^-$-transfer steps (see also Chap. 8).

Scheme 4.41 shows the reaction of Ph_2SiH_2 with CO_2 under imidazolium–carbene catalysis [178–180].

4.10.3 Insertion into C–H Bonds

The insertion of CO_2 into C–H bonds (4.36), despite its undoubtedly great industrial interest for

$$R - H + CO_2 \rightarrow R - CO_2H \tag{4.36}$$

the direct synthesis of carboxylic acids from hydrocarbons, is not documented by an abundant literature. The only industrial exploitation of such reaction is the Kolbe–Schmitt synthesis of hydroxybenzoic acids [181, 182] carried out by reacting CO_2 with Group 1 derivatives of phenol under forcing conditions (Scheme 4.42).

Such reaction has been re-examined more recently [183] and some elements of knowledge have been added to its otherwise not yet completely elucidated reaction mechanism.

A case related to the above is represented by the reaction with CO_2 of σ-PhO-Mn (II) and π(PhO-)Mn_2 systems. The former gives insertion into the Mn–O bond as expected to afford Mn–OC(O)OPh

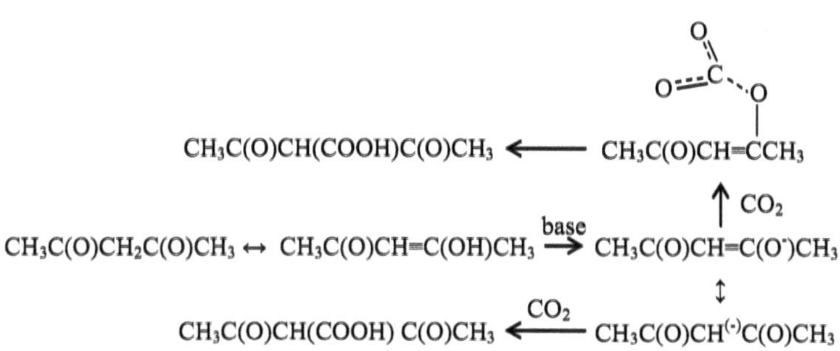

Scheme 4.43 CO_2 insertion into activated C–H bonds promoted by a base

$$CH_3COCH_2COCH_3 \xrightarrow{ZnS,\ h\nu} CH_3COCH(COOH)HCOCH_3 + CH_3COCH_2COCH_2\text{-}COOH$$

$$Ru@ZnS + CO_2 + light \rightarrow CO_2^{\cdot -}$$

$$CO_2^{\cdot -} + acac \rightarrow CH_3COCH^{\cdot}COCH_3 + CH_3COCH_2COCH_2^{\cdot} + CH_3COCH{=}CO^{\cdot}CH_3 + HOCO^{\cdot}$$

Scheme 4.44 Photocatalytic C–H activation and CO_2 insertion

derivatives, whereas the latter inserts CO_2 into the C–H bond affording very selectively Mn–OC_6H_4(*p*-COOH), *p*-hydroxobenzoic acid [184]. The insertion of CO_2 into the activated C–H methylenic bond of acetylacetone ($CH_3COCH_2COCH_3$, acac) has been reported by Chiusoli [185] to occur under base catalysis (4.37).

$$\begin{aligned} &CH_3COCH_2COCH_3 + CO_2 + PhONa \\ &\quad \rightarrow CH_3COCH(COONa)COCH_3 + PhOH \end{aligned} \tag{4.37}$$

In this case, the reaction is promoted by the tautomerism –CH_2–CO– $\leftrightarrow$ –CH=COH– which makes possible the extraction of the proton by the base. The resulting anion –CH=$CO^{(-)}$– in equilibrium with –$CH^{(-)}$–CO– makes possible the CO_2 trapping with formation of a C–CO_2 bond (Scheme 4.43).

A similar reaction is also promoted by imidazolium carbene carboxylates [186] as shown in (4.38), where the activated-H migrates to imidazole and CO_2 forms a C–C bond with acac.

$$R'R''Im = CO_2 + acac\text{-}H + MX \rightarrow R'R''ImH^+X^- + acac\text{-}CO_2M \tag{4.38}$$

More recently, the direct insertion of CO_2 into the C–H bond of acac has been shown to occur under photocatalysis (Scheme 4.44) [187]. The species identified as the initiator of the reaction is the $CO_2^{\bullet -}$ radical anion. It extracts an H atom either from the enolic form or from acac itself. Coupling of COOH with the acac-radical afford the acids.

Interestingly, under photocatalysis two different isomeric acids are obtained that imply the CO_2 insertion into the activated methylenic C–H bond but also into the more inert C–H bond of the terminal –CH_3 moiety, which was not affected under chemical catalysis.

This field deserves more research effort as the exploitation of the insertion into the C–H bond has great industrial potential for the synthesis of the –CO_2H moiety, which today is obtained through quite tortuous routes based on oxidation reactions which are often neither selective nor C-efficient.

4.11 Conclusions

The study of the reaction of CO_2 with several sets of E–X bonds has contributed to knowledge about the reactivity of the heterocumulene under different conditions, opening the path to the development of new metal-promoted synthetic methodologies and to new catalytic processes. The fundamental studies described in this chapter have great importance for the development of CO_2-chemistry, in its many aspects, from its use as a building block with incorporation of the entire heterocumulene into organic compounds (Chaps. 5 and 6) to its use as carbon source in reduction reactions for the production of molecules characterized by higher energy, either C1 or Cn molecules (Chaps. 7 and 8).

The basic concepts presented in this chapter are examples of application of the general principles described in the previous Chaps. 1 and 2, and represent the pillars on which CO_2 conversion is based: they are encountered and applied very often in reactions described in the following chapters.

Several aspects of the reactivity of the heterocumulene still need investigation, and new aspects will be discovered as our knowledge progresses.

References

1. Sneeden RPA (1982) Reactions of carbon dioxide. In: Wilkinson G, Stone FGA, Abel EW (eds) Comprehensive organometallic chemistry, vol 8. Pergamon, New York, pp 225–283
2. Eisenberg DC, Norton JR (1991) Hydrogen-atom transfer reactions of transition-metal hydrides. Isr J Chem 31:55–66
3. Labinger JA (1992) Nucleophilic reactivity of transition metal hydrides. In: Dedieu A (ed) Transition metal hydrides, Chap 10. VCH, New York
4. Kristjanadottir SS, Norton JR (1992) Acidity of hydrido transition metal complexes in solution. In: Dedieu A (ed) Transition metal hydrides, Chap 9. VCH, New York
5. Aresta M, Dibenedetto A, Pápai I, Schubert G, Macchioni A, Zuccaccia D (2004) Behaviour of $[PdH(dppe)_2]X$ ($X = CF_3SO_3^-$, SbF_6^-, BF_4^-) as proton or hydride donor: relevance to catalysis. Chem Eur J 10:3708–3716
6. Hawackery J, Lehn J-M, Zeissel R (1984) Electrocatalytic reduction of carbon dioxide mediated by $Re(dipy)(CO)_3Cl$. J Chem Soc Chem Commun 328–329
7. Hawackery J, Lehn J-M, Zeissel R (1986) Photochemical and electrochemical reduction of CO_2 to CO mediated by (2,2′ bipyridine)tricarbonyl-rhenium(I) and related complexes as homogeneous catalysts. Helv Chim Acta 69:1990–2012
8. Bradley MG, Roberts DA, Geoffroy GL (1981) Photogeneration of reactive $ReH(diphos)_2$. Its reversible coordination of CO_2 and activation of aromatic C-H bonds. J Am Chem Soc 103:379–384
9. Sullivan BP, Meyer TJ (1984) Photoinduced irreversible insertion of CO_2 into a metal-hydride bond. J Chem Soc Chem Comm 1244–1245
10. Sullivan BP, Krist K, Guard HE (1993) Formation of formate by insertion into a metal hydride bond. In: Electrochemical and electrocatalytic reactions of carbon dioxide. Elsevier, Amsterdam, pp. 128–129
11. Kolomnikov IS, Gusev AI, Alexandrov GG, Loboeva TS, Strutchov Yu T, Volpin MA (1973) Structure of the product formed in the reaction of carbon dioxide with ruthenium hydride complexes. J Organomet Chem 59:349–351

12. Darensbourg MY, Carlton EA (1987) Anionic transition metal hydrides. Adv Organomet Chem 27:1–47
13. DuBois DL, Berning DE (2000) Hydricity of transition metal hydrides and its role in CO_2 reduction. Appl Organomet Chem 14(12):860–862
14. Aresta M, Dibenedetto A, Angelini A (2014) Catalysis for the valorization of exhaust carbon: from carbon dioxide to chemicals, materials and fuels. Technological use of CO_2. Chem Rev 114:1709–1743
15. Inoue Y, Izumida H, Sasaki Y, Hashimoto H (1976) Catalytic fixation of carbon dioxide to formic acid by transition metal complexes under mild conditions. Chem Lett 863–864
16. Jessop PG, Ikariya T, Noyori R (1995) Homogeneous hydrogenation of carbon dioxide. Chem Rev 95:259–272
17. Leitner W (1995) Carbon dioxide as a raw material: the synthesis of formic acid and its derivatives from CO_2. Angew Chem Int Ed Engl 34:2207–2221
18. Jessop PG, Joo F, Tai C-C (2004) Recent advances in the homogeneous hydrogenation of carbon dioxide. Coord Chem Rev 248:2425–2442
19. Federsel C, Jackstell R, Beller M (2010) State-of-the-art catalysts for hydrogenation of carbon dioxide. Angew Chem Int Ed 49:6254–6257
20. Leitner W, Dinjus E, Gassner F (1998) In: Cornils B, Herrmann WA (eds) Aqueous-phase organometallic catalysis concepts and applications. Wiley-VCH, Weinheim, p 486
21. Kovacs G, Schubert G, Joo F, Pápai I (2006) Theoretical investigation of catalytic HCO_3^- hydrogenation in aqueous solutions. Catal Today 115:53–60
22. Engel DC, Versteeg GF, van Swaaij WPM (1997) Chemical equilibrium of hydrogen and aqueous solutions of 1:1 bicarbonate and formate salts with a common cation. Fluid Phase Equilib 135:109–136
23. Moret S, Dyson P, Laurenczy G (2014) Direct synthesis of formic acid from CO_2 by hydrogenation in acidic media. Nat Commun 5:1–7
24. Aresta M, Gobetto R, Quaranta E, Tommasi I (1992) A bonding-reactivity relationship for Ni $(PCy_3)_2(CO_2)$: a comparative solid-state-solution nuclear magnetic resonance study (^{31}P, ^{13}C) as a diagnostic tool to determine the mode of bonding of CO_2 to a metal center. Inorg Chem 31:4286–4290
25. Tanaka K, Ooyama D (2002) Multi-electron reduction of CO_2 via Ru-CO_2 -C(O)OH, -CO, -CHO, and CH_2OH species. Coord Chem Rev 226:211–218
26. DuBois DL, Miedaner AR, Curtis Haltiwanger RC (1991) Electrochemical reduction of carbon dioxide catalyzed by $[Pd(triphosphine)(solvent)](BF4)_2$ complexes: synthetic and mechanistic studies. J Am Chem Soc 113:8753–8764
27. Ishida H, Tanaka K, Morimoto M, Tanaka T (1986) Isolation of intermediates in the water gas shift reactions catalyzed by $[Ru(bpy)_2(CO)Cl]^+$ and $[Ru(bpy)_2(CO)_2]^{2+}$. Organometallics 5: 724–730
28. Dubois DL (1997) Development of transition metal phosphine complexes as electrocatalysts for CO_2 and CO reduction. Comments Inorg Chem 19:307–325
29. Cámpora J, Palma P, del Río D, Álvarez E (2005) CO insertion reactions into the M – OH bonds of monomeric nickel and palladium hydroxides. Reversible decarbonylation of hydroxycarbonyl palladium complex. Organometallics 23(8):1652–1655
30. Bertini I, Banci Luchinat C, Monnanni R (1987) The enzyme carbonic anhydrase. In: Aresta M, Forti G (eds) Carbon dioxide as a source of carbon: biochemical and chemical use. NATO Advanced Study Institute, Pugnochiuso
31. Palmer DA, Van Eldik R (1983) The chemistry of metal carbonate and CO_2 complexes. Chem Rev 83:651–731
32. Zhang X, Van Eldik R, Koike T, Kimura E (1993) Kinetics and mechanism of hydration of CO_2 and dehydration of HCO_3^- catalyzed by a Zn(II) complex of 1,5,9-tetraazadodecane. Inorg Chem 32:5749–5755
33. Zhang X, Van Eldik R (1995) A functional model for carbonic anhydrase. Thermodynamic and kinetic study of a tetraazacyclododecane complex of Zn(II). Inorg Chem 34:5606–5614

34. Huang D, Maklyents OV, Tan LL, Lee SC, Rybak-Akimova EV, Holm RH (2011) Kinetics and mechanistic study of an extremely rapid carbon dioxide fixation reaction. Proc Natl Acad Sci USA 108(4):1222–1227
35. Flynn BR, Vaska L (1973) Carbon dioxide fixation leading to stable molecular bicarbonato complexes of d^8 metals. J Am Chem Soc 95:5081–5082
36. Truscott BJ, Nelson DJ, Alexandra MZ, Slawin AMZ, Nolan SP (2014) CO_2 fixation employing an iridium(I)-hydroxide complex. Chem Commun 50:286–288
37. Lohr TL, Piers WE, Parvez M (2013) Reversible insertion of carbon dioxide into Pt(II)–hydroxo bonds. Dalton Trans 42:14742–14748
38. Aresta M, Schloss MV (eds) (1990) Enzymatic and model carboxylation and reduction reactions for carbon dioxide utilization. NATO Advanced Study Institute, Riva dei Tessali
39. Volpin ME (1972) The reaction of organometallic compounds of transition metals with molecular nitrogen and CO_2. Pure Appl Chem 30(3):607–626
40. Albano P, Aresta M, Manassero M (1980) Interaction of carbon dioxide with coordinatively unsaturated rhodium(I) complexes with the ligand 1,2-bis(diphenylphosphino)ethane. Inorg Chem 19:1069–1070
41. Allen OR, Dalgarno SJ, Field DL, Jensen P, Willis AC (2009) Insertion of CO_2 into the Ru–C bonds of cis- and trans-Ru(dmpe)$_2$Me$_2$(dmpe=Me$_2$PCH$_2$CH$_2$PMe$_2$). Organometallics 28:2385–2390
42. Jessop PG (2007) Homogeneous hydrogenation of carbon dioxide. In: Handbook of homogeneous hydrogenation, Chap 17. Wiley-VCH, Weinheim, pp 489–511
43. Darensbourg D, Kyran SJ, Yeung AD, Bengali AA (2013) Kinetic and thermodynamic investigation of CO_2 insertion into the Ru-Me and Ru-H bonds. An experimental and computational study. Eur J Inorg Chem 4024–4031
44. Darensbourg DJ, Holteamp MW (1995) Catalytic activity of zinc(II) phenoxides which possess readily accessible coordination sites. Copolymerization and terpolymerization of epoxides and carbon dioxide. Macromolecules 28:7577–7579
45. Super M, Berluche E, Costello C, Beckman E (1997) Copolymerization of 1,2-epoxycyclohexane and carbon dioxide using carbon dioxide as both reactant and solvent. Macromolecules 30:368–372
46. Super M, Beckman E (1998) Copolymerization of CO_2 and cyclohexene oxide. J Macromol Symp 127:89–108
47. Cheng M, Lobkovsky EB, Coates GW (1998) Catalytic reactions involving C1 feedstocks: new high-activity Zn(II)-based catalysts for the alternating copolymerization of carbon dioxide and epoxides. J Am Chem Soc 120:11018–11019
48. Beckman E (1999) Making polymers from carbon dioxide. Science 283:946–947
49. Darensbourg DJ, Holteamp MW, Struck GE, Zimmer MS, Niezgoda SA, Rainey P, Robertson JB, Draper JD, Reibenspies JH (1999) Catalytic activity of a series of Zn(II) phenoxides for the copolymerization of epoxides and carbon dioxide. J Am Chem Soc 121:107–116
50. Darensbourg DJ, Wildeson JR, Yarbrough JC, Reibenspies JH (2000) Bis 2,6-difluorophenoxide dimeric complexes of zinc and cadmium and their phosphine adducts: lessons learned relative to carbon dioxide/cyclohexene oxide alternating copolymerization processes catalyzed by zinc phenoxides. J Am Chem Soc 122:12487–12496
51. Cheng M, Moore DR, Reczek JJ, Chamberlain BM, Lobkovsky BE, Coates GW (2001) Single-site β-diiminate zinc catalysts for the alternating copolymerization of CO_2 and epoxides: catalyst synthesis and unprecedented polymerization activity. J Am Chem Soc 123:8738–8749
52. Cheng M, Darling NA, Lobkovsky EB, Coates GW (2000) Enantiomerically-enriched organic reagents *via* polymer synthesis: enantioselective copolymerization of cycloalkene oxides and CO_2 using homogeneous, zinc-based catalysts. Chem Commun 2007–2008
53. Eberhardt R, Allmendinger M, Luinstra GA, Rieger B (2000) The ethylsulfinate ligand: a highly efficient initiating group for the zinc β-diiminate catalyzed copolymerization reaction of CO_2 and epoxides. Organometallics 22:211–214

54. Inoue S (2000) Immortal polymerization: the outset, development, and application. J Polym Sci A 38:2861–2871
55. Kruper WJ, Dellar DV (1995) Catalytic formation of cyclic carbonates from epoxides and CO_2 with chromium metalloporphyrinates. J Org Chem 60:725–727
56. Mang S, Cooper AI, Colclough ME, Chauhan N, Holmes A (2000) Copolymerization of CO_2 and 1,2-cyclohexene oxide using a CO_2-soluble chromium porphyrin catalyst. Macromolecules 33:303–308
57. Allen SD, Moore DR, Lobkovsky EB, Coates GW (2002) High-activity, single-site catalysts for the alternating copolymerization of CO_2 and propylene oxide. J Am Chem Soc 124: 14284–14285
58. Darensbourg DJ, Lewis SJ, Rodgers JL, Yarbrough JC (2003) Carbon dioxide/epoxide coupling reactions utilizing Lewis base adducts of zinc halides as catalysts. Cyclic carbonate versus polycarbonate production. Inorg Chem 42:581–589
59. Darensbourg DJ, Yarbrough JC (2002) Mechanistic aspects of the copolymerization reaction of carbon dioxide and epoxides, using a chiral salen chromium chloride catalyst. J Am Chem Soc 124:6335–6342
60. Eberhardt R, Allmendinger M, Rieger B (2003) DMAP/Cr(III) catalyst ratio: the decisive factor for poly(propylene carbonate) formation in the coupling of CO_2 and propylene oxide. Macromol Rapid Commun 24:194–196
61. Darensbourg DJ, Yarbrough JC, Ortiz C, Fang CC (2003) Comparative kinetic studies of the copolymerization of cyclohexene oxide and propylene oxide with carbon dioxide in the presence of chromium salen derivatives. In situ FTIR measurements of copolymer vs cyclic carbonate production. J Am Chem Soc 125:7586–7591
62. Darensbourg DJ, Sanchez KM, Reibenspies JH, Rheingold AL (1989) Synthesis, structure, and reactivity of zerovalent group 6 metal pentacarbonyl aryl oxide complexes. Reactions with carbon dioxide. J Am Chem Soc 111:7094–7103
63. Simpson RD, Bergman RG (1992) A dramatic difference in the reactivities of alkoxido- and aryloxidorhenium complexes in insertion reactions. Angew Chem Int Ed Engl 31:220–223
64. Darensbourg DJ, Lee WZ, Phelps AL, Guidry E (2003) Kinetic study of the insertion and deinsertion of carbon dioxide into *fac*-$(CO)_3$(dppe)MnOR derivatives. Organometallics 22: 5585–5588
65. Mandal SK, Ho DM, Orchin M (1991) A convenient synthesis of the manganese and rhenium alkoxides and phenoxides *fac*-$(CO)_3$(dppe)MOR ($R = CH_3$, C_2H_5, C_6H_5; dppe=1,2-bis (diphenylphosphino)ethane). X-Ray structure of *fac*-$(CO)_3$(dppe)$ReOC_6H_5$. Inorg Chem 30:2244–2248
66. Mandal SK, Ho DM, Orchin M (1993) Reaction of electrophiles with manganese(I) and rhenium(I) alkoxide complexes: reversible absorption of atmospheric carbon dioxide. Organometallics 12:1714–1719
67. Aresta M, Dibenedetto A, Pastore C (2003) Synthesis and characterization of $Nb(OR)_4$[OC (O)OR] (R = Me, Et, Allyl) and their reaction with the parent alcohol to afford organic carbonates. Inorg Chem 42:3256–3261
68. Elmas S, Subhani MA, Vogt H, Leitner W, Muller T (2013) Facile insertion of CO_2 into metal-phenoxide bond. Green Chem 15:1356–1360
69. Hayward PJ, Blake DM, Nyman CJ, Wilkinson G (1969) Some novel peroxocarbonate complexes of platinum (II). J Chem Soc D Chem Commun 17:987–988
70. Hayward PJ, Blake DM, Wilkinson G, Nyman CJ (1970) Reactions of peroxobis(triphenylphosphine) platinum (II) and analogs with carbon dioxide, carbon disulfide, and other unsaturated molecules. J Am Chem Soc 92:5873–5878
71. Dahlenburg L, Prengel C (1984) Alkyl and aryl compounds of iridium and rhodium. 18. Oligophosphine ligands. 6. Reactivity of some alkyls and aryls of rhodium and iridium toward carbon dioxide. Facile formation and X-ray structural characterization of the peroxocarbonato complex [cyclic]-*mer*-$\overline{Rh(4MeC_6H_4)[OOC(O)O}][PhP(CH_2CH_2CH_2PPh_2)_2]$. Organometallics 3:934–936

72. Aresta M, Quaranta E, Ciccarese A (1985) Rhodium (I) promoted activation of carbon dioxide. C_1 Mol Chem 1:267–281
73. Aresta M, Tommasi I, Quaranta E, Fragale C, Mascetti J, Tranquille M, Galan F, Fouassier M (1996) Mechanism of formation of peroxocarbonates $\overline{Rh(OOC(O)O})(Cl)P_3$ and their reactivity as oxygen transfer agents mimicking monooxygenases. The first evidence of CO_2 insertion into the O-O bond of Rh(η^2-O_2) complexes. Inorg Chem 35:4254–4260
74. Aresta M, Quaranta E, Tommasi I, Mascetti J, Tranquille M, Borowiak M (1998) Formation of peroxocarbonates from $L_3Rh(O_2)Cl$ and $L_2Ni(CO_2)$: a unique reaction mechanism with carbon dioxide insertion into the O-O bond. Stud Surf Sci Catal 114:677–680
75. Borowiak MA, Jamroz MH, Dobrowolski JC, Bajdor K, Kazimirski JK, Mascetti J, Quaranta E, Tommasi I, Aresta M (2001) Application of the impulse oscillation model for modeling the formation of peroxocarbonates *via* carbon dioxide reaction with dioxygen transition metal complexes. A comparison with the experimental results obtained for Rh (η^2-O_2)ClP_3 [P = phosphane ligand]. J Mol Catal A 165:45–54
76. Standard bond energies. Department of Chemistry, Michigan State University. http://www.cem.msu.edu/~reusch/OrgPage/bndenrgy.htm
77. Aresta M, Tommasi I, Dibenedetto A, Fouassier M, Mascetti J (2002) Mechanism of formation of the peroxocarbonate complex $(PCy_3)_2Ni(CO_4)$ from solid $(PCy_3)_2Ni(CO_2)$ and dioxygen: an example of solid-state metallorganic reaction involving CO_2 deco-ordination and reinsertion into the O-O bond of $(PCy_3)_2Ni(O_2)$. Reactivity of the peroxocarbonate complex towards olefins in the solid state and in solution. Inorg Chim Acta 330:63–71
78. Yin X, Moss JR (1999) Recent developments in the activation of carbon dioxide by metal complexes. Coord Chem Rev 181:27–59
79. Behr A (1988) Carbon dioxide as an alternative C_1 synthetic unit: activation by transition metal complexes. Angew Chem Int Ed Engl 27:661–678
80. Walther D, Ruben M, Rau S (1999) Carbon dioxide and metal centres: from reactions inspired by nature to reactions in compressed carbon dioxide as solvent. Coord Chem Rev 182:67–100
81. Schenk S, Notni J, Kohn U, Wermann K, Anders E (2006) Carbon dioxide and related heterocumulenes at zinc and lithium cations: bioinspired reactions and principles. Dalton Trans 4191–4206
82. Belli Dell'Amico D, Calderazzo F, Labella L, Marchetti F, Pampaloni G (2003) Converting carbon dioxide into carbamato derivatives. Chem Rev 103:3857–3897
83. Quaranta E, Aresta M (2010) The chemistry of N-CO_2 bonds: synthesis of carbamic acids and their derivatives, isocyanates, and ureas. In: Aresta M (ed) Carbon dioxide as chemical feedstock. Wiley-VCH, Weinheim, pp 121–167
84. Matson EM, Fanwick PE, Bart SC (2011) Formation of trivalent U-C, U-N, and U-S bonds and their reactivity towards carbon dioxide and acetone. Organometallics 30:5753–5762
85. Chisolm MH, Extine M (1975) Carbon dioxide exchange reactions involving transition-metal N,N-dimethylcarbamato compounds: reversible insertion of carbon dioxide into transition-metal-nitrogen σ bonds. J Chem Soc Chem Commun 438–439
86. Chisolm MH, Extine M (1977) Reactions of transition metal-nitrogen σ bonds. 4. Mechanistic studies of carbon dioxide insertion and carbon dioxide exchange reactions involving early transition-metal dimethylamido and N, N-dimethylcarbamato compounds. J Am Chem Soc 99:792–802
87. Chisolm MH, Extine M (1977) Reactions of transition metal-nitrogen σ bonds. 3. Transition metal N, N-dimethylcarbamates. Preparation, properties, and carbon dioxide exchange reactions. J Am Chem Soc 99:782–792
88. Remko M, Rode BM (1995) Ab initio study of decomposition of carbamic acid and its thio and sila derivatives. J Mol Struct (Theochem) 339:125–131
89. Andrés J, Moliner V, Krecbl J, Silla E (1994) Comparison of several semiempirical and ab initio methods for transition state structure characterization. Addition of CO_2 to $CH_3NHCONH_2$. J Phys Chem 98:3664–3668

90. Andrés J, Moliner V, Krecbl J, Silla E (1993) A theoretical study of the addition mechanism of carbon dioxide to methylamine. Modelling CO_2–biotin fixation. J Chem Soc Perkin Trans 2:521–523
91. Andrés J, Moliner V, Krecbl J, Silla E (1993) A theoretical study of the effect of basis sets on stationary structures for the addition of carbon dioxide to methylamine: a relation among geometries, energy status, and electronic structures. Int J Quantum Chem 45:433–444
92. Chakraborty AK, Bischoff KB, Astarita G, Damewood JR (1988) Molecular orbital approach to substituent effects in amine-CO_2 interactions. J Am Chem Soc 110:6947–6954
93. Jamróz MH, Dobrowolski JC, Borowiak MA (1997) Ab initio study on the 1:2 reaction of CO_2 with dimethylamine. J Mol Struct 404:105–111
94. Jamróz MH, Dobrowolski JC, Borowiak MA (1999) The CO_2 with dimethylamine reaction: ab initio predicted vibrational spectra. J Mol Struct 482–483:633–637
95. Jamróz MH, Dobrowolski JC, Borowiak MA (2000) Theoretical IR spectra of the (2:1) ammonia–carbon dioxide system. Vib Spectrosc 22:157–161
96. Masuda K, Ito Y, Horiguchi M, Fujita H (2005) Studies on the solvent dependence of the carbamic acid formation from ω-(1-naphthyl)alkylamines and carbon dioxide. Tetrahedron 61:213–229
97. Hampe EM, Rudkevich DM (2003) Exploring reversible reactions between CO_2 and amines. Tetrahedron 59:9619–9625
98. Wittmann K, Wisniewski W, Mynott R, Leitner W, Kranermann CL, Rische T, Eilbracht P, Kluwer S, Ernsting JM, Elsevier CJ (2001) Supercritical carbon dioxide as solvent and temporary protecting group for Rhodium-catalyzed hydroaminomethylation. Chem Eur J 7:4584–4589
99. Furstner A, Ackermann L, Beck K, Hori H, Koch D, Langemann K, Liebl M, Six C, Leitner W (2001) Olefin metathesis in supercritical carbon dioxide. J Am Chem Soc 123:9000–9006
100. Xie X, Liotta CL, Eckert CA (2004) CO_2-protected amine formation from nitrile and imine hydrogenation in gas-expanded liquids. Ind Eng Chem Res 43:7907–7911
101. Fischer H, Gyllenhaal O, Vessmann J, Albert K (2003) Reaction monitoring of aliphatic amines in supercritical carbon dioxide by proton nuclear magnetic resonance spectroscopy and implications for supercritical fluid chromatography. Anal Chem 75:622–626
102. Dijkstra ZJ, Doornbos AR, Weyten H, Ernsting JM, Elsevier CJ, Keurentjes JTF (2007) Formation of carbamic acid in organic solvents and in supercritical carbon dioxide. J Supercrit Fluids 41:109–114
103. Kayaki Y, Suzuki T, Ikariya T (2008) Utilization of N, N-dialkylcarbamic acid derived from secondary amines and supercritical carbon dioxide: stereoselective synthesis of Z alkenyl carbamates with a CO_2-soluble ruthenium–$P(OC_2H_5)_3$ catalyst. Chem Asian J 3:1865–1870
104. Aresta M, Ballivet-Tkatchenko D, Bonnet MC, Faure R, Loiseleur H (1985) Synthesis and structural characterization of $Co(NO)_2[PhP(OCH_2CH_2)_2NH]Cl$: a novel carbon dixide carrier. J Am Chem Soc 107:2994–2995
105. Aresta M, Ballivet-Tkatchenko D, Belli Dell'Amico D, Bonnet MC, Boschi D, Calderazzo F, Faure R, Labella L, Marchetti F (1985) Isolation and structural determination of two derivatives of the elusive carbamic acid. Chem Commun 1099–1100
106. Jamróz MH, Dobrowolski JC (2002) Theoretical IR spectra and stability of carbamic acid complexes. Vib Spectrosc 29:217–221
107. Jamróz MH, Dobrowolski JC, Rode JE, Borowiak MA (2002) Comparison of calculated structural parameters and infrared spectra with experimental data for dimeric dibenzyl carbamic acid. J Mol Struct (Theochem) 618:101–108
108. Aresta M, Quaranta E (1992) Role of the macrocyclic polyether in the synthesis of *N*-alkylcarbamate esters from primary amines, CO_2 and alkyl halides in the presence of crown-ethers. Tetrahedron 48:1515–1530
109. Darensbourg DJ, Frost BJ, Larkins DL (2001) An experimental and theoretical investigation of the carbon dioxide insertion process into the tungsten-nitrogen bond of an anionic W (0) complex. Inorg Chem 40:1993–1999

110. Park S, Rheingold AL, Roundhill DM (1991) Synthesis and reaction chemistry of monomeric and dimeric amide complexes of platinum(II). Organometallics 10:615–623
111. Hartwig JF, Bergman RG, Andersen RA (1991) Insertion reactions of CO and CO_2 with ruthenium benzyl, arylamido, and aryloxide complexes: a comparison of the reactivity of ruthenium-carbon, ruthenium-nitrogen, and ruthenium-oxygen bonds. J Am Chem Soc 113: 6499–6508
112. Cowan RL, Trogler WC (1989) Syntheses, reactions and molecular structures of *trans*-hydrido(phenylamido)bis(triethylphosphine)platinum(II) and *trans*-hydridophenoxobis (triethylphosphine)platinum(II). J Am Chem Soc 111:4750–4761
113. Legzdins P, Rettig SJ, Ross KJ (1994) Competitive reactivity of W-C, W-N, and W-O bonds at the Cp*W(NO) fragment: insertion reactions of tert-butyl isocyanide, *p*-tolyl isocyanate, and carbon disulfide. Organometallics 13:569–577
114. Andersen RA (1979) Dialkylbis[bis(trimethylsilyl)amido]zirconium(IV) and -hafnium(IV). Preparation and reaction with carbon dioxide and *tert*-butylisocyanide. Inorg Chem 18: 2928–2932
115. Kloppenburg L, Petersen JL (1996) Facile conversion of an appended silylamido to a silyloxy ligand *via* isocyanate elimination. Synthesis of {[(C_5Me_4)$SiMe_2O$]Zr(η^2-O_2CMe) (μ-O_2CMe)}$_2$ *via* the carboxylation of [(C_5Me_4)$SiMe_2$(N-*t*-Bu)]$ZrMe_2$. Organometallics 15:7–9
116. Nieves JS, Royo P (2001) Insertion of carbon dioxide and isocyanide into tantalum-amide and tantalum-methyl bonds. J Organomet Chem 621:299–303
117. Ward BD, Orde G, Clot E, Cowley AR, Gade LH, Mountford P (2005) Reactions of neutral and cationic diamide-supported imido complexes with CO_2 and other heterocumulenes: issues of site selectivity. Organometallics 24:2368–2385
118. Dubberley SR, Friedrich A, Willman DA, Mountford P, Radius U (2003) Synthesis and reactivity of calix[4]arene-supported group 4 imido complexes. Chem Eur J 9:3634–3654
119. Guiducci AE, Boyd CL, Clot E, Mountford P (2009) Reactions of cyclopentadienyl-amidinate titanium imido compounds with CO_2: cycloaddition-extrusion vs. cycloaddition-insertion. Dalton Trans 5960–5979
120. Stewart CA, Dickie DA, Tang Y, Kemp RA (2011) Insertion reactions of CO_2, OCS and CS_2 into the Sn-N bond of $(Me_2N)_2Sn$: NMR and X-ray structural characterization of the products. Inorg Chim Acta 376:73–79
121. Inoue S, Yokoo Y (1972) Reaction of organoaluminum coordination compounds with carbon dioxide. Bull Chem Soc Jpn 45:3651–3653
122. Chang C-C, Srinivas B, Mung-Liang W, Wen-Ho C, Chiang MY, Chung-Sheng H (1995) Fixation of CO_2 by a series of ethynyl-bridged polynuclear aluminum-magnesium complexes. Synthesis, characterization, and crystal structures of [Me_2Al(μ-i-Pr_2N)$_2$ Mg (μ-C≡CR)]$_2$ (R = C_6H_5, C_6H_4-p-CH_3, t-Bu, $SiMe_3$), [Me_2Al(μ-Et_2N)$_2$ Mg(μ-C≡CC_6H_5)]$_2$, {(Me_2Al)$_2$[μ-OOC(i-Pr_2N)]$_2$}, and {(Me_2Al)$_2$[(μ-OOC(i-Pr_2N))$_2$]$_2$ Mg}. Organometallics 14:5150–5159
123. Chang CC, Ameerunisha MS (1999) Chemistry of the organodiamides of magnesium, aluminum and mixed Mg-Al systems. A review of the heterocumulene reactivity on the Mg-Al centers. Coord Chem Rev 189:199–278
124. Ruben M, Walther D, Knake R, Gorls H, Beckert R (2000) Fixation of carbon dioxide by oxalic amidinato magnesium complexes: structures and reactions of trimetallic magnesium carbamato and related complexes. Eur J Inorg Chem 1055–1064
125. Tang Y, Zakharov LN, Rheingold AL, Kemp RA (2004) Insertion of carbon dioxide into Mg-N bonds. Structural characterization of a previously unknown η^2 chelation mode to magnesium in magnesium carbamates. Organometallics 23:4788–479
126. Davies RP, Raithby PR, Snaith R (1996) A likely intermediate during the CO_2-induced activation of 2-alkylindoles toward electrophilic substitution: structure of a unique tetramer formed by joining two boat-shaped $(LiOCO)_2$ rings. Organometallics 15:4355–4356

127. Kennedy AR, Mulvey RF, Oliver DF, Robertson SD (2010) Lithium and aluminum carbamato derivatives of the utility amide 2,2,6,6,-tetramethylpiperidide. Dalton Trans 39: 6190–6197
128. McCowan CS, Groy TL, Caudle MT (2002) Synthesis, structure, and preparative transamination of tetrazinc carbamato complexes having the basic zinc carboxylate structure. Inorg Chem 41:1120–1127
129. Tang Y, Kassel WS, Zakharov LN, Rheingold AL, Kemp RA (2005) Insertion of carbon dioxide into Zn-N bonds. Syntheses and structure of tetrameric and dimeric alkylzinc carbamato complexes. Inorg Chem 44:359–364
130. Domide D, Kaifer E, Mautz J, Walter O, Behrens S, Himmel H-J (2008) Synthesis and characterisation of some new zinc carbamate complexes formed by CO_2 fixation and their use as precursors for ZnO particles under mild conditions. Eur J Inorg Chem 3177–3185
131. Nilsson Lill SO, Kohn U, Anders E (2004) Carbon dioxide fixation by lithium amides: DFT studies on the reaction mechanism of the formation of lithium carbamates. Eur J Org Chem 2868–2880
132. Himmel H-J (2007) On the mechanism of CO_2 insertion into the Mg-N bond of mono- and dinuclears magnesium compound: a quantum chemical study. Z Anorg Allg Chem 633: 2191–2198
133. Himmel H-J (2007) CO_2 fixation by alkyl zinc amides: a quantum chemical study motivated by recent experimental results. Eur J Inorg Chem 675–683
134. Lihs FJ, Caudle MT (2002) Kinetics and mechanism for CO_2 scrambling in a *N*-carboxyimidazolidone analogue for N^1-carboxybiotin. J Am Chem Soc 124:11334–11341
135. McCowan CS, Caudle MT (2005) Evidence for unimolecular CO_2 elimination in C-N metathesis reactions of carbamatozinc complexes $Zn_4O(O_2CAm)_6$ (Am = *N*-diethylamino, *N*-piperidyl, *N*-pyrrolidyl). Dalton Trans 238–246
136. Dureen MA, Stephan DW (2010) Reactions of boron amidinates with CO_2 and CO and other small molecules. J Am Chem Soc 132:13559–13569
137. Horley GA, Mahon MF, Molloy KC (2002) Synthesis and characterization of novel homoleptic N, N-dialkylcarbamato complexes of antimony: precursors for the deposition of antimony oxides. Inorg Chem 41:5052–5058
138. Aresta M, Quaranta E (1988) Reactivity of phosphacarbamates: transfer of the carbamate group promoted by metal assisted electrophilic attack at the carbon dioxide moiety. J Org Chem 53:4153–4154
139. Aresta M, Quaranta E (1992) Alkali-metal-assisted transfer of carbamate group from phosphocarbamates to alkyl halides: a new easy way to alkali-metal carbamates and to carbamate esters. J Chem Soc Dalton Trans 1893–1898
140. Yamazaki N, Tomioka T, Higashi F (1975) Trisubstituted ureas and thioureas from hexaalkylphosphorous triamides, primary amines, and carbon dioxide or carbon disulfide. Synthesis 384–385
141. Breederveld H (1962) The reaction of dialkylaminosilanes with carbon dioxide and with carbon disulphide. Rec Trav Chim 81:276–278
142. Cavell RG, Griend LV (1983) Formation and dynamic exchange processes in methyltris (trifluoromethyl)(dimethylcarbamato-O, O′)phosphorus(V). Inorg Chem 22:2066–2070
143. Aresta M, Ballivet-Tkatchenko D, Bonnet M (1985) The role of a bifunctional P, N ligand in CO_2 coordination. Nouv J Chim 9:321–323
144. Walther D, Gebbardt P, Fischer R, Kreher U, Gorls H (1998) Complexes of the bis (trimethylsilyl)-benzamidinato ligand '*siam*': synthesis and X-ray structures of $(siam)_2M$, (*siam*)(*siam*H)MX (M = Ni, Pd), $(siam)_2MnI$ and $(siam)ReO_3$, and their reactivity towards CO_2. Inorg Chim Acta 281:181–189
145. Sita LR, Babcock JR, Xi R (1996) Facile metathetical exchange between carbon dioxide and the divalent Group 14 bisamides $M[N(SiMe_3)_2]_2$ (M = Ge and Sn). J Am Chem Soc 118: 10912–10913

146. Habereder T, Noth H, Paine RT (2007) Synthesis and reactivity of new bis(tetramethylpiperidino)(phosphanyl)-alumanes. Eur J Inorg Chem 4298–4305
147. Bennet MA, Castro J, Edwards AJ, Kopp MR, Wenger E, Willis AC (2001) Preparation and reactivity of nickel(0) complexes with η^2-coordinated alkynylphosphines. Organometallics 20:980–989
148. Buhro WE, Chisholm MH, Martin JD, Huffmann JC, Folting K, Streib WE (1989) Reactions involving carbon dioxide and mixed amido-phosphido dinuclear compounds: $M_2(NMe_2)_4(PR_2)(M{\equiv}M)$, where $M = Mo$ and W. Comparative study of the insertion of carbon dioxide into metal-nitrogen and metal-phosphorus bonds. J Am Chem Soc 111: 8149–8156
149. Buhro WE, Chisholm MH, Folting K, Huffmann JC (1989) Phosphinecarboxylate ligands formed by insertion of carbon dioxide into metal-phosphido bonds. Preparation and structural characterization of tetrakis(di-*tert*-butylphosphinecarboxylato)dimolybdenum. Inorg Chem 26:3087–3088
150. Murakami M, Ishida N, Miura T (2006) Solvent and ligand partition reaction pathways in nickel-mediated carboxylation of methylenecyclopropanes. Chem Commun 643–645
151. Behr A, Thelen G (1984) Carbon-carbon coupling of carbon dioxide, cyclic hydrocarbon and nickel(0). C_1 Mol Chem 1:137–153
152. Inoue Y, Hibi T, Satake M, Hashimoto H (1979) Reaction of methylenecyclopropanes with carbon dioxide catalysed by palladium(0) complexes. Synthesis of five-membered lactones. J Chem Soc Chem Commun 982–982
153. Binger P, Weintz HJ (1984) Reaktionen der methylencyclopropane, VIII1). Palladium(0)-katalysierte darstellung ungesättigter γ-lactone aus methylencyclopropan und kohlendioxid. Chem Ber 117:654–665
154. Aresta M, Quaranta E, Ciccarese A (1985) Rh(I) promoted activation and fixation of carbon dioxide. C_1 Mol Chem 2:267–281
155. Dou XY, He LN, Yang ZZ, Wang JL (2010) Catalyst-free process for the synthesis of 5-aryl-2-oxazolidinones *via* cycloaddition reaction of aziridines and carbon dioxide. Synlett 14:2159–2163
156. Phung C, Ulrich RM, Ibrahim M, Tighe NTG, Lieberman DL, Pinhas AR (2011) The solvent-free and catalyst-free conversion of an aziridine to an oxazolidinone using only carbon dioxide. Green Chem 13:3224–3229
157. Hancock MT, Pinhas AL (2003) A convenient and inexpensive conversion of an aziridine to an oxazolidinone. Tetrahedron Lett 44:5457–5460
158. Du Y, Wu Y, Liu AH, He LN (2008) Quaternary ammonium bromide functionalized polyethylene glycol: a high efficient and recyclable catalyst for selective synthesis of 5-aryl-2-oxazolidinones from carbon dioxide and aziridines under solvent-free conditions. J Org Chem 73:4709–4712
159. Yang ZZ, He LN, Peng SY, Liu AH (2010) Lewis basic ionic liquids-catalyzed synthesis of 5-aryl-2-oxazolidinones from aziridines and CO_2 under solvent-free conditions. Green Chem 12:1850–1854
160. Mu WH, Chasse GQ, Fang DC (2008) High level *ab initio* exploration on the conversion of carbon dioxide into oxazolidinones: the mechanism and regioselectivity. J Chem Phys A 112: 6708–6714
161. Sudo A, Morioka Y, Sanda F, Endo T (2004) *N*-Tosyl aziridine, a new substrate for chemical fixation of carbon dioxide *via* ring expansion reaction under atmospheric pressure. Tetrahedron Lett 45:1363–1365
162. Wu Y, Liu G (2011) Organocatalyzed cycloaddition of carbon dioxide to aziridines. Tetrahedron Lett 52:6450–6452
163. Phung C, Pinhas AR (2010) The high yield and regioselective conversion of an unactivated aziridine to an oxazolidinone using carbon dioxide with ammonium iodide as the catalyst. Tetrahedron Lett 51:4552–4554

164. Miller AW, Nguyen SB (2004) (Salen)chromium (III)/DMAP: an efficient catalyst system for the selective synthesis of 5-substituted oxazolidinones from carbon dioxide and aziridines. Org Lett 6:2301–2304
165. Wu Y, He LN, Du Y, Wang JQ, Miao CX, Li W (2009) Zirconyl chloride: an efficient recyclable catalyst for synthesis of 5-aryl-2-oxazolidinones from aziridines and CO_2 under solvent-free conditions. Tetrahedron 65:6204–6210
166. Watile RA, Bagal DB, Patil YP, Bhanage BM (2011) Regioselective synthesis of 5-aryl-2-oxazolidinones from carbon dioxide and aziridines using $Br^-Ph_3{}^+PPEG_{600}P^+Ph_3Br^-$ as an efficient, homogenous recyclable catalyst at ambient conditions. Tetrahedron Lett 52: 6383–6387
167. Watile RA, Bagal DB, Deshmukh KM, Dhake KP, Bhanage BM (2011) Polymer supported diol functionalized ionic liquids: an efficient, heterogeneous and recyclable catalyst for 5-aryl-2-oxazolidinones synthesis from CO_2 and aziridines under mild and solvent free conditions. J Mol Catal A 351:196–203
168. Jiang HF, Ye JW, Qi CR, Huang LB (2010) Naturally occurring α-amino acid: a simple and inexpensive catalyst for the selective synthesis of 5-aryl-2-oxazolidinones from CO_2 and aziridines under solvent-free conditions. Tetrahedron Lett 51:928–932
169. Qi C, Ye J, Zeng W, Jiang H (2010) Polystyrene-supported amino acids as efficient catalyst for chemical fixation of carbon dioxide. Adv Synth Catal 352:1925–1933
170. Soga K, Hosoda S, Nakamura H, Ikeda S (1976) A new synthetic route to 2-oxazolidones. J Chem Soc Chem Commun 617–617
171. Kawanami H, Ikushima Y (2002) Regioselectivity and selective enhancement of carbon dioxide fixation of 2-substituted aziridines to 2-oxazolidinones under supercritical conditions. Tetrahedron Lett 43:3841–3844
172. Zhou H, Wang YM, Zhang WZ, Qu JP, Lu XB (2011) *N*-Heterocyclic carbene functionalized MCM-41 as an efficient catalyst for chemical fixation of carbon dioxide. Green Chem 13: 644–650
173. Tascedda P, Duñach R (2000) Electrosynthesis of cyclic carbamates from aziridines and carbon dioxide. Chem Commun 449–450
174. Fontana F, Chen CC, Aggarwal V (2011) Palladium-catalyzed insertion of CO_2 into vinylaziridines: new route to 5-vinyloxazolidinones. Org Lett 13:3454–3457
175. Ihata O, Kayaki Y, Ikariya T (2005) Aliphatic poly(urethane-amine)s synthesized by copolymerization of aziridines and supercritical carbon dioxide. Macromolecules 38:6429–6434
176. Pinkes JR, Steffey BD, Vites JC, Cutler AR (1994) Carbon dioxide insertion into the iron-zirconium and ruthenium-zirconium bonds of the heterobimetallic complexes $Cp(CO)_2MZr(Cl)Cp_2$: direct production of the μ-$\eta^1(C)$:$\eta^2(O, O')$-CO_2 compounds $Cp(CO)_2MCO_2Zr(Cl)Cp_2$. Organometallics 13(1):21–23
177. Riduan SN, Zhang Y (2010) Recent developments in carbon dioxide utilization under mild conditions. Dalton Trans 39:3347–3357
178. Ying JY, Nurhanna S, Riduan B (2011) Carbon dioxide reduction. European Patent 2340119 A1
179. Song JJ, Gallou F, Reeves JT, Tan Z, Yee NK, Senanayake CH (2006) Activation of TMSCN by N-heterocyclic carbenes for facile cyanosilylation of carbonyl compounds. J Org Chem 71:1273–1276
180. Suzuki Y, Bakar A, Muramatsu K, Sato M (2006) Cyanosilylation of aldehydes catalyzed by *N*-heterocyclic carbenes. Tetrahedron 62:4227–4231
181. Kolbe H (1860) Ueber Synthese der Salicylsäure. Just Lieb Ann Chem 113:125–127
182. Schmitt R, Burkard E (1877) Ueber Naphtolcarbonsäuren. Chem Ber 20:2699–1702
183. Dinjus E, Leitner W (1995) New insights into the palladium-catalysed synthesis of δ-lactones from 1,3-dienes and carbon dioxide. Appl Organomet Chem 9(1):43–50
184. Aresta M, Tommasi I, Dileo C, Dibenedetto A, Narracci M, Ziolkowski J, Jezierski A (2001) Synthesis and spectroscopic (^{1}H NMR, ESR) characterization of new aryloxy-Mn(II) complexes: steric control over *O*- vs. phenyl-π-coordination of ArO^-ligands (ArO^-=$C_6H_5O^-$,

4-methyl-$C_6H_4O^-$, 3,5-dimethyl-$C_6H_3O^-$, 2,6-di-*tert*-butyl-$C_6H_3O^-$, 2,6-dimethyl-$C_6H_3O^-$) to the "Mn(II)Cp" moiety, and their reactivity with carbon dioxide. Can J Chem 79(5–6): 570–577
185. Bottaccio G, Chiusoli GP, Alneri E, Marchi M, Lana G (1977) Process for the carboxylation of organic substrates with carbon dioxide. US Patent 4032555A
186. Aresta M, Tkatchenko I, Tommasi I (2003) In: Rogers RD, Seddon KR (eds) Ionic liquids as green solvents: progress and prospects, vol 856, ACS Symposium Series. ACS, Washington DC, p 93
187. Barran T, Dibenedetto A, Aresta M, Kruczała K, Macyk W (2014) Photocatalytic carboxylation of organic substrates with carbon dioxide at zinc sulfide with deposited ruthenium nanoparticles. ChemPlusChem 79:708–715

Chapter 5
Interaction of CO_2 with C–C Multiple Bonds

Reactions with Olefins, Cumulated and Conjugated Dienes, and Alkynes

Abstract In the presence of low-valent transition metal centers, CO_2 can react with unsaturated organic substrates, such as alkenes, alkynes, and conjugated and cumulated dienes, to give carboxylated products. This chapter focuses on the main mechanistic features of these carboxylation processes. A key step of these transformations is the metal-promoted oxidative coupling reaction between CO_2 and the unsaturated substrate. The mechanistic details of this step are highlighted and the relevance and role of CO_2 coordination to metal center in these reactions is argued.

5.1 Introduction

Under suitable conditions, CO_2 can interact with C–C multiple bonds of alkenes, conjugated dienes, cumulenes, and alkynes [1–9]. The interaction results in the coupling (see Sect. 5.2) of the organic substrate with the heterocumulene. These reactions are key steps for the functionalization of the above substrates, as they provide a powerful synthetic tool for the formation of a new C–C bond, and they open a further route, in addition to the insertion reactions (see also Chap. 4), to the incorporation of the heterocumulene into a variety of new carboxylated products (acids, esters, lactones, pyrones, etc.) often characterized by complex molecular structures and widely used in several fields (pharmaceuticals, cosmetics, etc.). The coupling reaction is promoted by a low-valent transition metal center, which undergoes oxidation. The carboxylation process may either occur stoichiometrically or proceed catalytically, depending on the nature of the organic substrate, the metal center, the ancillary ligands, and, more generally, the reaction conditions.

This chapter discusses the basic mechanistic aspects of these carboxylation processes and brings into focus the mechanistic details of the oxidative coupling reaction between CO_2 and the unsaturated substrate (olefin, conjugated diene, cumulene, alkyne) with the purpose of highlighting, whenever possible, the relevance and role of CO_2 coordination to the metal center in these transformations.

M. Aresta et al., *Reaction Mechanisms in Carbon Dioxide Conversion*,
DOI 10.1007/978-3-662-46831-9_5

5.2 Oxidative Coupling with CO_2

In the presence of a suitable transition metal center, CO_2 can react with a variety of unsaturated substrates, A=B, to form five-membered metallaheterocycles, as shown in Scheme 5.1a. This process implies the coupling of the CO_2 molecule with an unsaturated substrate A=B and the formation of three new σ-bonds, M–A, B–C, and O–M. The two electrons that are required to form these bonds are provided by the metal center, whose formal oxidation state increases by two units (5.1). Remarkably, in the known examples, the oxygen of CO_2 binds to the

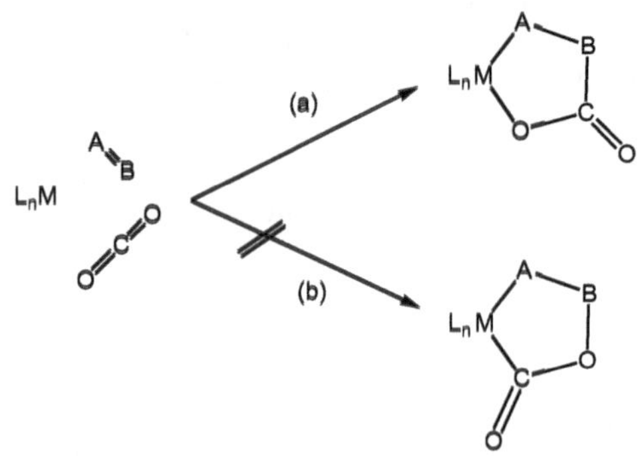

Scheme 5.1 Oxidative coupling with CO_2

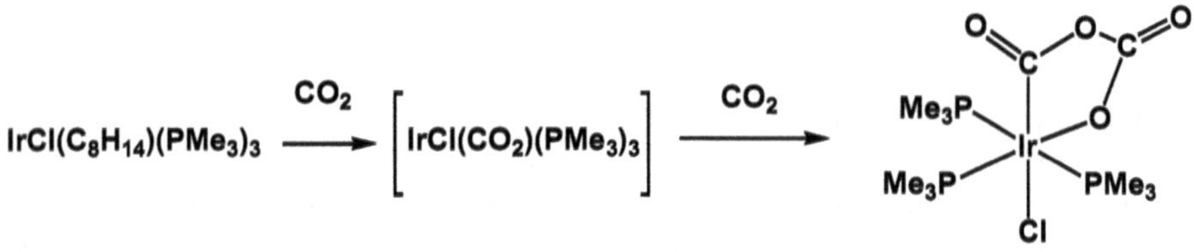

Scheme 5.2 Head-to-tail homocoupling of CO_2 promoted by $IrCl(C_8H_{14})(PMe_3)_3$ (adapted from [17])

metal, not the carbon (Scheme 5.1b). The coupling reaction is often promoted by electron-rich Ni(0) complexes,

$$M^{n+} \rightarrow M^{(n+2)+} + 2e^- \tag{5.1}$$

but examples involving low-valent complexes of other transition metals (Ti, Zr, Ta, Mo, Fe, Mn, Pd, Rh, Pt, etc.) are also known [1–10]. The nature of A=B may range among a large variety of unsaturated substrates. In most cases, however, A=B is an unsaturated hydrocarbon (olefin, diene, cumulene, alkyne, η^2-aryne) and the coupling reaction (B is a carbon atom) leads to the formation of a new C–C bond

[1–10]. These reactions are discussed in detail later. Relatively fewer studies have dealt with the oxidative coupling of CO_2 with nitroso- [11, 12] or diaza-compounds [10] and hetero-olefins, such as imines [1–10, 13], carbodiimides [14], azadienes [1–9, 15], or aldehydes [1–9, 16]. The latter processes lead to the formation of a new C–N (A=B: PhNO, R–N=N–R, RCH=NR, RN=C=NR, RCH=N–CH=CHR) or C–O bond (A=B: RCH=O), depending on the nature of the substrate A=B.

The role of the unsaturated substrate A=B can be played even by CO_2 itself [17]. The "head-to-tail" homocoupling of two CO_2 molecules at the Ir(I) center of $IrCl(C_8H_{14})(PMe_3)_3$ was described by Herskovitz [17] in 1976 and gives the metallacyclic bis-CO_2 adduct $IrCl(C_2O_4)(PMe_3)_3$ (Scheme 5.2).

Mechanistically, the formation of metallacycle (Scheme 5.1a) might result from a "three-center" concerted reaction involving the metal, the substrate, and CO_2, but it is unlikely that the three reactants in Scheme 5.1 might react in one step. More likely, the oxidative coupling reaction may proceed through a stepwise pathway, involving either the coordination of CO_2 to the metal center, followed by the reaction of the CO_2 adduct with the substrate (pathway (i); 5.2a, b),

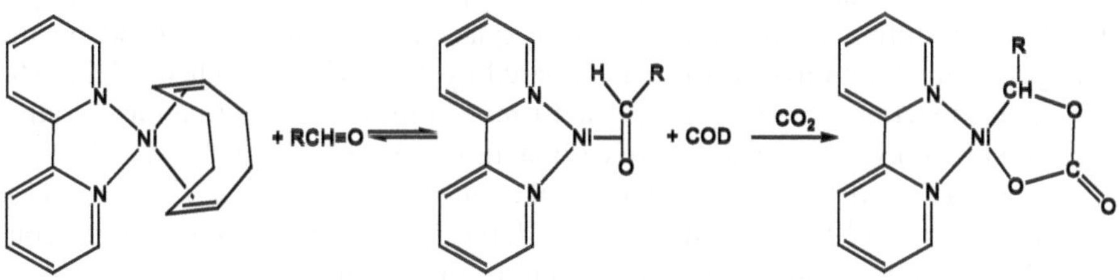

Scheme 5.3 Reaction of Ni(bpy)(cod) with propionaldehyde and CO_2 (adapted from [16])

or the interaction of CO_2 with the coordinated substrate A=B (pathway (ii); 5.3a, b).

$$L_nM + CO_2 \rightarrow L_nM(CO_2) \tag{5.2a}$$

$$L_nM(CO_2) + A{=}B \longrightarrow L_n\overline{M(A\text{-}B\text{-}C(O)O})\tag{5.2b}$$

$$L_nM + A = B \rightarrow L_nM(A = B) \tag{5.3a}$$

$$L_nM(A{=}B) + CO_2 \longrightarrow L_n\overline{M(A\text{-}B\text{-}C(O)O})\tag{5.3b}$$

Whatever pathway, (i) or (ii), may be followed, the role played by possible intermediates such as $L_nM(A{=}B)(CO_2)$ on the path leading to the final product, $L_n\overline{M(A\text{-}B\text{-}C(O)O})$,, should be defined too. These points are relevant to the general question whether or not the coordination of CO_2 to the metal center is a necessary prerequisite for the functionalization of the substrate A=B, and thus to the role of transition metal complexes in CO_2 fixation. The elucidation of these mechanistic details by means of experimental methods is not an easy task, however, and reliable answers have often been obtained only by modeling these processes by means of quantomechanical methods.

Pathway (ii) has been documented experimentally in a number of cases. For instance, the metallacyclic complex $Rh[ON(Ph)C(O)O](4\text{-}MeC_6H_4)[t\text{-}BuP(CH_2CH_2CH_2PPh_2)_2]$ was prepared by pressuring (3.5 MPa) a toluene solution of PhNO and $Rh(4\text{-}MeC_6H_4)[t\text{-}BuP(CH_2CH_2CH_2PPh_2)_2]$ with CO_2 [11]. The nitrosobenzene adduct $Rh(4\text{-}MeC_6H_4)(PhNO)[t\text{-}BuP(CH_2CH_2CH_2PPh_2)_2]$ is a possible intermediate of the rhodium-assisted CO_2/PhNO coupling reaction. In fact, this species, which can be isolated from mixtures of the tolyl–rhodium(I) complex and nitrosobenzene, interacted with carbon dioxide to give the carboxylated metallacycle complex. An analogous reaction was also described by Cenini, who observed the reversible oxidative coupling reaction of carbon dioxide with nitrosobenzene η^2-coordinated at platinum(0) [12].

The reaction of the nickel(0) complex Ni(bpy)(cod) (bpy = 2,2′-bipyridine; cod = 1,4-cyclooctadiene) with propionaldehyde and CO_2 has been the subject of a detailed kinetic study [16]. It has been shown that carbon dioxide did not react with Ni(bpy)(cod) directly but with the nickel propionaldehyde complex Ni(bpy)(RCH=O) resulting from the reversible displacement of the cod ligand by the aldehyde, with formation of a five-membered cyclic complex (Scheme 5.3). The cyclization reaction with CO_2 proceeded irreversibly according to an associative mechanism, implying the insertion of carbon dioxide into the Ni–O bond of Ni(bpy)(RCH=O). The activation parameters for the latter process were $\Delta H^{\ddagger} = 43 \pm 6$ kJ/mol and $\Delta S^{\ddagger} = -129 \pm 19$ J/(mol K). The large negative activation entropy confirms the oxidative coupling step, with bond formation and charge

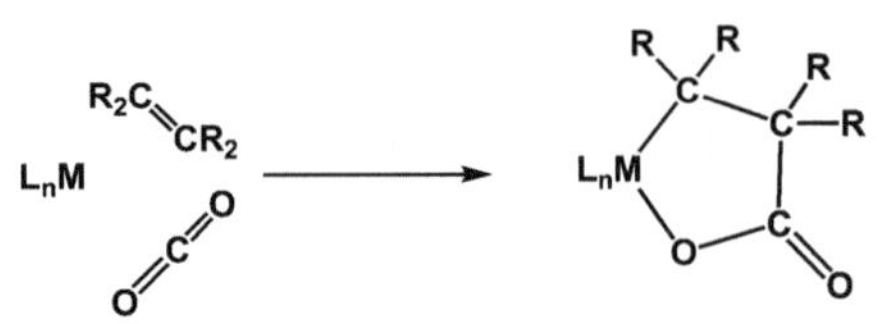

Scheme 5.4 Formation of metallacycles from olefins and carbon dioxide

creation in the transition state. It is assumed that the carbon dioxide molecule approaches the Ni(bpy)(RCH=O) complex and a four-center transition state is formed as shown in Structure 5.1.

Structure 5.1 Ni-promoted oxidative coupling of RCH=O with CO_2: proposed transition state

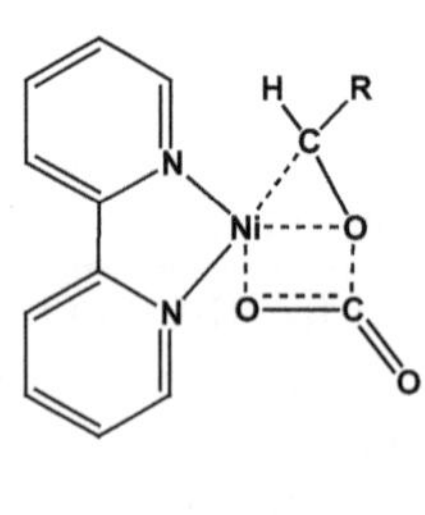

The formation of $IrCl(C_2O_4)(PMe_3)_3$ from $IrCl(C_8H_{14})(PMe_3)_3$ and CO_2 (Scheme 5.2) necessarily implies the intermediacy of a CO_2-Ir adduct. The mechanistic details of this reaction are not known. The authors assume that $IrCl(C_2O_4)(PMe_3)_3$ is formed from $IrCl(C_8H_{14})(PMe_3)_3$ via stepwise uptake of two CO_2 molecules with concomitant displacement of cyclooctene. The reaction is believed to be initiated by addition of the nucleophilic $IrCl(C_8H_{14})(PMe_3)_3$ to a CO_2 molecule, with one of the two (more basic) oxygens binding a second CO_2 molecule through C. Cyclization via Ir–O bond formation affords the product.

5.3 Carboxylation of Olefins

Stoichiometric carboxylation of olefins has been achieved by using a variety of low valent transition metal (Ti, Zr, V, Mo, W, Rh, Ni, Fe) complexes [1–10, 18–26]. Metallacyclocarboxylates have been generated and/or isolated by oxidative coupling of alkenes with CO_2 in the presence of Ni(0) [1–10], or Ti [18–20], Zr [1–9, 21], V [22], or Rh [23] reagents (Scheme 5.4). In the presence of Ni(0) the reaction may proceed reversibly depending on the nature of the ancillary ligands at the metal center: only N- or P-chelating ligands, such as TMEDA (*N,N,N′,N′*-tetramethyl-1,2-ethylenediamine), bpy, dcpe (1,2-bis(dicyclohexylphosphino) ethane), and dppe (1,2-bis(diphenylphosphino)ethane), were consistent with a reversible path, whereas P–N chelating ligands, such as $(alkyl)_2P–CH_2CH_2–N(alkyl)_2$, promoted the carboxylation reaction irreversibly. In agreement with the reversibility of the reaction, it was possible to replace, by other alkenes or alkynes, the olefin which has undergone coupling to form a five-membered ring metallacycle [1–10]. The formation of the nickelaheterocycles proceeds under kinetic control and the yield of the thermodynamically more stable final product depends on temperature and reaction time. Complexes with a primary carbon bound to nickel were thermodynamically more stable than those with a secondary carbon–nickel bond.

A few $L_2\overline{Ni(CH_2CH_2C(O)O})$ nickelacycles have also been structurally characterized (L = DBU (1,8-diazabicyclo[5.4.0]undec-7-ene); L_2 = bpy, $Ph_2P(O)CH_2PPh_2$) [24, 25]. Complex $(DBU)_2\overline{Ni(CH_2CH_2C(O})O)$, prepared from $Ni(cod)_2$ and DBU in the presence of CO_2/C_2H_4, underwent further ethene insertion to give a seven-membered ring complex, which, by subsequent β-H elimination and acid hydrolysis, gave pentenoic acids [25]. The Ni–C bond in $(DBU)_2\overline{Ni(CH_2CH_2C(O)O})$ was also available for insertion reactions with functionalized olefins and dienes, allowing the synthesis, after acid hydrolysis, of various acid products.

Coupling between ethene and CO_2 can also be induced by Fe(0) complexes [26]. Remarkably, these systems can promote ethylene coupling with two CO_2 molecules, affording a dicarboxylate which was esterified by means of acid hydrolysis in methanol. The nature of the ancillary ligands had a marked influence on the product distribution. In the presence of PMe_3, methyl malonate was obtained exclusively, whereas methyl succinate was formed preferentially in the presence of dcpe.

Despite the huge synthetic work, the mechanism of the coupling reaction of CO_2 with olefins (Scheme 5.4) in the presence of Ni(0) complexes has not been matter of extensive experimental studies. The hypothesis that coordinated CO_2 might react with olefin to afford a carboxylate metallacycle did not find sound support in the experiment. For instance, Aresta et al. found that $Ni(\eta^2\text{-}CO_2)(PCy_3)_2$, in the solid state, did not react with ethene even after 3 h at room temperature, whereas, in solution, a substitution reaction occurred easily with the formation of Ni $(PCy_3)_2(C_2H_4)$ [27–29]. The CO_2 substitution with ethene can be explained considering either (1) olefin attack to $Ni(\eta^2\text{-}CO_2)(PCy_3)_2$ in the state in which the Ni–C bond is loosened (see Scheme 2.3(iii)) with subsequent CO_2 elimination, or (2) the reaction of ethene with "$Ni(PCy_3)_2$" formed upon CO_2 dissociation. It is worth noting that, in toluene, under nitrogen, $Ni(\eta^2\text{-}CO_2)(PCy_3)_2$ is in equilibrium with the dinitrogen complex "$Ni(PCy_3)_2(N_2)$." The mechanism involving phosphine dissociation seems to be unlikely, as NMR spectroscopy did not reveal any free phosphine when $Ni(\eta^2\text{-}CO_2)(PCy_3)_2$ was dissolved in the presence of ethene. However, whatever the substitution reaction mechanism may be, the above findings clearly show that ethene is more tightly bound to "$Ni(PCy_3)_2$" than CO_2, in accordance with the results of a few early theoretical calculations carried out for the model fragment "$Ni(PH_3)_2$" [30]. This conclusion also agrees with the fact that, for instance, five-membered ring complexes $L_2Ni(CH_2CH_2C(O)O)$,, obtained from alkenes and CO_2, eliminate CO_2 upon heating to give the corresponding (ligand)Ni (0)–olefin complex, which can be isolated as crystals [10, 31].

More recently, the mechanistic details of nickel(0)-assisted oxidative coupling of CO_2 with C_2H_4 have been studied by Papai et al. by means of density functional calculations [32]. The basic question addressed in this theoretical study was whether the intermediacy of a carbon dioxide complex is a general requirement in CO_2–C_2H_4 coupling reactions. The calculations have been carried out for the model reaction (5.4), and have shown that this reaction is initiated by the addition

$$\text{Ni(cdt)} + \text{bpy} + CO_2 + C_2H_4 \longrightarrow \text{(bpy)}Ni(CH_2CH_2C(O)O) + \text{cdt} \quad (\text{cdt} = 1,5,9\text{-cyclododecatriene}) \tag{5.4}$$

of chelating bpy to Ni(cdt) to form a ternary $Ni(\eta^2\text{-cdt})(bpy)$ complex, which is followed by an olefin exchange that gives the $Ni(bpy)(C_2H_4)$ intermediate. The formation of the ethene complex $Ni(bpy)(C_2H_4)$ via the replacement of the η^2-coordinated cdt ligand was favored energetically by 8.4 kcal/mol. The ligand exchange for CO_2 to give complex $Ni(bpy)(CO_2)$ was far less likely, because the corresponding substitution reaction was predicted to be endothermic by 1.4 kcal/ mol. The

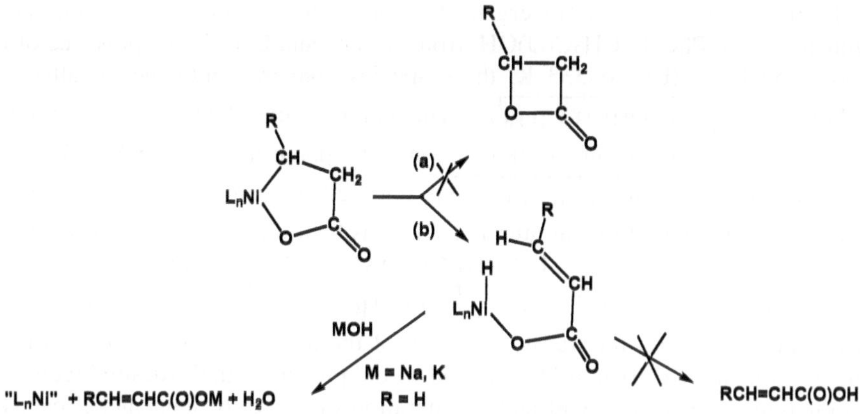

Scheme 5.5 Potential reaction pathways for $L_nNi(CH_2CH_2C(O)O)$

difference found for the stability of these complexes is consistent with the previously mentioned experimental observation that nickel–CO_2 complexes can easily undergo exchange reactions with olefins. The saturated 18e$^-$ Ni(bpy)(CO_2)(C_2H_4) complex, which was originally assumed to act as a precursor for the C–C bond formation [31], was also identified as a local minimum in the calculations. This complex, however, seems to be a fairly labile species, as it lies 7.1 kcal/mol higher in energy than "Ni(bpy)(C_2H_4) + CO_2," and the barrier towards this dissociation was calculated to be only 0.8 kcal/mol. It is therefore quite unlikely that complex Ni (bpy)(CO_2)(C_2H_4) may be an intermediate in the investigated reaction. The calculations showed that C–C bond formation in the present reaction occurred in a single step from the reaction of the Ni(bpy)(C_2H_4) intermediate with an incoming CO_2 molecule, supporting the fact that the formation of a "mixed" ethene-carbon dioxide complex is not a prerequisite for the CO_2–C_2H_4 coupling reaction. Nevertheless, the Ni(0) center has a prominent role in this reaction, because the C–C coupling is initiated by the metal-induced activation of CO_2 which allows the oxidative coupling with the coordinated ethene. Analogous conclusions were also achieved when dcpe or substituted bipyridines were selected as ancillary ligands in place of bpy. A thorough investigation of the role of the co-ligand showed that the two minima associated with the C–C bond formation and the related activation barrier depended heavily on the nature of the chelating ligands.

Scheme 5.5 shows two possible ways through which a nickelacycle carboxylate may, in principle, further react, once formed. Formation of β-lactone, which may form from $L_nNi(CH_2CH_2C(O)O)$ by reductive elimination (route a), has never been observed experimentally. On the other hand, the nickela-lactone may generate, through a β-H shift reaction, a nickel-hydrido-acrylate species, which may reductively eliminate the unsaturated acid RCH=CH_2C(O)OH (route b).

Catalytic synthesis of acrylic acids by direct coupling of olefins with CO_2 has attracted the interest of chemists since the early 1980s. Nevertheless, to date, this goal still remains elusive. Hoberg [33] reported the stoichiometric synthesis of cynnamic acid, $PhCH{=}CH_2C(O)OH$, from styrene and CO_2, in the presence of Ni $(cod)_2$ and DBU (L). At 333 K, these species converted into the metallacycle carboxylate $L_nNi(PhCHCH_2C(O)O)$.. The latter species liberated the saturated acid $PhCH_2CH_2C(O)OH$ upon acidification, but, upon heating to 358 K and further acid hydrolysis, $L_nNi(PhCHCH_2C(O)O)$ released cynnamic acid in high yield (>70 %). Formation of the unsaturated acid was believed to involve a Ni-hydrido-acrylate intermediate species, $L_nNi(H)(OC(O)CH_2{=}CHPh)$, which, at higher temperature (358 K), formed from $L_nNi(PhCHCH_2C(O)O)$ by β-H elimination. According to Hoberg, the transfer reaction of the hydride ligand from the nickel center to the imine carbon of DBU ligand would prevent both the desired reductive elimination of cynnamic acid and regeneration of the zerovalent nickel catalyst from taking place, thus inhibiting the synthesis of the unsaturated acid to proceed catalytically.

The synthesis of acrylic acid from ethene and CO_2 over a $Ni(cod)(DBU)_2$ catalyst has been modeled by using density functional theory (DFT) by Buntine and coworkers [34]. The overall catalytic reaction to form acrylic acid through CO_2 and ethene coupling can be described as having three main stages. Exothermic formation of the five-membered nickelacycle (−17.2 kJ/mol) drives the initial steps of the reaction and includes overcoming a barrier of +121.8 kJ/mol. This barrier corresponds to a transition structure characterized by the attack of CO_2 on nickel-coordinated ethene, in agreement with the previously discussed results of Papai. From the nickela-lactone, elongation of the Ni–O bond precedes β-H elimination through a transition structure located at +147.4 kJ/mol, resulting in a nickel-hydrido-acrylate species in which the acrylate moiety is bound through both oxygens (+21.3 kJ/mol). The final reductive elimination of acrylic acid might occur via a number of routes, the lowest energy of which takes place via a five-centered transition structure (+104.1 kJ/mol) involving the coupling of the hydride ligand with an unbound oxygen of the acrylate ligand. The calculations showed that the three main steps (oxidative coupling, β-H shift, reductive elimination) exhibited substantial energy barriers, but not so large that would be expected to prohibit the overall catalytic process. The catalyst deactivation reaction (H^- transfer to DBU), proposed by Hoberg to justify the lack of observed catalytic activity, is subject to a moderate barrier (+150.2 kJ/mol), but the proposed product lies +102.6 kJ/mol above the reactants, thereby ruling out this type of pathway as the cause of the non-catalytic activity. It is plausible that thermodynamic rather than kinetic constraints may prevent the overall reaction from taking place to a significant extent, in accordance with the solvation-corrected Gibbs free energy of the coupling reaction in question, which has been calculated to be unfavorable by +42.7 kJ/mol. In this regard, it is of interest to note that acrylic acid has been reported to react with Ni $(cod)_2$ (1 equiv.) in the presence of PCy_3 (1 equiv.) to give the nickela-lactone species $Cy_3PNi(CH_2CH_2C(O)O)$ [35].

The body of experimental evidence indicates that, in most of CO_2-alkene coupling reactions, the metalla-lactone structure represents a thermodynamic sink, as the metallacycle was found to be the final product. Only recently the conversion of nickela-lactones into acrylate species has been documented experimentally in an unambiguous way. This process can be induced either by ligand substitution with a suitable chelating agent and subsequent chelate activation [36] or by addition of strong electrophiles [37]. Walther et al. [36] have reported the conversion of nickelacycle carboxylates into nickel-hydrido-acrylates in reactions with 1,2-bis(diphenylphosphino)methane (dppm). These hydridoacrylates then evolved to give dimeric phosphido–Ni complexes bearing a bridging acrylate ligand. Moreover, Rieger succeeded in obtaining modest yields of methyl acrylate by methylation of nickela-lactones with MeI [37]. This result is quite intriguing as it emphasizes that alkyl esters of acrylic acids can be released more easily than the corresponding acids from nickela-lactones. A previous example of liberation of alkyl acrylates from the coordination sphere of a metal center was first documented by Aresta et al. [38] by reacting Pd complexes coordinating a preformed methoxycarbonyl moiety with ethene (Scheme 5.6). Remarkably, these Pd-complexes,

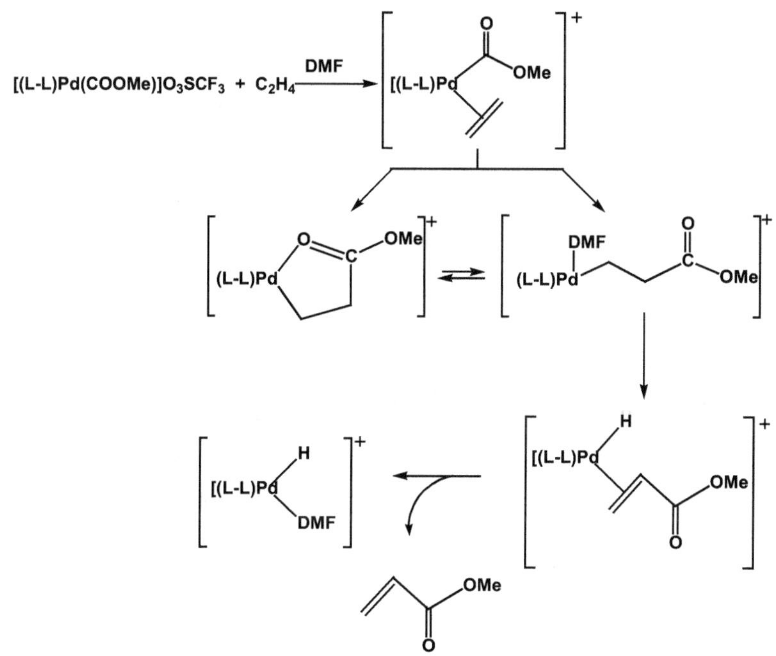

Scheme 5.6 Proposed mechanism for methyl acrylate formation from [(L–L)Pd(COOMe)] O_3SCF_3 (L–L=bpy, dppe, 2-(2-diphenylphosphino)ethylpyridine) and ethane (adapted from [38])

in the presence of CO_2 and ethene, also promoted the coupling of the heterocumulene with the olefin with facile release of ethyl acrylate.

The reductive elimination of acrylic acid from a hydrido–acrylate metal complex (see Scheme 5.5, right) is a kinetically (and thermodynamically) hindered process and never observed. It closely resembles the elimination of formic acid from hydrido formate metal complexes LnM(H)OCHO. In the latter case, the elimination process is in general facilitated by using an appropriate "shuttle" of the hydride (amines R_3N are used that are transformed into ammonium cations R_3NH^+) or of the formate anion (Group 1 metal cations are candidates, with K^+ as the most suited). The use of a $K^+(Na^+)$ shuttle has been investigated by Limback [39] who has reported a catalytic (also if limited to a few cycles) formation of $CH_2{=}CHCOOM$ (M = K, Na) by using Ni(0) catalysts in the presence of MOH (Scheme 5.5, left). This reaction has an industrial interest as acrylates are used as adsorbents.

Nevertheless, metal-hydrido-acrylate species can be more straightforwardly obtained from ethene and CO_2 by suitably changing the metal center. Carmona et al. found that a few Mo and W complexes possessed very similar chemistry and promoted the coupling of ethene and CO_2, affording dimeric (Mo, W) or monomeric (W) hydridoacrylate complexes which, unfortunately, were unable to eliminate acrylic acid (Structure 5.2) [40, 41]. For instance, in diethyl ether, the reactions

Structure 5.2 Dimeric and monomeric hydridoacrylate complexes

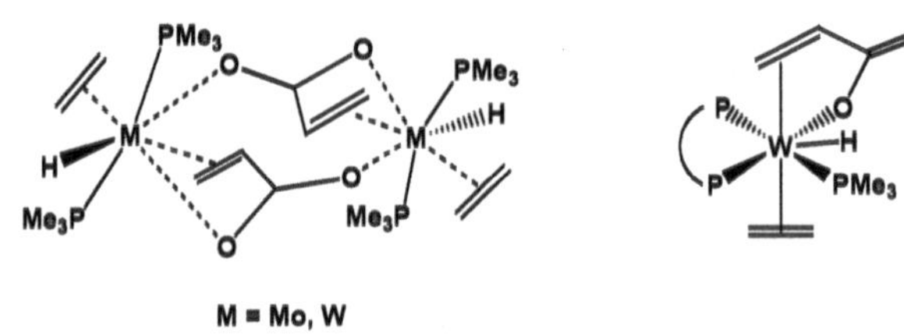

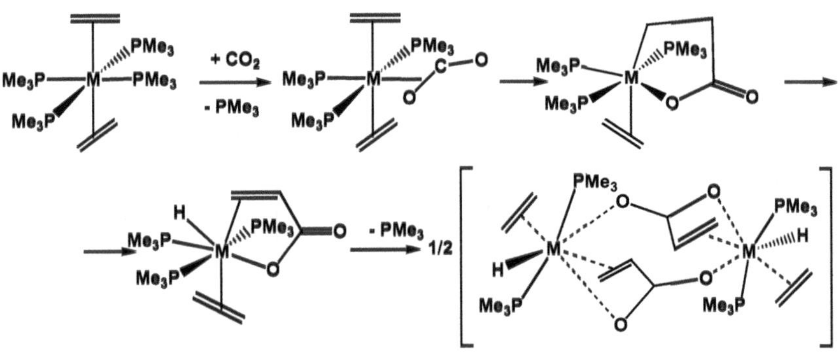

Scheme 5.7 Proposed mechanism for the formation of μ-$(CH_2{=}CHCOO)_2[HM(C_2H_4)(PMe_3)_2]_2$ (adapted from [42])

of the *trans*-$M(C_2H_4)_2(PMe_3)_4$ complexes (M = Mo, W) with CO_2, under mild conditions ($P_{CO_2} = 1$ atm ; room temperature or below), afforded μ-

$(CH_2{=}CHCOO)_2[HM(C_2H_4)(PMe_3)_2]_2$ binuclear hydrido–acrylate complexes (5.5).

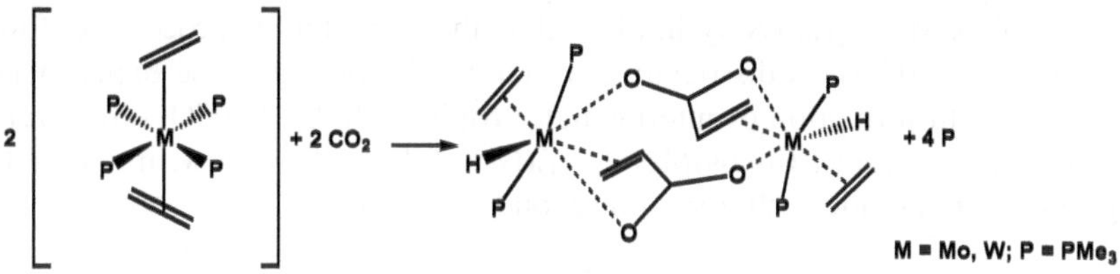

(5.5)

Scheme 5.7 summarizes the mechanism postulated by the authors for the genesis of the acrylate derivatives. The first step is the substitution of PMe_3 by CO_2 with formation of an intermediate undetected species, "$M(C_2H_4)_2(CO_2)(PMe_3)_3$," followed by oxidative coupling of C_2H_4 and CO_2 to afford a five-membered metalla-lactone, which could undergo β-H elimination to produce the hydrido-acrylate structure.

The consistency of the proposed mechanism has been supported by a few DFT calculations, which have been carried out for the specific case where M is molybdenum [42]. The substitution of one PMe_3 ligand by CO_2 takes place through a dissociative pathway. The substitution is slightly exothermic and leads to the complex $Mo(C_2H_4)_2(CO_2)(PMe_3)_3$, which is the key intermediate in the CO_2–C_2H_4 coupling process. The coupling takes place in a coplanar arrangement of the reacting ligands via a five-center transition state, resulting in a nearly planar metalla-lactone species. Differently from nickel (see above), coordination of CO_2 molecule to the metal center is a necessary step to achieve the CO_2–alkene coupling, because the formation of the new C–C bond is promoted by a simultaneous metal → CO_2 and metal → C_2H_4 electron donation mechanism. The formation of the metalla-lactone species is thermodynamically allowed, and the activation barrier has been estimated to be 11 kcal/mol. The five-membered metallacycle was found to be able to distort and adopt a structure involving β-agostic interactions. The agostic complex is also a crucial intermediate along the reaction path, because it facilitates the formation of the metal-hydride species through a β-hydrogen transfer process. The hydride formation is an exothermic step, with a related energy barrier (10 kcal/mol) comparable to that of the C–C coupling. The hydrido-acrylate monomer is already well below the energy of the reactants, but further stabilization is achieved by the dimerization process.

The mechanism of carbon dioxide functionalization with ethene at a molybdenum center has been investigated experimentally by Bernskoetter and Tyler [43], by reacting the zerovalent molybdenum complex, *trans*-(triphos)$Mo(C_2H_4)(N_2)_2$ (triphos = $(Ph_2PCH_2CH_2)_2PPh$), with CO_2. This reaction afforded the bridging acrylate hydride complex, $[(triphos)Mo\text{-}(H)(OC(O)CH{=}CH_2)]_2$ (5.6). Monitoring in situ the addition of CO_2 to a benzene-d_6 solution of the bis(dinitrogen)–molybdenum(0)–ethene complex by NMR spectroscopy revealed the formation of a long-lived reaction intermediate characterized as (triphos)$Mo(C_2H_4)(CO_2)$. The latter complex formed rapidly (≈15 min) from (triphos)$Mo(C_2H_4)(N_2)_2$ in the presence of excess CO_2 and converted slowly to the acrylate hydride species [(triphos)Mo(H)(OC(O)CH=CH_2)]$_2$ over several hours. The mechanism of carbon dioxide substitution with (triphos)Mo(C_2H_4)(N_2)$_2$ was explored. The coordinative saturation of complex (triphos)Mo(C_2H_4)(N_2)$_2$ implies that the substitution process to form (triphos)Mo (C_2H_4)(CO_2) is dissociative in nature. Accordingly, the experiments are consistent with formation of intermediate (triphos)Mo(C_2H_4)(CO_2) by reversible dinitrogen loss from (triphos)Mo(C_2H_4)(N_2)$_2$ followed by slow, irreversible, binding of carbon dioxide with loss of a second N_2 ligand.

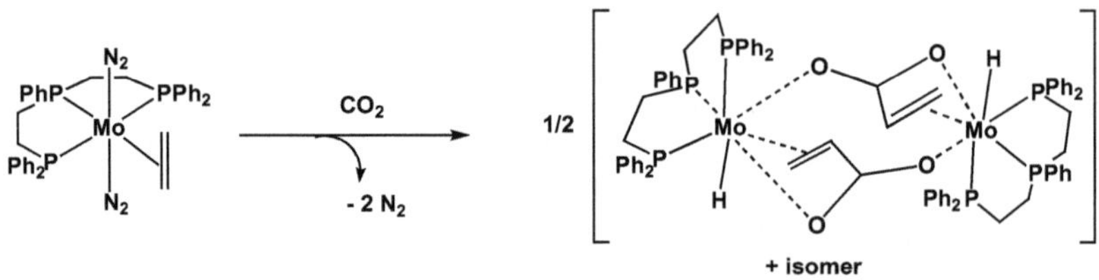

(5.6)

The observation of (triphos)Mo(C_2H_4)(CO_2) on the path to C–C bond formation between CO_2 and ethene provides the first experimental evidence that simultaneous coordination of both ethene and carbon dioxide at a metal center is a prerequisite for coupling the two unsaturates. This behavior is somewhat different from what was described to occur at nickel centers (see above). It cannot be excluded that the requirement of precoordination of both CO_2 and olefin, although being less stringent for late transition metal centers such as Ni, may likely be general for early transition metal centers.

The possibility of generating cleanly the intermediate complex (triphos)Mo (C_2H_4)(CO_2) was further exploited to study the kinetics and mechanism of molybdenum-mediated acrylate formation from CO_2 and ethene, for which an observed rate constant of $3.8(3) \times 10^{-5}\ s^{-1}$ was measured at 296 K corresponding to an activation energy of 23.5(2) kcal/mol. Kinetic analysis of the conversion of (triphos)Mo(C_2D_4)(CO_2) provided a slightly lower value for the observed rate constant ($3.1(5) \times 10^{-5}\ s^{-1}$ at 296 K) and revealed an isotopic effect of 1.2(2) for (triphos)Mo(C_2H_4)(CO_2) vs (triphos)Mo(C_2D_4)(CO_2) at 296 K. This small isotope effect is not consistent with a kinetic profile implying, as the rate-limiting step, β-hydride elimination from a metalla-lactone complex, which is expected to produce a more significant primary isotope effect. Measuring rate constants of the reaction over a 40 K temperature range allowed the computation of activation parameters for acrylate formation of $\Delta S^{\ddagger} = 1(7)$ eu and $\Delta H^{\ddagger} = 24(3)$ kcal/mol. The very low value of $\Delta S^{\ddagger}$ implicates a unimolecular transition structure for the rate-determining step, which, along with the small isotope effect for k_{C2H4}/k_{C2D4}, is consistent with oxidative C–C bond formation as the slow step during acrylate formation.

To date, examples of catalytic carboxylation of C=C double bonds of monoolefins still remains quite limited. A few early reports, claiming Rh-catalyzed carboxylation of ethene to propionic acid

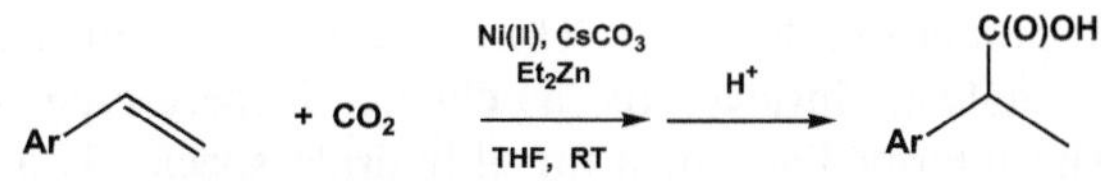

Scheme 5.8 Nickel-catalyzed reductive carboxylation of styrenes

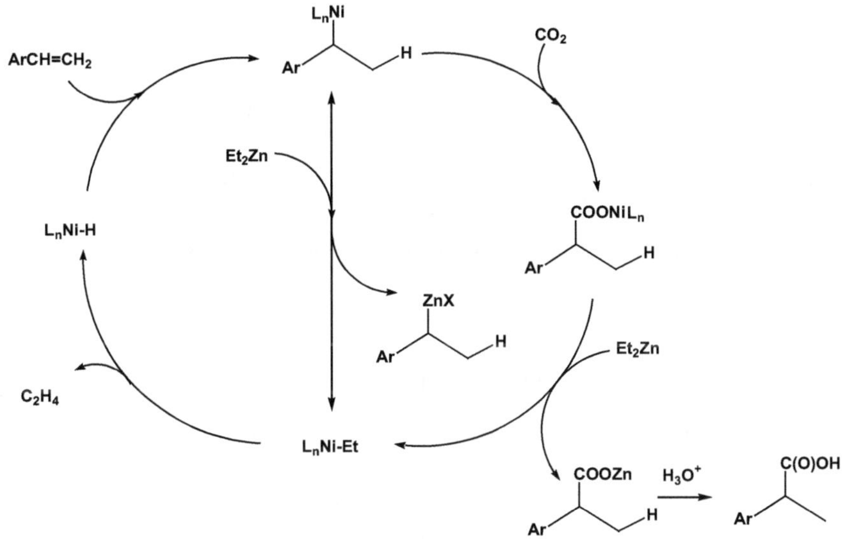

Scheme 5.9 Proposed mechanism for nickel-catalyzed hydrocarboxylation of styrenes (adapted from [48])

and ethyl propionate [44] or Ru-promoted carboxylation of propene to butyric esters [45] should be considered with caution as genuine examples of carboxylative processes, because, in both cases, there is no real proof that the carboxyl group derives from CO_2 [5, 6]. Duñach [46] extended to olefin carboxylation an electrochemical method successfully developed for catalytic incorporation of CO_2 into alkynes (see Sect. 5.4 for further details) and based on the use of a catalytic amount of a Ni(II) complex as the catalyst precursor. Later on, significant breakthrough in catalytic carboxylation of monoolefins has been achieved by Rovis and coworkers [47], who developed a protocol for nickel-catalyzed reductive carboxylation of styrenes under a CO_2 atmosphere. The latter study provided the first example of direct hydrocarboxylation of a single olefin. The reaction was carried out in THF, under mild conditions (room temperature, 0.1 MPa CO_2 pressure), in the presence of catalytic amounts (10 mol%) of $Ni(cod)_2$ or $Ni(acac)_2$ (acac = acetylacetonate), a base additive (20 mol%), and an organozinc reagent ($ZnEt_2$, 2.5 equiv.). The method proved to be effective for a variety of styrene derivatives with electron-withdrawing or neutral substituents and was tolerant of a variety of functional groups. The reaction was regioselective, as the α-carboxylated regioisomer

(Scheme 5.8) was produced selectively. The suggested mechanism implies a nickel-hydride active catalyst, instead of a classic Hoberg-type oxanickelacycle (Scheme 5.9). The intermediacy of a metal-hydride species is a peculiarity also distinguishing other metal-catalyzed hydrocarboxylation reactions [48–51]. Insertion of a styrene moiety into the nickel–hydride bond provides a benzyl nickel species which undergoes carboxylation. Subsequent *trans*-metalation of the reaction intermediate with Et_2Zn yields the hydrocarboxylation product and releases the precatalyst.

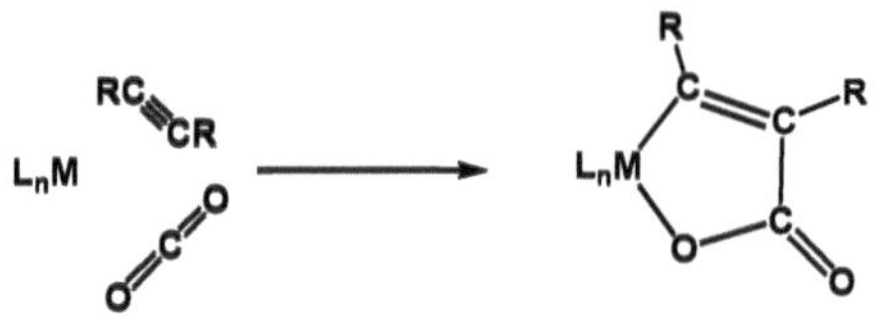

Scheme 5.10 Formation of oxametallacyclopentenones from alkynes and CO_2

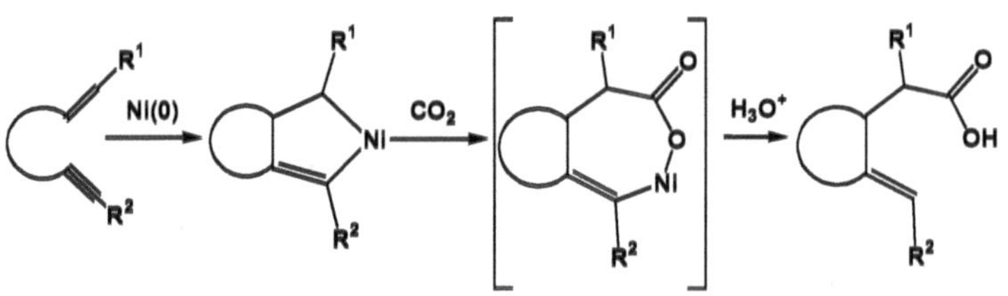

Scheme 5.11 Nickel-mediated carboxylative cyclization of enynes (adapted from [52])

5.4 Carboxylation of Alkynes

Alkynes and CO_2 can react stoichiometrically in the presence of Ni(0) complexes stabilized by P- or N-ancillary ligands (dcpe, dppe, bpy, TMEDA, DBU, etc.) to afford unsaturated pentaatomic oxanickelacycles (Scheme 5.10) [1–10]. Oxametallacyclopentenones have also been obtained by reaction of CO_2 with Ni- or Ta-aryne [1] complexes or by reacting the heterocumulene with alkyne complexes of Ti(II) or Zr(II) [1, 3, 19, 20].

Hoberg [1–10] first emphasized the synthetic potential of oxametallacyclopentenones, showing that they are useful starting materials or intermediates for the synthesis of unsaturated cyclic anhydrides (for instance, substituted maleic anhydrides, which can be obtained from the oxanickelacycles by reductive carbonylation with CO) or unsaturated carboxylic acids, which can be released upon protonolysis. The stoichiometric coupling reaction has also proved to be useful for the synthesis of unsaturated carboxylic acids through carboxylation and cyclization of enynes [48, 52, 53]. The latter reaction seems to proceed through oxidative cycloaddition of the α,ω-enyne moiety to the zero-valent nickel complex, regioselective insertion of carbon dioxide at the C(sp^3)–nickel bond, and hydrolysis of the resulting oxanickelacycle intermediate (Scheme 5.11).

The stoichiometric alkyne/CO_2 coupling reaction (Scheme 5.10) was adapted to work catalytically with respect to Ni, using chemical or electrochemical approaches.

Duñach [54, 55] reported the catalytic synthesis of a number of α,β-unsaturated carboxylic acid Mg salts from alkynes and CO_2 in the presence of catalytic amounts of L_nNi(II)-complexes (L = bpy, etc.), using a sacrificial Mg anode in a single-compartment electrochemical cell in which the catalytically active L_nNi(0)-reagent was catalytically regenerated at the cathode. The oxanickelacycle intermediate (Scheme 5.10; L_n = bpy) was isolated from the reaction mixture in the case of 4-octyne. Ni(II) was liberated from the metallacycle and recycled in the presence of Mg^{2+} ions and an appropriate proton source such as $(NBu_4)BF_4$ (supporting electrolyte) or DMF (reaction solvent). In this process, the nickel-magnesium exchange reaction is the key step to explain nickel catalysis. In fact, when the electrolysis was run in a two-compartment cell in the absence of magnesium ions, the

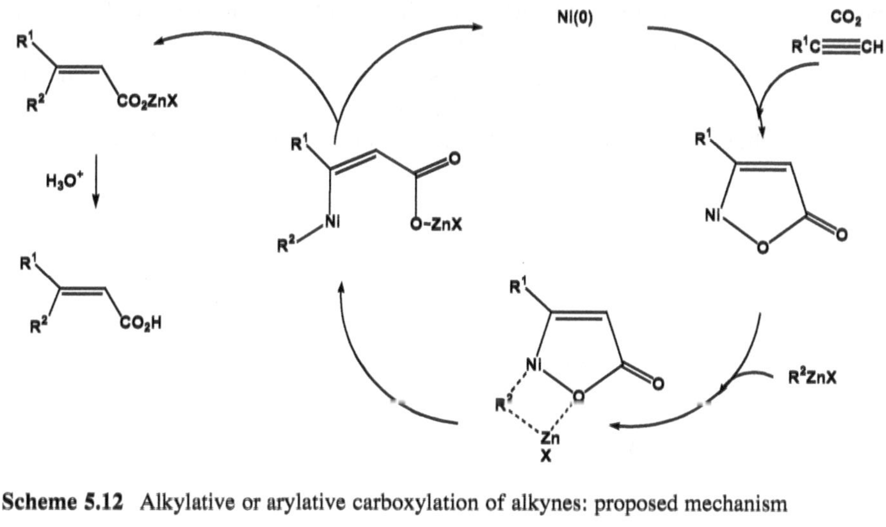

Scheme 5.12 Alkylative or arylative carboxylation of alkynes: proposed mechanism

reaction stopped at the stage of the metallacycle formation. The nickel species was not recycled and, therefore, the reaction occurred stoichiometrically.

Using stoichiometric amounts of an alkylzinc reagent and catalytic amounts of Ni(0)/DBU, Mori was able to prepare a number of β-alkylated or arylated α,-β-unsaturated carboxylic acids from alkynes and CO_2 (Scheme 5.12) [56, 57]. Through a *trans*-metalation reaction with the oxanickelacyclopentenone intermediate, the alkyl zinc reagent effected the reductive elimination of the product from the nickel center with concomitant regeneration of the active zerovalent catalyst.

Catalytic carboxylation of internal alkynes with CO_2 has been achieved in the presence of Ni [50] or Cu [51] catalysts, and through a hydrocarboxylative pathway, not involving the intermediacy of a metallaoxacyclopentenone, but, as previously noted for the hydrocarboxylation of alkenes (see Sect. 5.3), implying the intermediate formation of an M–H catalytically active species, which may easily be generated in the reaction mixture using diethylzinc [50] or hydrosilanes [51] as hydride transfer agent.

In principle, coupling reactions involving terminal alkynes and carbon dioxide can generate α,β-unsaturated carboxylic acids, which may be β- or α-substituted according to whether the alkyne is carboxylated at the 1- or 2-position (Scheme 5.13). Carboxylation at the 1-position of the alkyne was found to be predominant when the coupling reaction was carried out in THF at 273 K and promoted by $Ni(DBU)_2$ generated in situ [56–58]. These results are in sharp contrast with those of Duñach et al. [54], who, using electrochemically generated $(bpy)_2Ni$ complex in DMF at 293–353 K, showed that product regioselectivity was dominated by the α-regioisomer, involving carboxylation at the 2-position of the alkyne. Even subtle variations in the ancillary ligand or changes in the electronic and steric nature of alkyne can markedly affect the regioselectivity of the process [59, 60]. Using the Ni-promoted alkylative carboxylation protocol mentioned above and a variety of trimethylsilyl(TMS)-substituted alkynes ($RC{\equiv}C{-}SiMe_3$), Mori observed the predominant formation of α-silyl-β,β′-dialkyl-substituted α,-β-unsaturated carboxylic acids, whereas, with *t*-Bu ($RC{\equiv}C{-}CMe_3$) and Ph substituted ($RC{\equiv}C{-}Ph$) alkynes, the alternative regioisomer [(*t*-Bu)MeC=C(R)CO_2Me, (Ph)MeC=C(R)CO_2Me] formed preferentially [56]. Moving from electron-

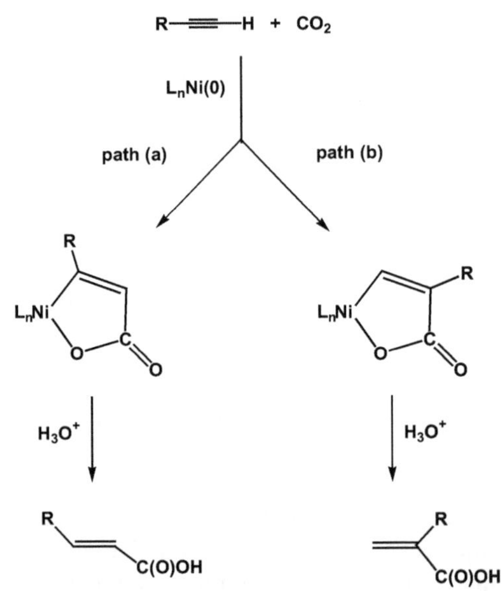

Scheme 5.13 Regioisomers from coupling terminal alkynes with CO_2

donating (4-OMe) to electron-withdrawing (4-CO_2Me) substituents on the Ph group increased preference for the pathway involving the carboxylation of the alkyne carbon bearing the aryl group.

Electronic effects of the substituents on the acetylenic carbons, as well as the nature of the ancillary ligands coordinated to nickel center, also markedly affect the rate of the coupling reaction. Saito [58] compared the carboxylation rates of both Ph–C≡CH and 4-MeO–C_6H_4–C≡CH with CO_2 in the presence of Ni(cod)$_2$/DBU (2 equiv.) and found that ethynylbenzene was carboxylated faster than 4-methoxy-1-ethynylbenzene, whereas Duñach noted that the reaction rate increased with the basicity of the ancillary ligand [59].

Only a limited number of theoretical studies have focused on the mechanism of oxidative coupling of alkynes with carbon dioxide [61–64]. The Ni(DBU)$_2$-mediated coupling reaction of alkynes with CO_2 has been modeled simultaneously and independently by two different groups using DFT approaches [63, 64]. Like the nickel-assisted ethylene/CO_2 coupling, the reaction pathway seems to proceed through an associative mechanism, involving, first, the η^2-coordination of the alkyne to zerovalent nickel with formation of $L_2Ni(\eta^2$-alkyne) (L = DBU [63] or 1,2-dimethyl-1,4,5,6-tetrahydropyrimidine selected as a model ligand (mDBU) for DBU [64]), which is followed by subsequent attack of the CO_2 carbon to one of the carbons of the coordinated alkyne. No evidence was found for a transition structure corresponding to attack of alkyne to nickel-bound CO_2. Analogously, the calculations did not support a dissociative pathway involving (1) the formation of a L_2Ni (η^2-CO_2)(η^2-alkyne) complex (18 e^-), (2) dissociation of one of the L ligands with generation of a LNi(η^2-CO_2)(η^2-alkyne) intermediate (16 e^-), (3) followed by the oxidative coupling step to afford the five-membered oxanickelacycle complex, which, finally, (4) re-coordinates L. The associative mechanism is preferred because more L ligands in the transition state make the coordinated alkyne substrate more electron-rich and more prone to electrophilic attack.

The issue of regioselectivity has also been addressed. Buntine's group [64], using a few terminal alkynes such as TMS–C≡C–H, *t*-Bu–C≡C–H, Ph–C≡C–H, and MeO-*p*-C_6H_4–C≡C–H calculated that the 3-R-substituted oxanickelacycle (path b, Scheme 5.13; L = mDBU, $n = 2$) was the thermodynamically preferred product for all the substrates investigated, because of the lower steric interaction between the alkyne-substituent and the ancillary ligand. However, the transition structure characterized by the attack of CO_2 on the alkyne carbon distal from the R-group was generally lower in energy, making the 2-R-substituted oxanickelacycle (path a, Scheme 5.13; L = mDBU, $n = 2$) the kinetically favored product. An exception to the above generalization was provided by the TMS-substituted alkyne, where the 3-R-substituted nickelacycle was calculated to be both kinetically and thermodynamically favored. In the latter case, the kinetic preference for path (b) over path (a) in Scheme 5.13 has been ascribed to the sheer size of the substituent, with the steric interaction between the incoming CO_2 moiety and the TMS group in path (b)-TS (transition state) being energetically more favorable than the steric hindrance between the TMS group and the ancillary ligand in path (a)-TS.

Li et al. [63], using $Ni(DBU)_2$ as promoter, focused on different model terminal alkynes, such as Me–C≡C–H, MeO–C≡C–H, and NC–C≡C–H, which exhibit a very different substituent effect. Electronic factors easily explain the results that NC–C≡CH gives the highest barrier whereas MeO–C≡CH gives the lowest barrier: the MeO-group makes the π-bonds of the substrate more electron-rich, whereas the CN-substituent makes the π-bonds more electron-poor. However, whatever the alkyne electronic properties are, the coupling reaction provided the same regioselectivity, because the 2-R-substituted nickelacycle (path a, Scheme 5.13; $L = DBU$, $n = 2$) was calculated to be the kinetically preferred regioisomer product for all the three substrates investigated, and also more stable from the thermodynamic point of view for R equal to OMe or CN. The calculations indicated that regioselectivity was dominated by steric repulsive interactions between the substituent at the alkyne carbon which forms the new carbon–carbon bond with the CO_2 carbon, and the unreacted C=O group of CO_2. For silyl-substituted internal alkynes, such as MeC≡CTMS, the additional steric repulsive interaction between the sterically demanding TMS substituent and the DBU ligand at the metal center also needs to be taken into account, when the regioselectivity is considered. In the latter case, both the additional steric effects and electronic factors favored the formation of compounds in which CO_2 couples with the silyl-substituted carbon, as also documented experimentally by Mori [56].

Replacing the DBU ancillary ligand with bpy resulted in higher activation energies [63, 64], making the bpy ligand a not so good ligand as DBU for the Ni-mediated coupling reactions, as observed early experimentally by Saito et al. [58] using THF as the reaction solvent ($T = 273$ K). The bpy ligand is less electron-donating than the DBU ligand. Consequently, the coordinated alkyne is not electron-rich enough and exhibits poorer nucleophilicity, leading to higher coupling barriers. According to the calculations in this case (see Scheme 5.13; $L_n = $ bpy, R = Me [63], Ph or *t*-Bu [64]) path (a) is kinetically favored over path (b), which leads to the thermodynamically more stable 3-R-substituted regioisomer. However, in the Ni/bpy system there is less driving force for the formation of the regioisomeric products relative to the encounter complex "(bpy)Ni(η^2-alkyne) + CO_2." The combination of a sterically demanding alkyne, such as *t*-BuC≡CH, with the bpy/Ni system even exhibits no driving force ($\Delta G = +0.7$ kJ/mol) for the formation of the kinetically preferred 2-R-susbstituted oxanickelacycle, and thus the thermodynamically preferred 3-R-substituted regioisomer would likely result.

Buntine [64] calculated that changing solvent from THF to DMF caused a decrease of activation barriers independent of the other parameters, without any preference for path (a) or (b) in Scheme 5.13 [L = DBU ($n = 2$) or bpy ($n = 1$)]. The latter finding is relevant to the results reported by Duñach (see above), who effectively used the electrochemically generated bpy/Ni(0) system as carboxylation catalyst, in DMF as reaction solvent [54, 55, 59]. The dominance of 3-R-substituted regioisomer products observed by Duñach [54, 55, 59] in her experiments may be related to the use of the bpy/Ni system which may disfavor the formation of sterically hindered 2-R-substituted products more than the DBU/Ni system. However, a different simpler explanation cannot be excluded, as the higher reaction temperatures (293–353 K) used by Duñach may help to favor the thermodynamically preferred 3-R-substituted product over the kinetically preferred 2-R-substituted regioisomer.

Internal or terminal alkynes can be catalytically converted into 2-pyrones by reaction with CO_2 in the presence of phosphine–Ni(0) complexes (5.7). In addition to 2-pyrones, cyclo-oligomers may also form as by-products.

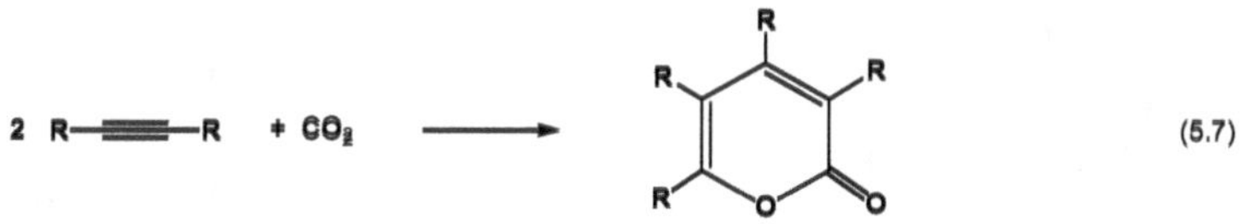

(5.7)

This reaction (5.7) was first reported by Inoue [65], using Ni(cod)$_2$/dppe as catalytic system, and further developed by Walther and coworkers [66], who set up an improved synthetic protocol capable of producing 2-pyrones with high yield and selectivity. The method involved the use of a THF-acetonitrile mixture as reaction solvent, in combination with monodentate basic phosphines with small cone angles (PEt_3, for instance). High yield and selectivity in pyrone have also been achieved using PMe_3 as ancillary ligand and supercritical carbon dioxide (sc-CO_2) as reactant and reaction medium [67]. Kishimoto [68] has described a new catalytic system, based on Ni(cod)$_2$ and $P(C_4H_9)_3$ or $P(C_8H_{17})_3$, which promoted the synthesis of a wide range of substituted pyrones with high yield (up to 98 %) and selectivity (up to 99 %) in compressed (4–14 MPa) CO_2. Tsuda et al. [8] significantly expanded the scope of reaction (5.7) using alkoxyalkynes or α,ω-diynes as substrates and developed interesting syntheses of highly functionalized cyclic compounds and novel polymeric materials. Later, the Ni(0)-catalyzed synthesis of 2-pyrones from CO_2 and diynes was reinvestigated and improved by Louie using N-heterocyclic carbenes (NHC) as ancillary ligands. A Ni(0)/NHC catalytic system was shown to promote effectively the pyrone synthesis at ambient temperature and 0.1 MPa CO_2 pressure [69].

An insight into the mechanism of the Ni-catalyzed formation of pyrone (5.7) has been gained by Hoberg [10, 31, 70, 71] and Walther [9, 72, 73]. According to the generally accepted mechanism (Scheme 5.14), the first step of the catalytic cycle is the formation of oxanickelacyclopentenone **A**. This intermediate can insert another alkyne molecule into the Ni–C(sp^2) bond to afford the oxanickelacycloheptadienone **B**, from which the pyrone product is liberated by reductive elimination. This mechanism finds sound support in the fact that, by using bidentate basic stabilizing ligands such as TMEDA, for instance, it was possible to isolate and characterize oxanickelacyclopentenone complexes, which have been shown to react with one more alkyne molecule affording oxanickelacycloheptadienone species, from which 2-pyrone can be released on heating. Comparative stoichiometric studies have not confirmed the hypothesis, initially proposed by Inoue [65], that pyrone synthesis might imply the preliminary coupling of two alkyne

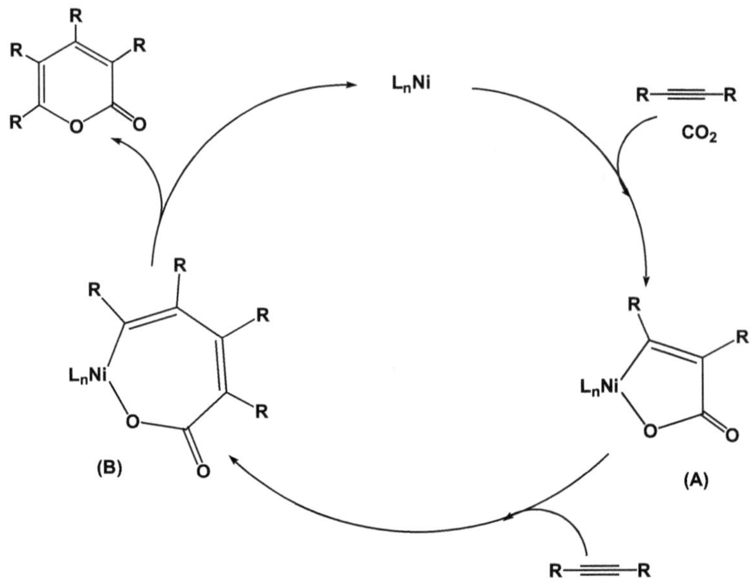

Scheme 5.14 Mechanism of Ni-catalyzed formation of 2-pyrones

molecules to form a nickelacyclopentadiene complex, followed by CO_2 insertion into the Ni–C bond or [4 + 2] cycloaddition to a CO_2 molecule. On the contrary, nickelacyclopentadiene species are involved as intermediates only in the alkyne cyclotrimerization.

Formation of cyclotrimer by-products emphasizes the ability of the catalytic system to catalyze not only the substrate/CO_2 hetero-coupling reaction but also homo-coupling processes. Homo-coupling can involve not only the substrate, as in the case of the formation of (cyclo)oligomers but also CO_2 itself. Walther et al. [72, 73] have shown that a competing reaction in the catalytic cyclooligomerization of hex-3-yne with CO_2 in the presence of Ni(TMEDA)(cod) is the formation of carbon monoxide and a nickel trimer with two (TMEDA)Ni(CO_3) units. The latter reaction pathway leads to catalyst deactivation. The reduction of CO_2 to CO probably proceeds through a nickelacycle intermediate analogous to that isolated with iridium (Scheme 5.2) involving the head-to-tail homocoupling of two CO_2 units.

2-Pyrone synthesis by cycloaddition of CO_2 to terminal alkynes (1-hexyne, 1-propyne) has also been investigated. This process can be catalytically promoted, albeit with low yield and selectivity, by Co [74] and Rh [75] complexes. Rh(dppe)(η^6-BPh_4), in acetonitrile, at 390 K, catalyzed the formation of 4,6-dimethyl-2-pyrone from 1-propyne and CO_2 (1 MPa) with a TON of 50 [75]. The Rh-catalyzed reaction has been proposed to proceed through a mechanism (Scheme 5.15) not involving an oxametallacycle intermediate species. The CO_2 insertion into the Rh–C(sp^2)-σ-bond of a Rh-alkenyl intermediate, obtained upon propyne dimerization, affords a linear unsaturated carboxylate which is converted into the pyrone.

The synthesis of 4,6-dimethyl-2-pyrone from 1-propyne and CO_2 (6 MPa, 403 K) was achieved by Pillai et al. using Rh_4- or Fe_2Rh_4-carbonyl cluster-derived catalysts, supported on selected metal oxides such as TiO_2, Al_2O_3, and ZrO_2 [76, 77]. Formation of 2-pyrone was accompanied by homocoupling by-products such as 1,3,5- and 1,2,4-trimethylbenzene. In the absence of CO_2 only the trimethylbenzenes were produced. FTIR studies indicated that a key intermediate in the formation of pyrone may be a monodentate carbonate absorbed on the catalyst surface [77].

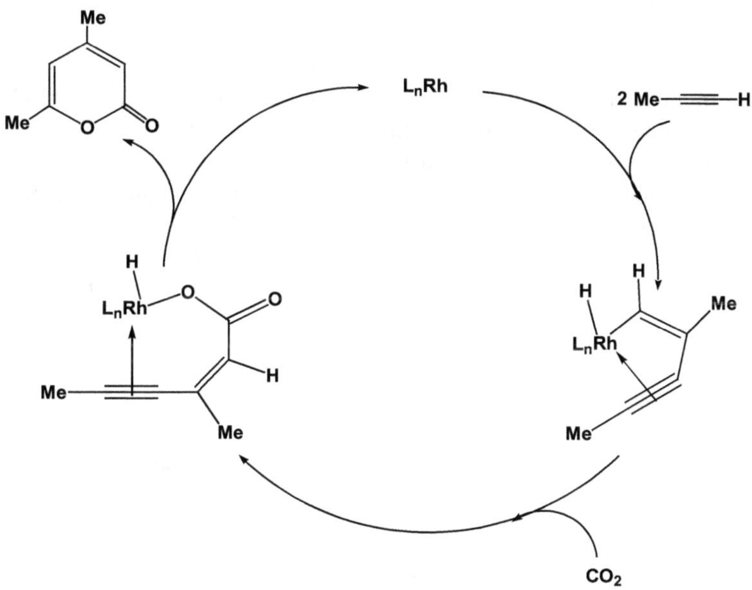

Scheme 5.15 Mechanism of Rh(dppe)(η^6-BPh_4)-catalyzed formation of 4,6-dimethyl-2-pyrone (adapted from [4])

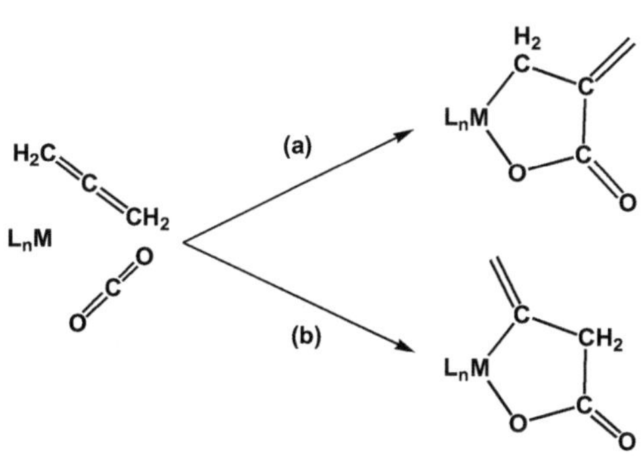

Scheme 5.16 Oxidative coupling allene-CO_2

5.5 Carboxylation of Allenes

Also allenes undergo stoichiometric 1:1 coupling reaction with CO_2 in the presence of $L_nNi(0)$ complexes (L = bpy, dcpe, dppe, DBU, PCy_3) [10, 78]. Pioneering studies by Hoberg [78] showed that the coupling reaction involves the carbon atom of CO_2 and the C(sp) atom of diene moiety and generates a conjugated five-membered nickelacyclic carboxylate, which shows a Ni–C(sp^3) bond (Scheme 5.16, route (a)). With substituted 1,2-dienes, such as $CH_2{=}C{=}CR^1R^2$ (R^1 = H, alkyl; R^2 = alkyl), two different five-membered oxanickelacycles were produced (Scheme 5.17). The metallacyclic carboxylates can be converted into the corresponding carboxylic acid methyl esters upon protonolysis with HCl/MeOH.

This stoichiometric chemistry has been revisited in the last few years [60, 79–82] and, for instance, applied to the synthesis of α-methylene-γ-hydroxy carboxylic acids and their derivatives (*cis*-β,γ-disubstituted α-methylene-γ-lactones) by means of nickel-mediated sequential addition of carbon dioxide and aldehydes into terminal or internal allenes [80, 81], or exploited for the preparation of β-hydroxycarboxylic acid derivatives by O_2-oxidation of the Ni-π-allyl species (5.8)

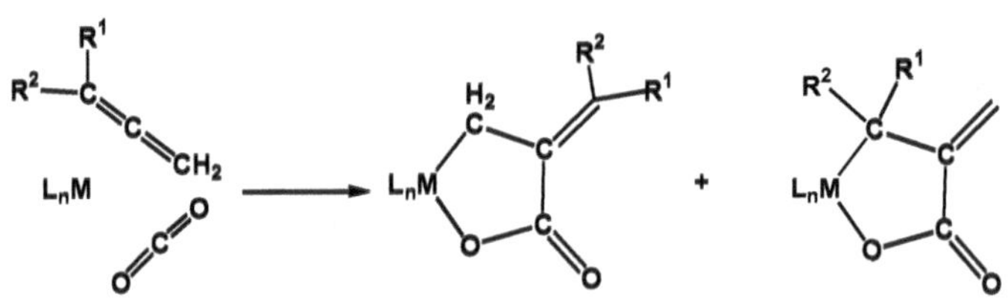

Scheme 5.17 Oxidative coupling of CO_2 with substituted allenes

in equilibrium with the oxanickelacycle intermediate generated by Ni(0)-mediated coupling of 1,2-dienes with CO_2 [82].

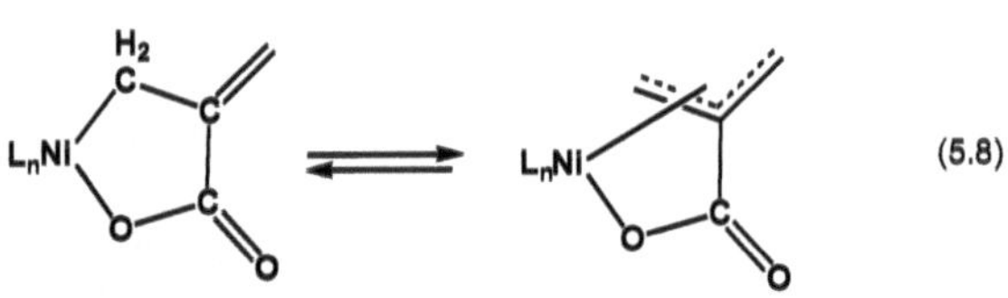

(5.8)

Unsaturated carboxylic acid magnesium salts have been obtained catalytically by reductive electrocarboxylation of a number of allenes with a $L_nNi(II)$-complex [L = bpy, dppe, PMDTA (*N,N,N′,N″,N″*-pentamethyldiethylene triamine)], as the catalyst, in the presence of a sacrificial magnesium anode [46, 83] (see also Sect. 5.3).However, regioselectivity was moderate. In addition to α,β-unsaturated acids, the process afforded vinyl acetic acid derivatives, involving carboxylation of a terminal carbon of allene moiety, and di-carboxylic acids. Catalytic double carboxylation of trimethylsilylallenes has also been achieved using Ni(0)/DBU as catalytic system and dimethyl zinc as *trans*-metalation/reducing agent [84].

The above catalytic processes are still based on Ni(0)-promoted oxidative coupling of substrate with CO_2 (Schemes 5.16 and 5.17). A new successful different approach, allowing the synthesis of β,γ-unsaturated carboxylic acids from allenes and CO_2, has been reported by Iwasawa [85, 86]. The protocol requires the use of a tridentate silyl pincer palladium hydride complex as the active catalyst in the presence of a suitable reducing agent (Scheme 5.18) and is based on the generation of a σ-allyl palladium species via hydropalladation of allene substrate, followed by its nucleophilic addition to CO_2. The catalyst precursor (X = trifluoromethanesulfonate in Scheme 5.18) is converted into a palladium hydride by *trans*-metalation with Et_3Al (or Et_2Zn), followed by β-elimination. Hydrometalation of the less sterically congested double bond of the allene moiety generates a σ-allyl Pd-complex. The tridentate structure of the silyl pincer ligand (P*Si*P) stabilizes the allyl group in the σ- rather than π-coordination mode and enhances the nucleophilicity of the σ-allyl palladium intermediate, because of the strong electron-donating nature and *trans*-influence of the silicon atom at the center of the tridentate pincer backbone. The σ-allyl palladium intermediate undergoes nucleophilic addition to CO_2 at the γ-position of palladium regioselectively to give, upon allylic transposition, a palladium carboxylate. The latter species can undergo a second *trans*-metalation and β-elimination with Et_3Al or Et_2Zn to regenerate the catalytically active Pd-hydride species and affords the aluminum or zinc salt of the β,γ-unsaturated carboxylic acid. The free acid was generated upon protonolysis. The pincer structure of the P*Si*P ligand helps to retard the liberation of Pd(0) and to avoid side reactions such as oxidative cyclization of allenes, usually promoted by Pd(0).

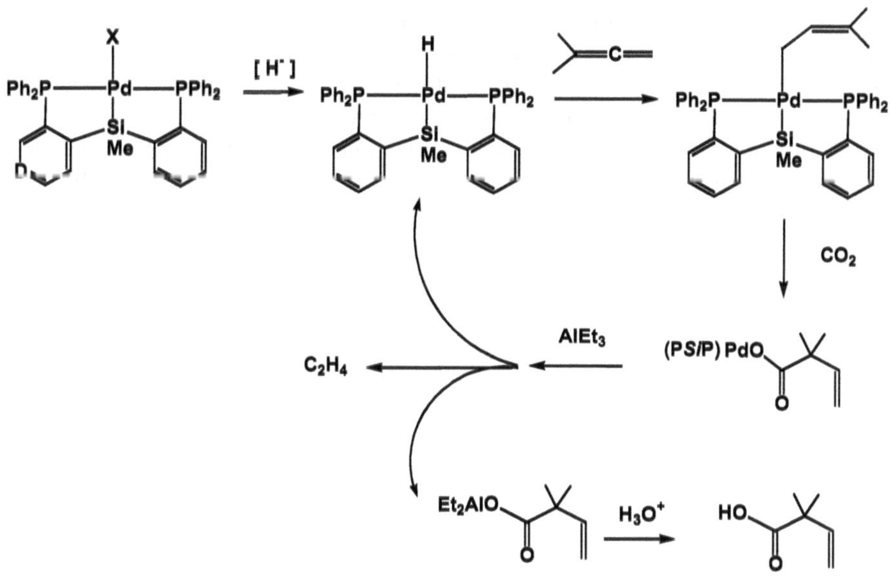

Scheme 5.18 Hydrocarboxylation of allenes with CO_2 catalyzed by silyl pincer-type palladium complex

To date, little information is available in the literature on catalytic co-oligomerization of allenes with CO_2. Allene and CO_2 can react in the presence of palladium catalysts to yield linear esters (**A** and **B**, Structure 5.3) as the main carboxylated products, together with lower amounts of 4,6-dimethyl-2-pyrone (**C**, Structure 5.3) and a mixture of allene linear oligomers and polymers [87].

Structure 5.3 Co-oligomerization allene-CO_2: linear and cyclic esters

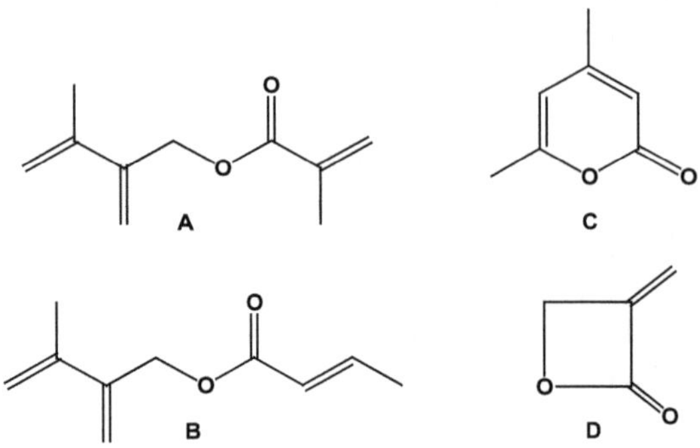

Pyrone formation is reminiscent of cyclooligomerization of alkynes (see Sect. 5.4), suggesting that allene might initially be isomerized to 1-propyne. However, an ad hoc experiment, involving propyne as the substrate, showed pyrone formation in very minor amounts and demonstrated that linear esters **A** and **B** (Structure 5.3) were not formed in this way. It has been suggested that the formation of the linear esters (and oligomers) may imply a hydrogen-transfer step which presumably proceeds through the intermediacy of a palladium hydride.

In aromatic solvents, allene and CO_2 reacted in the presence of Rh(dppe)(η^6-BPh_4) to yield pyrone **C** (Structure 5.3) besides 1.2.1.2-polyallene and a small amount of oligomers [88]. On the basis of the lack of reactivity of 1-propyne with CO_2, under conditions similar to those used for allene, the possibility that pyrone formation could imply a preliminary isomerization of allene to the alkyne can also be excluded in this case. It is worth noting that the pyrone formation requires an

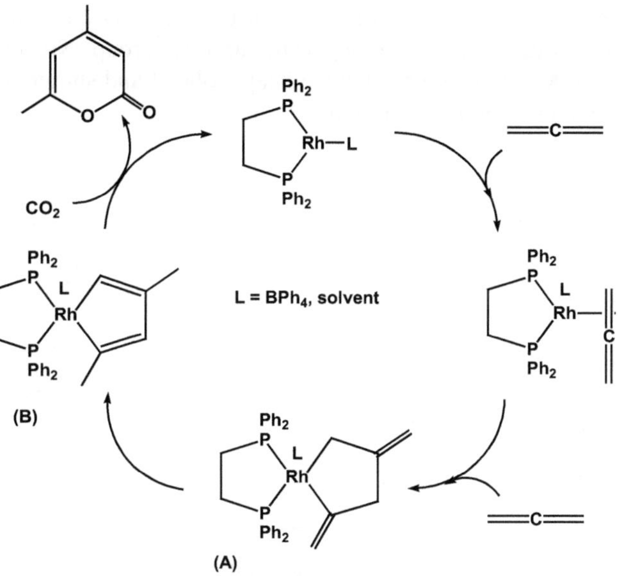

Scheme 5.19 Rh-catalyzed formation of 4,6-dimethyl-2-pyrone

initial head-to-tail (H–T) coupling of two allene units, which may be promoted by the metal center if two *cis* positions are made available in the coordination sphere of rhodium, which is coordinatively saturated in Rh(dppe)(η^6-BPh_4). This may occur at the working temperature (373 K), which favors the change of coordination sphere around the metal center, making possible the coordination of substrate followed by the H–T homocoupling step (Scheme 5.19). Subsequent isomerization of rhodacycle **A** (Scheme 5.19) to metallacycle **B** and reaction of the latter species with CO_2 can reasonably account for the formation of 4,6-dimethyl-2-pyrone. Carrying out the reaction in the presence of a base, used as a co-catalyst, promoted the formation of the linear ester **A** (Structure 5.3) and 1.2.2.1-polyallene, both requiring head-to-head coupling of allene units. The presence of a base co-catalyst caused a modification of the catalytic system because of formation of complex (dppe)Rh–Ph by phenyl transfer from boron to rhodium [89, 90]. Homolytic cleavage of the Rh–phenyl bond can afford rhodium radical species which have been proposed to initiate the observed H–H polymerization of substrate, as well as the formation of the precursor of the linear ester **A** (Structure 5.3). The formation of the linear ester **A** requires an external source of hydrogen, which may come from solvent. Accordingly, a significant solvent effect was observed and, in contrast with toluene, dichloromethane favored the formation of homopolymers.

Nickel(0) catalyzed cycloaddition reaction of CO_2 with allene has been reported to form 3-methyl-4-methylene-2-pentene-5-olide in 20 % yield [91]. Scheme 5.20 (L = dppe) summarizes the proposed mechanism. Insertion of CO_2 into a nickelacyclopentane intermediate (**A**, in Scheme 5.20) affords a π-allyl-nickel carboxylate, from which 3,4-dimethylenepentane-5-olide can be reductively eliminated.

Isomerization of the latter species to the final product may be driven by the more extended conjugation of diene moiety with carbonyl group. Nickelacyclopentane species, such as **A** in Scheme 5.20, have been isolated and shown to form π-allyl-nickel carboxylates upon reaction with CO_2 [87].

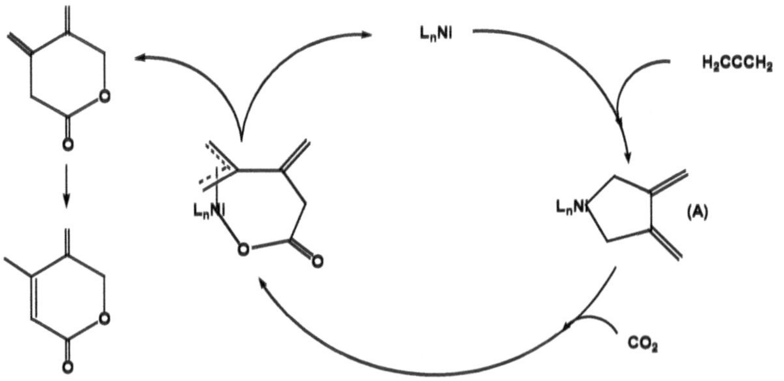

Scheme 5.20 Ni-catalyzed cycloaddition of allene with CO_2

A different regiochemistry has been observed by Tsuda [92] in the cycloaddition of CO_2 with methoxyallene. The cooligomerization process was catalyzed by $Pd_2(dba)_3 \cdot CHCl_3/(n\text{-}Bu)_2PCH_2CH_2Py$ (dba = dibenzylideneacetone; Py = 2-pyridyl). The reaction afforded (*E*)-5-methoxy-2-(methoxymethylene)-4-methylene-5-olide regio- and stereo-specifically. The methoxy substituent plays a crucial role in this reaction, as 1,2-pentadiene did not produce any co-oligomer with CO_2 in the presence of the same catalytic system. It has been proposed that the electron-donating methoxy group may favor the formation of a palladium metallacyclic intermediate (Structure 5.4) by conjugation with the electron-withdrawing carbon of the carbonyl group.

Structure 5.4 Pd-promoted co-oligomerization methoxyallene-CO_2: proposed metallacyclic intermediate

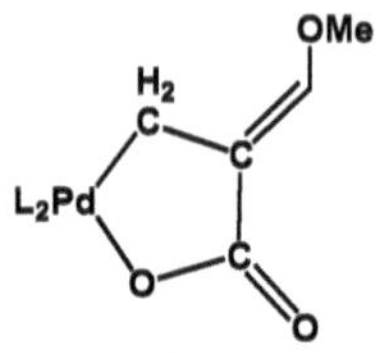

The reaction of allene with carbon dioxide, using *trans*-$[RhCl(C_2H_4)P(i\text{-}Pr)_3]_2$ as catalyst, has allowed an unprecedented reactivity to be emphasized [93]. Complex $[RhCl(C_2H_4)P(i\text{-}Pr)_3]_2$ is an active catalyst of allene H–T polymerization. Extended studies by Aresta's group have shown that the incidence of this process can be controlled either working at high temperatures, or at low temperature but in presence of H_2, which, by acting as a mass regulator, can repress the polymer formation in favor of oligomeric products. With a view to favoring allene-CO_2 coupling, avoiding the formation of polymers, the allene/CO_2 cooligomerization reaction has been investigated under CO_2 pressure (6 MPa), in the presence of H_2 (0.03–0.1 MPa), at temperatures just above room temperature (300–310 K). Under the above conditions, complex $[RhCl(C_2H_4)P(i\text{-}Pr)_3]_2$ promoted, albeit in very low yield, the formation of four-membered lactone 3-methylene-2-oxetanone (**D**, Structure 5.3), which is the product of formal [2 + 2] addition of allene to CO_2. It should be noted that this product was never observed when allene was reacted with CO_2 in absence of H_2. Moreover, it was not found when working in the presence of hydrogen at temperatures higher than 320 K. The formation of the new heterocyclic product can be related to a modification of the starting catalyst $[RhCl(C_2H_4)P(i\text{-}Pr)_3]_2$, which in the presence of hydrogen has been shown to convert easily into $[RhH_2(Cl)P(i\text{-}Pr_3)]_2$. The new hydride species is believed to react with allene to afford a reactive intermediate which is able to couple with CO_2.

The non-catalyzed combination of the two cumulenes has been investigated from a theoretical point of view [93–95]. The calculations have shown that CO_2 and allene may undergo a [2 + 2] addition reaction affording 3-methylene-2-oxetanone, as well as diketene. Energetics for the formation of both species are not very favorable, with 3-methylene-2-oxetanone being the isomer highest in energy. The Rh-hydride catalysis may circumvent the energy barrier and allow the formation of the lactone through a different reaction pathway.

5.6 Carboxylation of Conjugated Dienes

Stoichiometric carboxylation of conjugated dienes with CO_2 has been carried out in the presence of both late (Ni, Fe) and early (Zr) low valent transition metal systems [1–10, 48, 96–102]. Depending on the nature of the metal center, ancillary ligands, diene substrate, and, more generally, working conditions, the carboxylation reaction may evolve in different ways and afford metal-carboxylate products implying 1:2 or 1:1 or 2:1 CO_2/1,3-diene coupling, which are useful synthones for further reactions with electrophiles. For instance, butadiene, which is η^4-coordinated to iron(0) in $(Me_3P)_3Fe(C_4H_6)$, can react with CO_2 to give, according to the experimental conditions, either the mono- or the 1,4-dicarboxylated product [10, 96]. Yasuda et al. [97] studied the reaction of CO_2 with a series of Zr-diene complexes of the type ZrL_2(*s-cis*-diene) and ZrL_2(*s-trans*-diene) ($L = \eta^5\text{-}C_5H_5$, $\eta^5\text{-}C_5Me_5$; diene = butadiene, isoprene). The course of the carboxylation reaction was found to be markedly affected by the bulkiness of the auxiliary ligands and the geometry (*s-cis*, *s-trans*) of ligated dienes.

However, most studies have dealt with Ni(0)-based systems [1–10, 48, 98–102]. The reactivity of conjugated dienes with Ni(0)-complexes and carbon dioxide is higher than that exhibited by monoenes. Accordingly, the reaction of Ni(bpy)(cod) with CO_2 and 1,3,7-octatriene yielded products which indicate that CO_2 attacks only at the diene moiety of octatriene [101].

In the presence of monodentate P-ligands, Ni(0) promoted the CO_2/1,3-diene coupling in a 1:2 ratio to yield a Ni(II)-C_9-carboxylate. This behavior reflects the well-recognized ability of Ni(0)-complexes bearing labile phosphane ligands to promote olefin oligomerization. In fact, the above carboxylation reactions are usually believed to imply, as the first step, the linkage of two diene units to form a bis-allyl-C_8 chain, followed by CO_2 insertion into a Ni-allyl σ-bond to afford Ni (II)-carboxylates, which may undergo further rearrangements reactions. By using P (O-*i*-Pr)$_3$ as ancillary ligand, Hoberg succeeded in obtaining 2-methylene-3-vinylcyclopentanecarboxylic acid from butadiene and CO_2 even catalytically, albeit in low yield and with a low turnover (TON = 29 cycles) [98]. Mori has enlarged the scope of the Ni(0)-promoted 1:2 CO_2/1,3-diene coupling reaction, using bis-1,3-dienes as the substrates [99]. She found that the system Ni(0)/PPh_3 catalyzed, under very mild conditions, the addition of CO_2 and organozinc to bis-1,3-dienes (Scheme 5.21). The reaction resulted in the carboxylative cyclization of the bis-1,3-diene moiety with high regio- and stereoselectivity. This process was further developed into an enantioselective asymmetric reaction by using a chiral P, O chelating ligand [100].

CO_2/diene coupling in a 1:1 ratio to give η^3-allyl-C_5-carboxylate Ni-complexes (Scheme 5.22) has been shown to be possible in the presence of P- or N-chelating ligands [1–10, 101]. A few η^3-allyl-C_5-carboxylate-Ni-complexes have also been structurally characterized [9]. Although η^3-coordination of the allyl moiety is thermodynamically more stable compared

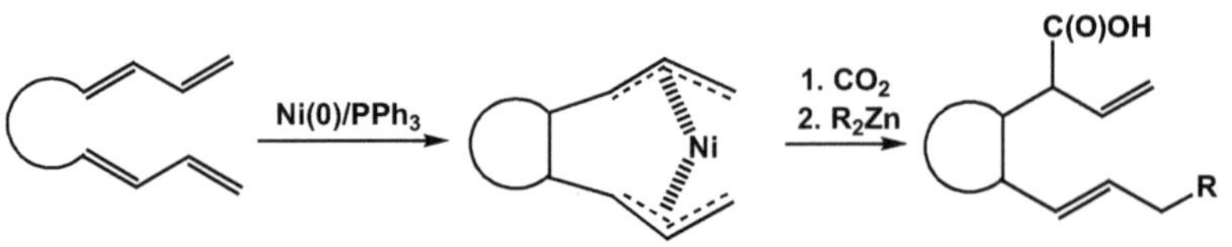

Scheme 5.21 Carboxylative cyclization of bis-1,3-dienes (adapted from [99])

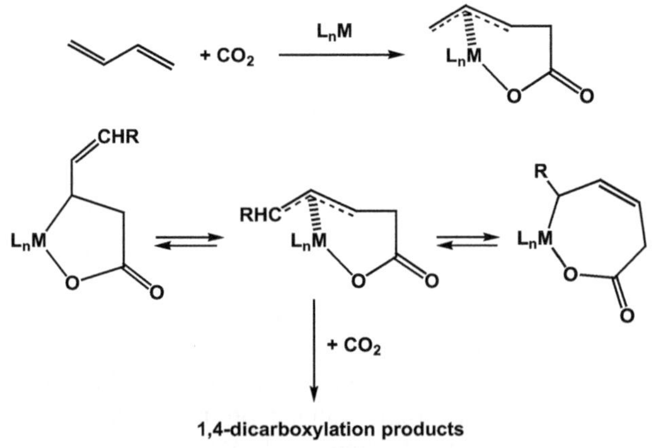

Scheme 5.22 Mono- and 1,4-dicarboxylation of 1,3 dienes with CO_2

to the η^1-bonding mode, an η^3–η^1 rearrangement of the oxo-π-allyl intermediate formed by diene-CO_2 oxidative coupling is also possible and can be envisaged to account for the formation of α,ω-dicarboxylate complexes, which have been reported in a few cases (Scheme 5.22; M = Fe [96], Zr [97], Ni [101]). Remarkably, only 1,4- and no 1,2-dicarboxylation has been observed. Mori [102] has shown that DBU is a superior ligand for nickel(0)-promoted oxidative coupling of dienes with CO_2. In the presence of the amidine ligand (2 equiv.), $Ni(cod)_2$ (1 equiv.) promoted the 1:1 oxidative coupling of 4-phenyl-1,3-butadiene with carbon dioxide under mild conditions (273 K, 0.1 MPa CO_2 pressure) in short times (4 h). The process can also run in the presence of an excess of organozinc reagents, acting as reducing and *trans*-metalation agents. Interestingly, depending on the nature of the organozinc compound, mono- or 1,4 dicarboxylated products can be obtained [102].

Schindler et al. carried out a kinetic analysis for the reaction of isoprene (ip) with carbon dioxide and Ni(bpy)(cod) [103]. Ni(bpy)(cod) did not react directly with CO_2 (see also Sect. 5.2), but reacted reversibly with isoprene to give Ni(bpy)(ip). The kinetic results support a mechanism where, in a fast preequilibrium step, one of the bonds between Ni and cod is cleaved and, in a following step, ip coordinates to the free coordination site (Scheme 5.23). At low temperatures Ni(bpy)(ip) can further react, through a consecutive equilibrium, with another ip molecule to give $Ni(bpy)(ip)_2$. However, in the presence of CO_2, the nickel–ip complex, Ni(bpy)(ip), reacts irreversibly with the heterocumulene to form the product complex (3-5-η^3)-3-methyl-3-pentenylato)nickel bipyridine (Scheme 5.23). The activation parameters for the latter step were calculated to be $\Delta H^{\ddagger} = 25 \pm 7$ kJ/mol and $\Delta S^{\ddagger} = -184 \pm 24$ J/(mol K). The large negative activation entropy indicates that the carboxylation reaction proceeds in an associative way, during which CO_2

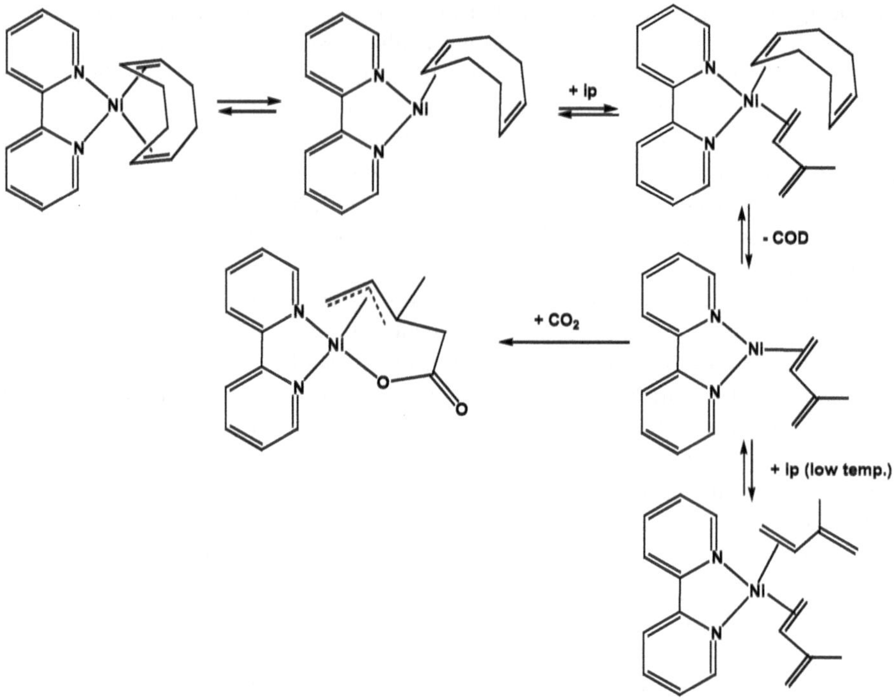

Scheme 5.23 Carboxylation of isoprene with CO_2 mediated by Ni(bpy)(cod): reaction mechanism (adapted from [103])

is inserted into a Ni–C bond in an oxidative coupling step. Therefore, also in this case, the carboxylation reaction seems to involve the activation of the substrate (ip) rather than CO_2. Only after diene activation can a reaction with CO_2 take place.

As already noted for alkenes, alkynes, and allenes, an alternative route to the one-to-one coupling of conjugated dienes with the heterocumulene, which allows bypassing of the oxidative coupling step between the two substrates, can be provided by the hydrocarboxylation pathway, involving hydrometalation of conjugated diene to give a π-allyl–metal complex, followed by reaction with CO_2 [104, 105]. Recently, the hydropalladation route explored by Iwasawa and coworkers for the catalytic carboxylation of allenes (see Sect. 5.5 and Scheme 5.18) has also been exploited successfully by the same research group for the catalytic carboxylation of conjugated 1,3-dienes, using the same PS*i*P-pincer type palladium complex described in Scheme 5.18 as catalyst precursor and $AlEt_3$ as reducing/*trans*-metalation agent. The protocol provided a very useful method for the synthesis of β,γ-unsaturated carboxylic acid derivatives through efficient one-to-one coupling of 1,3-dienes with CO_2 [105].

In the presence of transition metal (Pd, Ni, Ru, Rh) catalysts, conjugated dienes can also cooligomerize with carbon dioxide. The first studies on catalytic cooligomerization of 1,3-dienes with CO_2 date back to the end of the 1970s and were carried out by Inoue [106] and Musco [107], who succeeded in synthesizing lactones and linear esters, albeit in low yield. Since then, several studies [1–9] have been conducted by different research groups with the aim of improving both yield and selectivity of these reactions and rationalizing the reaction mechanism.

The best results in terms of activity and selectivity have been achieved using 1,3-butadiene as the substrate in the presence of Pd-based catalytic systems. Behr's group [6, 108–112] studied in detail the cooligomerization of butadiene with CO_2, throwing light on the conditions leading to the synthesis of δ-lactone 2-ethylidene-6-heptene-5-olide (**1**, Scheme 5.24), the most representative of the products which can form in this process. The formation of **1** is accompanied by that of other by-products, such as 2-ethylidene-5-heptene-4-olide (**2**) and its conjugated isomer 2-ethyl-2,4-heptadiene-4-olide (**3**), linear esters **4** and **5**, and diene oligomers (Scheme 5.24). The distribution of the products depends on the working conditions and is markedly affected by temperature. For instance, using the catalytic system Pd $(acac)_2/P(i\text{-}Pr)_3$ [6], it was found that the formation of lactone **1** was very selective at low temperatures (333–343 K), but yields were low. The selectivity decreased with increasing temperature because of formation of esters and diene oligomers. However, above 403 K only oligomers of butadiene were observed. The synthesis of lactone **1** has been optimized and developed to miniplant scale [110–112]. The selectivity to the δ-lactone has been increased above 95 % by recycling of by-products, reaching an overall butadiene conversion of 45 %. Moreover, solvent, unreacted starting materials, and catalyst can also be recycled, affording a stable process which is ready for industrial exploitation. δ-Lactone **1** is a richly functionalized molecular system which may find practical utilization in numerous synthetic applications [109]. This compound has proved to be a versatile synthone for further reactions, leading to the synthesis in high yields of interesting products, such as acids, alcohols or diols, aldehydes, amino acids or amines, esters, silanes, and even polymers.

Scheme 5.24 summarizes the catalytic cycle leading to the formation of δ-lactone **1**, as well as the other observed by-products. The cycle also applies to Ni- and Ru-based catalysts, which, however, exhibited poor catalytic activity.

The catalytically active species is a palladium(0) complex stabilized by a tertiary phosphine, which is formed initially in situ from a palladium(II) precursor. Very interesting yields and selectivities to δ-lactone have been achieved by using catalytic systems based on $Pd(acac)_2$ and phosphines with good σ-donor properties and bulky substituents, such as $P(i\text{-}Pr)_3$ or PCy_3. It is worth noting that these ligands, which are very active in the synthesis of δ-lactone, contain a CHR_2 group adjacent to the phosphorus atom. It has been suggested that this group may be involved in the activation of the heterocumulene [6].

The first step is butadiene dimerization by oxidative coupling of two diene molecules to give a bis-η^3-allylpalladium complex **A** which may exist in equilibrium with the η^3,η^1 isomer **A′**. Direct coupling of CO_2 with a diene molecule is not a competitive reaction at this stage, as the diene dimerization is faster at the Pd (0) center. Complex **A′** can eliminate 1,3,7-octatriene or insert CO_2 into the Pd–C σ-bond to give an allylic C_9-Pd-carboxylate **B**. The latter species can undergo reductive elimination to give δ-lactone **1** or convert into γ-lactone **2**, which can isomerize to the more conjugated γ-lactone **3**. The reductive elimination step affording δ-lactone **1** is reversible, as **1** can oxidatively add to Pd and regenerate the allylic Pd-carboxylate **B** from which γ-lactones **2** and **3** can form. Accordingly, the same catalytic systems able to promote the formation of **1** from butadiene and CO_2 also catalyze the isomerization of δ-lactone **1** to γ-lactones **2** and **3**. The latter reaction, which is somewhat facilitated at higher temperatures and catalyst concentrations, indicates that, under the conditions used in the cooligomerization reaction (353–363 K; 8–15 h), δ-lactone **1** is a kinetic product, as the formation of this species is kinetically preferred over that of the thermodynamically more stable γ-lactones **2** and **3**. The key intermediate in the catalytic cycle is the allylic C_9-palladium-carboxylate **B**. This species is also an intermediate in the formation of the linear esters because it can react with two other butadiene molecules to give **4** and **5**. Accordingly, a mixture of the esters **4** and **5** has been obtained by reacting δ-lactone **1** with an excess of butadiene in the presence of the catalytic system Pd $(acac)_2/P(i\text{-}Pr)_3$.

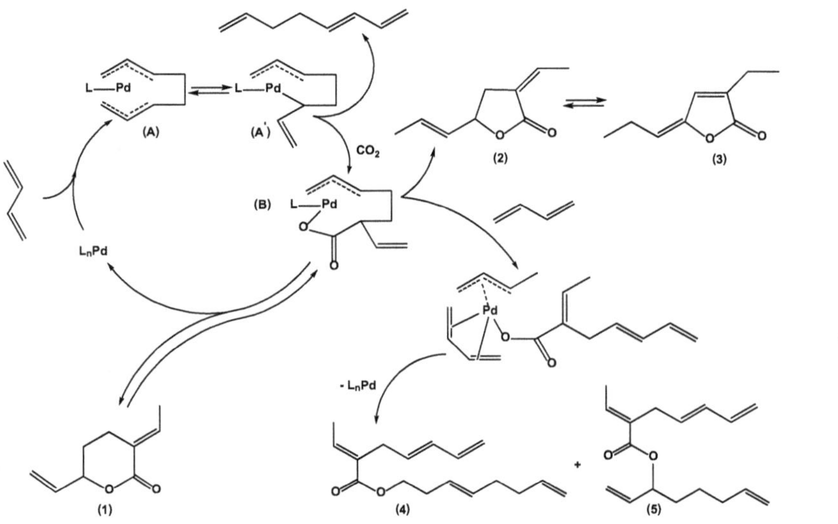

Scheme 5.24 Pd-catalyzed cooligomerization of butadiene with CO_2: reaction mechanism (adapted from [109])

The presence of nitrile groups in the reaction medium markedly facilitates the formation of δ-lactone **1**. In fact, the cooligomerization process proceeded satisfactorily in nitrile solvents, such as acetonitrile, propionitrile, and benzonitrile, whereas butadiene oligomers formed mainly when solvents such as toluene, acetone, and THF were employed. It has been proposed that nitrile molecules can stabilize the catalytically active Pd(0)/phosphine system through the formation of a labile $Pd(NCR)_x(PR_3)_y$ complex. The nitrile ligands are less tightly bound to the metal center than phosphine ligands and, therefore, can more easily generate free coordination sites for the coordination of diene. Based on this "nitrile-effect," Pitter and coworkers have designed and prepared new hemi-labile phosphino-nitrile ligands, $R_2P(CH_2)_n$–CN, which increased the versatility of the catalytic system, allowing one to carry out the catalytic process in solvents other than nitriles or also in solvent-free conditions [113, 114]. Behr and coworkers have recently explored the use of organic carbonates as reaction solvents as an alternative to toxic acetonitrile for the cooligomerization of butadiene with CO_2 [112]. Cyclic carbonates, such as ethylene carbonate, propylene carbonate, butylene carbonate, and glycerol carbonate, facilitated the formation of lactone **1** much more than linear carbonates, such as dimethyl- or diethyl carbonate. Very interesting conversions were observed, with selectivities depending on the size of the substituents in the carbonate molecule.

The cooligomerization process can also be promoted by Rh catalysts [108, 115]. However, in the presence of neutral or ionic complexes of Rh(I) or Rh(III) as catalyst precursors (for instance, the system $(C_2H_4)_2$Rh(acac)/PEt_3), γ-lactone 2-ethyl-2,4,9-undecatriene-4-olide **(6)** was isolated, in

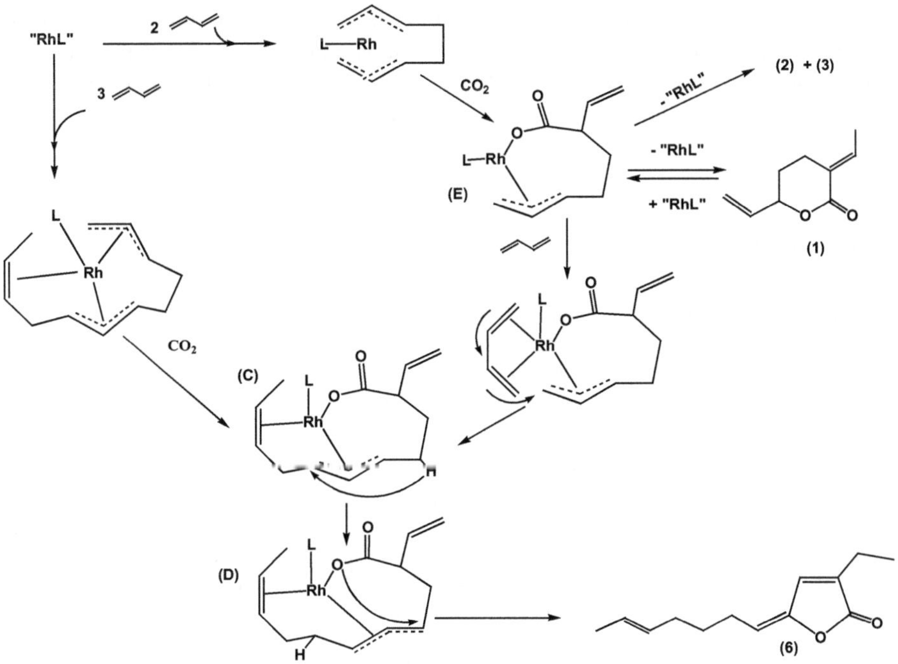

Scheme 5.25 Rh-catalyzed cooligomerization of butadiene with CO_2: plausible reaction mechanisms

which three molecules of butadiene are combined with one molecule of CO_2 (5.9). The cooligomerization reaction proceeded in acetonitrile with low yield (5–8 %) and TON, under relatively severe conditions (393 K, 24 h). C_9-Lactones **1–3** (see

Scheme 5.24) and diene oligomers also formed as by-products. Remarkably, C_{13}-γ-lactone **6** can be obtained in much higher yield from the C_9-δ-lactone **1** and butadiene in the presence of $(C_2H_4)_2Rh(acac)/PEt_3$.

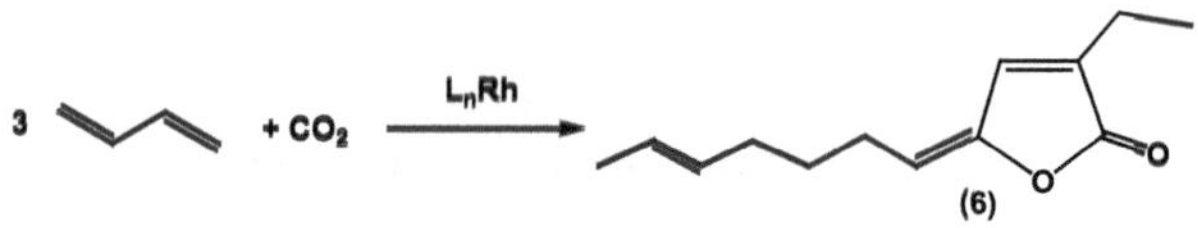

(6) (5.9)

(5.9)

A few plausible reaction pathways for the formation of C_{13}-γ-lactone **6** have been summarized in Scheme 5.25. A C_{12}-bis-η^3-allyl-complex is initially formed by reaction of the starting Rh-complex with three butadiene molecules. This intermediate may further insert CO_2 into the terminal allyl moiety to give a C_{13}-carboxylate-Rh-complex **C**. Upon H-shift, **C** can isomerize to intermediate **D**, from which C_{13}-γ-lactone **6** can be released by reductive elimination. However, a parallel path may also be operative wherein a C_8-bis-allyl Rh-species is initially formed. This species can react, to a certain extent, with CO_2 to give the C_9-carboxylate-Rh-complex **E**, which can further react with the diene to give the final product. It is worth noting that the C_9-Rh-carboxylate **E** is analogous to intermediate **B** in Scheme 5.24. Therefore, the intermediacy of **E** can easily explain not only the formation of the observed carboxylated by-products **1-3**, but also the formation of γ-lactone **6** from δ-lactone **1** and butadiene (Scheme 5.25).

References

1. Yin X, Moss JR (1999) Recent developments in the activation of carbon dioxide by metal complexes. Coord Chem Rev 181:27–59
2. Pandey KK (1995) Reactivities of carbonyl sulfide (COS), carbon disulfide (CS_2) and carbon dioxide (CO_2) with transition metal complexes. Coord Chem Rev 140:37–114
3. Leitner W (1996) The coordination chemistry of carbon dioxide and its relevance for catalysis: a critical survey. Coord Chem Rev 153:257–284
4. Aresta M, Quaranta E, Tommasi I, Giannoccaro P, Ciccarese A (1995) Enzymatic versus chemical carbon dioxide utilization. Part I. The role of metal centres in carboxylation reactions. Gazz Chim Ital 125:509–538
5. Braunstein P, Matt D, Nobel D (1988) Reactions of carbon dioxide with carbon-carbon bond formation catalyzed by transition metal complexes. Chem Rev 88:747–764
6. Behr A (1988) Carbon dioxide activation by metal complexes. VCH Verlagsgesellschaft mbH, Weinheim
7. Louie J (2005) Transition metal-catalyzed reactions of carbon dioxide and other heterocumulenes. Curr Org Chem 9:605–623
8. Tsuda T (1995) Utilization of carbon dioxide in organic synthesis and polymer synthesis by the transition metal-catalyzed carbon dioxide fixation into unsaturated hydrocarbons. Gazz Chim Ital 125:101–110
9. Walther D (1987) Homogeneous catalytic reactions of carbon dioxide with unsaturated substrates, reversible CO_2-carriers, and transcarboxylation reactions. Coord Chem Rev 79: 135–174

10. Hoberg H (1987) Stoichiometric reactions of C-C bond formation promoted by metal systems. In: Aresta M, Forti G (eds) Carbon dioxide as a source of carbon. D. Reidel, Dordrecht, pp 275–293
11. Dahlenburg L, Prengel C (1986) Organometallic compounds of iridium and rhodium. XXX. Oxidative coupling of carbon dioxide and nitrosobenzene by $Rh(4\text{-}MeC_6H_4)$ [*t*-BuP $(CH_2CH_2CH_2PPh_2)_2$]. Inorg Chim Acta 122:55–57
12. Cenini S, Porta F, Pizzotti M, Crotti C (1985) Insertion of carbon dioxide, of CO_2-like molecules, and of other unsaturated compounds into the platinum–nitrogen bond of [Pt $(PPh_3)_2(PhNO)$]. J Chem Soc Dalton Trans 1:163–168
13. Graët A, Sinault L, Fusaro MB, Vallet A-L, Seu C, Kilgore JL, Baum MM (2010) Reactivity of the $Ni(0)\text{-}CO_2$-imine system: new pathway to vicinal diamines. Organometallics 29:1997–2000
14. Walther D, Dinjus E, Herzog V (1983) Aktivierung von kohlendioxid an übergangsmetalizentren: die oxydative kopplung von CO_2 mit carbodiimiden am elektronenreichen nickel (0). Z Chem 23:188
15. Walther D, Dinjus E, Herzog V (1984) Aktivierung von kohlendioxid an übergangsmetallkomplexen: topo- und regioselektivität der metallaringschlußreaktion mit substraten, die azomethin- und olefingrappen enthalten, am elektronenreichen nickel(0)-komplexrump. Z Chem 24:298–299
16. Geyer C, Dinjus E, Schindler S (1998) Reactions of aldehydes with carbon dioxide at nickel (0) centers. A detailed kinetic analysis. Organometallics 17:98–103
17. Herskovitz T, Guggeberger LJ (1976) Carbon dioxide coordination chemistry. The structure and some chemistry of the novel CO_2 addition product $IrCl(C_2O_4)(PMe_3)_3$. J Am Chem Soc 98:1615–1616
18. Cohen SA, Bercaw JE (1985) Titanacycles derived from reductive coupling of nitriles, alkynes, acetaldehyde, and carbon dioxide with bis(pentamethylcyclopentadienyl)(ethylene)-titanium(II). Organometallics 4:1006–1014
19. Six Y (2002) The reaction of dialkoxytitanacyclopropanes and dialkoxytitana-cyclopropenes with carbon dioxide. J Chem Soc Perkin Trans 1:1159–1160
20. Six I (2003) Titanium-mediated carboxylation of alkynes with carbon dioxide. Eur J Org Chem 2003:1157–1171
21. Yamashita K, Chatani N (2005) Cp_2ZrCl_2-mediated three-component coupling reactions of CO_2, ethylene (or alkynes), and electrophiles leading to carboxylic acid derivatives. Synlett 6:919–922
22. Hessen B, Meetsma A, Van Bolhuis F, Teuben JH, Helgesson G, Jagner S (1990) Chemistry of carbon monoxide free monocyclopentadienylvanadium(I) alkene and alkyne complexes. Organometallics 9:1925–1936
23. Aresta M, Quaranta E (1993) Synthesis, characterization and reactivity of $Rh(bpy)(C_2H_4)Cl$: a study on the reaction with C_1 molecules (CH_2O, CO_2) and $NaBPh_4$. J Organomet Chem 463:215–221
24. Hipler B, Döring M, Dubs C, Görls H, Hübler T, Uhlig E (1998) Bildung und strukturen von nickelacyclen des typs $(LL')\overline{NiCH_2CH_2C(O)O}$. Z Anorg Allg Chem 624:1329–1335
25. Hoberg H, Peres Y, Krüger C, Tsay Y-H (1987) A 1-oxa-2-nickela-5-cyclopentanone from ethene and carbon dioxide: preparation, structure, and reactivity. Angew Chem Int Ed Engl 26:771–773
26. Hoberg H, Jenni K, Angermund K, Krüger C (1987) CC-linkages of ethene with CO_2 on an iron(0) complex—synthesis and crystal structure analysis of $[(PEt_3)_2Fe(C_2H_4)_2]$. Angew Chem Int Ed Engl 26:153–155
27. Aresta M, Quaranta E, Tommasi I (1988) Reduction of co-ordinated carbon dioxide to carbon monoxide via protonation by thiols and other Broensted acids promoted by Ni-sistems: a contribution to the understanding of the mode of action of the enzyme carbon monoxide dehydrogenase. J Chem Soc Chem Commun 450–452

28. Aresta M, Gobetto R, Quaranta E, Tommasi I (1992) A bonding-reactivity relationship for Ni $(PCy_3)_2(CO_2)$: a comparative solid state-solution nuclear magnetic resonance study (^{31}P, ^{13}C) as a diagnostic tool to determine the mode of binding of CO_2 to a metal center. Inorg Chem 32:4286–4290
29. Aresta M, Dibenedetto A (2002) Key issues in carbon dioxide utilization as a building block for molecular organic compounds in the chemical industry. ACS Symp Ser 809:54–70
30. Dedieu A, Bo C, Ingold F (1990) Theoretical studies on carbon dioxide organometallic reactivity. In: Aresta M, Schloss JV (eds) Enzymatic and model reactions for carbon dioxide utilization. Kluwer, Dordrecht, pp 23–42
31. Hoberg H, Schaefer D, Burkhart G, Krüger C, Romao MJ (1984) Nickel(0)-induzierte C-C-verknüpfung zwischen kohlendioxid und alkinen sowie alkenen. J Organomet Chem 266: 203–224
32. Papai I, Schubert G, Mayer I, Besenyei G, Aresta M (2004) Mechanistic details of nickel(0)-assisted oxidative coupling of CO_2 with C_2H_4. Organometallics 23:5252–5259
33. Hoberg H, Peres Y, Milchereit A (1986) C-C Verknüpfung von alkenen mit CO_2 an nickel(0): herstellung von zimtsäure aus styrol. J Organomet Chem 307:C38–C40
34. Graham DC, Mitchell C, Bruce MI, Metha GF, Bowie JH, Buntine MA (2007) Production of acrylic acid through nickel-mediated coupling of ethylene and carbon dioxide—a DFT study. Organometallics 26:6784–6792
35. Yamamoto T, Igarashi K, Komiya S, Yamamoto A (1980) Preparation and properties of phosphine complexes of nickel-containing cyclic amides and esters $[(PR_3)_nNiCH_2CH(R^1)COZ\ (Z=NR^2, O)]$. J Am Chem Soc 102:7448–7456
36. Fischer R, Langer J, Malassa G, Walther D, Gorls H, Vaughan G (2006). A key step in the formation of acrylic acid from CO_2 and ethylene: the transformation of a nickelalactone into a nickel-acrylate complex. Chem Commun 2510–2512
37. Bruckmeier C, Lehenmeyer MW, Reichardt R, Vagin S, Rieger B (2010) Formation of methyl acrylate from CO_2 and ethylene via methylation of nickelalactones. Organometallics 29:2199–2202
38. Aresta M, Pastore C, Giannoccaro P, Kovacs G, Dibenedetto A, Papai I (2007) Evidence of spontaneous release of acrylates from a transition metal complex upon coupling ethene or propene with a carboxylic moiety or CO_2. Chem Eur J 13:9028–9034
39. Lejkowski ML, Lindner R, Kageyama T, Bódizs GE, Plessow PN, Müller IB, Schäfer A, Rominger F, Hofmann P, Futter C, Schunk SA, Limbach M (2012) The first catalytic synthesis of an acrylate from CO_2 and an alkene—a rational approach. Chem Eur J 18: 14017–14025
40. Alvarez R, Carmona E, Galindo A, Gutierrez E, Marin JM, Monge A, Poveda ML, Ruiz C, Savariault JM (1989) Formation of carboxylate complexes from the reactions of CO_2 with ethylene complexes of molybdenum and tungsten. X-Ray and neutron diffraction studies. Organometallics 8:2430–2439
41. Galindo A, Pastor A, Perez PJ, Carmona E (1993) Bis(ethylene)complexes of molybdenum and tungsten and their reactivity towards CO_2. New examples of acrylate formation by coupling of ethylene and carbon dioxide. Organometallics 12:4443–4451
42. Schubert G, Papai I (2003) Acrylate formation via metal-assisted C-C coupling between CO_2 and C_2H_4: reaction mechanism as revealed from density functional calculations. J Am Chem Soc 125:14847–14858
43. Bernskoetter WH, Tyler BT (2011) Kinetics and mechanism of molybdenum-mediated acrylate formation from carbon dioxide and ethylene. Organometallics 30:520–527
44. Lapidus AL, Pirozhkov SD, Koryakin AA (1978) Catalytic synthesis of propionic acid by the carboxylation of ethylene by carbon dioxide. Bull Acad Sci USSR, Div Chem Sci (Engl Transl) 27(12):2513–2515
45. Besecke S, Schroder G (1981) Carboxylic acid esters from olefins. DE 2948888A1. Chem Abstr 95, 97107m

46. Dérien S, Clinet J-C, Duñach E, Périchon J (1992) Electrochemical incorporation of carbon dioxide into alkenes by nickel complexes. Tetrahedron 48:5235–5248
47. Williams CM, Johnson JB, Rovis T (2008) Nickel-catalyzed carboxylation of styrenes using CO_2. J Am Chem Soc 130:14936–14937
48. Riduan SN, Zhang Y (2010) Recent developments in carbon dioxide utilization under mild conditions. Dalton Trans 39:3347–3357
49. Zhang Y, Riduan SN (2011) Catalytic hydrocarboxylation of alkenes and alkynes with CO_2. Angew Chem Int Ed 50:6210–6212
50. Li S, Yuan W, Ma S (2011) Highly regio- and stereoselective three-component nickel-catalyzed *syn*-hydrocarboxylation of alkynes with diethyl zinc and carbon dioxide. Angew Chem Int Ed 50:2578–2582
51. Fujihara T, Xu T, Semba K, Terao J, Tsuji Y (2011) Copper-catalyzed hydrocarboxylation of alkynes using carbon dioxide and hydrosilanes. Angew Chem Int Ed 50:523–527
52. Takimoto M, Mizuno T, Mori M, Sato Y (2006) Nickel-mediated cyclization of enynes under an atmosphere of carbon dioxide. Tetrahedron 62:7589–7597
53. Takimoto M, Mizuno T, Sato Y, Mori M (2005) Nickel-mediated carboxylative cyclization of enynes. Tetrahedron Lett 46:5173–5176
54. Dérien S, Duñach E, Périchon J (1991) From stoichiometry to catalysis: electroreductive coupling of alkynes and carbon dioxide with nickel-bipyridine complexes. Magnesium ions as the key for catalysis. J Am Chem Soc 113:8447–8454
55. Dérien S, Clinet J-C, Duñach E, Périchon J (1993) Activation of carbon dioxide: nickel-catalysed electrochemical carboxylation of diynes. J Org Chem 58:2578–2588
56. Shimizu K, Takimoto M, Sato Y, Mori M (2005) Nickel catalyzed regioselective synthesis of tetrasubstituted alkene using alkylative carboxylation of disubstituted alkynes. Org Lett 7: 195–197
57. Mori M (2007) Regio- and stereoselective synthesis of tri- and tetrasubstituted alkenes by introduction of CO_2 and alkylzinc into alkynes. Eur J Org Chem 2007:4981–4993
58. Saito S, Nakagawa S, Koizumi T, Hirayama K, Yamamoto Y (1999) Nickel-mediated regio- and chemoselective carboxylation of alkynes in the presence of carbon dioxide. J Org Chem 64:3975–3978
59. Labbé E, Duñach E, Périchon J (1988) Ligand-directed reaction products in the nickel-catalyzed electrochemical carboxylation of alkynes. J Organomet Chem 353:C31–C36
60. Aoki M, Kaneko M, Izumi S, Ukai K, Iwasawa N (2004) Bidentate amidine ligands for nickel (0)-mediated coupling of carbon dioxide with unsaturated hydrocarbons. Chem Commun 2568–2569
61. Sakaki S, Mine K, Taguchi D, Arai T (1993) Formation of the oxanickelacyclopentene complex from nickel(0), carbon dioxide, and alkyne. An *ab initio* MO/SD-CI study. Bull Chem Soc Jpn 66:3289–3299
62. Sakaki S, Mine K, Hamada T, Arai T (1995) Formation of the oxanickelacyclopentene complex from nickel(0), carbon dioxide, and alkyne. An *ab initio* MO/SD-CI Study. Part II. Reactivity and regioselectivity of hydroxyacetylene. Bull Chem Soc Jpn 68:1873–1882
63. Li J, Jia G, Lin Z (2008) Theoretical studies on coupling reactions of carbon dioxide with alkynes mediated by nickel(0) complexes. Organometallics 27:3892–3900
64. Graham DC, Bruce MI, Metha GF, Bowie JH, Buntine MA (2008) Regioselective control of the nickel-mediated coupling of acetylene and carbon dioxide. J Organomet Chem 693: 2703–2710
65. Inoue Y, Itoh Y, Kazama H, Hashimoto H (1980) Reaction of dialkyl-substituted alkynes with carbon dioxide catalyzed by nickel(0) complexes. Incorporation of carbon dioxide in alkyne dimers and novel cyclotrimerization of the alkynes. Bull Chem Soc Jpn 53:3329–3333
66. Walther D, Schönberg H, Dinjus E, Sieler J (1987) Aktivierung von kohlendioxid an übergangsmetallzentren: selektive cooligomerisation mit hexin(−3) durch das katalysator-system acetonitril/trialkylphosphan/nickel(0) und struktur eines nickel(0)-komplexes mit side-on gebundenem acetonitril. J Organomet Chem 334:377–388

67. Jessop PJ, Leitner W (eds) (1999) Chemical synthesis using supercritical carbon dioxide. Wiley-VCH, Weinheim
68. Kishimoto Y, Mitani I (2005) Solvent-free synthesis of 2-pyrone from alkynes and carbon dioxide catalyzed by Ni(1,5-cyclooctadiene)$_2$/trialkylphosphine catalysts. Synlett 14:2141–2146
69. Tekavec TN, Zuo G, Simon K, Louie J (2006) An in situ approach for nickel-catalyzed cycloaddition. J Org Chem 71:5834–5836
70. Hoberg H, Schaefer D, Burkhart G (1982) Oxanickelacyclopenten-derivate, ein neuer typ vielseitig verwendbarer synthone. J Organomet Chem 228:C21–C24
71. Hoberg H, Schaefer D (1982) Modellkomplexe des nickels für die [2+2+2′]-cycloaddition von alkinen mit kohlendioxid. J Organomet Chem 238:383–387
72. Kempe R, Sieler J, Walther D, Reinhold J, Rommel K (1993) Aktivierung von CO_2 an übergangsmetallzentren: zum ablauf der CO_2-reduktion an nickel(0)-fragmenten. Z Anorg Allg Chem 619:1105–1110
73. Walther D, Bräunlich G, Kempe R, Siele J (1992) Aktivierung von CO_2 an übergangsmetallzentren: zum ablauf der homogen-katalytischen bildung von 2-pyron aus kohlendioxid und hex-3-in an nickel(0)-fragmenten. J Organomet Chem 436:109–119
74. Diversi P, Ingrosso G, Lucherini A, Malquori S (1987) Cobalt-catalyzed cyclotrimerization of alkynes and heterocumulenes. J Mol Catal 40:267–280
75. Albano P, Aresta M (1980) Some catalytic properties of Rh(diphos)(η^6-BPh_4). J Organomet Chem 190:243–246
76. Pillai SM, Ohnishi R, Ichicawa M (1990) Cycloaddition of carbon dioxide to propyne over supported Rh_4 and Fe_2Rh_4 carbonyl cluster-derived catalysts. J Chem Soc Chem Commun 246–247
77. Pillai SM, Ohnishi R, Ichicawa M (1992) A mechanistic study of cycloaddition of carbon dioxide to propyne over supported Rh_4 and Fe_2Rh_4 carbonyl cluster-derived catalysts. React Kinet Catal Lett 48:201–208
78. Hoberg H, Oster BW (1984) Nickel(0)-induzierte CC-verknüpfung zwischen 1,2-dienen und kohlendioxid. J Organomet Chem 266:321–326
79. Takimoto M, Kawamura M, Mori M (2004) Nickel-mediated regio-and stereoselective carboxylation of trimethylsilylallene under an atmosphere of carbon dioxide. Synthesis 2004:791–795
80. Takimoto M, Kawamura M, Mori M (2003) Nickel(0)-mediated sequential addition of carbon dioxide and aryl aldehydes into terminal allenes. Org Lett 5:2599–2601
81. Takimoto M, Kawamura M, Mori M, Sato Y (2011) Nickel-promoted carboxylation/cyclization cascade of allenyl aldehyde under an atmosphere of CO_2. Synlett 2011:1423–1426
82. Aoki M, Izumi S, Kaneko M, Ukai K, Takaya J, Iwasawa N (2007) Ni(0)-promoted hydrocarboxylation of 1,2-dienes by reaction with CO_2 and O_2. Org Lett 9:1251–1253
83. Dérien S, Clinet J-C, Duñach E, Périchon J (1990) Coupling of allenes and carbon dioxide catalyzed by electrogenerated nickel complexes. Synlett 1990:361–364
84. Takimoto M, Kawamura M, Mori M, Sato Y (2005) Nickel-catalyzed regio-and stereoselective double carboxylation of trimethylsilylallene under an atmosphere of carbon dioxide and its application to the synthesis of chaetomellic acid A anhydride. Synlett 2005:2019–2022
85. Takaya J, Iwasawa N (2008) Hydrocarboxylation of allenes with CO_2 catalyzed by silyl pincer-type palladium complex. J Am Chem Soc 130:15254–15255
86. North M (2011) Synthesis of β, γ-unsaturated acids from allenes and carbon dioxide. Angew Chem Int Ed 48:4104–4105
87. Dohring A, Jolly PM (1980) The palladium catalyzed reaction of carbon dioxide with allene. Tetrahedron Lett 21:3021–3024
88. Aresta M, Ciccarese A, Quaranta E (1985) Head to head and head to tail coupling of allene and co-condensation with carbon dioxide promoted by 1,2-bis(diphenylphosphino)ethane(η^6-tetraphenylborate)rhodium C_1. Mol Chem 1:283–295

89. Aresta M, Quaranta E, Tommasi I (1993) Spectroscopic characterization of (diphos)Rh(η^6-$PhBPh_3$) in solution and study on its reactivity as a total phenylating agent. Formation of C-C and C-C-C bonds by reaction with aldehydes. Gazz Chim Ital 123:271–278
90. Aresta M, Quaranta E, Tommasi I, Dérien S, Duñach E (1995) Tetraphenylborate anion as a phenylating agent: chemical and electrochemical reactivity of BPh_4–Rh-complexes toward mono- and dienes and carbon dioxide. Organometallics 14:3349–3356
91. Sasaki Y (1989) Nickel-catalyzed cyclo-cooligomerization of allene with carbon dioxide. J Mol Catal 54:L9–L12
92. Tsuda T, Yamamoto T, Saegusa T (1992) Palladium-catalyzed cycloaddition of carbon dioxide with methoxyallene. J Organomet Chem 429:C46–C48
93. Aresta M, Dibenedetto A, Papai I, Schubert G (2002) Unprecedented formal [2 + 2] addition of allene to CO_2 promoted by $[RhCl(C_2H_4)(P^iPr_3)]_2$: direct synthesis of the four membered lactone α-methylene-β-oxiethanone. The intermediacy of $[RhH_2Cl(P^iPr_3)]_2$: theoretical aspects and experiments. Inorg Chim Acta 334:294–300
94. Rode JE, Dobrowolski JCZ (2006) Reaction paths of the [2 + 2] cycloaddition of X=C=Y molecules (X, Y≡S or O or CH_2). *Ab initio* study. J Phys Chem A 110:207–218
95. Dobrowolski JCZ, Jamroz MH, Borowiak MA, Quaranta E, Aresta M (2006) Theoretical IR and Raman spectra of diketene and its 3-methylene isomer. Vibr Spectrosc 22:19–28
96. Hoberg H, Jenni K, Krüger C, Raabe E (1986) CC-Kupplung von CO_2 und butadien an eisen (0)-komplexen–ein neuer weg zu α, ω-dicarbonsäuren. Angew Chem 98:819–820
97. Yasuda H, Okamoto T, Matsuoka Y, Nakamura A, Kai Y, Kanehisa N, Kasai N (1989) Diverse reaction courses in the controlled carbometalation of heterocumulenes with zirconium-diene complexes and molecular structures of carbon dioxide, isocyanate and ketene 1:1 and 1:2 inserted compounds. Organometallics 8:1139–1152
98. Hoberg H, Gross S, Milchereit A (1987) Nickel(0)-catalyzed production of a functionalized cyclopentanecarboxylic acid from 1,3-butadiene and CO_2. Angew Chem Int Ed Engl 26: 571–572
99. Takimoto M, Mori M (2002) Novel catalytic CO_2 incorporation reaction: nickel catalyzed regio- and stereoselective ring-closing carboxylation of bis-1,3-dienes. J Am Chem Soc 124: 10008–10009
100. Takimoto M, Nakamura Y, Kimura K, Mori M (2004) Highly enantioselective catalytic carbon dioxide incorporation reaction: nickel-catalyzed asymmetric carboxylative cyclization of bis-1,3-dienes. J Am Chem Soc 126:5956–5957
101. Behr A, Kanne U (1986) Nickel complex induced C-C linkage of carbon dioxide with trienes. J Organomet Chem 317:C41–C44
102. Takimoto M, Mori M (2001) Cross-coupling reaction of oxo-π-allylnickel complex generated from 1,3-diene under an atmosphere of carbon dioxide. J Am Chem Soc 123:2895–2896
103. Geyer C, Schindler S (1998) Kinetic analysis of the reaction of isoprene with carbon dioxide and a nickel(0) complex. Organometallics 17:4400–4405
104. Gao Y, Iijima S, Urabe H, Sato F (1994) Carbon dioxide fixation by $Cp_2(\eta^3$-allyl)Ti complexes generated by various dienes. Inorg Chim Acta 222:145–153
105. Takaya J, Sasano K, Iwasawa N (2011) Efficient one-to-one coupling of easily available 1,3-dienes with carbon dioxide. Org Lett 13:1698–1701
106. Inoue Y, Sasaki Y, Hashimoto H (1976) Reaction of carbon dioxide with butadiene catalysed by palladium complexes. Synthesis of 2-ethylidenehept-5-en-4-olide. J Chem Soc Chem Commun 605–606
107. Musco A (1980) Co-oligomerization of butadiene and carbon dioxide catalysed by tertiary phosphine-palladium complexes. J Chem Soc Perkin Trans 1:693–698
108. Behr A, He R, Juszak KD, Krueger C, Tsay YH (1986) Possibilities of controlling transition metal-catalyzed reactions of 1,3-dienes with carbon dioxide. Chem Ber 119:991–1015
109. Behr A, Henze H (2011) Use of carbon dioxide in chemical syntheses via a lactone intermediate. Green Chem 13:25–39

110. Behr A, Becker M (2006) The telomerisation of 1,3-butadiene and carbon dioxide: process development and optimisation in a continuous miniplant. Dalton Trans 4607–4613
111. Behr A, Bahke P, Becker M (2004) Palladium-katalysierte telomerisation von kohlendioxid mit butadien im labor- und miniplantmaßstab. Chem Ing Tech 76:1828–1832
112. Behr A, Bahke P, Klinger B, Becker M (2007) Application of carbonate solvents in the telomerisation of butadiene with carbon dioxide. J Mol Catal A Chem 267:149–156
113. Pitter S, Dinjus E (1997) Phosphinoalkyl nitriles as hemilabile ligands: new aspects in the homogeneous catalytic coupling of CO_2 and 1,3-butadiene. J Mol Catal A Chem 125:39–45
114. Buchmüller K, Dahmen N, Dinjus E, Neumann D, Powietzka B, Pitter S, Schön J (2003) Control of homogeneously catalyzed reactions by phase equilibria. Green Chem 5:218–223
115. Behr A, He R (1984) Rhodium catalyzed reaction of butadiene and carbon dioxide. J Organomet Chem 276:C69–C72

Chapter 6
Reaction Mechanisms in the Direct Carboxylation of Alcohols, Polyols, Cyclic Ethers, and Cyclic Amines to Afford Monomeric Compounds and Polymeric Materials

Abstract This chapter deals with the utilization of CO_2 in the carboxylation of alcohols, diols, polyols, and epoxides to create a variety of compounds such as linear carbonates, cyclic monomeric carbonates, and polycarbonates. Homogeneous, heterogenized, and heterogeneous catalysts are described. The problem of "water elimination" is considered and routes for water-trapping discussed. DFT calculations used to support the reaction mechanism are presented with the identified transition states relevant to various mechanistic scenarios.

6.1 Utilization of Organic Carbonates and Conventional Synthetic Routes

Molecular acyclic (Structure 6.1a) and cyclic (Structure 6.1b) organic carbonates are expected to experience a considerable expansion of their market during the coming years because of their use in the chemical (as solvent and reagent [1]), pharmaceutical (as intermediates [2, 3]), and polymer industries [4] (Fig. 6.1).

The lower members of the acyclic carbonate family, namely dimethyl-carbonate (DMC) and diethylcarbonate (DEC), have been proposed as components of gasoline [6]. Considering the percentage they can reach in gasoline (up to 5 7 %), it is foreseeable that their free market may face a growth of more than two orders of magnitude from the actual 0.1 Mt $year^{-1}$ [7, 8]. It is useful to point out that most of the DMC actually produced is for captive use in the industry of polycarbonates: only ~10 % is on the market. To satisfy market demand, new synthetic technologies need to be developed that are not based on the use of phosgene (6.1), which is banned in several countries and cannot be transported [9, 10].

$$2ROH + COCl_2 + 2Base \rightarrow (RO)_2CO + 2BaseHCl \qquad (6.1)$$

Safer and environmentally friendly synthetic methodologies, either CO-, CO_2-, or urea-based routes, are available in the scientific and patent literature. Some of them

M. Aresta et al., *Reaction Mechanisms in Carbon Dioxide Conversion*,
DOI 10.1007/978-3-662-46831-9_6

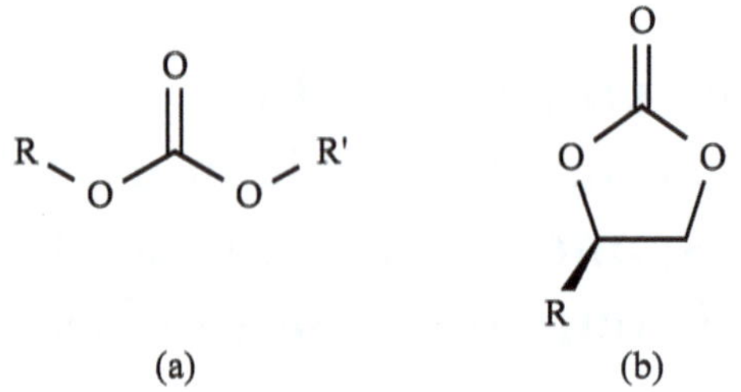

Structure 6.1 (a) linear carbonate; (b) cyclic carbonate

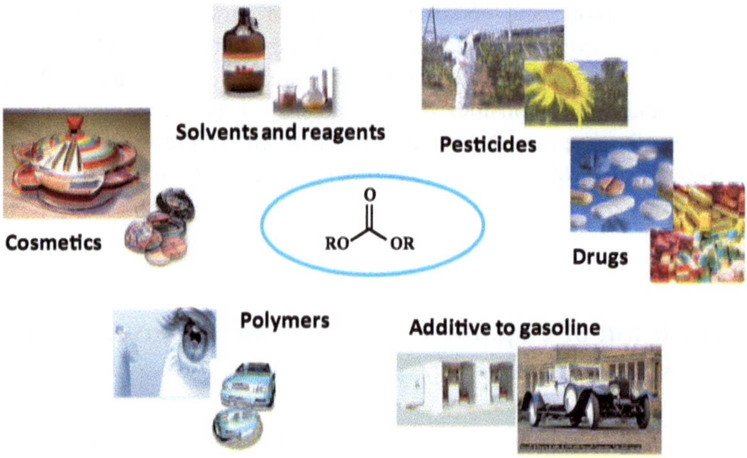

Fig. 6.1 Uses of dialkylcarbonates. Reprinted with permission from [5]

are on stream (oxidative carbonylation of methanol) or close to commercialization (urea chemistry). Noteworthy, the oxidative carbonylation better suits methanol than higher alcohols. The technology has been developed by ENIChem [11–15] (6.2) and UBE [16] (6.3), each having drawbacks that have prevented the development to a large scale exploitation.

$$2CH_3OH + 1/2O_2 + CO \xrightarrow{Cu/Pd} (CH_3O)_2CO + H_2O \qquad (6.2)$$

$$2CH_3OH + 1/2O_2 + CO \xrightarrow{Pd/NO} (CH_3O)_2CO + H_2O \qquad (6.3)$$

The trans-esterification of other carbonates is also used (6.4) [17], which mainly uses ethene carbonate (6.4) prepared from ethene epoxide (the industrial process needs improvement) and CO_2 (6.5) [18, 19].

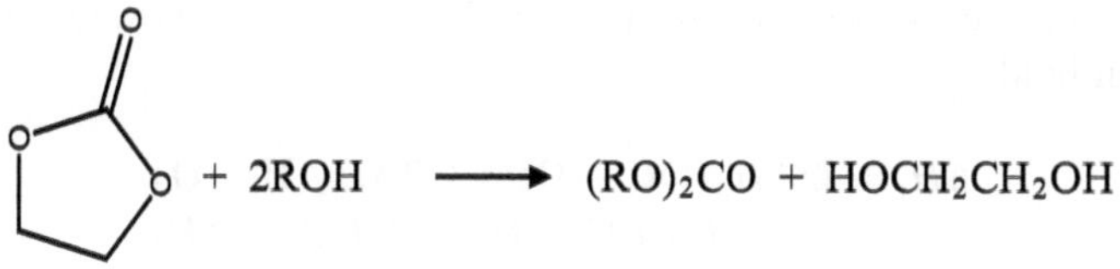

(6.4)

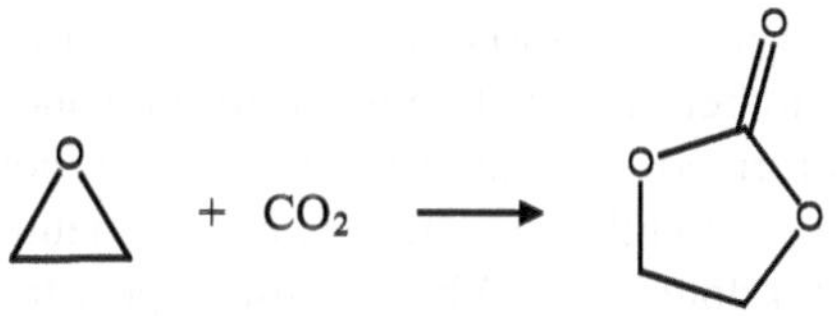

(6.5)

The direct carboxylation of alcohols (6.6) and the reaction of alcohols with urea (used as an active form of CO_2) (6.7) represent appealing alternatives. The former has some kinetic and thermodynamic barriers, the latter raises NH_3 recovery issues and product separation problems.

$$\begin{gathered}2ROH + CO_2 \rightarrow (RO)_2CO + H_2O \\ R = Me,\ Et,\ Allyl\end{gathered} \tag{6.6}$$

$$2ROH + H_2NCONH_2 \rightarrow (RO)_2CO + 2NH_3 \tag{6.7}$$

6.2 Direct Carboxylation of Alcohols

Because of their Gibbs free energy (see below), reaction (6.6) is shifted to the left much more than reaction (6.7). As a matter of fact, the literature shows that the equilibrium concentration of DMC in reaction (6.6) (R = Me) is less than 1 %, depending on the reaction temperature [20], whereas in reaction (6.7) it can easily exceed 20 %. Other alcohols, such as ethanol and allyl alcohol [21], show an analogous behavior. Unfavorable thermodynamics is not the only drawback of reaction (6.6): the water formed in the reaction pushes the equilibrium to the left and affects the catalyst structure and activity [20, 22, 23] in both homogeneous and heterogeneous processes; water traps have been used, either chemical- or process-based, and are discussed in Sect. 6.2.2.3.

6.2.1 Thermodynamic and Kinetic Issues

Let us start with a general consideration of (6.6). It formally implies the following elementary reactions:

1. Breaking of two O–H bonds in two alcohol molecules
2. Formation of two O–H bonds in water
3. Breaking of two C–O bonds in CO_2
4. Formation of two CO bonds in the carbonate

All together, we can summarize the ΔH change as in (6.8) where the affected bonds are in bold:

$$\Delta H_{\mathrm{r}} = [2\Delta H_{\mathrm{f}}(\mathrm{OC} - \mathbf{OR}) + 2\Delta H_{\mathrm{f}}(\mathbf{H} - \mathbf{O})] \\ - [2\Delta H_{\mathrm{d}}(\mathbf{RO} - \mathbf{H}) + \Delta H_{\mathrm{d}}(\mathbf{C} = \mathbf{O})], \tag{6.8}$$

where r is the reaction; f the formation of a bond; and d the dissociation of bond.

Because the same number of O–H bonds are formed and cleaved, and two O–C bonds are cleaved and formed, despite different kinds of bonds being implied, one can foresee that the overall enthalpy effect is almost neutral and depends on the difference of energy of a double C=O bond with respect to two single C–O bonds and by the different dissociation-formation energy of O–H bonds in water and alcohols. If aromatic compounds are used, then the presence of the aromatic ring with respect to the aliphatic group makes a difference to enthalpy. Conversely, one can foresee that the entropic factor may play a more significant role than ΔH, as gaseous CO_2 is converted into condensed phase CO_2-containing products, and the effect is the same for all reactions, should it be either an aliphatic or an aromatic derivative. In summary, the enthalpy effect is driven by the nature of the alcohol, although the entropic effect is driven by CO_2.

If a liquid alcohol is reacted with gaseous CO_2 to afford liquid carbonates and liquid water, reaction (6.6) has a negative enthalpy of reaction if R is an alkyl group. Conversely, when R is an aromatic moiety, ΔH is positive (see Table 6.1). By applying the equation that correlates the thermodynamic quantities: Gibbs free energy-enthalpy-entropy of reaction-temperature (6.9), one can calculate the value of ΔG, considering that ΔS is equal to ~34 kcal mol^{-1} for the various reactions. Temperature also plays a role; an increase makes the product ($T\Delta S$) more negative and ΔG becomes more positive.

$$\Delta G = \Delta H - T\Delta S \tag{6.9}$$

As a result of the unfavorable thermodynamics, the equilibrium concentration of the formed carbonate turns out to be quite low, often around 1 % or even less, depending on the reaction temperature, which usually is in the range 400–450 K for a metal-catalyzed reaction. Keeping the temperature as low as possible is a key issue in this reaction.

Table 6.1 ΔH° and ΔG° for the synthesis of organic carbonates from alcohols and CO_2

R group	ΔH° at 298 K kcal mol^{-1}	ΔG° at 298 K kcal mol^{-1}
CH_3	−4.0	6.0
C_2H_5	−3.8	6.2
CH_2=CH–CH_2	−3.9	6.1
C_6H_5	12.1	22.1

Adapted with permission from [21]. Copyright (2003) American Chemical Society

However, super-active catalysts are necessary to carry out the reaction at the lowest temperature, hopefully in the range 300–350 K or lower. Another approach used, discussed later, is the coupling of two reactions: the carboxylation of alcohols and a strongly exergonic reaction that together may produce an overall negative ΔG. The exergonic reaction often used is the hydration of an organic substrate, such as (1) multiple C–N bonds (cyanides, carbodiimides); (2) stressed rings; (3) reactive acetals or ketals or orthoesters. Obviously such partner reactions need to be associated in equimolar amounts to the carboxylation and the effect is that one often gains in conversion of alcohols but loses in post-conversion treatment, as the separation of the products may not be easy and requires energy. Additionally, the formed carbonates (which are good alkylating or carboxy-alkylating agents; see below) can react with the new formed species with an overall loss of selectivity, thus possibly producing no net benefit. It is worth recalling that the CO_2-based chemistry targets sustainability and reduction of C-footprint of synthetic processes: such targets cannot be missed. To reduce the effect of the entropic factor, the carboxylation reaction can be carried out in supercritical CO_2 ($scCO_2$). To be effective, the reactive system must be in a single phase. Alternatively, high CO_2 pressure is used using liquid alcohols as reaction medium. In fact, if one operates at a pressure of 300 MPa, the reaction is shifted to the right: such operative conditions are not sustainable.

It has been mentioned that water trapping using reactive species may help the thermodynamics of the system but can generate negative effects on the composition of the reaction mixture. Water can also be trapped by using chemico-physical techniques, which do not modify the reactive system, as discussed in Sect. 6.2.2.3.

From the kinetic point of view, it must be said that the bond-cleavages involved in the reactive systems are all quite energetic [24]. This results in a global process which has kinetic barriers to overcome. The use of a catalyst can influence such parameter. It has been said in Chap. 4 that CO_2 inserts quite rapidly in an M–O bond: therefore the formation of an M–OR bond has an overall kinetic relevance (through activation of the RO–H bond and M–OR bond formation) although the insertion of CO_2 into the M–OR bond to afford M–OC(O)OR is rapid. The following reaction of cleavage of the M–OC(O)OR bond by reaction with a second molecule of alcohol and formation of M–OH and $(RO)_2C{=}O$ also plays a role in the overall kinetics. The M–OH species has then to be converted again into M–OR for further reaction to occur. Alternatively, it is possible that the M–OR species is formed in the direct reaction of M–OC(O)OR with two molecules of alcohols [25], an aspect discussed in this chapter.

6.2.2 Reaction Mechanism

In a catalyst-free process, the direct carboxylation of alcohols requires, a priori, two distinct steps as categorized in Scheme 6.1. In (*i*) an alcohol molecule is activated by a base and an alkoxo-moiety is generated which interacts (*ii*) with CO_2 affording the hemi-carbonate, $RO{-}C(O)O^-$. The acid activation of the second molecule of alcohol (*iii*) produces the alkyl moiety which reacts with the hemi-carbonate and generates the dialkylcarbonate. Interestingly, the role of acid and base can be played by a catalyst. Various kinds of catalysts are discussed in the following sections and the known mechanisms are presented. The pathway presented in Scheme 6.1 is not followed in all catalyzed processes and its occurrence depends on the properties of the catalyst used.

6.2.2.1 Homogeneous and Heterogenized Catalysts

Tin-Based Catalysts Soluble metallorganic species of Sn(IV), namely *n*-dibutyl-dialkoxostannanes, *n*-$Bu_2Sn(OR)_2$ (R = methyl, ethyl, *n*-butyl) have been used as catalysts in the synthesis of DMC or DEC. Early reports [26] say that, at 423 K under 2.8 MPa CO_2 pressure, DMC was produced using dicyclohexylcarbodiimide as a water chemical trap. Sn-catalysts were reported to show 100 % selectivity towards DMC [27].

$$\text{ROH} + \text{base} \rightarrow \text{RO}^- + \text{base-H}^+ \qquad \text{(i)}$$

$$\text{RO}^- + \text{CO}_2 \rightarrow \text{RO-CO}_2^- \qquad \text{(ii)}$$

$$\text{ROH} + \text{acid} \rightarrow \text{R}^+ + \text{acidOH}^- \qquad \text{(iii)}$$

$$\text{RO-CO}_2^- + \text{R}^+ \rightarrow (\text{RO})_2\text{CO} \qquad \text{(iv)}$$

$$\text{Base-H}^+ + \text{acid-OH}^- \rightarrow \text{base} + \text{acid} + \text{H}_2\text{O} \qquad \text{(v)}$$

Scheme 6.1 Putative reaction path for the carboxylation of alcohols. Adapted with permission from [28]. Copyright (2006) Springer Science + Business

Ethanol was also converted into DEC using the same catalyst, but with a lower conversion yield which can be explained in terms of the lower acidity of ethanol ($pK_a = 15.9$) with respect to methanol ($pK_a = 15.5$). Ballivet and coworkers have carried out studies [22, 23, 29–31] on the structural characterization of tin compounds, in solution and in the solid-state after the first cycle of reaction, and have defined the structure of the species formed from $(CH_3)_2Sn(OCH_3)_2$ **1**, *n*-$Bu_2Sn(OCH_3)_2$ **2**, and *n*-$Bu_2Sn[OCH(CH_3)_2]_2$ **3** used as putative catalysts. Other authors have used the same systems with different alkyl groups (see below). The compounds isolated by Ballivet et al. from the reaction mixture are described as resting species in the catalytic cycle (Structure 6.2).

Compound **A** is the product of monoinsertion of CO_2 into Sn–OR bonds of **2**. DFT calculations have shown that the double CO_2 insertion on **2** to afford *n*-$Bu_2Sn[OC(O)OCH_3]_2$ is thermodynamically possible [33], but such compounds were not isolated nor their presence ascertained in the catalytic runs. When heated, **A** converts into **B** and DMC via an intramolecular methyl-transfer mechanism. Species **B** is very stable because of the formation of O–Sn–O bonds. The formation of DMC requires that **B** reacts with external methanol, affording **C**. Moving from **A** to **C**, the nuclearity of the complex increases from 2 to 4 and 10 Sn atoms with the

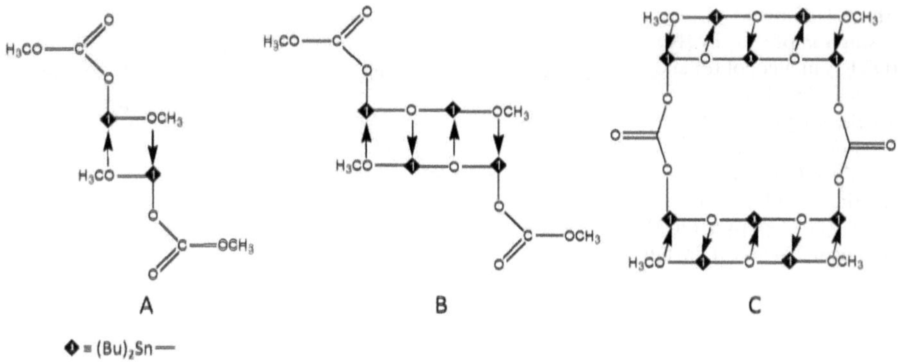

Structure 6.2 Reprinted from [32]. Copyright (2014) Elsevier Inc

formation of extended Sn–O–Sn linkages. Eventually **C** can be recycled to afford **A** which restarts the cycle: **C** is not the active catalyst. In **A**, **B**, and **C** the tin centers are always penta-coordinated. The overall kinetics of the DMC formation is slow because of the complex interconversion mechanism of the metal species active in catalysis, which requires the cleavage and formation of energetic bonds, as mentioned above. Sc-CO_2 has been used as solvent in such reactions [30, 31].

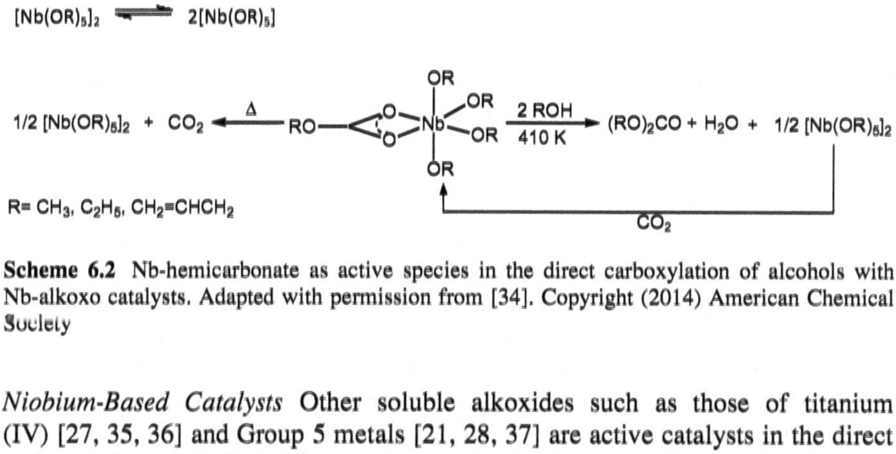

Scheme 6.2 Nb-hemicarbonate as active species in the direct carboxylation of alcohols with Nb-alkoxo catalysts. Adapted with permission from [34]. Copyright (2014) American Chemical Society

Niobium-Based Catalysts Other soluble alkoxides such as those of titanium (IV) [27, 35, 36] and Group 5 metals [21, 28, 37] are active catalysts in the direct carboxylation of alcohols. The reaction mechanism has been elucidated with penta-alkoxo species of Group 5 elements [21, 28, 37] combining DFT and experimental studies. The overall reaction mechanism is depicted in Scheme 6.2.

The niobium alkoxides are dimeric species, and they need to dissociate into the monomeric species before CO_2 can react. Such reaction depends on the solvent used: alcohols favor the dissociation more than benzene. This is because of the capacity of alcohols to assist the Nb–O bond cleavage in the dimer by coordinating to Nb. However, alcohols can bind to the coordinatively unsaturated monomeric Nb-alcoxo species formed upon dissociation, preventing their re-association. Such

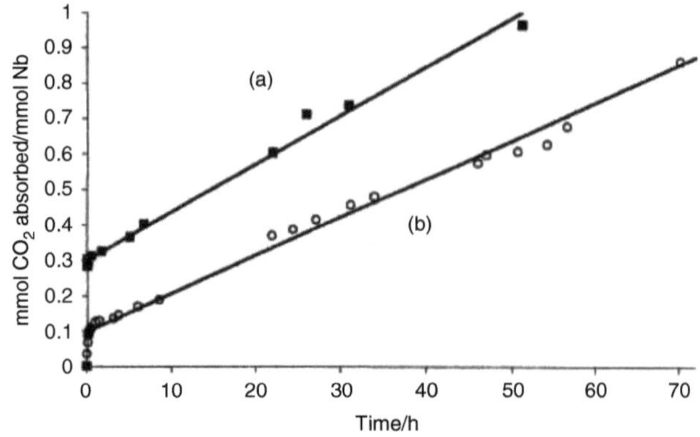

Fig. 6.2 Kinetics of absorption of CO_2 by $[Nb(OR)_5]_2$ in alcohol (a) and benzene (b). The accelerating effect of alcohols is evident (R=CH_3, C_2H_5, allyl). Reprinted with permission from [21]. Copyright (2003) American Chemical Society

pentacoordinated species can react with CO_2 to afford the chelating hemicarbonate $Nb(OR)_4(OCO_2R)$ as thermodynamic product and the monodentate as kinetic product and real catalyst [21, 28].

The curve of absorption of CO_2 by the Nb-complexes in alcohol clearly shows an induction time because of the dissociation of the dimer. Interestingly, the dissociation is slower in benzene than in alcoholic solutions (k_{obs1} value of 1.15×10^{-5} s^{-1} in ethanol and 0.97×10^{-5} s^{-1} in benzene), accounting for the solvating effect of the solvent (Fig. 6.2). The curve also shows that the reaction is stopped when a molar ratio $CO_2/Nb = 1$ is reached, suggesting that only a mono-alkylhemicarbonate species is formed. The reaction is first order in the monomeric Nb-alkoxo complex and first order in CO_2 (6.10) with a value $k_{obs} = 1.19 \times 10^{-2}$ L mol^{-1} s^{-1} which includes the K_{diss} of the dimer.

$$\text{rate}_{carb} = k_{2obs}\left[\text{Nb(OEt)}_5\right][\text{CO}_2] \tag{6.10}$$

From the determined k_{diss} and k_{obs}, k_{carb} was calculated and found equal to ca. 10^3, a value in line with what is reported in Sect. 4.4.

The active species is the Nb-monoalkylcarbonate which was isolated. The alkylcarbonate exists as a stable chelating species in the solid state, as shown by the FTIR data: bands at 1600 [ν_{asym}(OCOO)], 1414, 1330 [ν_{sym} (OCOO)] and 807 cm^{-1}. The IR data also suggest, for the ethyl carbonate group, an η^2-coordination to the metal center through two O atoms.

In fact, the experimental value of $\Delta\nu = (\nu_{as} - \nu_{sym})$ for the OCOO moiety has been found to be 270 cm^{-1}, which is the correct value reported for bridging or chelating OCOO moieties [38]. In benzene solution, the hemicarbonate is monomeric and monodentate [21], pointing thus to a chelating hemicarbonate in the solid state. The conversion of the two forms (Scheme 6.3) is rapid. The formation of the monodentate form is essential for catalysis to occur with dialkylcarbonates formation.

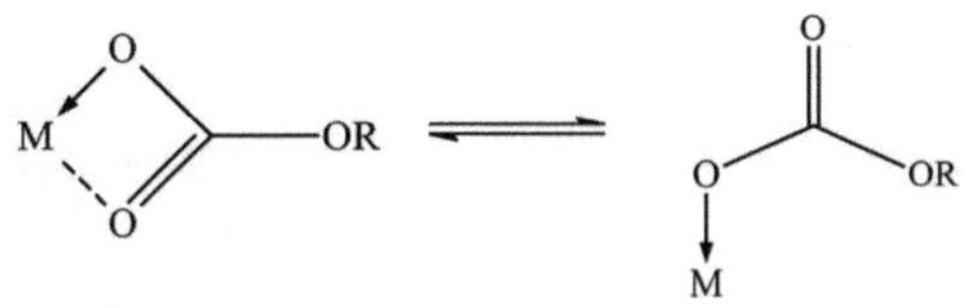

Scheme 6.3 Reversible interconversion of the chelating (solid state)-monodentate(solution) forms of the alkylcarbonate moiety on Nb(V). Adapted with permission from [21]. Copyright (2003) American Chemical Society

Interestingly, when the solid $Nb(OR)_4(OCO_2R)$ complex was heated, CO_2 was released (Scheme 6.1, left) and no formation of dialkylcarbonates was observed [21]. Conversely, heating in the presence of alcohols, dialkylcarbonates were produced, giving back the starting alkoxide complex. Nb–alkoxides thus show a different behavior with respect to Sn–dialkoxo complexes.

Ruling out the intramolecular transfer of the methyl moiety, the formation of dimethylcarbonate requires the intervention of external methanol. The reaction of the $(CH_3O)_4Nb(OCO_2CH_3)$ **3** in Scheme 6.4 with methanol can in principle occur using one or two molecules of methanol, **A** and **B** in Scheme 6.4.

Two approaches were used to decide which is the most likely route: (1) DFT calculations to model the reaction and find the TS energy for the various possible processes; (2) an experimental study on the influence of the molar ratio alcohol–Nb–CO_2 on the formation of carbonates.

It is worth emphasizing that, in principle, tri-molecular reactions (in the present case, implying two molecules of methanol and the Nb-hemicarbonate moiety as depicted in the green path in Scheme 6.4) are not entropically favored with respect to bi-molecular reactions (one molecule of methanol reacting with one Nb-hemicarbonate, fuchsia path in Scheme 6.4). Nevertheless, it has been calculated by DFT that the interaction of two molecules of methanol with a molecule of Nb-hemicarbonate has more favorable ΔH_R that may balance out the negative entropic contributions [21, 28]. Using two (or more) molecules of alcohol in such a solvolysis reaction bears to less stressed structures for the TS than when using a single molecule: the energy difference can be as large as 25 kcal mol^{-1} (see Sect. 6.2.2.3).

Scheme 6.4 shows the three possible routes of generation of DMC from Nb-methylcarbonate. The blue route is ruled out by the fact that in the absence of methanol DMC is not formed, the green route implies one molecule of methanol per Nb, and the red route implies two molecules of methanol per Nb.

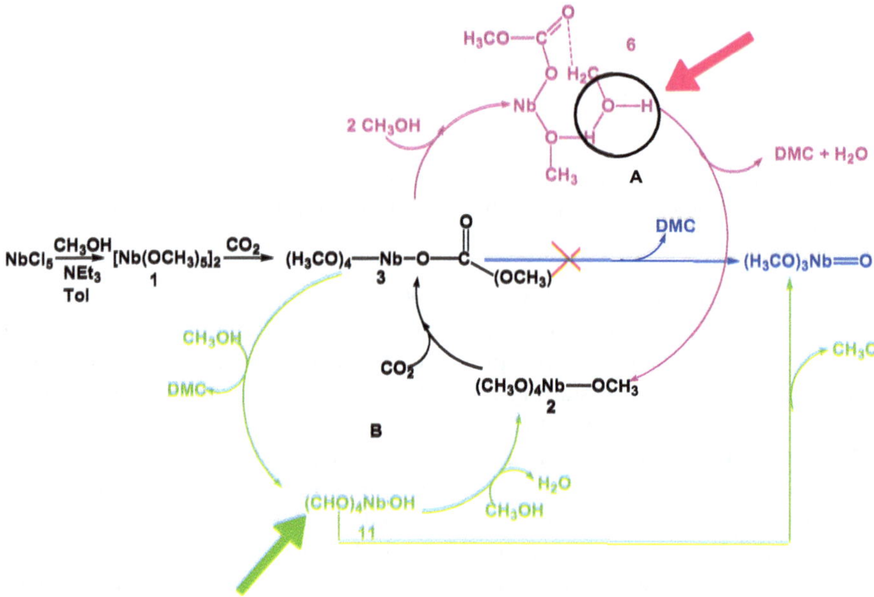

Scheme 6.4 Reaction mechanism of Nb-methylcarbonate with zero, one, or two molecules of methanol. Adapted with permission from [28]. Copyright (2006) Springer Science + Business

DFT calculations (Fig. 6.3) clearly show that the kinetically most favored route involves two molecules of methanol per Nb center. This theoretical finding is supported by the experimental observation that increasing the molar ratio CH_3OH/Nb over 1 increases the rate of formation of DMC from $(CH_3O)_4Nb–OC(O)OCH_3$ [25].

Importantly, the catalysts discussed above, similar to the Sn catalysts, require a quite anhydrous medium as in the presence of water the $[O{=}Nb(OR)_3]_2$ species is formed, inactive in catalysis [21, 28, 37, 39]. The equilibrium concentration of the dialkylcarbonates is in the range 0.5–2 %, depending on the reaction conditions and the alcohols used.

Several other catalysts have been used, but details about the nature of active species and about the reaction mechanism are lacking in the literature. We mention here the work done by Sakakura [40]. Scheme 6.5 shows the catalytic cycle proposed by the authors who have coupled Sn-systems to water-trapping organics.

The proposed mechanism, recalls the fundamental steps proposed by Ballivet (see above) also if multinuclear Sn specie are only mentioned here but not isolated. Again, carbon dioxide is inserted into the non-bridging Sn–O bond of dimeric $[R_2Sn(OMe)_2]_2$ to afford a dimeric methoxo-bridged dinuclear structure bearing two carbonate moieties, each on a different Sn atom. Such a dimer does not undergo a second CO_2 insertion, even at high CO_2 pressure (up to 6.0 MPa). The thermolysis of the dinuclear carbonato-complex results in the formation of DMC in a reasonable yield and a multinuclear Sn complex which reacts with methanol to afford the starting dimer. The hemicarbonato complex is said to lose CO_2 readily under an N_2 atmosphere, even at room temperature, to regenerate the methoxo complex. The organic water traps allow the catalysts to be used for several cycles.

Heterogenized Catalysts As discussed above, once equilibrium is reached, the catalyst must be separated from the reaction mixture and recovered, if one wishes to reuse it or water traps are used which increase the complexity of the system.

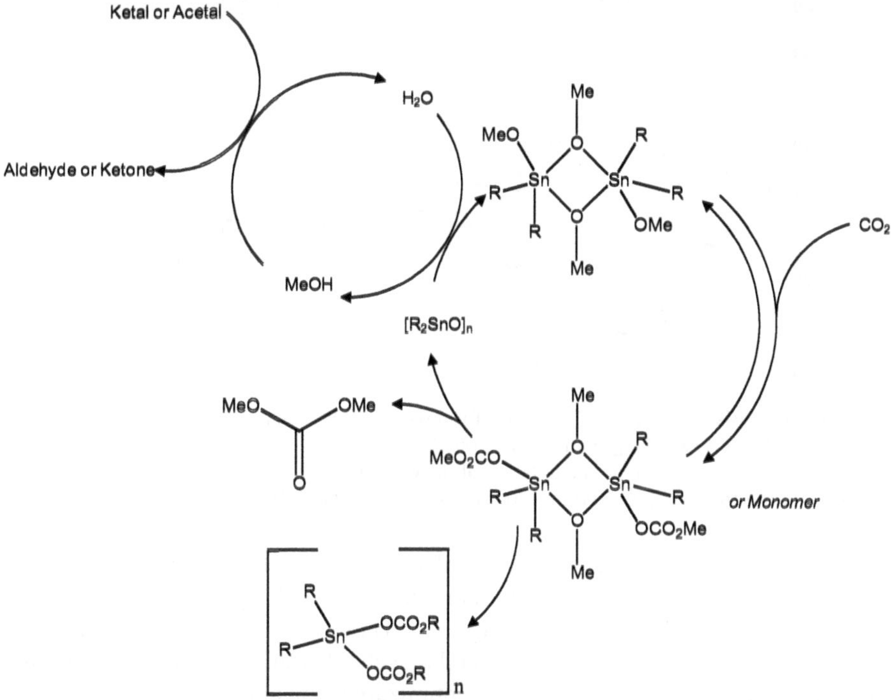

Scheme 6.5 Mechanism of formation of dialkyl carbonates proposed by Sakakura. Adapted with permission from [40]. Copyright (2000) Elsevier

The recovery of the soluble monomeric catalyst (the Nb case) implies the separation by evaporation of the alcohols. In the case of Sn, oligomers are formed which slow down the process, but would be easily separated as they are not soluble. Although the Nb catalysts maintain their activity for several cycles (Fig. 6.4), the low concentration of dialkylcarbonates makes the recovery procedure an expensive operation. This has suggested the preparation and use of heterogenized catalysts for an easier recovery (separation by filtration). The heterogenization can be implemented in several different ways and can influence the activity of the catalyst. The complexes can be supported on an inert material (inorganics such as silica [42], alumina, aluminosilicates, etc.), inglobated into inert matrices (organic or inorganic) or into cavities of a polymeric structure (organic polymers [39] or zeolites), or linked to the surface of a catalyst with a "tether" (tethered catalysts [43]). The latter technique is interesting because it leaves the catalytic center quite unaltered from the point of view of the coordination sphere around the metal and avoids strong interactions with the support which may change the properties of the metal center. The immobilization of Sn complexes on mesoporous SBA-15 affords more

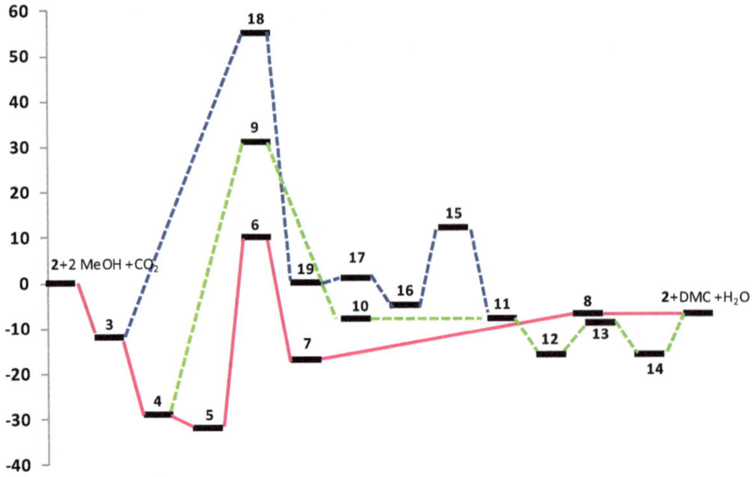

Fig. 6.3 Comparison of the energy profiles for three possible mechanisms of formation of DMC from $[Nb(OMe)_4(OCOOMe)]_2$ with implication of zero (*blue line*), one (*green line*), and two (*pink line*) methanol molecules in the methylation of the $CH_3OC(O)O^-$ moiety bonded to Nb. Adapted with permission from [25]. Copyright (2006), Springer Science + Business

resistant catalysts than the starting molecular Sn catalysts [44]. Examples of catalysts tethered on polymers are represented by polystyrene-grafted [39, 43] organo-tin and Nb-alkoxo species (Fig. 6.7). They have been used in a batch or a flow reactor and were shown to have good recyclability and long activity. Although easily recyclable, these systems need optimization for maximizing their activity. The reaction mechanism has not been discussed in detail, but at least for Nb-catalysts it seems to repeat the one observed with homogeneous catalysts, as the condition of the metal center is not changed in the two cases (Structure 6.3). Conversely, for the Sn-tethered complexes it is hard to think that the same behavior observed with soluble complexes can hold.

In fact, the Sn-centers in the supported catalyst are no longer free to undergo association through Sn–O–Sn bridges, as they are quite far away and in a different environment. This aspect requires further investigation for clarifying the reaction mechanism which operates in such cases.

6.2.2.2 Heterogeneous Catalysts

Taking into account what has been discussed above, it appears that heterogeneous catalysts would be the best candidates to drive the carboxylation of alcohols. In this paragraph their use is discussed and the known aspects of the reaction mechanisms reported. The catalysts are divided into two classes: single-metal and multi-metal catalysts.

Single-Metal Catalysts Metal oxides such as CeO_2 [17, 46], ZrO_2 [47], and TiO_2 [48]which are non-soluble in the reaction medium (heterogeneous) have been used

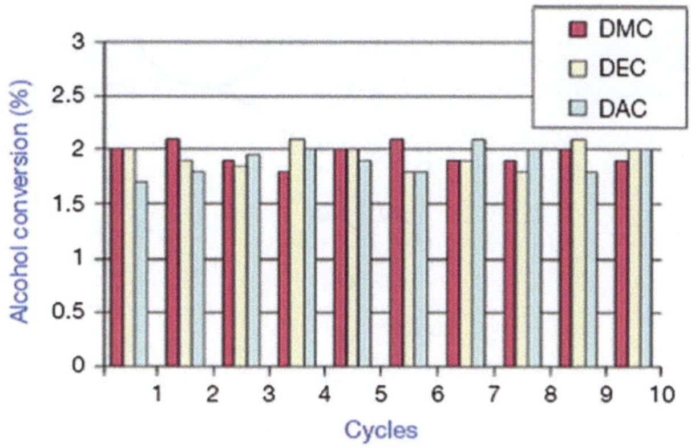

Fig. 6.4 Activity of the recycled Nb(V)-catalyst. Reprinted from [41]. Copyright (2013) Elsevier Ltd

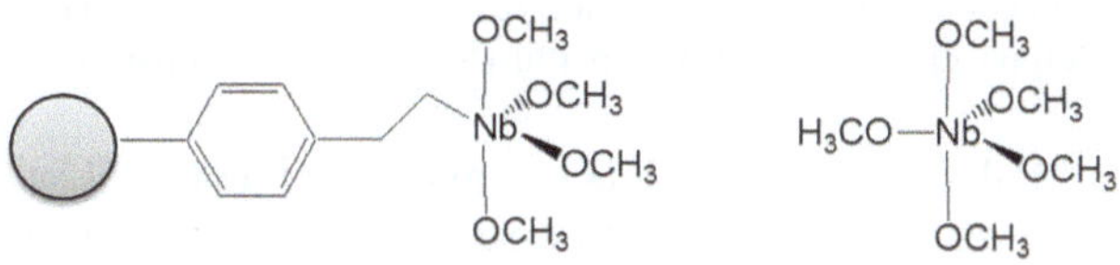

1 2

Structure 6.3 Reprinted from [43]

in the direct carboxylation of alcohols (methanol and ethanol). ZrO_2-based materials were active in the temperature range 400–440 K, showing a selectivity towards DMC close to 100 % [17]. In order to increase the DMC yield, ZrO_2 was modified with H_3PO_4 [49, 50] or used as the mixed oxide $Ce_{0.2}Zr_{0.8}O_2$ [51].

The attempt to design active catalysts is justified by the fact that if they were active at lower temperature, the thermodynamics of the process would be improved and the equilibrium conversion yield increased to more usable values. Therefore, modification of the oxides with phosphoric acid has been carried out in an attempt to modify the acid/basic ratio of active sites. Very often, it has been observed that an increase of catalyst activity is coupled with a lowering of the selectivity: the conversion yield is improved but the lower selectivity raises the problem of the cost of separation of the products. Examples of such unfortunate coincidences are given for the following catalysts. Whereas ZrO_2 is used at 440 K with a low yield (<1 %) and selectivity close to 100 %, $H_3PW_{12}O_{40}/ZrO_2$ is active at 373 K, but dimethylether-DME and CO are formed as by-products [49, 50, 52–54].

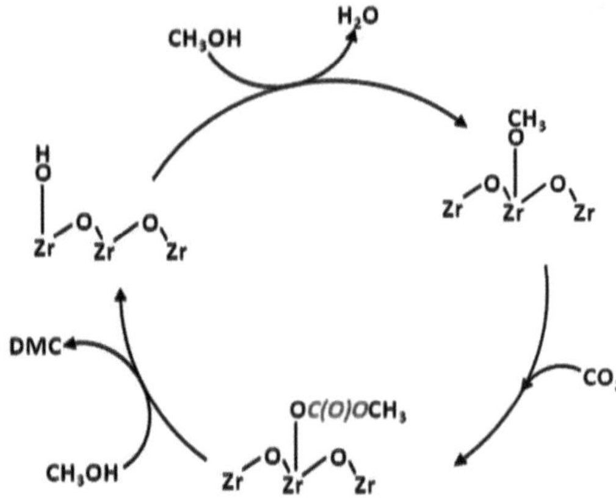

Scheme 6.6 Postulated key surface species involved in the DMC formation from CH_3OH and CO_2 on ZrO_2. Adapted with permission from [55]. Copyright (2010) John Wiley and Sons

$H_3PW_{12}O_{40}/Ce_{0.1}Ti_{0.9}O_2$ provides a higher DMC yield at 443 K (2.5 %) [53, 54] with respect to ZrO_2 with a selectivity much lower than 100 % which correlates with the higher number of acid and basic sites. Unsupported $H_3PMo_{12}O_{40}$ and its copper salt $Cu_{1.5}PMo_{12}O_{40}$ are active at temperatures as low as 313–333 K under atmospheric pressure of CO_2 [56] but the DMC selectivity does not exceed 23 %. DME, formaldehyde, and methyl formate are the other identified products.

The benefit of a higher conversion is, thus, almost cancelled out by the loss in selectivity, which implies energy-intensive post reaction separation processes which makes the carboxylation of alcohols a net CO_2 emitter, more than a CO_2 user.

Jung and Bell using ZrO_2, have proposed the reaction sequence summarized in Scheme 6.6 [57] based on NMR, IR, and Raman spectroscopic studies.

Surface Zr–O–Zr–OH groups were proposed to react with methanol to afford water and the Zr–O–Zr–O–CH_3 moiety. The catalyst behaves here as an acid. Then CO_2 insertion occurs, converting the methoxy fragment into the hemicarbonate $CH_3OC(O)O$–Zr–O–Zr species. A subsequent reaction with CH_3OH is proposed to lead to DMC. The question may arise whether a Brønsted or Lewis acid is required. Observing that when the oxide was loaded with H_3PO_4 the activity of the catalyst was increased [49, 50], it has been suggested that Brønsted acid sites are more important than Lewis ones [52–54]. Single-metal oxides show an interesting catalytic activity, but a drawback is represented by their deactivation. The reasons for such deactivation were investigated in the case of CeO_2. Figure 6.5 (blue curve) shows that, after the first cycle, the activity decreases, reaching a marginal conversion of methanol/alcohol in a few cycles.

The degradation of commercial CeO_2 was investigated through BET and XPS studies [58] and it was shown that the loss of activity was caused by two factors: (1) the decrease of the active surface during the reaction (BET decreased from 8.8 to 1.6 $m^2\,g^{-1}$) and (2) the reduction of the oxidation state of 18 % of surface cerium atoms from Ce(IV) to Ce(III), a species inactive in catalysis. The size of the

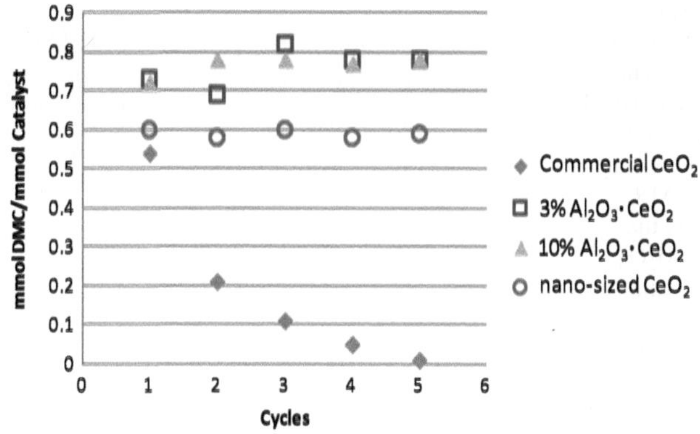

Fig. 6.5 Comparison of the behavior of commercial micro-sized ceria (*blue*) with that of nano-sized ceria (2–10 nm, *empty circles*) and nano-sized Al_2O_3-modified ceria (*empty squares* 3 % alumina; *triangles*, 10 % alumina). The DMC yield in five consecutive cycles of reaction remains constant. In each cycle the same amount (50 mg) of catalyst (CeO_2 or modified ceria) was placed in a reactor with 4 mL of MeOH, at 408 K for 3 h under 5.0 MPa of CO_2. Adapted with permission from [45]. Copyright (2010) Elsevier

particles of the catalyst also plays a key role: nano-sized ceria (5–10 nm) is more active than micro-sized particles and more resistant to deactivation [58].

Mixed Oxides To improve the resistance and life of the CeO_2 catalyst, mixed oxides were prepared and studied. When commercial ceria was loaded with 3–10 % alumina (Fig. 6.5, upper curves) the catalyst was stabilized and had a much longer life [58].

Such improvement was because Al_2O_3, which is bound to the surface, as demonstrated by the insurgence of a signal at 40 ppm produced by a "pentacoordinated dispersed" Al species (Fig. 6.6) in the solid state ^{27}Al NMR spectrum [58], or dissolved in the very first layers of ceria, prevents O-transfer to species bound to the ceria surface and, thus, prevents the Ce(IV) to Ce(III) reduction, as proven by XPS studies [45, 58, 59]

XPS spectra of samples of CeO_2 at $t = 0$ and after use in catalysis show the insurgence of signals caused by Ce(III), whereas the Al-loaded samples do not show any Ce(III) signal after use in several catalytic runs. The reaction mechanism was ascertained by a combination of spectroscopic (IR and multinuclear NMR) experimental tests and DFT calculations. High temperature DRIFT studies on the catalyst allowed the collection of information on the reactions occurring at the surface under the different operative conditions and hemicarbonate species were characterized, also showing that the Al/Ce mixed oxide, unlike CeO_2, does not oxidize methanol [45, 59] to formate. ^{13}C NMR was useful to demonstrate both the formation of *S*-OCH_3 (*S* is the catalyst surface) moieties (signal at 48.5 ppm; Fig. 6.7c) and the surface conversion of such moieties into *S*-OC(O)OCH_3 (signal at 54.2 ppm; Fig. 6.12c) upon admission of CO_2.

The neat catalysts have no ^{13}C signals (Fig. 6.7a). The low intensity signal of the carbon of the hemicarbonate group –OC(O)O– is located at 160.5 ppm. By

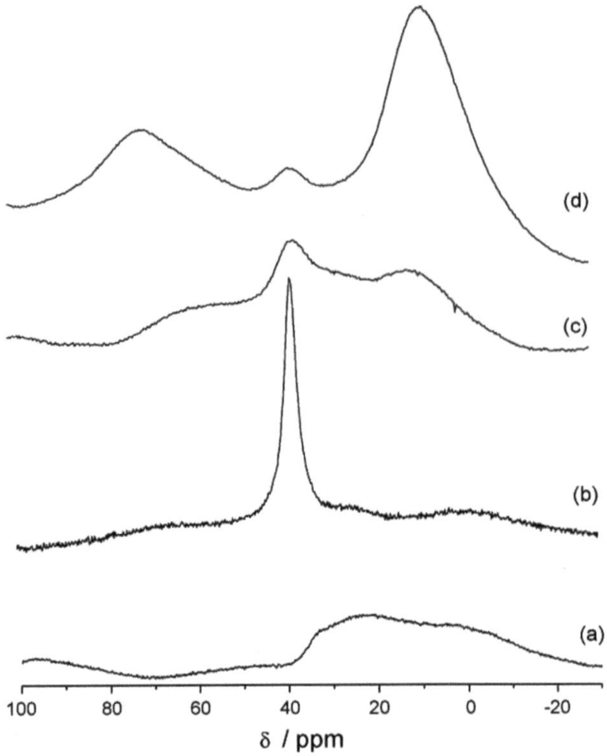

Fig. 6.6 ^{27}Al-NMR spectrum of solid Al(x%) CeO_2(x = 3–20 %). The peak at 40 ppm is ascribed to "pentacoordinated dispersed" Al. Increasing the percentage of Al_2O_3 above 10 % causes the separation of the Al_2O_3 phase. Reprinted with permission from [58]. Copyright (2008) Elsevier

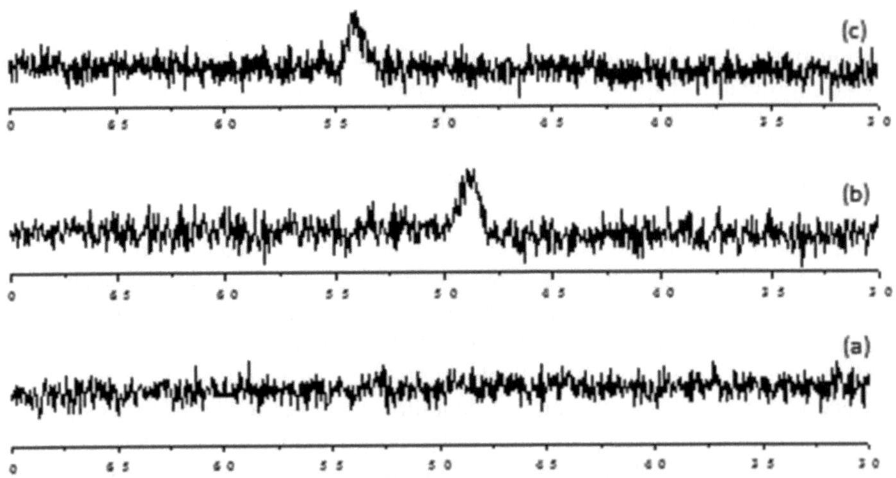

Fig. 6.7 Solid state ^{13}C-NMR spectrum of Al_2O_3(10 %)–CeO_2: (**a**) neat; (**b**) after treatment with methanol at 400–403 K; (**c**) sample (**b**) after admission of CO_2. Reprinted with permission from [45]. Copyright (2008) Elsevier

carefully choosing the time of exposure to CO_2 of the surface bearing the methoxo moieties, it was possible to observe the signals of both the methoxo and hemicarbonate group in the spectrum of the same sample.

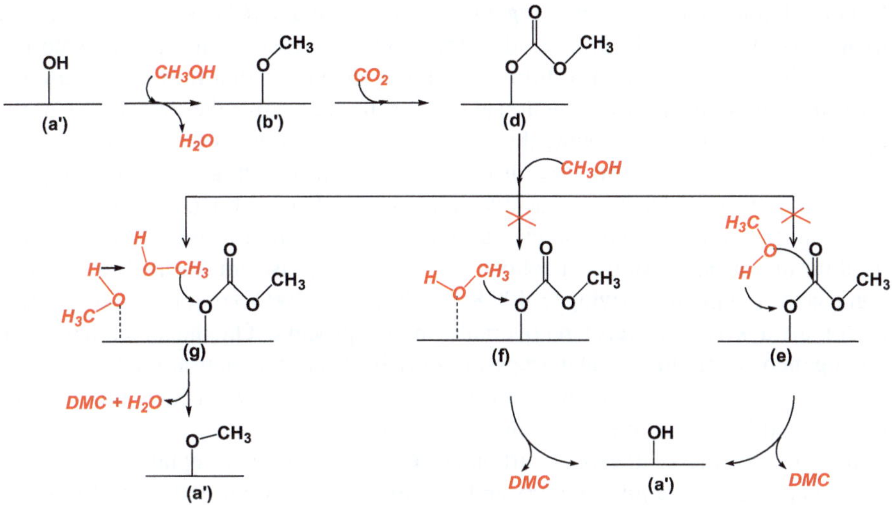

Scheme 6.7 Steps for the formation of DMC from methanol and CO_2 via gas-phase methanol attack onto surface-bound $OCOOCH_3$. Gas-phase species are in *red*, *arrows* show bonds that are formed. Reprinted with permission from [45]. Copyright (2008) Elsevier

The formation of DMC was then followed by combining DRIFT, NMR, and GC-MS studies.

On the basis of what has been discussed above, the methyl transfer can in principle occur through the interaction of the *S*-$OC(O)OCH_3$ group with either a methoxo moiety standing nearby on the surface of the catalyst or methanol present in the gas phase. The former hypothesis was ruled out with specific studies in the absence of gas-phase methanol (Scheme 6.7). Al_2O_3–CeO_2 bearing both the –OCH_3 and the –$OC(O)OCH_3$ moieties was heated in absence of MeOH and DMC was not revealed. Conversely, when methanol was admitted, the formation of DMC was observed.

A DFT study has confirmed that two molecules of methanol are involved in the release of DMC and routes *(e)* and *(f)* in Scheme 6.6 are ruled out. Also, the dependence of the rate of formation of DMC on the concentration of methanol was confirmed by experiments which did show that increasing the amount of methanol above 1:1 molar ratio $CH_3OH/CH_3OC(O)O^-$ causes an increase of the rate of reaction, as the energy is lowered.

Al_2O_3- and Nb_2O_5-loaded ceria were compared in the carboxylation of ethanol. It was found that Nb_2O_5 3 %/CeO_2 mixed oxides are 50 % more active than the relevant Al/Ce systems [45, 59, 60].

6.2.2.2.1 Water Removal Techniques

Water removal for shifting equilibrium in (Eq. 6.6) to the right and preserving the catalysts has been attempted using both chemical water traps and physical means. Chemical traps can be either organic species (acetals [61], ketals [62], easily hydratable bonds such as cyanides [63], and imides) or inorganic materials (zeolite-based materials). The former have the advantage that they can be used at the same temperature of the reaction and make more negative the overall ΔG, as the hydration is exergonic, allowing the reaction to occur at lower temperature and with much better yields (see the case discussed below in the Organic Promoters paragraph). Zeolites need to be used at low temperature (also 240 K) [64], otherwise their acidic surface may decompose the carbonates at temperatures of >400 K, usually used for the carboxylation of alcohols: this makes their use not energetically convenient as the temperature swing (>400 K → 240 K → >400 K) requires energy which is not always (completely) recoverable in the process. On the other hand, for homogeneous (organic) water traps the drawback is that they are soluble in the reaction medium and, thus, need to be recovered from the reaction mixture and regenerated for re-use, otherwise stoichiometric use makes the direct carboxylation of alcohols for the synthesis of carbonates economically unacceptable. Such post-reaction processing requires energy and increases the economic and energetic cost of the process. Moreover, it frequently happens that the addition of such species results in a loss of selectivity, as parallel reactions may occur which consume the formed carbonate. It is useful to recall at this point that dialkyl-, alkylaryl-, or diaryl-carbonates behave as alkylating (arylating) or alkoxy (aryloxy) carbonylating agents [65]. Among the organic species used as water traps, we mention acetals (e.g., 2,2-dimethoxypropane, DMP) [61, 66], ketals [62], cyanides [63–66], and butene oxide [63, 64, 66–71]. Among inorganic water traps we mention molecular sieves of 3 or 4 Å [72, 73]. A specific case of organic water traps, dicyclohexylcarbodiimide-DCC, which act as promoters of the reaction are further discussed in Sect. 6.2.2.3.

Pervaporation Membrane for Water Removal The use of pervaporation membranes for water removal proposes a serious alternative which allows the water separation to operate at the same reaction temperature without any loss in selectivity. The Nb_2O_5 3–10 %/CeO_2 catalyst has been used in a flow reactor for several days coupled to a pervaporation membrane for water elimination [60] (Fig. 6.8).

The use of supercritical conditions allowed a higher conversion of ethanol to be reached with respect to using liquid ethanol pressurized with CO_2. The use of pervaporation membranes is by far the best performing technology for water separation as it minimizes post-reaction separation costs and allows working in a continuous mode.

With respect to homogeneous organic water traps, membranes avoid any post-reaction processing of the reaction mixture and, with respect to inorganic traps, they avoid the swift temperature change. With respect to both organic and inorganic water traps, membranes save the energy necessary for the recycling of the hydrated species.

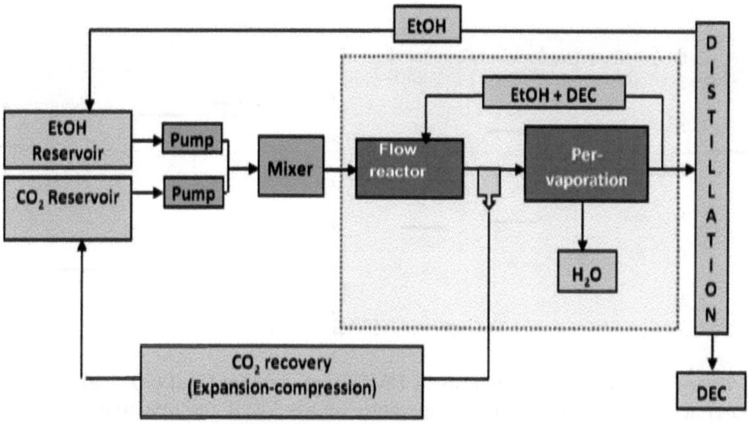

Fig. 6.8 Setup used for the carboxylation of ethanol in single phase *sc*-conditions, using a flow reactor and an inorganic ceramic pervaporation membrane for water separation. Reprinted with permission from [60]. Copyright (2012) John Wiley and Sons

6.2.2.3 Organic Promoters

Dicyclohexylcarbodiimide, CyN=C=NCy (DCC), is known to react with protic compounds with addition of the latter across one of the C=N bonds. Water, for example, affords the relevant dicyclohexylurea CyHN–C(O)–NHCy (DCU), a reaction strongly exergonic ($\Delta G = -40\ \text{kcal mol}^{-1}$).

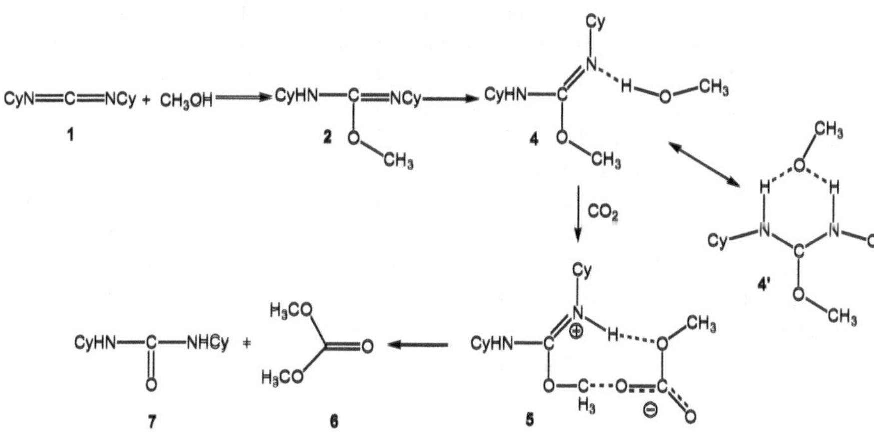

Scheme 6.8 Reaction mechanism for the formation of dimethyl carbonate from CO_2 and methanol in the presence of DCC. Reprinted with permission from [25]. Copyright (2005) American Chemical Society

Such reaction has been used for a long time for capturing water formed in esterification reactions and for shifting the equilibrium to the right [25]. However,

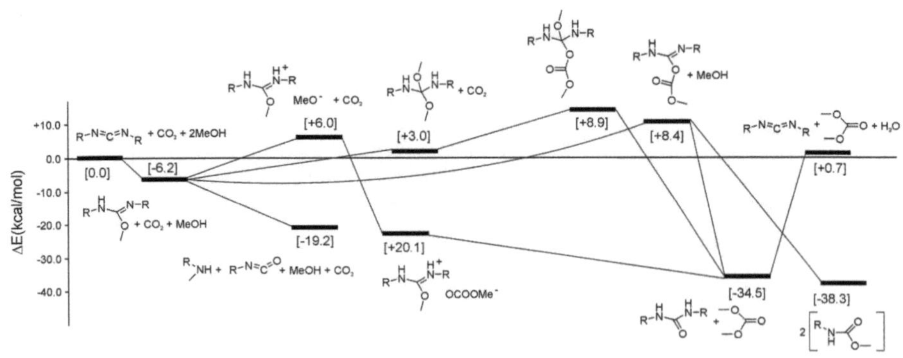

Fig. 6.9 Energy profile for the synthesis of DMC from methanol and CO_2 under DCC catalysis (*lower part*) at 330 K. Increasing the temperature above 340 K the selectivity of the process is decreased as new compounds (carbamates and isocyanates) are formed (*upper part*). Reprinted with permission from [74]. Copyright (2014) Springer

DCC was used as a water trap in the transition metal catalyzed synthesis of organic carbonates from alcohols and CO_2 [26]. It was shown later on [25] that DCC is able to promote the synthesis of organic carbonates, even under much milder conditions than transition metal systems, as justified by the quite negative hydration energy, making its use as a water trap in conjunction with metal systems unseemly. The reaction mechanism has been demonstrated (Scheme 6.8) and DFT calculations have given the mechanistic details and the TSs energies (Fig. 6.9) [25, 74].

The first step (addition of methanol to DCC) is formally analogous to the addition of water to DCC. It may occur with the implication of one or two (or even more) methanol molecules. Scheme 6.8 shows that the activation energy of methanol addition is significantly reduced in the trimolecular reaction.

Once isourea **2** is formed, a second alcohol molecule interacts via H-bonding with the iminic nitrogen of **2** affording **4** (see Scheme 6.7) which is in equilibrium with its more stable form **4′**, isolated in the solid state [28, 74]. The H-bond formed between the methanol molecule and the iminic C=N bond increases the nucleophilicity of the O-atom, enabling a favorable interaction with CO_2 and giving rise to the formation of the hemicarbonate, "ion-paired" with isourea (**5** in Scheme 6.7). The migration of the methyl group of isourea to the hemicarbonate anion $^-OC(O)OCH_3$ affords DMC **6** and urea **7**. The net stepwise reactions are shown in (6.11a) and (6.11b), and the overall reaction is shown in (6.12).

$$\mathrm{CyHNC(OCH_3) = NCy + HOCH_3 \overset{CO_2}{\leftrightarrows} CyHNC^+(OCH_3) - NHCy\ {}^-OC(O)OCH_3} \tag{6.11a}$$

$$\mathrm{CyHNC^+(OCH_3) - NHCy\ {}^-OC(O)OCH_3 \rightarrow CyHNC(O)NHCy + (CH_3O)_2CO} \tag{6.11b}$$

$$\mathrm{CyN = C = NCy + 2\,CH_3OH + CO_2} \\ \rightarrow \mathrm{CyHNC(O)NHCy + (CH_3O)_2CO} \tag{6.12}$$

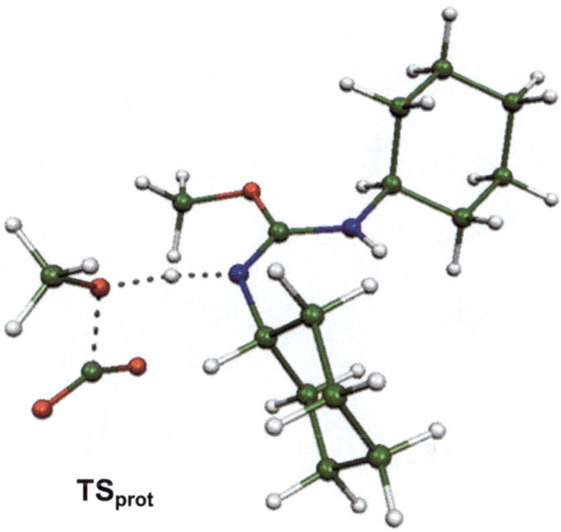

Scheme 6.9 Transition state of isourea protonation taking place in concert with C–O bond formation. Reprinted with permission from [74]. Copyright (2014) Springer

DFT calculations allowed one to ascertain that the precursor state of reaction (6.11a) is a three-center adduct involving an initial weak $CH_3OH{\cdots}CO_2$ interaction (the energy is 2.4 kcal mol^{-1}) which is enhanced by the $CH_3O-H{\cdots}N \rightarrow CH_3O{\cdots}H-N$ hydrogen shift (**4**, Scheme 6.7). As already said, the H-bonding increases the negative charge on the methoxo-oxygen, causing the strengthening of the O–CO_2 bond. When CH_3OD, instead of CH_3OH, was reacted with isourea and CO_2, an isotopic effect equal to 1.8 was observed which confirms that the H (D) transfer is the rate-determining step [25]. The concerted mechanism of the proton shift and the O–C bond formation is clearly borne out by the structure of the transition state **TS**$_{\mathbf{prot}}$, in which the proton is between nitrogen and oxygen, whereas the O–CH_3 bond distance is 1.79 Å and CO_2 is bent (Scheme 6.9).

However, the presence of CO_2 in the reaction medium converts the alkoxo- into the hemicarbonate-species which is more stable than the former because of charge delocalization over the two oxygen atoms of the $ROCOO^-$ group. Calculations carried out on the interaction of a methoxo anion with carbon dioxide show that CO_2 addition to CH_3O^- anion occurs spontaneously (with no energy barrier), and the formation of the CH_3OCOO^- anion is exothermic by 40.4 kcal mol^{-1}, underlying the importance of stabilization introduced by the presence of CO_2. The protonation of the iminic-N of isourea weakens its O–CH_3 bond as indicated by the O–C bond distances in **4** and **5** (1.43 vs 1.46 Å, respectively). This favors the methyl transfer to the –OC(O)O*Me* moiety, a step represented by $\mathbf{TS_{methyl}}$ (Scheme 6.10).

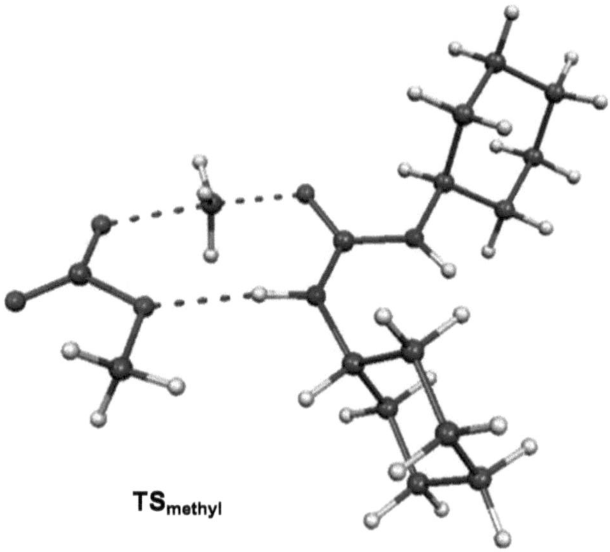

Scheme 6.10 Transition state of methyl migration. Reprinted with permission from [74]. Copyright (2014) Springer

The *quasi* planar CH_3 group is positioned between the two O-atoms, and the N–H···O hydrogen bond is much weaker than in the ion-pair. Because the migrating CH_3 group carries a positive charge, both components of **5** become neutral, and the final state is characterized as an H-bonded complex formed between the *cis-trans* isomers of urea and DMC (7···6). The relative energies of the identified stationary points are summarized in Fig. 6.10. Because of the ionic nature of **5**, the solvent effects play an important role in the energetics, and therefore the solvation energies for the five stationary points were estimated in terms of the PCM model, and corrected for solvent effects. As Fig. 6.11 shows, the two models provide rather different energy profiles.

Although the gas-phase calculations predict the ion-pair to be well above (+12.7 kcal mol^{-1}) the isourea-methanol-CO_2 adduct, this step clearly becomes energetically permitted (−7.1 kcal mol^{-1}) in the solvated model causing, therefore, the energy of $\mathbf{TS_{methyl}}$ shift to a much lower value.

The energy barriers for the two steps of the reaction are predicted to be 9.8 and 14.5 kcal mol^{-1} (from −7.1 to 7.4 kcal mol^{-1}) with the solvated model, which are reasonably low, and one may conclude that the two-step reaction mechanism represents a conceivable low energy route between *isourea-methanol-CO_2* and *DMC + DCU*. Furthermore, these results are consistent with the fact that the reaction takes place faster in polar solvents, justifying the theoretical observation

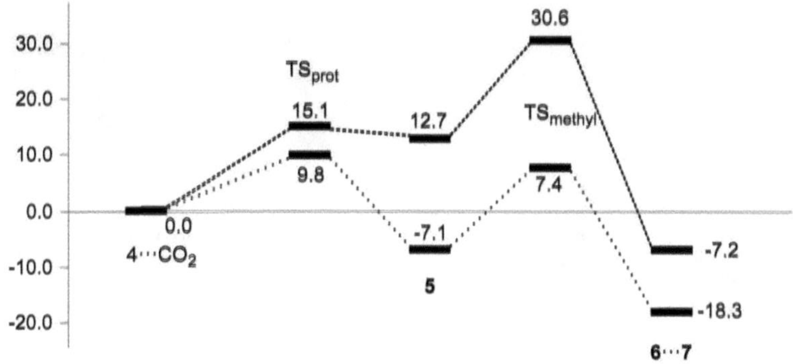

Fig. 6.10 Computed energy profile for the **4**···CO_2 → **7**···**6** reaction pathway as obtained by the gas phase (*dashed*) and solvated (*dotted*) models. The dielectric constant of the polarizable medium was set to $e = 32.6$ to model methanol as a solvent. Energies (*y* axis) are in kcal mol^{-1}. Reprinted with permission from [25]. Copyright (2005) American Chemical Society

that the stability of the ion-pair intermediate and the gap of the proton shift appear to be sensitive to the polarity of the solvent.

Attempts to find experimental evidence with NMR and IR studies successfully supported the above mechanism. **4′** could be clearly identified *via* both its ^1H and ^{13}C NMR spectra in CD_3CN at 273 K which, respectively, showed the equivalence of the two aminic protons in solution, as depicted in Scheme 6.11, and the equivalence of the N-bonded carbon atoms of the cyclohexyl rings which are non-equivalent in **3**. If CO_2 is added to a CD_3CN solution containing **3** and methanol in equimolar amounts, a new signal located at 3.41 ppm appears in the ^1H NMR spectrum of the solution because of the proton of the methyl group of the hemi-carbonate moiety. At 300 K no further change is observed within minutes, but heating this solution to 330 K causes the fast appearance in the ^1H NMR spectrum of the typical DMC signal at 3.70 ppm.

IR cannot clearly confirm the formation of **4:4′**, but when an acetonitrile solution of methanol and isourea is exposed to CO_2, a solid is slowly formed at 273 K, showing in its IR spectrum bands at 1 530 and 1 283 cm^{-1}, attributed to the symmetrical and asymmetrical carbonyl stretching of the methylcarbonate anion $CH_3OC(O)O-$.

The position of the bands and the difference of 302 cm^{-1} between the *sym* and *asym* carbonyl stretching frequency confirm that the methylcarbonate is not monodentate and in the solid state behaves as a μ^2-O,O bidentate ligand [21], generating the stable form **5′** shown in Scheme 6.12. **5′** in the solid state is unable to give the formation of DMC at 330 K. Conversely, after dissolving it in CH_3CN and heating the solution at 330 K, a new IR signal centered at 1 755 cm^{-1} appears (C=O asymmetrical stretching because of the DMC).

The GC-MS analysis of the solution confirms the formation of DMC. The reaction of isourea **3** with methanol and CO_2 goes to completion with 100 % conversion into DMC at 330 K. Above such temperature the reaction is more complex as other products such as carbamates are formed (Fig. 6.9).

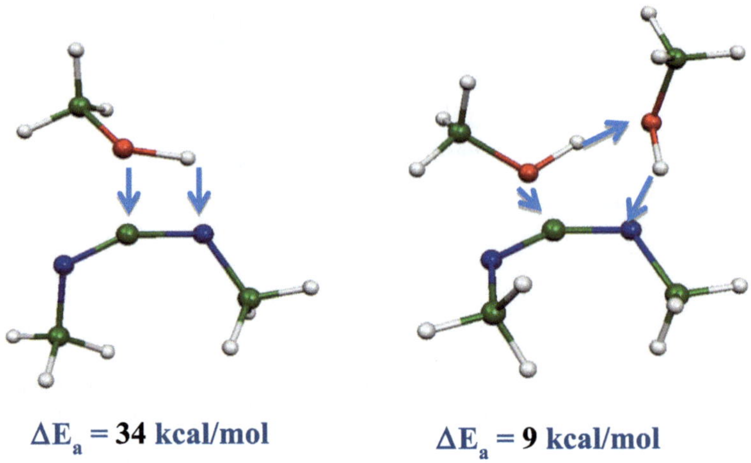

Fig. 6.11 Computed energy barriers of the addition of methanol to DCC using one (*left*) or two (*right*) molecules of methanol per DCC (energies are relative to separated species). Reprinted with permission from [74]. Copyright (2014) Springer

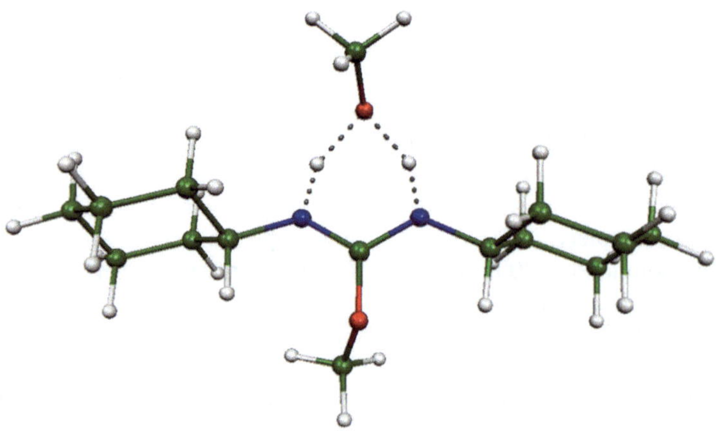

Scheme 6.11 Transition state between two asymmetric structures of complex **4′**. Reprinted with permission from [74]. Copyright (2014) Springer

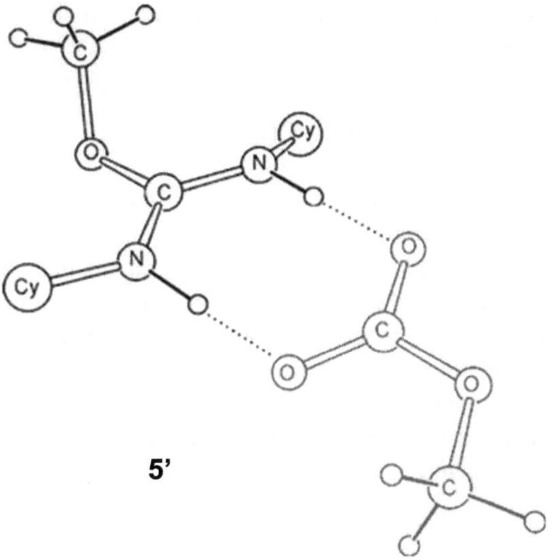

5'

Scheme 6.12 Optimized structure of the bidentate ion-pair **5′** formed between the protonatedsc conditions isourea and the hemi-carbonate anion (bond lengths in Å). Reprinted with permission from [74]. Copyright (2014) Springer

B3LYP/6-311++G** calculations reveal that **5′** is 19.8 kcal mol^{-1} more stable than **5**, which is attributed to the formation of the double H-bond and also to the internal structural rearrangement of the protonated isourea moiety (Fig. 6.9). However, the data above clearly demonstrate that DCC is a promoter of the carboxylation of alcohols to dilakylcarbonates at room temperature, with a ΔG of ca. −38 kcal mol^{-1}, because of the formation of DCU.

DCC may, thus, promote the reaction of alcohols with carbon dioxide to occur under temperature and pressure conditions that are much milder than those encountered for metal-catalyzed reactions as reported above. In fact, the latter require 420–470 K in CO_2-pressurized alcohols (>5 MPa) or under supercritical conditions (sc-conditions) for reaching a ca. 1–2 % concentration of carbonate at equilibrium at best. However, it does not make sense to use DCC as a water trap in combination with metal systems as DCC itself promotes the carboxylation reaction in milder conditions than metal systems. Favorable thermodynamics is generated by coupling of the carboxylation reaction with the formation of DCU from DCC, a thermodynamically favored reaction, and brings about a quantitative conversion of the DCC–alcohol–CO_2 mixture when all reagents are in a stoichiometric ratio 1:2:1 as shown in (6.12). An interesting additional piece of information is that the resulting DCU can be easily recovered as it is insoluble in the reaction medium (different from other organic water traps whose hydrated forms are soluble), and reconverted back into DCC [75], a finding which may support the practical application of such synthetic methodology for the synthesis of carbonates of alcohols bearing thermally unstable moieties that require low reaction temperatures. Conversely, DCC is less suited with alcohols such as methanol or ethanol, which would be used at a >10 Mt year^{-1} scale.

6.2.2.4 Urea as an Active Form of CO_2

The unfavorable thermodynamics of the direct carboxylation of alcohols has pushed efforts to find alternative routes for the conversion of alcohols into the relevant carbonates. An interesting possibility is represented by the use of urea which is formed from CO_2 and NH_3. Urea can be considered an active form of CO_2 (Fig. 6.12). Its formation ΔG is some 200 kJ mol^{-1} less negative than that of CO_2. However, in principle, its reactions with alcohols would have less negative thermodynamics than the reaction of CO_2 and, consequently, the equilibrium positions should be shifted towards the right, reaching higher conversion at the equilibrium.

The synthesis of urea and its reaction with alcohols are reported in (6.13)–(6.15):

$$2NH_3 + CO_2 \rightarrow (H_2N)_2CO \quad (6.13)$$

$$ROH + (H_2N)_2CO \rightarrow ROC(O)NH_2 + NH_3 \quad (6.14)$$

$$ROC(O)NH_2 + ROH \rightarrow (RO)_2CO + NH_3 \quad (6.15)$$

The alcoholysis of urea using alcohols and polyols has been known for a long time [76] to proceed in two quite distinct steps. Reaction (6.14), bearing to the urethane, at least with the lower homologues of the aliphatic alcohols, is a thermal reaction and does not require any catalyst [76–78]. The second step is a transition metal-catalyzed reaction. Ammonia formed in the two latter reactions can be recovered and reacted again with CO_2 to reform urea, thus resulting in a net consumption of

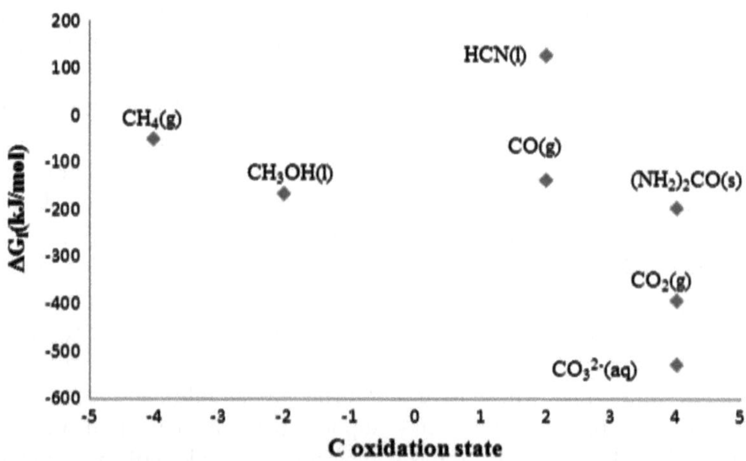

Fig. 6.12 Free energy of formation of C1 species. Adapted with permission from [34]. Copyright (2014), American Chemical Society

CO_2 in the carboxylation, or used for other synthetic purposes, such as in amination of organic substrates [79].

If reaction (6.15) is carried out at a temperature above 400 K, then the reactive system becomes much more complex. In fact, at 408 K the decomposition of urea (6.16) takes place, producing HNCO and NH_3 able to react with the catalysts.

$$(H_2N)_2CO \rightarrow HNCO + NH_3 \tag{6.16}$$

The decomposition has been clearly demonstrated by Sun for the reaction of urea with methanol [78]. Moreover, as the linear carbonate formed is an excellent alkylating agent, *N*-alkylation of urea and alkyl carbamate may occur. However, reactor engineering and catalyst design are of primary importance to circumvent secondary reactions. A catalyst quite active in alcoholysis of urea is ZnO. Under the reaction conditions it converts into $Zn(NCO)_2(NH_3)_2$, a soluble form [80]. Therefore, ZnO is the precursor of the catalysts and not the catalyst itself: reaction (6.15) proceeds under homogeneous catalysis. A point is that the solubilization of Zn does not allow its easy recovery, with some practical problems. Research in this area is still progressing to shed light on the real catalytic species and its implication in catalysis. Sun et al., on the basis of DFT studies, have proposed that alcohol ROH adds to $Zn(NCO)_2(NH_3)_2$ forming $Zn(NCO)(NH–COOR)(NH_3)_2$ which then further reacts with alcohol to afford the carbonate [81, 82]. A different mechanism has been discovered by the Aresta group who found that the urethane ($H_2NC(O)OR$) coordinates to the metal center through the carbonyl moiety and undergoes further reaction with alcohol to afford dialkylcarbonates and NH_3 reforming the original catalyst $Zn(NCO)_2(NH_3)_2$ [83]. The latter mechanism is supported by the fact that alkylcarbamate can be used as starting materials instead of urea. Therefore, although it is clear that $Zn(NCO)_2(NH_3)_3$ is the real catalyst, the mechanism it works by is still under investigation: possibly, different mechanisms operate in different experimental conditions. It is also well ascertained that in order to shift the equilibrium to the right, continuous elimination of NH_3 is benign. Catalytic distillation has been applied to the synthesis of dimethyl carbonate by urea methanolysis [84] with variable results: although the concept of catalytic distillation is simple, its practical application is quite complex, implying process modeling, thermodynamics of the fluid phase equilibria, and optimization of the chemical kinetics. It may work well if products have quite different boiling points and do not form azeotropic mixtures.

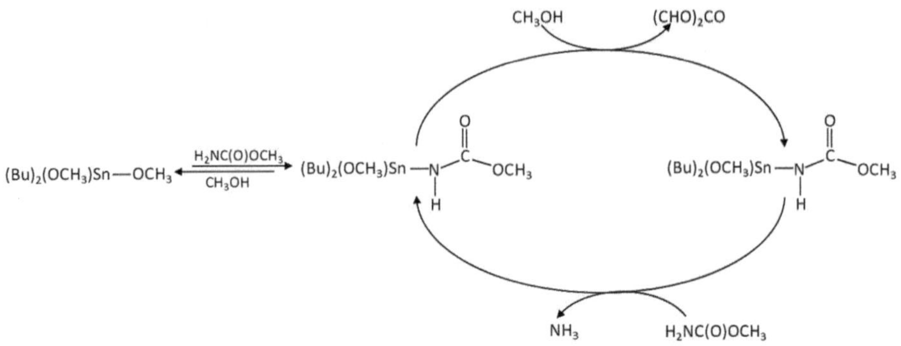

Scheme 6.13 Reaction pathway proposed for the formation of DMC from methyl carbamate and methanol with n-$Bu_2Sn(OCH_3)_2$ precursors. Reprinted from [32]. Copyright (2014) Elsevier Inc

The use of soluble organotin compounds [85–87], metal oxides other than ZnO, e.g., CaO, MgO, ZrO_2, PbO_2 [80, 88, 89], ionic liquids [90, 91], and mixed oxides [92–95] has been reported. Dibutyltin(IV) soluble precursors give interesting results. Correlations between catalyst structure and activity have been demonstrated for tin catalysts [96]. A series of di-n-butyl tin(IV) compounds have been synthesized, characterized by NMR and IR spectroscopies, and screened for methyl carbamate methanolysis at 463 K. The key reactions proposed are depicted in Scheme 6.13.

6.3 Direct Carboxylation of Diols and Polyols

The direct carboxylation of diols and polyols has been attempted with variable results. Diols have been investigated for a long time with some success. A variety of synthetic techniques have been used. Among effective catalysts we recall: organic bases (15.3 % yield, 100 % selectivity). Catalyst: TBD, 1,5,7-triazabiciclo [4.4.0dec-5-ene] in the presence of ammonium carbonate and acetonitrile as a water trap [97], and the use of electrogenerated heterocyclic carbenes [98]. The above reaction suffers thermodynamic restrictions and water traps have been used to shift the equilibrium. Recently, CeO_2 has been used for the direct carboxylation of propanediol to afford propenecarbonate with yields >60 % in the presence of 2-CN-pyridine as water scavenger [100].

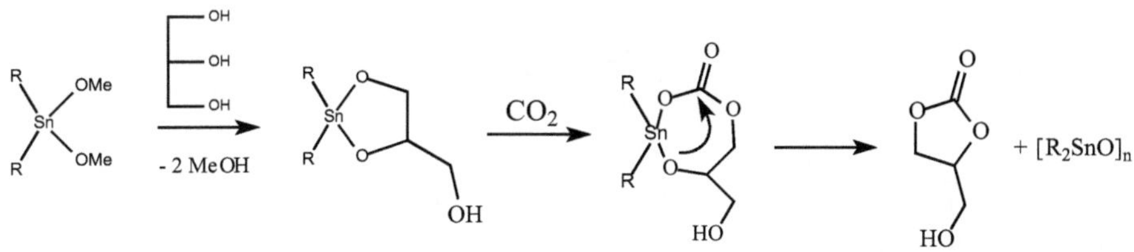

Scheme 6.14 Proposed reaction mechanism for the direct carboxylation of glycerol to glycerol carbonate. Reprinted with permission from [99]. Copyright (2006) Elsevier

Glycerol has been shown to react with CO_2 under n-Bu_2SnCl_2 or n-Bu_2SnO catalysis to afford ca. 6 % glycerol carbonate, which is the equilibrium concentration as shown by investigating the reaction at growing catalyst loadings [99]. The proposed reaction mechanism is shown in Scheme 6.14. The tin glycerato complex has been shown to oligomerize and separate. The oligomer does not react extensively with CO_2. Therefore the insertion of CO_2 was proposed to take place on the monomer before the oligomer was formed (or also concurrently with the first glycerol–O–Sn bond formation). The elimination brings about the formation of an oligomeric Sn-oxo complex which must be reactivated by glycerol. Such reaction has later been modeled using DFT and thermodynamic calculations using the Peng–Robinson equation of state and Mathias–Klotz–Prausnitz mixing rule [101]. It has been found that the reaction is only slightly exothermic ($\Delta H° = -7.38$ kJ mol^{-1}) and presents low equilibrium constants ($K_{eq298} = 7.71 \times 10^{-6}$ and $K_{eq423} = 3.26 \times 10^{-6}$) which justify the conversion yield of ca. 6 % reported above. The thermodynamic quantities above are of the same order of magnitude as those reported above for linear carbonates.

Also in this case, the use of urea is of great help for increasing the conversion [102]. Up to 90 % glycerol carbonate recovery has been reported using γ-Zr-phosphate as catalyst [103, 104]. The reaction mechanism was elucidated through kinetic and spectroscopic studies and proposed to be as reported in Scheme 6.15 [105].

The reaction conditions play a key role for having a 100 % selectivity. The temperature must be kept below 408 K: at higher temperatures the conversion rate was increased but the selectivity was drastically decreased because of the occurrence of reaction (6.16) which drove other concurrent conversions of glycerol [105] to afford oxazolidinones (Scheme 6.16).

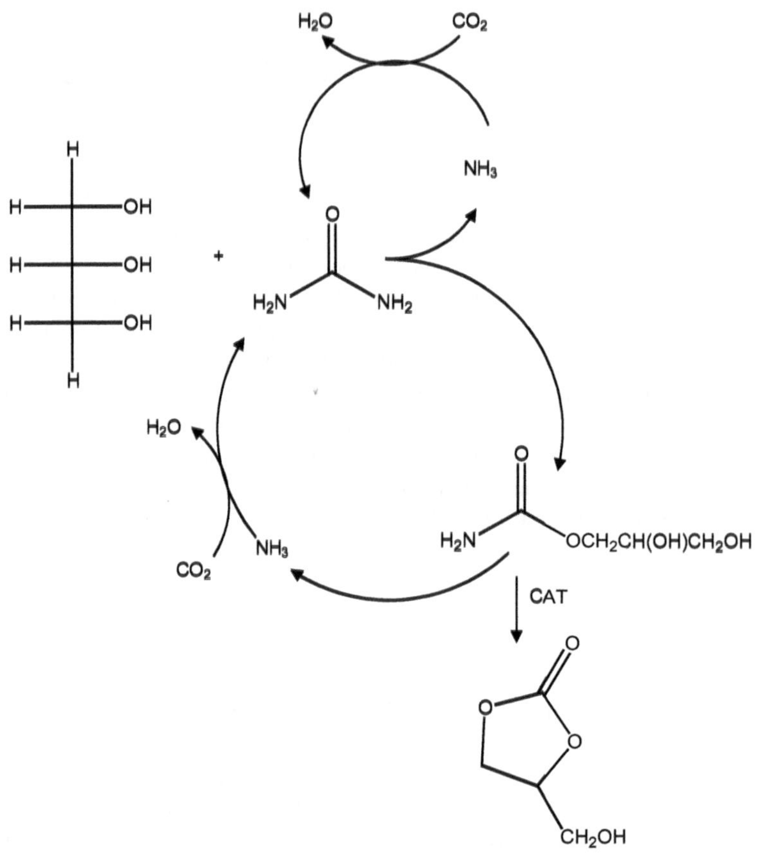

Scheme 6.15 Reaction mechanism in the glycerolysis of urea to afford glycerol carbonate at 408 K. Reprinted with permission from [103]. Copyright (2009) Elsevier

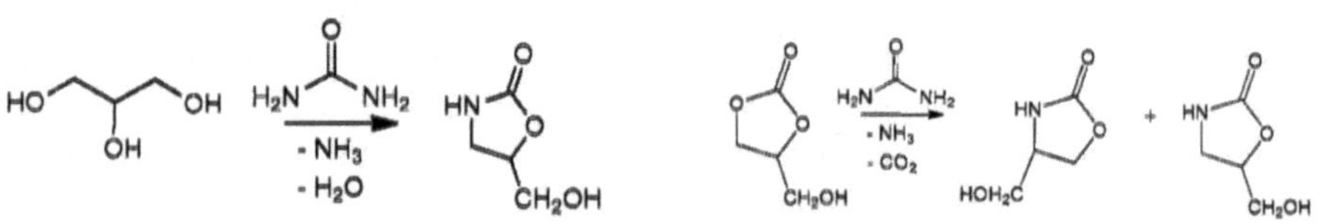

Scheme 6.16 Conversion of glycerol or glycerol carbonate into oxazolidinones above 420 K. Reprinted with permission from [105]. Copyright (2012) John Wiley and Sons

The produced glycerol carbonate has been used as platform molecule for the synthesis of several chemicals, including epichlorohydrin [106].

6.4 Oxidative Carboxylation of Olefins to Afford Cyclic Carbonates

The oxidative carboxylation of olefins appears to be an interesting approach to the synthesis of cyclic carbonates from cheap and easily available reagents such as olefins, non-O_2-free CO_2, and air (6.17) mostly under heterogeneous catalysis.

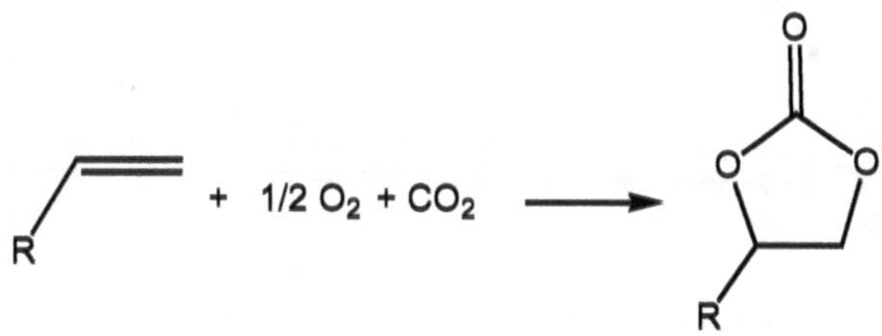

(6.17)

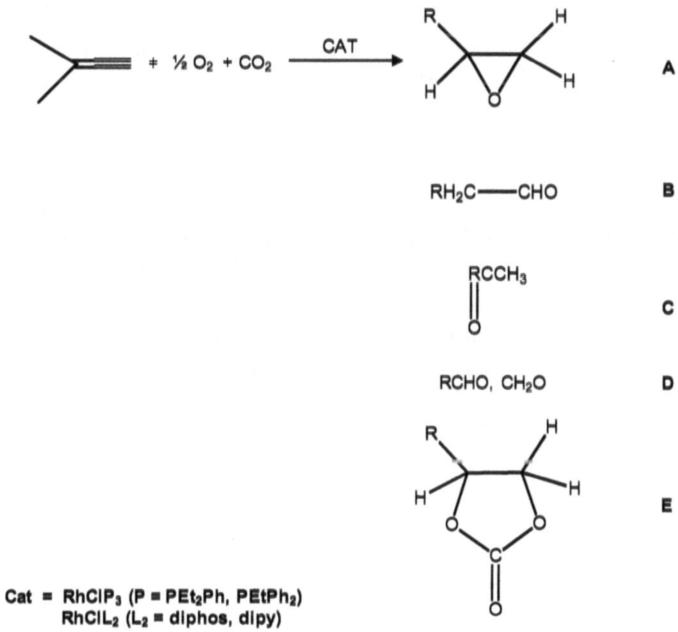

Scheme 6.17 Product distribution in the catalytic oxidation of olefins. Reprinted with permission from [107]. Copyright (2002) Elsevier

The direct oxidative carboxylation of olefins [108–110] couples two processes, namely (1) the epoxidation of the olefins and (2) the carbonation of the epoxide, occurring in the same reactor. Interestingly, it has been shown that CO_2 modulates the oxidant properties of O_2 [111].

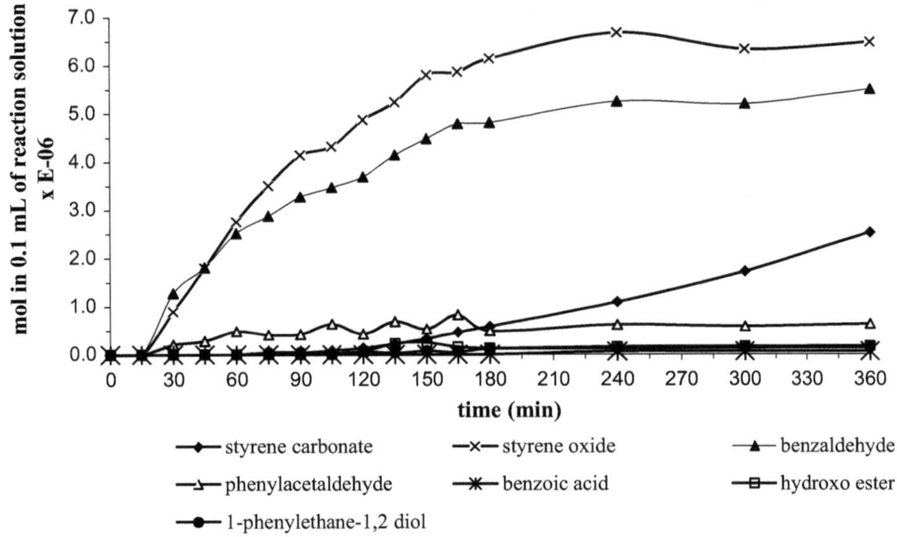

Fig. 6.13 Kinetics of products formation in the oxidative carboxylation of styrene at 393 K under metal oxides catalysis. Styrene oxide is formed in a slight excess with respect to benzaldehyde with a selectivity of ca. 55–60 % with respect to styrene. Reprinted with permission from [107]. Copyright (2002) Elsevier

The reaction can go via "one-oxygen transfer" or "two-oxygen transfer" to the olefin (Scheme 6.17). The former affords the epoxide (A), target product, or its isomers aldehyde (B) and ketone (C). The epoxide is the precursor of the carbonate (E).

The latter route splits the olefin into two aldehydes (RCHO and CH_2O, D) which in turn, depending on the reaction conditions, can be oxidized to the relevant acids.

The epoxide has been quantified in the reaction mixture [107] which leads one to infer that the reaction is a two-step process in which the carboxylation of the epoxide is the rate-determining step under the reaction conditions. Figure 6.13 shows the formation of the epoxide and carbonate in the one-pot oxidative carboxylation.

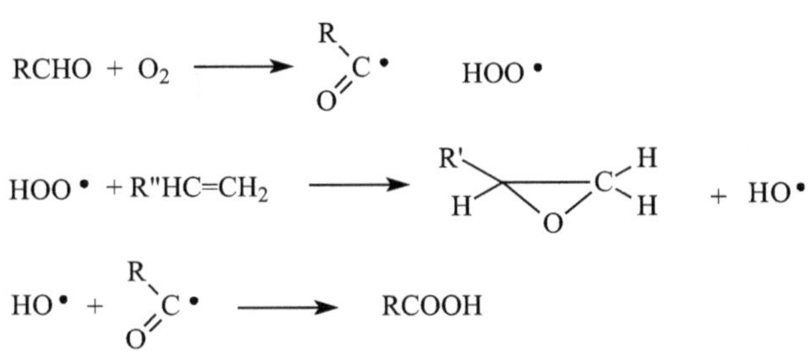

Scheme 6.18 Role of aldehydes in promoting the formation of epoxides from olefins and O_2. Reprinted with permission from [107]. Copyright (2002) Elsevier

The reaction is quite complex and the splitting of styrene to afford benzaldehyde on the one hand is a negative fact as it subtracts the reagent, on the other hand it is beneficial as aldehydes (Scheme 6.18) are known to promote the epoxidation of olefins via formation of peroxo groups [107].

The reaction mechanism of the overall process can, thus, be distinguished in two steps, the first relating to the formation of the epoxide and the second to the carboxylation of the epoxide (Scheme 6.19). Interestingly, it has been found that the carboxylation step can be favored by the use of nucleophiles (halides are common agents, or the solvent itself) which helps in stabilizing intermediate cationic species, as shown in Scheme 6.19. The elimination of the epoxide from the metal site competes with its carboxylation: therefore, such stabilizers may favor the carboxylation over ring-closure to afford the epoxide and, thus, facilitate the formation of the carbonate.

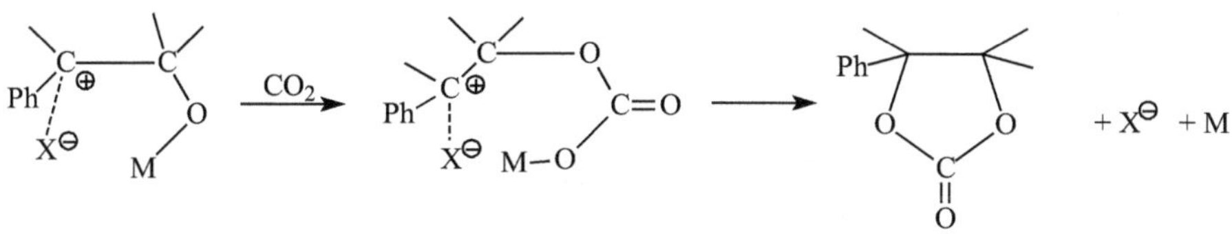

Scheme 6.19 Role of nucleophiles in the stabilization of open forms of the epoxide linked to a catalyst. Reprinted with permission from [107]. Copyright (2002) Elsevier

Only a few examples are reported in the literature of the direct carbonation of olefins, namely the direct functionalization of propene [112, 113] and styrene [108]. Such an approach has a drawback represented by the addition of dioxygen across the C=C double bond with formation of aldehydes or of the relevant acids. Using Rh-complexes the active species in the epoxidation of the olefin has been shown to be the peroxocarbonate moiety [114]. Using differently labeled O_2 molecules such as $^{16(18)}O–^{18(16)}O$ it has been possible to build peroxo carbonates bearing a well defined O-isotope of the peroxo group linked to a metal atom, namely $M–^{16}O–^{18}O$ or $M–^{18}O–^{16}O$ (see Sect. 4.4). However, it has been possible to show that the metal-bound O-atom of the peroxo group is transferred to an oxophile such as an olefin [115].

Several other oxidants have been used for the oxidation of olefins, such as organic hydroperoxides or hydrogen peroxide [107], which have limitations in their market volume not comparable with that of cyclic carbonates if the latter have to be used as monomers for polymers or in other large scale uses. The goal must be the direct use of dioxygen (or even better air) under controlled conditions which may block the radical reactions involving O_2 and olefins.

It has been shown that the cleavage of the olefin double bond can be avoided using a two-reactor system ((6.18) and (6.19)) in which a metal oxide transfers one lattice oxygen to an olefin with formation of the olefin oxide and is then re-oxidized using dioxygen in a separate reactor [116]. A single reactor can be used which is flushed with the olefin to get the epoxide and then with the correct amount of O_2 to regenerate the oxide.

$$M_xO_y + \text{propene} \rightarrow M_xO_{y-1} + \text{propene oxide} \tag{6.18}$$

$$M_xO_{(y-1)} + \frac{1}{2}\,O_2 \rightarrow M_xO_y \tag{6.19}$$

In this way, the direct contact of O_2 with the olefin is prevented and the radical process of addition across the double bond is avoided. Reactions (6.18) and (6.19) are slow and the selectivity towards the epoxide in reaction (6.18) strongly depends on the catalyst preparation, the nature of the metal, and the reaction temperature. Using propene, the formation of the epoxide is in concurrence with the formation of acetone and propionaldehyde. Moreover, depending on the preparation of the metal oxide, the same catalyst can push the reaction to the formation of acroleine or even to the total oxidation of propene to CO_2 and water [117]. If, instead of the only olefin, a mixture of olefin and CO_2 is admitted on the catalyst in its oxidized form, the carbonate is formed which can be recovered by condensation and the excess olefin recycled.

6.5 Carboxylation of Cyclic Ethers

Cyclic ethers are substrates of great interest as sources of cyclic monomeric carbonates or even as co-monomers for the production of polymers through ring-opening carboxylative polymerization. Their reactivity is driven by different parameters, such as:

1. The ring size and the ring stress energy: three-, four-, five-, and six-membered ethers have a quite different reactivity (see Table 6.2)
2. The substituents on the ring: groups which may produce a different charge localization on O or ethereal Cs have an influence on the rate of opening of the ring

Table 6.2 Properties of cyclic ethers

Name	Number of atoms in the ring	Ring-strain energy (kJ mol^{-1})	Basicity (pK_b)	Backbone
Ethene oxide	3	115.36		Linear
Propene oxide	3	114.2	15.7	Linear
Epichlorohydrin	3	*115*	16.7	Linear
Vinylcyclohexene	3	*113*	15.5	Linear
Styrene oxide	3	*102*	16.4	Linear
Indene oxide	3	*95*	15.2	Alicyclic
Cyclopentene oxide	3	*89*	15.4	Alicyclic
Cyclohexeneoxide	3	*92*	15.5	Alicyclic
Oxetane	4	106.4	13.2	Cyclic
Tetrahydrofuran	5	28.8	13.5	Cyclic

Values in italics indicate a lack of confidence. Adapted with permission from [118]. Copyright (2013) Elsevier

3. The ring backbone: an ethereal ring on a linear alkyl chain (e.g., propeneoxide) has a different reactivity compared to the same size ring on an alicyclic structure (e.g., cyclohexeneoxide)

These features influence the overall reaction rate and may also govern the propagation of the insertion reaction and its regularity in the case of the copolymerization reaction. Table 6.2 presents some properties of cyclic ethers. Such properties have long been known to influence the polymerization process [119]. The ring strain energy has been calculated using ab initio methods [120] or by other methods [121]. The basicity of cyclic ethers (Table 6.2) has been determined experimentally [118] by using an adaptation of the method used for the determination of the basicity of amines [122, 123].

The carboxylation of epoxides (6.20) has been known since 1943 but has been exploited only much more recently. It may afford either cyclic carbonates or polymers, depending on the catalyst used [124–126] and the reaction conditions [127–138].

A large variety of catalysts has been used, some of them in equimolar amounts or suffering a very short life.

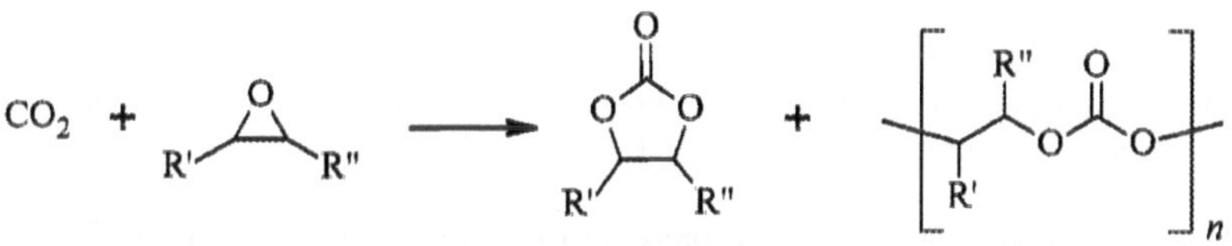

(6.20)

6.5.1 Synthesis of Monomeric Cyclic Carbonates

Organometallic species and metal complexes [139], classical Lewis acids [140], metal phthalocyanines [141], metal halides [142], also coupled to β-cyclodextrin (β-CD), a hydrogen bonding agent that accelerates the ring-opening in the absence of organic solvents [143], polyoxometalates with ammonium salts [144], and H-heterocyclic carbenes [145] all catalyze the ring opening of epoxides. The *salen* or analogous ligands based on an "$N_2 + O_2$" set of donor atoms have been largely used with metals such as: aluminum [146, 147], nickel [146], copper [146, 148, 149], zinc [146, 148, 149], magnesium [149], cobalt [146–148], chromium [146, 147, 150], tin [146, 147, 151], manganese [146, 147] or ruthenium [147]. Recently, bromine (or *N*-bromosuccinimide, NBS) has been reported as an effective catalyst for the carboxylation of epoxides [152] in the presence of bases. Heterogeneous catalysts [153, 154], including metal oxides [155–157], are interesting catalysts characterized by a longer life with respect to homogeneous systems. Amides, such as dimethylformamides (DMF) or dialkylacetamides (DAA), often used as solvents, promote by themselves the carboxylation of epoxides [108], perhaps to a limited extent. The synthesis of optically active carbonates has been carried out and, interestingly, metal oxides can catalyze the carboxylation of optically active epoxides with total retention of configuration [110, 147], most probably through a double inversion mechanism. The resolution of a racemic mixture of the epoxides was not very successful, with an *ee* of the order of 22 %, using Nb(IV) complexes with optically active (N, O, P as donor atoms) ligands, because of the de-anchoring of the ligand from the metal center at the reaction temperature, as shown by an NMR study [110]. Microwaves have been shown to promote the formation of carbonate in IL [159, 160]. Exhaustive reviews on the carboxylation of epoxides have recently been published [146, 158, 161–168].

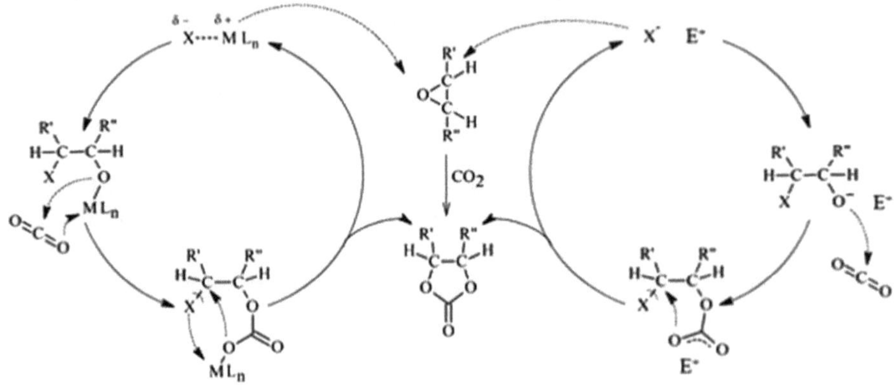

Scheme 6.20 Ring opening of epoxides and coupling to CO_2 promoted by nucleophiles or electrophiles. Adapted with permission from [158]. Copyright (2010) Elsevier Ltd

The mechanism of the carboxylation passes through several steps, such as: ring opening, CO_2 coupling to the opened ether structure, back-biting to form the cyclic carbonate, and elimination with regeneration of the catalyst. The ring opening is driven in different directions according to the catalyst employed, as shown in Scheme 6.20. An electrophile (or Lewis acid) attacks the O-atom and the carbocation formed is stabilized by interaction with the nucleophilic species X^- (often a halide anion). An insertion into the E–O bond (usually fast if the nucleophile is a metal cation; see Sect. 4.4) generates the open form of the carbonate. Back-biting and cyclization produce the carbonate. Conversely, a nucleophile (halide ion, quite often) attacks the C of the epoxide and makes the open carbanionic form of the epoxide (stabilized by interaction with an electrophile) ready for an attack at the C of CO_2. Ring closure by back-biting and elimination of the catalyst produces the carbonate. As a matter of fact, although the ring opening can be started by a nucleophile or an electrophile, catalysis requires a Lewis acid–base pair. Ammonium salts of various structural complexities play such a role. A metal most likely produces an attack at oxygen with formation of a "MOR$^+$" moiety stabilized by a nucleophile (halide ion) [109]. An alternative to monometallic systems are bimetallic catalysts [169, 170] which have also been shown to be quite active in the presence of tetrabutylammonium salts (bromide) (Scheme 6.21). Of interest is the kinetic law discovered by the authors (6.21).

$$\text{rate} = k_{\text{obs}}[\text{epoxide}]^1 \text{ where}$$
$$k_{\text{obs}} = k[CO_2]^1[\text{metal cat}]^1[Bu_4NBr]^1[Bu_3N]^1 \quad (6.21)$$

On the basis of such a kinetic law, the authors have proposed the reaction mechanism depicted in Scheme 6.22. The presence in the kinetic equation of both the ammonium ion and the free amine is quite singular. The role of the latter is to interact with Lewis acids, such as coordinatively unsaturated metals or other species.

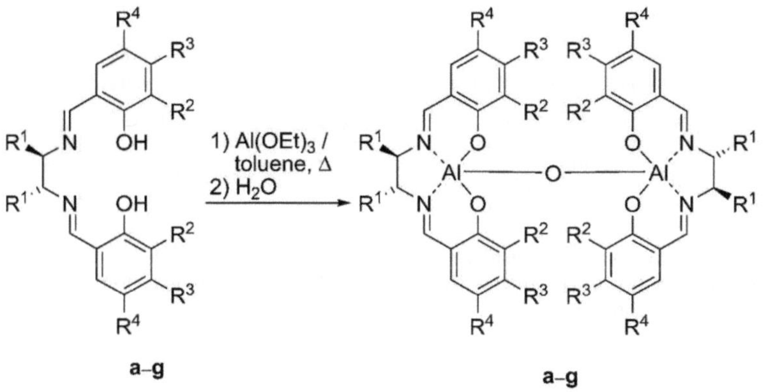

a: R^1-R^1 = $(CH_2)_4$; R^2 = R^4 = *t*Bu; R^3 = H
b: R^1-R^1 = $(CH_2)_4$; R^2 = R^3 = H; R^4 = *t*Bu
c: R^1-R^1 = $(CH_2)_4$; R^2 = R^3 = R^4 = H
d: R^1-R^1 = $(CH_2)_4$; R^2 = *t*Bu; R^3 = H; R^4 = NO_2
e: R^1-R^1 = $(CH_2)_4$; R^2 = *t*Bu; R^3 = H; R^4 = OMe
f: R^1-R^1 = $(CH_2)_4$; R^2 = R^4 = H; R^3 = OMe
g: R^1 = H; R^2 = R^4 = *t*Bu; R^3 = H

Scheme 6.21 Bimetallic Al–O–Al catalysts used in the carboxylation of epoxides. Adapted with permission from [169]. Copyright (2010) John Wiley and Sons

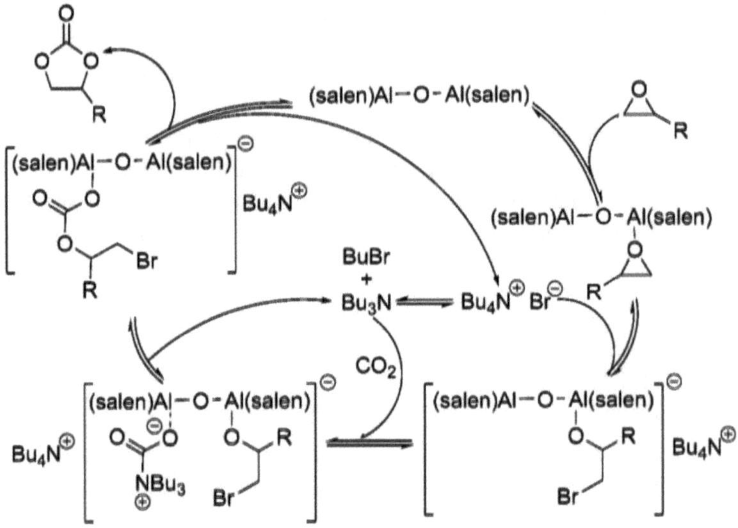

Scheme 6.22 Reaction mechanism for the carboxylation of epoxides under the action of a bimetallic-Al-complex, Bu_4NBr and Bu_3N. Adapted with permission from [169]. Copyright (2010) John Wiley and Sons

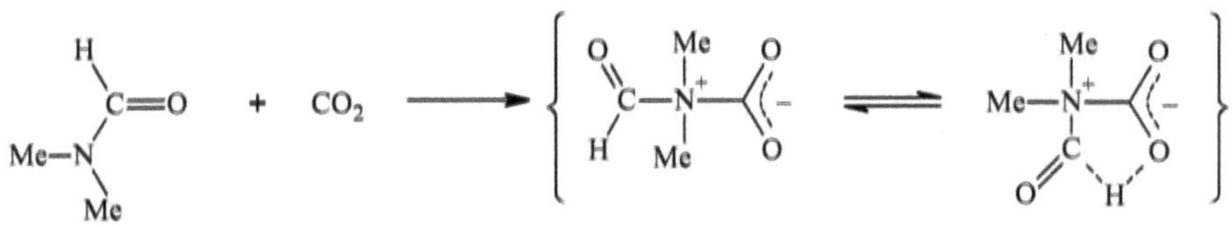

Scheme 6.23 Interaction of dimethylformamide with CO_2. Adapted with permission from [109]. Copyright (2003) Elsevier Science BV

In the case of Br_2 promoted carboxylation [171], Br^+ is supposed to attack the oxygen of the epoxide whereas Br^- stabilizes the carbocation. This is a particular case, as it is known that bromine in the presence of bases affords Br^- and BrO_3^- or BrO^-, according to the conditions.

That solvents may play a key role in the carboxylation of epoxides was proposed by Dibenedetto et al. [109] who showed that both dimethylformamide and diacetamide may also catalyze the carboxylation of epoxides if at a low extent (Scheme 6.23). The active role of DMF has been confirmed in the case of the bromine catalyzed carboxylation of epoxides [171]. DMF has been proposed to attack CO_2 (Scheme 6.23) and prompt it to a reaction with epoxides.

Sc-CO_2 is often used as solvent and reagent and is considered to be an economically viable and ecologically benign reaction medium. It has several advantages such as no flammability, lack of toxicity, absence of a gas-liquid phase boundary, and possible simplifications in workup. The advantages of using Sc-CO_2 alone [172] or in combination with IL [173, 174] have been described. Amides play a positive role because of participation either in the ring opening of the epoxide or in a preliminary coordination of CO_2 [109].

6.5.2 Synthesis of Polycarbonates

Polycarbonates have long been produced by using phosgene. This field is now moving to new synthetic strategies based on: (1) the direct copolymerization of olefin-oxides and CO_2 (propene carbonate, Novomer), (2) the use of CO_2 substitutes (organic carbonates formed from CO_2) if the epoxide is not easily produced (this is the case of BP-A), or (3) the polymerization of preformed monomeric carbonates, as represented in Scheme 6.24. The latter would produce a very regular polymer characterized by 50 % CO_2 and co-monomer.

The carboxylation of epoxides may afford either monomers, as discussed above, or polycarbonates: for example, Al-porphyrin complexes [175, 176] or Zn-compounds [177] promote the formation of polycarbonates. The pioneering studies of Inoue [178] and Kuran [179] have opened the route to the investigation of the copolymerization of CO_2 and epoxides. The key issue here is to master the alternate insertion epoxide-CO_2.

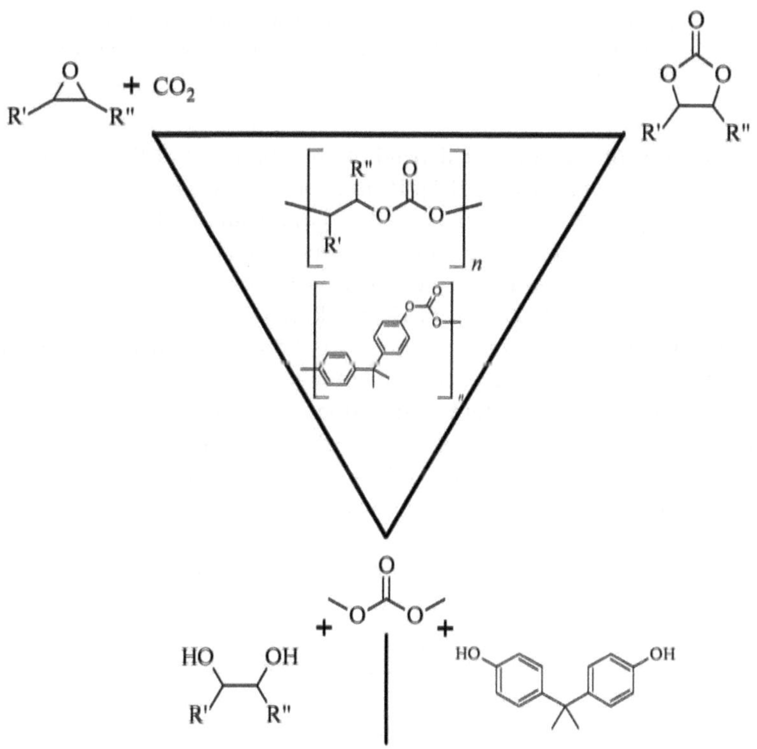

Scheme 6.24 Routes to polycarbonates: copolymerization of CO_2 and epoxides (*top left*), copolymerization of dialkylcarbonates and diols (*bottom*), polymerization of pure cyclic monomers (*top right*)

Table 6.3 Parameters that influence the rate of the polymerization and the regularity of the alternate insertion

Parameter	Role
Metal center	Metals such as Al, Co, Cr, Mn, Zn, and Sn are active in the form of soluble complexes. A single site is necessary for catalysis to occur. In general such complexes have an octahedral structure with the active site at one of the axial positions. The *trans* ligand has an influence in labilizing the M–O or M–C bond and favoring the insertion
Ligands	A tetradentate ligand is usually required, of the type of salen or salan, with a N_2+O_2 set of binding atoms. Also N_4 sets have been used. The rigidity of the N_2+O_2 set of atoms may play a role. Ornamental groups of various bulkiness and electron donor properties also play a key role. Dicarboxylato ligands have shown some usability as Zn complexes. Ligands must force a planar geometry around the metal center and control the insertion at the apical position. Chiral ligands may regulate the insertion of asymmetric olefins
Cyclic ether	The facility of opening of the ring may speed the reaction. The structure of the ether is also important. An ethereal link in a linear structure behaves differently from alicyclic ethers
Temperature	Low temperatures in the range 298–330 K produce polymers, higher temperatures tend to form molecular cyclic carbonates
Solvent	The micropolarity of the solvent may help to stabilize polar or ionic forms and be beneficial to the reaction. For example, solvents with donor atoms (halides, nitrogen, oxygen) may be useful in the stabilization of the open form of the ether
Co-catalyst	A co-catalyst is used to start the reaction or stabilize intermediates. Halides are usually good starters and large ammonium cations play a role in the stabilization of intermediate forms

The properties of the polymer depend on the regularity of the incorporation of CO_2 and the epoxide [125, 180, 181]. Bifunctional Co-salen complexes have been used as promoters of the alternating copolymerization of epoxides and CO_2 [182]. Co(III) complexes with tetraamidomacrocyclic [183] and salen-type ligands have been demonstrated to be very efficient in the copolymerization of epoxides and CO_2 with a TOF of 15 000 h^{-1} [184], working with a substrate/catalyst ratio of 50 000. Double metal cyanide systems also show an interesting activity [185] (TOF = 3 856 h^{-1}). It must be recalled that the polymer molecular mass is important but the regularity of the alternate insertion is also very important. Therefore, a high rate of insertion must be coupled to a perfect alternate insertion for producing polymers that have good properties. Often, polymers with average properties are obtained because of a poor alternate insertion.

The use of epoxides characterized by a different molecular structure may be of interest as polymers with different structural properties may be obtained. Therefore, an issue of great interest is the discovery of new routes to the production of epoxides using dioxygen instead of hydrogen peroxide in the epoxidation of olefins.

From the short discussion above, it comes out that a number of parameters influence the "perfect" polymerization. They can be identified as metal center, ligands, cyclic ether, co-catalyst, temperature, solvent. Table 6.3 summarizes the role of each parameter and presents the influence on the polymerization.

The reaction mechanism for the copolymerization is initiated as shown in Scheme 6.24, left and once the first coupling epoxide-CO_2 is performed, the growth of the chain should proceed with a perfect alternate insertion of epoxide and CO_2.

During the chain growth, a back-biting elimination mechanism may occur in several ways [186] which brings about the formation of cyclic carbonate, as proposed by Darensbourg (Scheme 6.25).

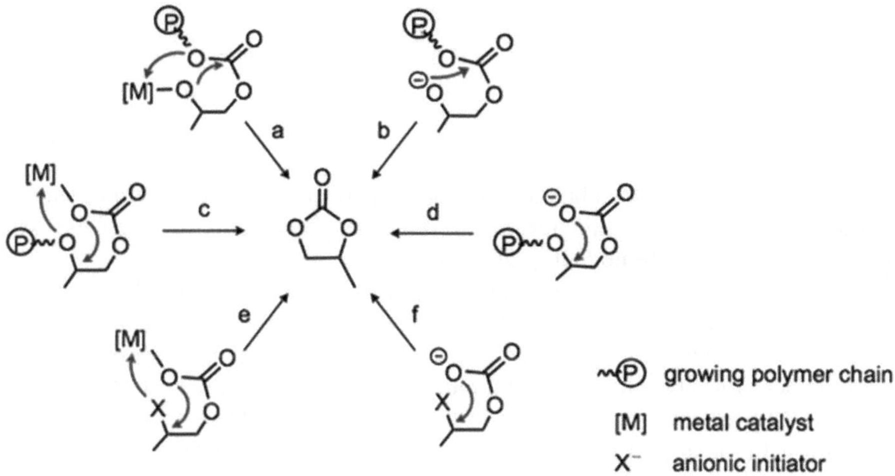

Scheme 6.25 Back-biting and elimination of cyclic carbonate in the chain growth. Reprinted from [186]. Copyright (2014) Elsevier

Polycarbonates are used in several different applications, from buildings to car manufacture, CDs and specialty optical. The market has seen a steady growth in the last years: the 2013 demand was over 4.5 Mt. PC demand is foreseen to grow by roughly 5 % per year on average by 2016 [187]. Other sources foresee a growth of the market by 6 % annually, although the PC demand by China is estimated to be higher than 8 % on a yearly basis. Such a forecast is based on the expansion of the use of PCs in the building sector and in the mobility sector with major growth in China [188, 189].

6.6 Formation of Polyurethanes: Carboxylation of Cyclic Amines

The formation of carbamates RR′N–COOR″ has been discussed in Chap. 4. There are several applications in the pharmaceutical, agrochemical, materials, and polymer industries. Dicarbamates are of particular interest as they are used as monomers for polymers. Among dicarbamates, those derived from primary amines are of great industrial interest as they can easily lose alcohol and produce isocyanates (6.22).

Diisocyanates are monomers for the production of polyurethanes (6.23) which find extensive industrial application as foams, insulating materials, packaging, etc.

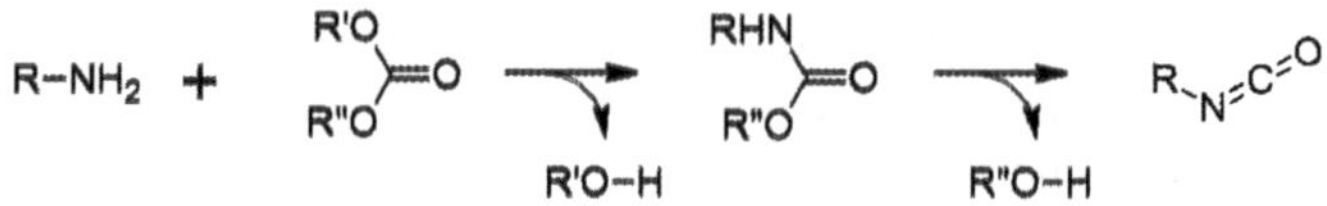

(6.22)

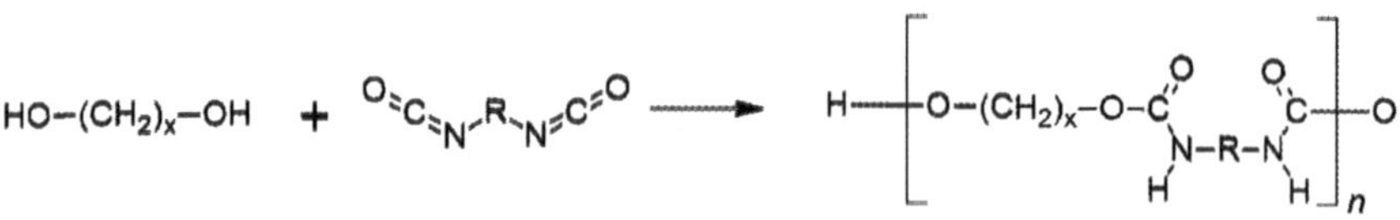

(6.23)

Polyurethanes can also be obtained via copolymerization of cyclic amines such as aziridines (analogous to epoxides) and azetidines (analogous to oxetanes) with CO_2, according to Scheme 6.26 [190, 191].

The interaction of such cyclic amines with metal centers and CO_2 to afford oxazolidinones (cyclic carbamates) has already been discussed in Sect. 4.9. Here we discuss the production of polymers, an application that has not received further development for more than 30 years, most likely because of the cost of the cyclic amines. Only recently a revamping in this area has brought us to reconsider the reaction with the aim of improving the quality of the polymer [192] and bringing close to one the ratio amine/CO_2 which is usually much higher than one because of the easier insertion of the aziridine with respect to CO_2. Polyurethanes are long-lasting materials and their synthesis from CO_2 would represent a way of storing CO_2 for a long time.

Very recently the cyclic carbamate structure has been generated by a route based on the reaction of glycerol carbonate (made from glycerol and CO_2 or urea; see below) and urea as a source of *N*-atoms for the substitution of a ring-oxygen with a ring-NH group. So, such a reaction uses a bio-derived C-skeleton and cheap co-reagents (CO_2 and urea) for the production of added value molecules such as the oxazolidinones. The direct reaction of glycerol with urea can produce both glycerol carbonate and oxazolidinones, according to the temperature (Scheme 6.27) [105]. The reaction mechanism of such a reaction that produces useful monomers has been fully elucidated as shown in Scheme 6.27. Noticeably, the oxazolidinones are interesting monomers for polymers.

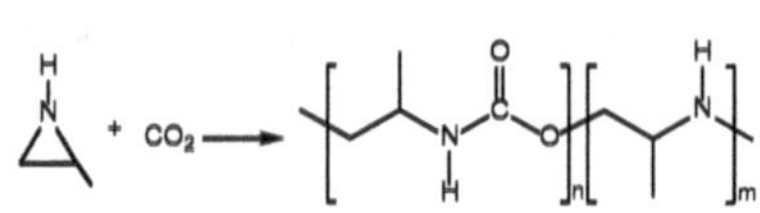

Scheme 6.26 Copolymerization of aziridines with CO_2. Reprinted from [181]

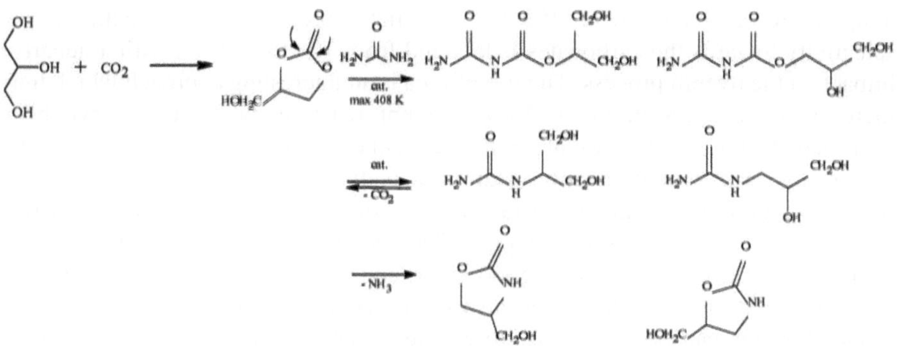

Scheme 6.27 Synthesis of oxazolidinones from glycerol, CO_2 and urea. Adapted with permission from [105]. Copyright (2012) John Wiley and Sons

6.7 Conclusions

The reaction mechanism for formation of dialkyl carbonates from alcohols and CO_2 has been investigated with different catalysts and shown to be a two-step process. The first step is always an alcohol base-activation that affords the alkoxo moiety RO^-. This species is promoted by a homogeneous or heterogeneous catalyst and affords a *cat*-OR species which undergoes CO_2 insertion to give the emicarbonate *cat*-OC(O)OR.

The latter undergoes alkylation with formation of the dialkylcarbonate, $(RO)_2CO$. The alkylation step has been demonstrated to occur in different ways depending on the catalyst used. In the DCC- and *n*-$Bu_2Sn(OCH_3)_2$-catalyzed processes for synthesis of dimethylcarbonate it has been shown that the methyl cation CH_3^+ is provided by a *cat*-OCH_3 species which is converted into either a C=O or an O–Sn–O moiety, respectively. None of the end compounds, DCU or Sn–O–Sn oligomers, are per se active in catalysis. In the homogeneous Nb-penta-alkoxo-catalyzed synthesis of dialkylcarbonates $(RO)_2CO$ (R=CH_3, C_2H_5, CH_2=CH–CH_2) the critical step is represented by a simultaneous acid- and base-activation of the alcohol molecule which concurrently affords the carbonate and regenerates the catalyst which can be isolated and reused. The same is true for heterogenized Nb-catalysts. Heterogeneous catalysts have been shown to be characterized by a mechanism that follows the scheme previously discussed up to the formation of the *S*-OC(O)OR moiety (*S* is the surface). The latter is converted into the organic carbonate by interaction with gas phase alcohols. At least in one case, namely the synthesis of diethlylcarbonate from ethanol and CO_2 using Nb-Ceria mixed oxides, the involvement of two molecules of ethanol in the latter step has been demonstrated. The data available so far seem to support the view that the acid-activation of methanol is the key step, and the involvement of Brønsted acids seems to boost the process more than Lewis acids. The use of hyperactive heterogeneous catalysts (loaded with H_3PO_4) which allow the reaction to occur at a lower temperature (310 instead of 450 K) does not represent a real benefit, as the selectivity towards the carbonates is lowered from 100 to <20 %, with a negative impact on the overall process. The use of urea is an interesting approach which may increase the conversion yield. Two different reaction mechanisms have been proposed. Cyclic carbonates are produced either by reaction of epoxides with CO_2 or by direct oxidative carboxylation of olefins, a reaction that needs control for avoiding olefin splitting under radical conditions caused by oxygen. The actual selectivity of 55–60 % is a starting point. The reaction of diols or polyols with urea also gives cyclic carbonates with good yields and selectivities. Cyclic amines react with CO_2 to afford oxazolidinones, also synthesized by reacting polyols with urea. Polycarbonates can be formed by three routes, namely the copolymerization of epoxides and CO_2 (a process that requires catalyst control for a regular 1:1 copolymerization), the reaction of diols with linear carbonates, and the polymerization of monomeric cyclic carbonates. The reaction mechanism has been investigated and critical parameters identified. The search for new catalysts and monomers for new polymers with new properties is a current hot topic. Cyclic amines can co-polymerize with CO_2 or oxazolidinones can be polymerized in a more regular way. This field is fast moving, considering the industrial interest existing in new materials with new properties which, eventually, can be used for chemical storage of CO_2.

References

1. Delledonne D, Rivetti F, Romano U (2001) Developments in the production and application of dimethyl carbonate. Appl Catal A Gen 221(1–2):241–251
2. Shaikh AAG, Sivaram S (1996) Organic carbonates. Chem Rev 96:951–976, and references therein
3. Tullo AH (2001) Fighting for position in polycarbonate. C&EN 79(15):15–16
4. Buysch HJ (1992) Carbonic acid esters. In: Ullmann's encyclopedia of industrial chemistry, vol A5. VCH, Weinheim, pp 197–201
5. Aresta M, Dibenedetto A, He LN (2012) Analysis of demand for captured CO_2 and products from CO_2 conversion, a report exclusively for members of the carbon dioxide capture and conversion (CO_2CC) program of the Catalyst Group Resources (TCGR)
6. Pacheco MA, Marshall CL (1997) Review of dimethyl carbonate (DMC) manufacture and its characteristics as a fuel additive. Energy Fuels 11(1):2–29
7. Aresta M, Dibenedetto A (2003) In: Aresta M (ed) Carbon dioxide: recovery and utilization. Kluwer, Dordrecht, pp 211–260
8. (2001) TEXACO study
9. Société Nationale des Poudres et Explosifs (1973) Continuous manufacture of alkyl carbonates. FR 2163884, 1973 Fr. Patent, 7 pp
10. Damle SB (2000) Carbonic and carbonochloridic esters. In: Othmer K (ed) Encyclopedia of chemical technology, vol 5, 4th edn. Wiley, New York, NY, pp 77–97
11. Romano U, Tesei R, Massi MM, Rebora P (1980) Synthesis of dimethyl carbonate from methanol, carbon monoxide, and oxygen catalyzed by copper compounds. Ind Eng Chem Prod Res Dev 19(3):396–403
12. Romano U (1993) Dimethyl carbonate and its production technology. Chim Ind Milan 75(4): 303–306

13. Perrotti E, Cipriani G (1974) Process for the preparation of esters of carbonic acid. US Patent 3846468
14. Romano U, Tesei R, Cipriani G, Micucci L (1980) Method for the preparation of esters of carbonic acid. US Patent 4218391
15. Matsuzaki T, Nakamura A (1997) Dimethyl carbonate synthesis and other oxidative reactions using alkyl nitrites. Catal Surv Jpn 1:77–88
16. Nishihira K, Tanaka S, Kodama K, Kaneko T (1992) Process for preparing diester of carbonic acid. Eur Pat Appl EP501507
17. Aresta M, Dibenedetto A, di Bitonto L (2014) Cerium-based binary and ternary oxides in the transesterification of dimethylcarbonate with phenol. ChemSusChem 7(4):1155–1161
18. Bhanage BM, Fujita S, Ikushima Y, Torii K, Arai M (2003) Synthesis of dimethyl carbonate and glycols from carbon dioxide, epoxides and methanol using heterogeneous Mg containing smectite catalysts: effect of reaction variables on activity and selectivity performance. Green Chem 5:71–75
19. Bhanage BM, Fujita S, Ikushima Y, Arai M (2001) Synthesis of dimethyl carbonate and glycols from carbon dioxide, epoxides, and methanol using heterogeneous basic metal oxide catalysts with high activity and selectivity. Appl Catal A Gen 219:259–266
20. Tomishige K, Sakaihori T, Ikeda Y, Fujimoto K (1999) A novel method of direct synthesis of dimethyl carbonate from methanol and carbon dioxide catalyzed by zirconia. Catal Lett 58: 225–229
21. Aresta M, Dibenedetto A, Pastore C (2003) Synthesis and characterization of $Nb(OR)_4[OC(O)OR]$ (R = Me, Et, Allyl) and their reaction with the parent alcohol to afford organic carbonates. Inorg Chem 42(10):3256–3261, and references therein
22. Choi J-C, Sakakura T, Sako T (1999) Reaction of dialkyltin methoxide with carbon dioxide relevant to the mechanism of catalytic carbonate synthesis. J Am Chem Soc 121:3793–3794
23. Ballivet-Tkatchenko D, Douteau O, Stutzmann S (2000) Reactivity of carbon dioxide with n-butyl(phenoxy)-, (alkoxy)-, and (oxo)stannanes: insight into dimethyl carbonate synthesis. Organomet 19:4563–4567
24. Sanderson RT (1976) Chemical bonds and bond energy. Academic, New York, NY
25. Aresta M, Dibenedetto A, Fracchiolla E, Giannoccaro P, Pastore C, Pàpai I, Schubert G (2005) Mechanism of formation of organic carbonates from aliphatic alcohols and carbon dioxide under mild conditions promoted by carbodiimides. DFT calculation and experimental study. J Org Chem 70(16):6177–6186
26. Isaacs NS, O'Sullivan B, Verhaelen C (1999) High pressure routes to dimethyl carbonate from supercritical carbon dioxide. Tetrahedron 55:11949–11956
27. Sakakura T, Saito Y, Okano M, Choi J-C, Sako T (1998) Selective conversion of carbon dioxide to dimethyl carbonate by molecular catalysis. J Org Chem 63:7095–7096
28. Aresta M, Dibenedetto A, Pastore C, Pàpai I, Schubert G (2006) Reaction mechanism of the direct carboxylation of methanol to dimethylcarbonate: experimental and theoretical studies. Top Catal 40(1–4):71–81
29. Ballivet-Tkatchenko D, Jerphagnon T, Ligabue R, Plasseraud L, Poinsot D (2003) The role of distannoxanes in the synthesis of dimethyl carbonate from carbon dioxide. Appl Catal A Gen 255:93–99
30. Ballivet-Tkatchenko D, Chambrey S, Keiski R, Ligabue R, Plasseraud L, Richard P, Turunen H (2006) Direct synthesis of dimethyl carbonate with supercritical carbon dioxide: characterization of a key organotin oxide intermediate. Catal Today 115:80–87
31. Kohno K, Choi J-C, Ohshima Y, Yili A, Yasuda H, Sakakura T (2008) Reaction of dibutyltin oxide with methanol under CO_2 pressure relevant to catalytic dimethyl carbonate synthesis. J Organomet Chem 693:1389–1392
32. Dibenedetto A, Angelini A (2014) Synthesis of organic carbonates. Adv Inorg Chem 66: 25–81

33. Ballivet-Tkatchenko D, Chermette H, Plasseraud L, Walter O (2006) Insertion reaction of carbon dioxide into Sn–OR bond. Synthesis, structure and DFT calculations of di- and tetranuclear isopropylcarbonato tin(IV) complexes. Dalton Trans 43:5167–5175
34. Aresta M, Dibenedetto A, Angelini A (2014) Catalysis for the valorization of exhaust carbon: from CO_2 to chemicals, materials, and fuels. Technological use of CO_2. Chem Rev 114(3): 1709–1742
35. Kizlink J, Pastucha I (1995) Preparation of dimethyl carbonate from methanol and carbon dioxide in the presence of Sn(IV) and Ti(IV) alkoxides and metal acetates. Collect Czech Chem Commun 60:687–692
36. Kohno K, Choi J-C, Ohshima Y, Yasuda H, Sakakura T (2008) Synthesis of dimethyl carbonate from carbon dioxide catalyzed by titanium alkoxides with polyether-type ligand. ChemSusChem 1:186–188
37. Dibenedetto A, Pastore C, Aresta M (2006) Direct carboxylation of alcohols to organic carbonates: comparison of the Group 5 element alkoxides catalytic activity. Catal Today 115:88–94
38. Kato M, Ito T (1985) Facile carbon dioxide uptake by zinc(II)-tetraazacycloalkane complexes. 1. Syntheses, characterizations, and chemical properties of (monoalkyl carbonato) (tetraaza-cycloalkane)zinc(II) complexes. Inorg Chem 24:504–505
39. Aresta M, Dibenedetto A, Nocito F, Pastore C (2008) Comparison of the behaviour of supported homogeneous catalysts in the synthesis of dimethylcarbonate from methanol and carbon dioxide: polystyrene-grafted tin-metallorganic species versus silesquioxanes linked Nb-methoxo species. Inorg Chim Acta 361:3215–3220
40. Sakakura T, Saito Y, Choi J-C, Sako T (2000) Synthesis of dimethyl carbonate from carbon dioxide: catalysis and mechanism. Polyhedron 19:573–576
41. Aresta M, Dibenedetto A, Angelini A (2013) From carbon dioxide to valuable products under homogeneous catalysis. In: Reedijk J, Poeppelmeier KBT (eds) Comprehensive inorganic chemistry II. Elsevier, Amsterdam, pp 563–586
42. Zhong SH, Kong LL, Li HS, Xiao XF (2002) Preparation of $Ti_2(OMe)_4/SiO_2$ catalyst and its reactivity for DMC synthesis from CO_2 and CH_3OH. Ranliao Huaxue Xuebao 30(5):454–458
43. Aresta M, Dibenedetto A, Nocito F, Angelini A, Gabriele B (2010) Synthesis and characterization of a novel polystyrene-tethered niobium methoxo species. Its application in the CO_2-based carboxylation of methanol to afford dimethyl carbonate. Appl Catal A Gen 387: 113–118
44. Fan B, Zhang J, Li R, Fan W (2008) In situ preparation of functional heterogeneous organotin catalyst tethered on SBA-15. Catal Lett 121:297–302
45. Aresta M, Dibenedetto A, Pastore C, Angelini A, Aresta B, Pápai I (2010) Influence of Al_2O_3 on the performance of CeO_2 used as catalyst in the direct carboxylation of methanol to dimethylcarbonate and the elucidation of the reaction mechanism. J Catal 269:44–52
46. Tomishige K, Yoshida Y, Arai Y, Kado S, Kunimori K (2006) Direct synthesis of organic carbonates from the reaction of CO_2 with methanol and ethanol over CeO_2 catalysts. Catal Today 115:95–101
47. Tomishige K, Ikeda Y, Sakaihori T, Fujimoto K (2000) Catalytic properties and structure of zirconia catalysts for direct synthesis of dimethyl carbonate from methanol and carbon dioxide. J Catal 192:355–362
48. Ma J, Sun N, Zhang X, Zhao N, Xiao F, Wie W (2009) A short review of catalysis for CO_2 conversion. Catal Today 148:221–231
49. Ikeda Y, Asadullah M, Fujimoto K, Tomishige K (2001) Structure of the active sites on H_3PO_4/ZrO_2 catalysts for dimethyl carbonate synthesis from methanol and carbon dioxide. J Phys Chem B 105:10653–10658
50. Ikeda Y, Sakaihori T, Tomishige K, Fujimoto K (2000) Promoting effect of phosphoric acid on zirconia catalysts in selective synthesis of dimethyl carbonate from methanol and carbon dioxide. Catal Lett 66:59–62

51. Tomishige K, Furusawa Y, Ikeda Y, Asadullah M, Fujimoto K (2001) CeO_2–ZrO_2 solid solution catalyst for selective synthesis of dimethyl carbonate from methanol and carbon dioxide. Catal Lett 76:71–74
52. Allaoui LA, Acuissi A (2006) Effect of the Brønsted acidity on the behavior of CO_2 methanol reaction. J Mol Catal A Chem 259:281–285
53. La KW, Song IK (2006) Direct synthesis of dimethyl carbonate from CH_3OH and CO_2 by $H_3PW_{12}O_{40}/Ce_xTi_{1-x}O_2$ catalyst. React Kinet Catal Lett 89:303–309
54. La KW, Jung JC, Kima H, Baeck SH, Song IK (2007) Effect of acid–base properties of $H_3PW_{12}O_{40}/Ce_xTi_{1-x}O_2$ catalysts on the direct synthesis of dimethyl carbonate from methanol and carbon dioxide: a TPD study of $H_3PW_{12}O_{40}/Ce_xTi_{1-x}O_2$ catalysts. J Mol Catal A Chem 269:41–45
55. Tkatchenko DB, Dibenedetto A (2010) Synthesis of linear and cyclic carbonates. In: Aresta M (ed) CO_2 as chemical feedstock. Wiley-VCH, Weinheim, p 178
56. Jiang C, Guo Y, Wang C, Hu C, Wu Y, Wang E (2003) Synthesis of dimethyl carbonate from methanol and carbon dioxide in the presence of polyoxometalates under mild conditions. Appl Catal A Gen 256:203–212
57. Jung KT, Bell AT (2001) An in situ infrared study of dimethyl carbonate synthesis from carbon dioxide and methanol over zirconia. J Catal 204:339–347
58. Aresta M, Dibenedetto A, Pastore C, Cuocci C, Aresta B, Cometa S, De Giglio E (2008) Cerium(IV)oxide modification by inclusion of a hetero-atom: a strategy for producing efficient and robust nano-catalysts for methanol carboxylation. Catal Today 137:125–131
59. Finocchio E, Daturi M, Binet C, Lavalley JC, Blanchard G (1999) Thermal evolution of the adsorbed methoxy species on $Ce_xZr_{1-x}O_2$ solid solution samples: a FT-IR study. Catal Today 52:53–63
60. Dibenedetto A, Aresta M, Angelini A, Ethiraj J, Aresta BM (2012) Synthesis, characterization, and use of Nb^V/Ce^{IV}-mixed oxides in the direct carboxylation of ethanol by using pervaporation membranes for water removal. Chem-A Eur J 18(33):10324–10334
61. Sakakura T, Saito Y, Choi J-C, Masuda T, Sako T, Oriyama T (1999) Metal-catalyzed carbonate synthesis from carbon dioxide and acetals. J Org Chem 64:4506–4508
62. Aresta M, Dibenedetto A, Di Leo C, Tommasi I, Amidio E (2003) The first synthesis of a cyclic carbonate from a ketal in sc-CO_2. J Supercrit Fluids 25:177–180
63. Honda M, Kuno S, Sonehara S, Fujimoto K, Suzuki K, Nakagawa Y, Tomishige K (2011) Tandem carboxylation-hydration reaction system from methanol, CO_2 and benzonitrile to dimethyl carbonate and benzamide catalyzed by CeO_2. ChemCatChem 3(2):365–370
64. Tomishige K, Kunimori K (2002) Catalytic and direct synthesis of dimethyl carbonate starting from carbon dioxide using CeO_2-ZrO_2 solid solution heterogeneous catalyst: effect of H_2O removal from the reaction system. Appl Catal A Gen 237:103–109
65. Carafa M, Quaranta E (2009) Synthesis of organic carbamates without using phosgene: carbonylation of amines with carbonic acid diesters. Mini-Rev Org Chem 6(3):168–183
66. Honda M, Kuno S, Begum N, Fujimoto K-I, Suzuki K, Nakagawa Y, Tomishige K (2010) Catalytic synthesis of dialkyl carbonate from low pressure CO_2 and alcohols combined with acetonitrile hydration catalyzed by CeO_2. App Catal A Gen 384(1–2):165–170
67. Honda M, Suzuki A, Noorjahan B, Fujimoto K-I, Suzuki K, Tomishige K (2009) Low pressure CO_2 to dimethyl carbonate by the reaction with methanol promoted by acetonitrile hydration. Chem Commun 30:4596–4598
68. Eta V, Arbvela PM, Leino AR, Kordàs TD, Salmi T, Murzoin DY, Perikkolas J (2010) Synthesis of dimethyl carbonate from methanol and carbon dioxide: circumventing thermodynamic limitations. Ind Eng Chem Res 49:9609–9617
69. Eta V, Mäki-Arvela P, Wärnå J, Salmi T, Mikkola J-P, Murzin DY (2011) Kinetics of dimethyl carbonate synthesis from methanol and carbon dioxide over ZrO_2–MgO catalyst in the presence of butylene oxide as additive. Appl Catal A Gen 404:39–46

70. Leino E, Mäki-Arvela P, Eränen K, Tenho M, Murzin DY, Salmi T, Mikkola JP (2011) Enhanced yields of diethyl carbonate via one-pot synthesis from ethanol, carbon dioxide and butylene oxide over cerium (IV) oxide. Chem Eng J 176–177:124–133
71. Leino E, Mäki-Arvela P, Eta V, Kumar N, Demoisson F, Samikannu A, Leino AR, Shchukarev A, Murzin DY, Mikkola J-P (2013) The influence of various synthesis methods on the catalytic activity of cerium oxide in one-pot synthesis of diethyl carbonate starting from CO_2, ethanol and butylene oxide. Catal Today 210:47–54
72. Wagner A, Haas W (1994) Process for producing dialkyl carbonate. WO Patent 022805
73. Cheong M, Kim S-C, Park JB (1997) Dimethyl carbonate synthesis via carbon dioxide activation in the presence of iodide catalysts. New J Chem 21:1143–1145
74. Aresta M, Dibenedetto A, Angelini A, Pàpai I (2014) Reaction mechanisms in the direct carboxylation of alcohols for the synthesis of acyclic carbonates. Top Catal 58(1):2–14
75. Aresta M, Dibenedetto A, Stufano P, Aresta BM, Maggi S, Papai I, Rokob TA, Gabriele B (2010) The solid state structure and reactivity of NbCl5 · (N, N-dicyclohexylurea) in solution: evidence for co-ordinated urea dehydration to the relevant carbodiimide. Dalton Trans 39: 6985–6992
76. Aresta M, Dibenedetto A, Devita C, Bourova OA, Chupakhin ON (2004) New catalysts for the conversion of urea into carbamates and carbonates with C1 and C2 alcohols. Stud Surf Catal 153:213–220
77. Zhao W, Peng W, Wang D, Zhao N, Li J, Xiao F, Wei W, Sun Y (2009) Zinc oxide as the precursor of homogenous catalyst for synthesis of dialkyl carbonate from urea and alcohols. Catal Commun 10:655–658
78. Wang H, Wang M, Zhao W, Wei W, Sun Y (2010) Reaction of zinc oxide with urea and its role in urea methanolysis. React Kinet Mech Catal 99:381–389
79. Dubois JL (2011) Amination of organic substrates method for the co-production of non-cyclic carbonates and amino acids. EP Patent 2137133B1
80. Wang M, Zhao N, Wei W, Sun Y (2005) Synthesis of dimethyl carbonate from urea and methanol over ZnO. Ind Eng Chem Res 44(19):7596–7599
81. Zhao W, Wang F, Peng W, Zhao N, Li J, Xiao F, Wei W, Sun Y (2008) Synthesis of dimethyl carbonate from methyl carbamate and methanol with zinc compounds as catalysts. Ind Eng Chem Res 47:5913–5917
82. Gaoa Y, Penga W, Zhaoa N, Wei W, Sun Y (2011) A DFT study on the reaction mechanism for dimethyl carbonate synthesis from methyl carbamate and methanol. J Mol Catal A Chem 35:29–40
83. Dibenedetto A, Angelini A, Fasciano S, Papai I, Curulla F, Aresta M (2014) The reaction mechanism in the ethanolysis of urea with transition metal based catalysts: DFT calculations and experiments. J CO2 Util 8:27–33
84. Wang M, Wang H, Zhao N, Wei W, Sun Y (2007) High-yield synthesis of dimethyl carbonate from urea and methanol using a catalytic distillation process. Ind Eng Chem Res 46(9): 2683–2687
85. Saleh RY, Michaelson RC, Suciu EN, Kuhlmann B (1994) Dialkyl isocyanato tin alcoholate catalysts and dimers thereof. US Patent 5,561,094
86. Ryu JY (2000) Catalyst for making dialkyl carbonates. US Patent 6010976 A
87. Lin H, Yang B, Sun J, Wang X, Wang D (2004) Kinetics studies for the synthesis of dimethyl carbonate from urea and methanol. Chem Eng J 103:21–27
88. Wang M, Wang H, Zhao N, Wei W, Sun Y (2006) Synthesis of dimethyl carbonate from urea and methanol over solid base catalysts. Catal Commun 7:6–10
89. Guo L, Zhao X, An H, Wang Y (2012) Catalysis by lead oxide for diethyl carbonate synthesis from ethyl carbamate and ethanol. Chin J Catal 33:595–600
90. Fan M-M, Wang H, Zhang P-B, Ni BQ (2012) Synthesis, characterization and catalysis performance of ionic liquid 1-butyl-3-methylimidazolium chlorozincate. Chin J Inorg Chem 28:1333–1337

91. Wang H, Lu B, Wang X, Zhang J, Cai QI (2009) Highly selective synthesis of dimethyl carbonate from urea and methanol catalyzed by ionic liquids. Fuel Proc Technol 90: 1198–1201
92. Joe W, Lee HJ, Hong UG, Anh YS, Song CJ, Kwon BJ, Song IK (2012) Urea methanolysis to dimethyl carbonate over ZnO–CeO_2–MO (MO: La_2O_3, Y_2O_3, Co_2O_3, Ga_2O_3, and ZrO_2) catalysts. J Ind Eng Chem 18:1730–1735
93. Joe W, Lee HJ, Hong UG, Anh YS, Song CJ, Kwon BJ, Song IK (2012) Synthesis of dimethyl carbonate from urea and methanol over ZnO(X)–$CeO_2(1 - X)$ catalysts prepared by a sol–gel method. J Ind Eng Chem 18:1018–1022
94. Wang D, Zhang X, Gao Y, Xiao F, Wei W, Sun Y (2010) Zn/Fe mixed oxide: heterogeneous catalyst for the synthesis of dimethyl carbonate from methyl carbamate and methanol. Catal Commun 11:430–433
95. Wang D, Zhang X, Zhao W, Peng W, Zhao N, Xiao F, Wei W, Sun Y (2010) Synthesis of dimethyl carbonate from methyl carbamate and methanol catalyzed by mixed oxides from hydrotalcite-like compounds. J Phys Chem Sol 71:427–430
96. Ryu JY, Gelbein AP (2001) Process and catalyst for making dialkyl carbonates. US Patent 6392078 B1
97. Huang S, Ma J, Li J, Zhao N, Wei W, Sun Y (2008) Efficient propylene carbonate synthesis from propylene glycol and carbon dioxide via organic bases. Catal Commun 9:276–280
98. Wu LX, Wang H, Tu Z-Y, Ding B-B, Xiao Y, Lu J-X (2012) Synthesis of cyclic carbonates from CO_2 and diols via electrogenerated N-heterocyclic carbenes. Int J Electrochem Sci 7: 11540–11549
99. Aresta M, Dibenedetto A, Nocito F, Pastore C (2006) A study on the carboxylation of glycerol to glycerol carbonate with carbon dioxide: the role of the catalyst, solvent and reaction conditions. J Mol Catal 257:149–153
100. Honda M, Tamura M, Nakao K, Suzuki K, Nakagawa Y, Tomishige K (2014) Direct cyclic carbonate synthesis from CO_2 and diol over carboxylation/hydration cascade catalyst CeO_2 with 2-cyano-pyridine. ACS Catal 4:1893–1896
101. Carrera G, Visak Z, Bogel-Lukasik R, Nunes Da Ponte M (2011) Thermodynamic studies for the synthesis of 1,2-glycerol carbonate from CO_2 and glycerol. ICCDU XI, Dijon-FR, September 2011, Book of Abstracts, OC62, p 81
102. Vieville C, Yoo JW, Palet S, Mouloungui Z (1998) Synthesis of glycerol carbonate by direct carbonatation of glycerol in supercritical CO_2 in the presence of zeolites and ion exchange resins. Catal Lett 56:245–247
103. Aresta M, Dibenedetto A, Nocito F, Ferragina C (2009) Valorization of bio-glycerol: new catalytic materials for the synthesis of glycerol carbonate via glycerolysis of urea. J Catal 268:106–114
104. Aresta M, Dibenedetto A, Nocito F, Dubois JL (2010) Synthesis process of polyol carbonate from polyols, conducted in using a solvent selective for polyols carbonates. WO Patent WO2010040786 A3
105. Dibenedetto A, Nocito F, Angelini A, Papai I, Aresta M, Mancuso R (2013) Catalytic synthesis of hydroxymethyl-2-oxazolidinones from glycerol or glycerol carbonate and urea. ChemSusChem 6(2):345–352
106. Dibenedetto A, Angelini A, Aresta M, Ethiraj J, Fragale C, Nocito F (2011) Converting wastes into added value products: from glycerol to glycerol carbonate, glycidol and epichlorohydrin using environmentally friendly synthetic routes. Tetrahedron 67:1308–1313
107. Aresta M, Dibenedetto A (2002) Carbon dioxide as building block for the synthesis of organic carbonates: behavior of homogeneous and heterogeneous catalysts in the oxidative carboxylation of olefins. J Mol Catal 182–183:399–409
108. Aresta M, Quaranta E, Ciccarese A (1987) Direct synthesis of 1,3-benzodioxol-2-one from styrene, dioxygen and carbon dioxide promoted by Rh(I). J Mol Catal 41:355–359
109. Aresta M, Dibendetto A, Gianfrate L, Pastore C (2003) Nb(V) compounds as epoxides carboxylation catalysts: the role of the solvent. J Mol Catal A Gen 204–205:245–252

110. Aresta M, Dibendetto A, Gianfrate L, Pastore C (2003) Enantioselective synthesis of organic carbonates promoted by Nb(IV) and Nb(V) catalysts. Appl Catal A Gen 255:5–11
111. Aresta M, Fragale C, Quaranta E, Tommasi I (1992) Carbon dioxide as modulator of the oxidative properties of dioxygen in the presence of transition metal systems. J Chem Soc Chem Commun 4:315–317
112. Dibenedetto A, Aresta M, Nocito F, Pastore C, Venezia AM, Chirykalova E, Kononenko VI, Shevchenko VG, Chupova IA (2006) Synthesis of cyclic carbonates from epoxides: use of reticular oxygen of Al_2O_3 or Al_2O_3-supported CeO_x for the selective epoxidation of propene. Catal Today 115:117–123
113. Dibenedetto A, Aresta M, Distaso M, Pastore C, Venezia AM, Liu C-J, Zhang M (2008) High throughput experiment approach to the oxidation of propene-to-propene oxide with transition-metal oxides as O-donors. Catal Today 137:44–51
114. Aresta M, Tommasi I, Quaranta E, Fragale C, Mascetti J, Tranquille M, Galan F, Fouassier M (1996) Mechanism of formation of peroxocarbonates $Rh(OOC(O)O)(Cl)P_3$ and their reactivity as oxygen transfer agents mimicking monooxygenases. The first evidence of CO_2 insertion into the O-O bond of $Rh(\eta^2\text{-}O_2)$ complexes. Inorg Chem 35:4254–4260
115. Aresta M, Quaranta E, Tommasi I, Mascetti J, Tranquille M, Borowiak M (1998) Formation of peroxocarbonates from $L_3Rh(O_2)Cl$ and $L_2Ni(CO_2)$: a unique reaction mechanism with carbon dioxide insertion into the O-O bond. Stud Surf Sci Catal 114:677–680
116. Aresta M, Dibenedetto A, Tommasi I (2000) Direct synthesis of organic carbonates by oxidative carboxylation of olefins catalyzed by metal oxides: developing green chemistry based on carbon dioxide. Appl Organomet Chem 14:799–802
117. Dibenedetto A, Aresta M. Unpublished results
118. Darensbourg DJ, Chung WC (2013) Relative basicities of cyclic ethers and esters. Chemistry of importance to ring-opening co- and terpolymerization reactions. Polyhedron 58:139–143
119. Tanaka Y (1967) Contribution of ring strain and basicity to reactivity of cyclic ethers in cationic copolymerization. J Macromol Sci A Chem 1(6):1059–1068
120. Dudev T, Lim C (1998) Ring strain energies from ab initio calculations. J Am Chem Soc 120: 4450–4458
121. Dill JB, Grenberg A, Liebman JF (1979) Substituent effects on strain energies. J Am Chem Soc 101:6814–6818
122. Gordy W, Stanford SC (1941) Spectroscopic evidence for hydrogen bonds: comparison of proton-attracting properties of liquids, III. J Chem Phys 9:204–214
123. Gordy W (1941) Spectroscopic evidence for hydrogen bonds: comparison of proton-attracting properties of liquids, IV. J Chem Phys 9:215–223
124. Coates GW, Moore DR (2004) Discrete metal-based catalysts for the copolymerization of CO_2 and epoxides: discovery, reactivity, optimization, and mechanism. Angew Chem Int Ed 43:6618–6639
125. Darensbourg DJ (2007) Making plastics from carbon dioxide: salen metal complexes as catalysts for the production of polycarbonates from epoxides and CO_2. Chem Rev 107: 2388–2410
126. Darensbourg DJ, Mackiewicz RM, Phelps AL, Billodeaux DR (2004) Copolymerization of CO_2 and epoxides catalyzed by metal salen complexes. Acc Chem Res 37:836–844
127. Limura N, Takagi M, Iwane H, Ookago J (1995) Production of propylene carbonate. Japanese Patent 07,267,944
128. Inoue K, Oobuko H, Kokai Tokkyo Koho (1995) Production of alkylene carbonate. Japanese Patent 07,206,846
129. Inoue K, Oobkubo H, Kokai Tokkyo Koho (1995) Production of alkylene carbonate. Japanese Patent 07,206,847
130. Inoue K, Oobkubo H (1995) Production of alkylene carbonate. Japanese Patent 07,206,848
131. Inaba M, Hasegawa K, Nagaoka H, Kokai Tokkyo Koho (1997) Production of alkylene carbonate. Japanese Patent 09,067,365

132. Ichikawa S, Iwane H, Kokai Tokkyo Koho (1997) Production of alkylene carbonate. Japanese Patent 09,235,252
133. Tojo M, Fukuoka S (1991) Production of alkylene carbonate. Japanese Patent 03,120,270 to Asahi Chem Ind
134. Bobyleva LI, Kryukov SI, Bobylev BN, Liakumovich AG, Surovstev A, Karpov OP, Akhmedyanova RA, Koneva SA (1992) Method for production of cyclic carbonates. Yaroslavskij Polit. Institut SU Patent 1,781,218
135. Mais FJ, Buysch HJ, Mendoza-Frohn C, Klausener A (1993) Method for the preparation of alkylene carbonates. EU Patent 543,249 to Bayer
136. Kuran W, Listos T (1994) Initiation and propagation reactions in the copolymerization of epoxide with carbon dioxide by catalysts based on diethylzinc and polyhydric phenol. Macromol Chem Phys 195:977–984
137. Sakai T, Kihara N, Endo T (1995) Polymer reaction of epoxide and carbon dioxide. Incorporation of carbon dioxide into epoxide polymers. Macromolecules 28:4701–4706
138. Sakai T, Tsutsumi Y, Ema T (2008) Highly active and robust organic–inorganic hybrid catalyst for the synthesis of cyclic carbonates from carbon dioxide and epoxides. Green Chem 10:337–341
139. Darensbourg DJ, Holtcamp MW (1996) Catalysts for the reactions of epoxides and carbon dioxide. Coord Chem Rev 153:155–174
140. Inoe K, Oobkubo H (1995) Kokai Tokkyo Koho. Japanese Patent 07,206,847
141. Marquis ET, Sanderson JR (1994) Texaco Chemical Co. US Patent 5,283,365
142. Sone M, Sako T, Kamisawa C (1999) Kokai Tokkyo Koho. Japanese Patent 11,335,372
143. Song J, Zhang Z, Han B, Hu S, Li W, Xie Y (2008) Synthesis of cyclic carbonates from epoxides and CO_2 catalyzed by potassium halide in the presence of β-cyclodextrin. Green Chem 10:1337–1341
144. Langanke J, Greiner L, Leitner W (2013) Substrate dependent synergistic and antagonistic effect of ammonium halide and polyoxometalate catalysts in the synthesis of cyclic carbonates from oleochemical epoxides and CO_2. Green Chem 15:1173–1182
145. Zhou H, Wang YM, Zhang WZ, Qu JP, Lu XB (2011) N-Heterocyclic carbene functionalized MCM-41 as an efficient catalyst for chemical fixation of carbon dioxide. Green Chem 13: 644–650
146. Decortes A, Castilla AM, Kleij AW (2010) Salen-complex-mediated formation of cyclic carbonates by cycloaddition of CO_2 to epoxides. Angew Chem Int Ed 49:9822–9837
147. North M, Pasquale R, Young C (2010) Synthesis of cyclic carbonates from epoxides and CO_2. Green Chem 12:1514–1539
148. Shen Y-M, Duan W-L, Shi M (2003) Chemical fixation of carbon dioxide catalyzed by binaphthyldiamino Zn, Cu, and Co salen-type complexes. J Org Chem 68:1559–1562
149. Lu X-B, Feng X-J, He R (2002) Catalytic formation of ethylene carbonate from supercritical carbon dioxide/ethylene oxide mixture with tetradentate Schiff-base complexes as catalyst. Appl Catal A 234:25–34
150. Paddock RL, Nguyen ST (2001) Chemical CO_2 fixation: Cr(III) salen complexes as highly efficient catalysts for the coupling of CO_2 and epoxides. J Am Chem Soc 123:11498–11499
151. Jing H, Edulji SK, Gibbs JM, Stern CL, Zhou H, Nguyen ST (2004) (Salen)tin complexes: syntheses, characterization, crystal structures, and catalytic activity in the formation of propylene carbonate from CO_2 and propylene oxide. Inorg Chem 43(14):4315–4327
152. Kozak JA, Wu J, Su X, Simeon F, Hatton TA, Jamison TF (2013) Bromine catalysed conversion of CO_2 and epoxides to cyclic carbonates under continuous flow conditions. J Am Chem Soc 135:18497–18501
153. Li Y, Zhao XQ, Wang Y (2005) Synthesis of dimethyl carbonate from methanol, propylene oxide and carbon dioxide over KOH/4A molecular sieve catalyst. Appl Catal A 279:205–208
154. Zhang X, Wei W, Sun Y (2005) International conference on carbon dioxide utilization. ICCDU VIII, Oslo

155. Yano T, Matsui H, Koike T, Ishiguro H, Fujihara H, Yoshihara M, Maeshima T (1997) Magnesium oxide-catalysed reaction of carbon dioxide with anepoxide with retention of stereochemistry. Chem Commun 12:1129–1130
156. Yamaguchi K, Ebitani K, Yoshida T, Yoshida H, Kaneda KJ (1999) Mg–Al mixed oxides as highly active acid–base catalysts for cycloaddition of carbon dioxide to epoxides. Am Chem Soc 121:4526–4527
157. Aresta M, Dibendetto A (2001) 221st ACS national meeting, San Diego, Abstract 220
158. Shibata I, Mitani I, Imakuni A, Baba A (2011) Highly efficient synthesis of cyclic carbonates from epoxides catalyzed by indium tribromide system. Tetrahedron Lett 52:721–723
159. Machac JR Jr, Marquis ET, Woodrum SA (2000) US Patent 654,438
160. Kuruppathparambil RR, Tharun J, Dongwoo K, Kathalikkattil AC, Park DW (2014) Microwave-assisted one pot-synthesis of amino acid ionic liquids in water: simple catalysts for styrene carbonate synthesis under atmospheric pressure of CO_2. Catal Sci Technol 4: 963–970
161. Riduan SN, Zhang Y (2010) Recent developments in carbon dioxide utilization under mild conditions. Dalton Trans 39:3347–3357
162. Song J, Zhang B, Jiang T, Yang G, Han B (2011) Synthesis of cyclic carbonates and dimethyl carbonate using CO_2 as a building block catalyzed by MOF-5/KI and MOF-5/KI/K_2CO_3. Front Chem Chin 6:21–30
163. Ulusoy M, Kilic A, Durgun Z, Tasci B, Cetinkaya J (2011) Silicon containing new salicylaldimine Pd(II) and Co(II) metal complexes as efficient catalysts in transformation of carbon dioxide (CO_2) to cyclic carbonates. J Organomet Chem 696:1372–1379
164. Liang S, Liu H, Jiang T, Song J, Yang G, Han B (2011) Highly efficient synthesis of cyclic carbonates from CO_2 and epoxides over cellulose/KI. Chem Commun 47:2131–2133
165. Buchard A, Kember MR, Sandeman KG, Williams CK (2011) A bimetallic iron(III) catalyst for CO_2/epoxide coupling. Chem Commun 47:212–214
166. Dengler JE, Lehenmeier MW, Klaus S, Anderson CE, Herdtweck E, Riege B (2011) A one-component iron catalyst for cyclic propylene carbonate synthesis. Eur J Inorg Chem 3:336–343
167. Kilic A, Ulusoy M, Durgun M, Tasci Z, Yilmaz I, Cetinkaya B (2010) Hetero- and homoleptic Ru(II) catalyzed synthesis of cyclic carbonates from CO_2; synthesis, spectroscopic characterization and electrochemical properties. Appl Organomet Chem 24:446–453
168. Ulusoy M, Sahin O, Kilic A, Buyukgungor O (2011) Multinuclear Cu(II) Schiff base complex as efficient catalyst for the chemical coupling of CO_2 and epoxides: synthesis, X-ray structural characterization and catalytic activity. Catal Lett 141:717–725
169. Clegg W, Harrington RW, North M, Pasquale R (2010) Cyclic carbonate synthesis catalysed by bimetallic aluminium-salen complexes. Chem Eur J 16:6828–6843
170. Castro-Osma JA, Alonso-Moreno C, Lara-Sánchez A, Martínez J, North M, Otero A (2014) Synthesis of cyclic carbonates catalysed by aluminium heteroscorpionate complexes. Catal Sci Technol 4:1674–1684
171. Eghbali N, Li C-J (2007) Conversion of carbon dioxide and olefins into cyclic carbonates in water. Green Chem 9:213–215
172. Kawanami H, Ikushima Y (2000) Chemical fixation of carbon dioxide to styrene carbonate under supercritical conditions with DMF in the absence of any additional catalysts. Chem Commun 21:2089–2090
173. Dupont J, De Souza RF, Suarez PAZ (2002) Ionic liquid (molten salt) phase organometallic catalysis. Chem Rev 102:3667–3692
174. Kawanami H, Sasaki A, Matsui K, Ikushima Y (2003) A rapid and effective synthesis of propylene. Chem Commun 7:896–897
175. Sugimoto H, Inoue SJ (2004) Copolymerization of carbon dioxide and epoxide. Polym Sci A Polym Chem 42:5561–5573

176. Sugimoto H, Ohtsuka H, Inoue S (2004) Alternating copolymerization of carbon dioxide and epoxide. Aluminum Schiff base complex – quartenary ammonium salt systems as novel initiators. Stud Surf Sci Catal 153:243–246
177. Super M, Berluche E, Costello C, Beckman E (1997) Copolymerization of 1,2-epoxycyclohexane and carbon dioxide using carbon dioxide as both reactant and solvent. Macromolecules 30:368–372
178. Inoue S (1987) In: Aresta M, Forti G (eds) Carbon dioxide as a source of carbon: chemical and biochemical uses, NATO-ASI Series C. Reidel Publishers, Dordrecht, p 331
179. Rokicki A, Kuran W (1981) The application of carbon dioxide as a direct material for polymer synthesis in polymerization and polycondensation. J Macromol Sci Rev Macromol Chem C21:135–136
180. Wu GP, Wei S-H, Ren WM, Lu X-B, Xu T-Q, Darensbourg DJ (2011) Perfectly alternating copolymerization of CO_2 and epichlorohydrin using cobalt(III)-based catalyst systems. J Am Chem Soc 133(38):15191–15199
181. Darensbourg DJ, Andreatta JR, Moncada AI (2010) Polymers from carbon dioxide: polyucarbonates, polythiocarbonates and polyurethanes. In: Aresta M (ed) CO_2 as chemical feedstock. Wiley-VCH, Weinheim, p 213
182. Li H, Niu Y (2011) Alternating copolymerization of CO_2 with propylene oxide and terpolymerization with aliphatic epoxides by bifunctional cobalt salen complex. Polym J 43:121–125
183. Gosh A, Ramidi P, Pulla S, Sullivan SZ, Collom SL, Gartia Y, Munshi P, Biris AS, Noll BC, Berry BC (2010) Cycloaddition of CO_2 to epoxides using a highly active Co(III) complex of tetraamidomacrocyclic ligand. Catal Lett 137:1–7
184. Kim BE, Varghese JK, Han YG, Lee BY (2010) Cobalt(III) complexes of various salen-type ligand bearing four quaternary ammonium salts and their reactivity for CO_2/epoxide co-polymerization. Bull Korean Chem Soc 31:829–834
185. Lee K, Ha JJY, Cao C, Park DW, Ha C-S, Kim I (2009) Effect of complexing agents of double metal cyanide catalyst on the copolymerizations of cyclohexene oxide and carbon dioxide. Catal Today 148:389–397
186. Darensbourg DJ (2014) Personal adventures in the synthesis of co-polymers from carbon dioxide and cyclic ethers. Adv Inorg Chem 66:1–23
187. (2012) After major downturn, global demand for polycarbonate growing again, Says IHS chemical report. IHS Online Pressroom, Englewood, CO, February 13, 2012. http://press.ihs.com/press-release/commodities-pricing-cost/after-major-downturn-global-demand-polycarbonate-growing-agai
188. Chemeurope.com (2011) Survey of polycarbonate market in China. Market Studies, Berlin. http://www.chemeurope.com/en/studies/13725/survey-of-polycarbonate-market-in-china.html
189. Demonstration plant for polycarbonates (PPC, PEC) polyols, Novomer, USA; Five commercial plants for BPA with 65 to 260 kt/y capacity, Asahi Kasei Corp., Japan; Polyols production and PPC with a capacity of 10 kt/y, Jinioong-Cas Chemail Co., China; Pilot plant for polyethercarbonates, Bayer Material Science; BASF: PPC
190. Soga K, Hosoda S, Nakamura K, Ikeda S (1976) A new synthetic route to 2-oxazolidones. J Chem Soc Chem Comm 16:617
191. Inoue S (1976) High polymers from carbon dioxide. Chemtech 6(9):588–594
192. Ihata O, Kayaki Y, Ikaryia T (2004) Synthesis of thermoresponsive polyurethane from 2-methylaziridine and supercritical carbon dioxide. Angew Chem Int Ed 43:717–719

Chapter 7
Carbon Dioxide Conversion in High Temperature Reactions

Abstract This chapter deals with high temperature reactions in which carbon dioxide (CO_2) is used either as dehydrogenation agent ("soft oxidant") or hydrogen user ("reduction to other C1 molecules"). The first part of the chapter covers reactions such as the oxidative coupling of methane, the dehydrogenation of alkanes to olefins, the dehydrogenation of ethylbenzene to styrene, and, finally, CO_2 reforming of methane.

The second part covers the reaction of CO_2 with hydrogen in processes such as the reverse water gas shift reaction and the synthesis of both methanol and dimethyl ether.

All processes discussed here already have an industrial exploitation or are related to processes on stream, so that a large set of data is available in the literature. This chapter illustrates the building-up of knowledge for demonstrating the reaction mechanism in processes in which very often several conversion routes of CO_2 coexist and the role of support and catalysts is essential for addressing the reaction in one or another direction.

7.1 Introduction

In recent years, significant efforts have been made towards the utilization of CO_2 in high temperature catalyzed reactions, such as dry reforming of methane (DRM), oxidative coupling of methane (OCM), and dehydrogenation of hydrocarbons (DH), in addition to the conversion of CO_2 into oxygen containing fuels such as methanol and CH_3OCH_3 (DME, dimethyl ether) [1]. Transformation of CO_2 into hydrogen-containing chemicals typically requires hydrogen or electroreductive processes implying e^- and H^+ transfer [2–5] (see Chap. 8). The use of CO_2 as oxidant in selective processes and C-source in the synthesis of methanol or other energy rich species such as DME are key reactions [6] that may substantially contribute to the conversions of large volumes of CO_2 contributing to greenhouse gases (GHG) mitigation, besides having an economic value.

Co-authored by Prof S. Kawi and Dr. Y. Kathiraser, ChBE Department, NUS, Singapore

M. Aresta et al., *Reaction Mechanisms in Carbon Dioxide Conversion*,
DOI 10.1007/978-3-662-46831-9_7

7.2 CO_2 as Oxidant

CO_2 activated by a catalyst is able to play a role as an oxidant or dehydrogenating agent and to replace steam and oxygen. Examples are the oxidative coupling reactions, the oxidative dehydrogenation, and reforming reactions. CO_2 has an important role in gasification of carbon deposits on a catalyst (thus avoiding catalyst deactivation), especially in reactions such as CO_2 (dry) reforming of methane (DRM) (see Sect. 7.1.4) whereby coke formation is a prevalent issue, especially with the widely used Ni-based catalysts.

7.2.1 OCM Promoted by CO_2

Natural gas with methane as the principal component is mainly utilized for heating, methanol and ammonia synthesis, or generation of electric power [7]. With the recent discovery of large reservoirs of shale gas, methane could soon increase its role as fuel and feedstock for the synthesis of chemicals [8], replacing oil. The conversion of methane via OCM into C_2 hydrocarbons such as ethane (7.1) and ethene (7.2) [8] is an interesting technology for the direct conversion of natural gas into molecules which can find applications in the energy sector (ethane) or in the chemical and polymer industry (ethene) [9].

$$2CH_4 + 0.5O_2 \rightarrow CH_3CH_3 + H_2O \quad \Delta H^\circ_{298K} = -42.4\,kJ\,mol^{-1} \tag{7.1}$$

$$2CH_4 + O_2 \rightarrow CH_2CH_2 + 2H_2O \quad \Delta H^\circ_{298K} = -67.4\,kJ\,mol^{-1} \tag{7.2}$$

C–C coupling is at the core of organic synthesis and serves as the foundation for generating more complex organic compounds from simpler ones [10]. Light olefins such as ethene and propene, are often used as building blocks in the production of chemicals.

OCM to C_2 species proceeds via a two-step process, namely: (1) C–H bond cleavage over the oxidized catalyst sites producing $CH_3\bullet$ or $CH_2\bullet$ radicals; and (2) coupling of the above radicals to produce ethane and ethene, respectively [11]. In the presence of O_2, some unselective gas-phase radical reactions which cause deep oxidation have a profoundly unfavorable impact on the OCM reaction, resulting in formation of CO_x [12].

However, despite its prospective cost-effectiveness, commercialization attempts at such high temperature reactions (T = 873–1 173 K) have not been successful so far because of the low C_2 yields. As a matter of fact, the current commercialized process for ethene production is based on the pyrolysis of naphtha or the steam-based dehydrogenation of ethane [8]. Returning to oxidative coupling reactions, to prevent loss of C_2, the utilization of an oxidant softer than O_2 is desirable [7, 8], which does not promote the formation of radicals. CO_2 can play such a role and offers interesting advantages over O_2 because utilization of CO_2 prevents deep combustion prevalent in O_2-mediated reactions, which lead to undesirable formation of various CO_x by-products. Second, the reaction between methane and CO_2 is more chemoselective and is mainly controlled by catalysis, because CO_2 would not induce gas-phase radical reactions [7]. In addition, the utilization of CO_2 with methane offers tremendous potential in terms of reducing the climate impact, because both CO_2 and methane are potent GHGs.

The main pathways for OCM with CO_2 as a "soft" oxidant are shown in (7.3) and (7.4) [13]:

$$2CH_4 + CO_2 \rightarrow CH_3CH_3 + CO + H_2O \quad \Delta H^\circ_{298K} = +62.1\,\mathrm{kJ\,mol^{-1}} \tag{7.3}$$

$$2CH_4 + 2CO_2 \rightarrow CH_2 = CH_2 + 2CO + H_2O \quad \Delta H^\circ_{298K} = +196.2\,\mathrm{kJ\,mol^{-1}} \tag{7.4}$$

The CO_2-OCM reaction is a complex heterogeneous process involving [7]:

1. Cleavage of C–H bond of the methane molecule on the solid surface
2. Dissociation of CO_2 to CO and oxygen atoms
3. Recombination of CH_3• or CH_2• radicals released from the surface
4. Oxidative or radical dehydrogenation of ethane to ethene

7.2.1.1 Catalytic Systems for CO_2-OCM

The principal requisite to target in designing the catalytic system for CO_2-mediated OCM reactions is the ability of the catalyst in activating both highly stable CO_2 and methane molecules, ensuring high product selectivity. The two C–O bonds in CO_2 require a high amount of energy for their dissociation (>200 kcal/mol) [14]. In general, the effective catalysts for conventional OCM reaction are strong basic oxides, which behave as *p*-type semiconductors at higher temperatures and possess oxygen anion mobility [15, 16]. They can be divided into four groups [15], namely: (1) highly basic pure oxides, of which the early members of the lanthanide oxide series (excluding CeO_2) are most promising; (2) basic oxides loaded with Group 1 or 2 cations (e.g., Li/MgO, Ba/MgO, and Sr/La_2O_3, as monophasic oxides); (3) a few transition metal oxides that contain Group 1 cations; and (4) any of the above-mentioned materials promoted with chloride ions. Previous works on OCM with O_2 as oxidant highlighted the importance of catalyst basicity [17, 18]. The Na_2WO_4/Mn/SiO_2-based catalyst has been widely studied for OCM [18, 19]. Using such a catalyst, Palermo et al. [18] obtained 33 % methane conversion with a C_2 selectivity of 80 %. They have attributed such interesting result to the presence of Na^+ which plays a dual role as both structural and chemical promoters and can convert the active and unselective amorphous silica carrier into a catalytically inert crystalline phase. The surface-bound WO_4 tetrahedron was found to be the catalytically active site (based on Raman data by Wang et al. [20]) in the Na_2WO_4-based catalyst. Research on this catalyst for the CO_2-OCM reaction has shown an interesting C_2

selectivity of 94 % at the reaction temperature of 1 093 K. Even though reactions at higher temperatures are more favourable for methane conversion, on increasing the temperature the bulk lattice oxygen of the catalyst becomes more activated, thus promoting deep oxidation of methane, which is detrimental to the selectivity towards the desired C_2 products. In fact, the selective oxidation of methane to C_2 species is facilitated by the surface lattice oxygen [21].

Some early studies on basic oxides such as MgO, Sm_2O under co-CO_2-fed OCM reactions were found to improve C_2 yield and selectivity, whereby the presence of CO_2 suppressed catalyst deactivation and decreased the amount of coke produced in the long run [22]. According to Lunsford et al. [23, 24], the enhanced selectivity in C_2 over Li/MgO could be attributed to the poisoning effect of CO_2 resulting in surface modification (i.e., carbonate formation), which promoted secondary reactions with the alkyl radicals. These findings were concurred by Wang and Ohtsuka [25] who studied the activity of Group 2 metals such as Ca, Sr, and Ba on Mn-based ternary mixed oxide. The improvement in selectivity for the Sr- and Ba-based ternary mixed oxides was attributed to the formation of $SrCO_3$ and $BaCO_3$ carbonates. These carbonates were postulated to react further with MnO_2 located on the boundaries forming $SrMnO_{2.5}$ and $BaMnO_{2.5}$, which can improve the methane conversion and selectivity to C_2 hydrocarbons. Even though carbonates have high decomposition temperatures (mostly above 1 100 K), the carbonates formed from strontium and barium in this catalyst system are most likely to possess greater reactivity in the presence of Mn, which resulted in improved catalytic performance.

Rare earth elements such as lanthanum oxide have also been tested for the CO_2-OCM reaction. Results from Chen et al. [26] demonstrate that the ternary oxide of La_2O_3–ZnO could give a C_2 yield of 2.8 % with high stability at 1 123 K. However, Xu et al. [27] found that by doping the appropriate composition of La_2O_3–ZnO mixed oxide with suitable amounts of K_2O, a significant improvement was found, with a C_2 yield more than six times higher (18.1 %). Recently, Oshima et al. [28] investigated the performance of La_2O_3–ZnO mixed oxide as catalyst for the CO_2-OCM reaction using electric field non-thermal plasma and they obtained a C_2 yield at 4.5 %, even at the low temperature of 423 K. In a plasma process, the high energy electron-impact dissociation of the C–H bond to radicals CH_x is the main process involved [29]. Compared to conventional fixed-bed reactors with high operating temperatures, Oshima et al. [28] obtained good performance, which was attributed to the synergetic effect between La cation, tetragonal ZrO_2, and the electric field. The utilization of electric field non-plasma techniques appears to be interesting and with much potential because the operation temperature can be drastically lowered. Besides, the utilization of an electric field facilitates a quick start-up, which is especially suited for smaller scale processes [28].

Redox catalysts such as ceria modified with Group 1 or 2 metal oxides are also effective for the CO_2-OCM reaction [30, 31]. The redox interchange between Ce^{4+} and Ce^{3+} oxidation states is crucial for CO_2 dissociation to CO and lattice oxygen species. The modification of ceria with metal oxides such as

CaO [31, 32] improves C_2 selectivity and, because of the basic nature of CaO, produces a large chemisorption of CO_2. In addition, the incorporation of bivalent Ca^{2+} (0.99 Å) cations into the CeO_2 lattice generates defect sites, which promote redox reactions between Ce^{4+} (0.97 Å) and Ce^{3+} (1.14 Å) [33] with improved methane conversion. He et al. [34] have prepared CeO_2/ZnO catalysts and observed that higher catalytic conversions were obtained using nanostructured catalysts, compared to the bulk counterparts prepared using conventional impregnation techniques. This is further support to the importance of nanostructured catalysts, which, because of their large surface areas, allow greater exposure of catalytic sites to the reactants. Previous investigations have also emphasized the importance of the redox properties of the catalysts, using PbO–MgO, whereby PbO supplies oxygen for the coupling reaction. It should be noted that PbO does not form stable carbonates under the operating conditions [35]. The redox properties and the participation of lattice oxygens from rare earth elements such as praseodymium (Pr) and terbium oxides were also found to produce a high conversion yield and C_2 selectivity up to 50 % for a stable CO_2-OCM catalytic performance (20 h test run) [36].

7.2.1.2 Reaction Mechanism for CO_2-OCM

Depending on the type of metal oxide used as catalyst, two reaction mechanisms are suggested for the CO_2-OCM process. One pathway is the adsorption of CO_2 over a defective metal oxide, followed by O-transfer from CO_2 to the oxide lattice with CO release. The active oxygen species selectively converts methane into the methyl radical which, upon recombination, gives ethane. Ethene would be formed through oxidative dehydrogenation of ethane [7]. The second possible mechanism involves the reaction of methane with lattice oxygen of the metal oxides, forming methyl radicals, which are responsible for the formation of C_2 hydrocarbons. In this process, the metal is partially reduced and the oxygen vacancies are replenished by CO_2 [36], which is converted into CO. This mechanism mainly applies to catalysts with redox centers such as Ce(IV), or even to those which possess an abundance of lattice oxygen vacancies, especially in a defective oxide structure. However, there is the possibility of combination of both pathways in this process.

Wang et al. [31] investigated the possible mechanism involved in the CO_2-OCM reaction using CaO–CeO_2 catalysts. Because ceria is a redox catalyst, they examined the possibility of its lattice oxygen species participating in the reaction of abstracting H from the methane molecule by carrying out the reaction in the presence and absence of CO_2. A lower selectivity in the absence of CO_2 proved that the heterocumulene plays an important role in the reaction. By coupling kinetic studies and CO_2 chemisorption data, they proposed the mechanism shown in Scheme 7.1.

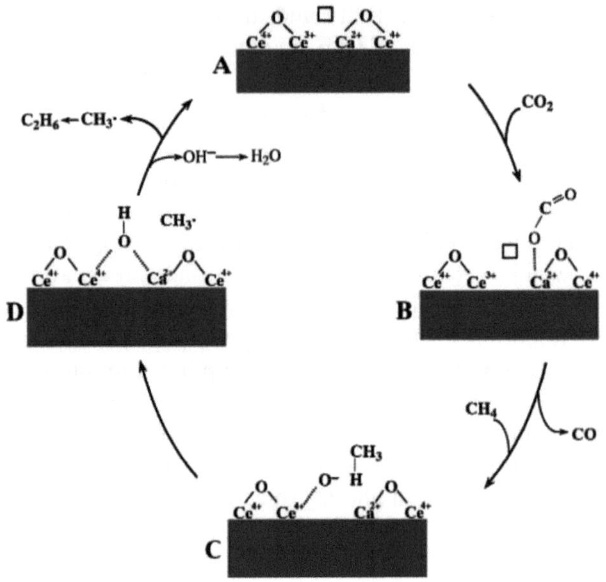

Scheme 7.1 Proposed mechanism for CO_2-OCM reaction over CaO–CeO_2 catalyst [31]

The authors proposed that CO_2 adsorbs on the catalyst by interacting with the acid Ca^{2+} sites at the outermost layer via the O-atom (B). The Ce^{3+} sites present in ceria activate CO_2 forming CO and "active oxygen" species present most likely as subsurface oxygen (C), responsible for the conversion of methane to methyl radical (D). Hence, the cooperation of Ca^{2+} and Ce^{3+} sites in solid solution of the catalyst leads to a greater chemisorption and activation of CO_2, which produces more active oxygen species for selective C_2 formation. As discussed above, the incorporation of Ca^{2+} into CeO_2 lattice (proven by XPS analysis) generates defect sites, which can also promote redox interchange between Ce^{4+} and Ce^{3+} [31].

Wang and Ohtsuka [25] have studied Sr(Ba)O–MnO_2 oxide ternary catalysts for the CO_2-OCM reaction. Sr(Ba)CO_3 and MnO were the main species at steady state of the reaction and were converted into $SrMnO_{2.5}$ after helium treatment. By correlating the reaction data, which show the presence of CO_2 and CO, with XPS data (Fig. 7.1), demonstrating the reduction of the oxidation state of Mn from mainly 4+ to 2+ after reaction (increased to 3+ after helium treatment), they proposed that the reaction had to take place according to (7.5). The locations at the boundaries between Sr carbonate and MnO_2 is most likely the focal reaction point.

$$2SrCO_3 + 2MnO \rightarrow 2SrMnO_{2.5} + CO + CO_2 \tag{7.5}$$

XRD confirmed the formation of $SrMnO_{2.5}$ after reaction. The latter was considered responsible for generating methyl radicals (7.6), which were coupled to generate

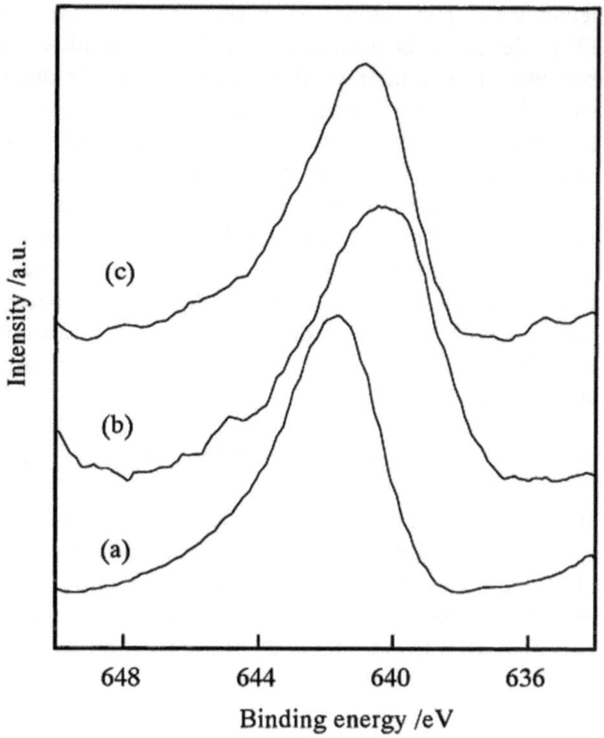

Fig. 7.1 Mn $2p^{3/2}$ XPS spectra for the Sr–Mn catalyst: (**a**) before reaction; (**b**) after reaction at 1,123 K; (**c**) after He treatment of (**b**) at 1,123 K [28]

ethane. Research in this area still needs development to reach exploitable catalyst systems.

$$2SrMnO_{2.5} + CH_4 \rightarrow 2SrO + 2MnO + CH_3\bullet + OH_{(a)} \tag{7.6}$$

7.2.2 Oxidative Dehydrogenation of Alkanes

The utilization of CO_2 in the oxidative dehydrogenation (ODH) of alkanes has received considerable attention in recent years. CO_2 has the potential to act as both a hydrogen acceptor and a "soft" oxidant. As co-fed gas in the oxidative dehydrogenation of light alkanes, it shifts the reaction equilibrium towards more olefin production [37, 38]. The ODH reaction is of great interest as it represents a straightforward reaction for the production of olefins such as ethene, propene, and butene from their respective saturated hydrocarbons. Such olefins are classified as building blocks in the petrochemical industry. Ethene, in particular, has a diversified derivatives portfolio and can be used for the production of plastics, polymers, fibers, and packaging materials [39, 40]. The dehydrogenation of light alkanes is reversible and, because of limitation by thermodynamics, the reaction is often carried out at high temperatures (>800 K). Industrially, steam cracking of hydrocarbons or naphtha is used for the synthesis of ethene, a very energy-consuming technology and prone to severe catalyst deactivation caused by coking [39]. Hence, it is desirable to utilize an oxidant capable of oxidizing, in the presence of a suitable catalyst, the hydrogen produced so as to shift the equilibrium. The use of molecular oxygen as oxidant can effectively reduce the reaction temperature. Nevertheless, undesirable factors affect this reaction such as the low selectivity because of the production of CO_x species from HCs, the need to remove excess heat of reaction, and the undesired reaction mixture flammability [41]. On the other hand, CO_2 has been found to serve as a medium for heat supply for the endothermic dehydrogenation reaction, as a diluent to enhance equilibrium conversion of alkanes and as a coke removal agent from the catalyst [38]. In addition, CO_2 inhibits deep oxidation of the reaction products, thus improving the olefin selectivity [42]. The generic reaction pathway for the CO_2-ODH is shown in (7.7):

$$C_nH_{n+2} + CO_2 \rightarrow C_nH_n + CO + H_2O \tag{7.7}$$

It should be noted that, in the dehydrogenation of higher alkanes, the olefins produced generally contain allylic hydrogens. The H-abstraction from the allylic position has a greater affinity towards production of coke precursors because of lower bond dissociation energy compared to the saturated substrate [43, 44]. This results in catalyst deactivation, which in turn leads to low olefin yields.

7.2.2.1 Catalytic Systems for CO_2-ODH of Alkanes

Notably, catalysts with redox properties, such as molybdenum-, chromium-, and vanadia-based catalysts, show high activity in various oxidative dehydrogenation reactions of hydrocarbons [45–48]. Factors influencing the reaction also include acid–base bifunctionality, which plays an important role in CO_2-mediated dehydrogenation reactions [49]. Both basic sites and Lewis-acid vacant sites are important for hydrocarbons activation [50]. In fact, an enhanced basicity results in an improved performance because of the rapid desorption of the electron-rich alkenes, whereas Lewis acid sites enhance the dehydrogenation process [51]. In addition, in the presence of CO_2 as feed, surface basicity favors the adsorption and reactivity of the acid CO_2 molecules [52] (see also previous chapters).

The best performing chromium-based catalysts for CO_2-ODH of light alkanes are those supported on mixed oxides, mesoporous silica, and zeolites [53–55] as well as diamond [43]. Such high surface area supports facilitate good catalyst dispersion, which is essential for high catalytic activity and stability. Cr-based catalysts supported on mesoporous supports such as zeolitic TS-1 [55], high silica H-ZSM-5 [56], and siliceous MCM-41 [53] were found to be highly active and selective for the CO_2-ODH reaction of ethane and propane. Despite their high catalytic activity, however, deactivation was observed caused by the reduction of the active chromium species on the support. In fact, as suggested by Takehira et al. [53], based on EXAFS data, tetrahedral Cr(VI) is coordinatively unsaturated and represents the active site for alkane dehydrogenation; bulk Cr(III) ions, with a coordinatively saturated and stable octahedral geometry, are much less active, if not inactive for alkane dehydrogenation [57–59]. The regeneration of the deactivated catalyst, i.e., reoxidation of Cr(III) to Cr(VI) species, can be performed with CO_2. However, the reduction–oxidation cycle involving Cr(VI) in tetrahedral coordination and Cr(III) in octahedral coordination and CO_2 has an important role in alkane dehydrogenation. Similar to its mesoporous counterpart, the catalytic activity of chromium supported on silica was found to correlate with the chromium dispersion, which is affected by the grafting of chromate species with various nuclearities on the silica surface [60]. Based on the investigation by Botavina et al. [60], DR-UV–vis studies indicate that Cr(VI) in the form of extended polychromates (less dispersed) results in a lower activity compared to the surface chromates with lower nuclearity. It is now clear that the nature of the support affects the distribution of the surface chromium species, ultimately playing a decisive role in catalytic activity [61].

Besides supporting the catalysts on mesoporous supports, it is also possible to incorporate the catalyst within mesoporous structures, up to a certain loading percentage. For example, Liu et al. [62] have investigated the performance in ethane dehydrogenation of Cr incorporated into MSU-1 (a type of mesoporous molecular sieve) and Cr supported on MSU-1, the former labeled as Cr-MSU-1 and the latter as Cr/MSU-1. The family of MSU-1-type mesoporous silica is less acidic than the SBA-15 or MCM-41 types and possesses uniform, three-dimensional, worm-like channels, favorable for the diffusion of molecules, in addition to its properties of high surface area [63]. The higher activity of the Cr-supported catalysts, especially at Cr loading >1 wt%, however, is coupled to a lower selectivity because of cracking reactions. This is attributed to a higher ratio of inactive polychromate species compared to the more selective monochromate species. In addition, with higher Cr loading (>8 wt%), the catalyst dispersion is severely affected, resulting in the decrease of Cr(VI) species and an increase in crystallite Cr_2O_3 species [57, 62]. More recently, Baek et al. [63] prepared highly dispersed chromium oxide catalysts supported on mesoporous silica using an electrostatically controlled assembly process via the $(N^0M^{n+})I^0$ pathway developed by Pinnavaia et al. [64]. This catalyst was found to exhibit 90 % selectivity towards propene formation with a yield of 30 % in the CO_2-ODH of propane. A steady state in terms of propane conversion was achieved within 25 h, and the deactivation observed was attributed to coke formation and reduction of active chromium sites.

CO_2 as oxidant (Scheme 7.2) is found to boost the regeneration of the reduced chromium sites formed during the dehydrogenation of propane (7.8) and the consecutive reduction with H_2 (7.9) [47], where CrO_x represents the oxidized form and CrO_{x-1} represents the reduced form of the catalyst).

$$C_3H_8 + CrO_x \leftrightarrows CrO_{x-1} + C_3H_6 + H_2O \quad (7.8)$$

$$CO_2 + CrO_{x-1} \leftrightarrows CO + CrO_x \quad (7.9)$$

$$H_2 + CrO_x \leftrightarrows CrO_{x-1} + H_2O \quad (7.10)$$

$$CO_2 + CrO_{x-1} \leftrightarrows CO + CrO_x \quad (7.11)$$

Scheme 7.2 CO_2-ODH of propane over chromium oxide catalytic sites

The effect of promoters such as Fe, Co, and Mn oxides on Cr-based Cr_2O_3/ZrO_2 nano-composite catalysts for CO_2-ODH of ethane was found to increase markedly the ethene selectivity [50]. In this study, the acid–base properties of the catalysts were found to have the most significant influence. Weak acid sites promote ethene formation through better ethane activation, whereas strong acid–base sites favor the side reactions of reforming and cracking of ethane [50]. Another important phenomenon discussed in this study is that the amount of various types of oxygen species, i.e., the lattice oxygen (O_I), oxygen in adsorbed CO_2 or carbonates (O_{II}), and adsorbed oxygen species (O_{III}), can influence the reaction rate and the selectivity towards ethene production. In fact, lattice oxygen species are mostly responsible for selective oxidation of hydrocarbons, whereas the adsorbed oxygen species contribute towards deep oxidation, leading towards more CO_x formation [50, 65]. Anyway, the amount of adsorbed CO_2 is quite important as it can promote the dry reforming of ethane, a concurrent reaction to ODH. It should be noted that catalysts such as $Na_2WO_4/Mn/SiO_2$ suitable for OCM reaction for ethene production are also active in the ODH of ethane [66]. Zhu et al. [67] have investigated the effect of Co addition to $Na_2WO_4/Mn/SiO_2$ for CO_2-ODH of ethane. However, likewise Ni, the presence of Co favors the reforming of ethane with CO production.

The catalytic properties of vanadium oxide depend on the acid–base nature of the supporting materials [68] and on the dispersion of the vanadium species on the support [69]. Park et al. [44] have successfully synthesized V_2O_5 supported on various mixed oxides such as SnO_2–ZrO_2 [44], CeO_2–ZrO_2, and TiO_2–ZrO_2 [37]. The SnO_2–ZrO_2-supported V-catalyst was found to have a good catalytic activity attributed to the high surface area and superior acid–base bifunctionality, with generation of more medium strength acid–base sites compared to vanadia supported on single oxides. Such optimal basicity was found to inhibit dehydrogenation of *n*-butenes to butadienes, besides precluding side reactions to occur because of the easy desorption of alkenes [70]. The reducibility of V_2O_5 is strongly affected by the nature of the support. In the case of CeO_2–ZrO_2-supported V_2O_5, despite the slightly lower conversion with respect to TiO_2–ZrO_2 and SnO_2–ZrO_2-supported catalysts, the total selectivity towards C4 olefins is much higher and this is attributed to the participation of larger amounts of lattice oxygen species and oxygen vacancies (which are replenished by CO_2) because of the formation of $CeVO_4$ species [37]. This is not surprising, because CeO_2–ZrO_2 supports are known for their redox nature because of the increased oxygen mobility, good thermal resistance, high oxygen-storage capacity, and facile oxidation/reduction of the Ce^{4+}/Ce^{3+} couple [71, 72]. However, when V_2O_5 is incorporated in chromium oxide catalysts, an enhanced redox activity is observed, because of the reduction of Cr^{6+} to Cr^{3+} and V^{5+} to V^{4+}, resulting in a higher ethene selectivity [73]. Nevertheless, comparing Cr and V–Cr catalysts, one can find that the latter produce a lower yield because they require a higher reaction temperature.

A well-dispersed catalyst results in a better stability for the ODH reaction and, therefore, mesoporous structured supports such as MCM-41 or SBA-15 are highly indicated. It has been found that impregnation of vanadium on mesoporous siliceous MCM-41 results in greater density of surface-active sites compared to amorphous silica because of the large pore diameters and the stronger interaction which exists between the active species and its support [74], hence the increasing interest in this type of support. In addition, SBA-15 types of mesoporous silica possess thicker walls (31–64 Å) compared to their MCM-41 counterparts, thus providing greater thermal and hydrothermal stability [69]. At predominantly low vanadia content (1–5 wt%), isolated tetrahedral vanadium oxide species containing terminal V=O groups well dispersed on the mesoporous silica have been identified to be the active sites. Excess vanadia loading leads to a larger presence of polymerized vanadium species on the silica surface, which may result in a negative effect on the catalytic performance in the dehydrogenation process [69]. Recently, mesoporous Ga_2O_3 and Ga_2O_3–Al_2O_3 prepared via nanocasting were identified as effective catalysts for the CO_2-ODH of propane [75]. At 832 K, propene was produced with a yield of 10–18 % and a high selectivity of 91–95 %. The higher stability of the mesoporous gallium oxide catalysts is an important fact, because the conventional gallium oxide-based catalyst investigated by the same research group had shown deactivation within 1 h of reaction time [76]. The new preparation method for Ga_2O_3-based catalysts can effectively provide a hard template, which suppresses the collapse of mesoporous structure typically occurring during gallium oxide recrystallization [75].

Several studies have investigated the CO_2-ODH of light alkanes, not only with CO_2 as the sole oxidant but also co-fed with oxygen [38]. Early work concentrated on the co-feed of CO_2 (considered as promoter) together with oxygen as a means to improve the dynamic phenomena at the surface of oxides in order to improve the yield of olefins via ODH reactions. This approach was believed to improve the efficiency of the catalyst because CO_2 acted as a modulator of the oxidative properties of oxygen [14, 77]. Among the beneficial effects arising from the introduction of CO_2 in the system are: (1) faster, more extended formation of oxycarbonate phase and its regeneration; (2) formation of a peroxocarbonate [77, 78] intermediate, which serves as a promoter for gas phase oxygen oxidation; (3) lower tendency for deep oxidation of hydrocarbons (whilst minimizing hot spot formation); and (4) decrease in coking effect [14].

Urlan et al. [38] showed that the ODH of *n*-butane over titanium pyrophosphate (TiP_2O_7) can be successfully improved by co-feeding CO_2 in the system. However, when CO_2 was used as the sole oxidant, the *n*-butane conversion was lower. Similarly, when CO_2 partial pressures were increased over V–Mg–O catalysts, the yield and selectivity of C_4 dehydrogenation products were found to show some improvement [79]. It was surmised that, when CO_2 was used as the sole oxidant, the competition with alkanes, significantly affected the adsorption of CO_2 on the catalyst surface and, thus, the reoxidation of the catalyst, leading to a drop in conversion [44].

Another group of catalysts, i.e., the molybdenum oxide-based, possess a chemistry comparable to that of vanadium oxide, because Mo can be present as MoO_x monomers, polymers, or MoO_3 crystallites, depending on the Mo loading, type of support, and catalyst preparation techniques and conditions [80]. For instance, a study on the effect of CO_2 for ODH over Mo_2C/SiO_2 has demonstrated that CO_2 is beneficial in oxidizing the Mo_2C forming Mo-oxycarbide at higher temperatures, with the Mo-oxycarbide species playing an important role in the activation of the alkane [46, 81]. Similar behavior can be found for other supports such as Al_2O_3 and TiO_2. In fact, Dury et al. [14] found that the introduction of just 3 % CO_2 for the ODH reaction of propane over $NiMoO_4$ leads to dissociation of CO_2 on the catalyst surface, inducing the oxidation of molybdenum suboxide at temperatures around 673–723 K. Experiments using $NiMoO_4$ in alkane oxidation have shown that the presence of molybdenum oxides as well as molybdates can enhance the catalytic properties [82, 83]. The oxygen species derived from the dissociation of CO_2 were found to play a more positive role in oxidation compared to molecular oxygen by cleaning and protecting the sites during the reaction [14, 83] and avoiding radical formation.

7.2.2.2 Reaction Mechanism for CO_2-ODH of Alkanes

Two major roles are proposed for CO_2 in the CO_2-ODH reaction, i.e., (1) CO_2 oxidizes the reduced catalytic sites during the dehydrogenation process, hence participating in a redox process and (2) CO_2 removes hydrogen formed in the dehydrogenation process via the reverse water gas shift (RWGS) reaction, thus shifting the reaction equilibrium towards product formation.

7.2.2.2.1 Redox Mechanism

The redox nature of the catalysts used for the ODH reactions benefit from CO_2 dissociation on catalyst sites producing surface oxygen species is parallel to the alkane dehydrogenation by CO_2. For example, based on the investigation conducted by Mimura et al. [56] for the CO_2-ODH of ethane over the reducible Cr supported on zeolite (Cr/ZSM-5), an important role of CO_2 in the Cr redox process in the catalytic cycle was highlighted. A comparison of the FTIR spectra of the fresh (A) catalyst and treated (B) catalyst (Fig. 7.2a) indicates that the reduction of the peak intensity at 925 cm^{-1} attributed to the Cr=O double bond can be related to the reaction with ethane. Such reduction is indicative of a reduced average oxidation state of the Cr species under the reaction condition and this can be related to a higher dehydrogenation rate by Cr=O species compared to the reoxidation rate

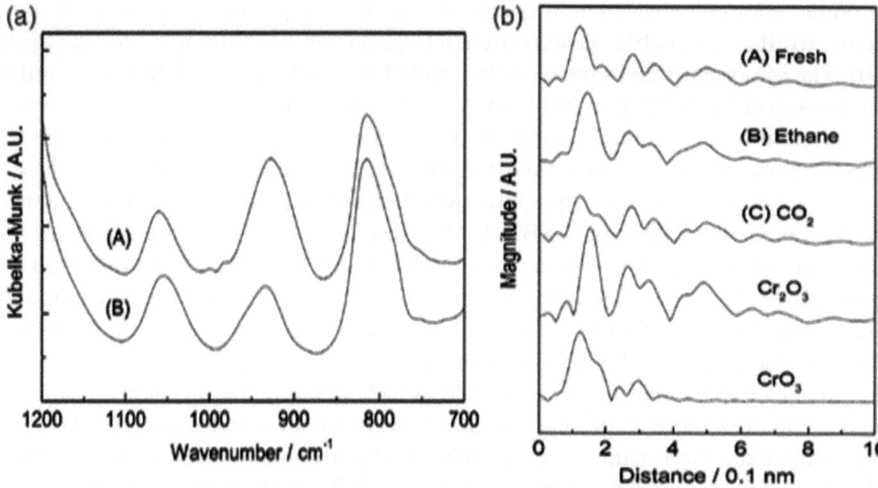

Fig. 7.2 (**a**) FTIR spectra of Cr/HZSM-5 catalyst: (A)-Fresh catalyst; (B)- After ethane/CO_2 treatment at 773 K for 30 min; (**b**) EXAFS data for Cr/HZSM-5 catalyst under various conditions at 773 K (treatment for 30 min) [56]

of the reduced Cr species by CO_2. The EXAFS data (Fig. 7.2b) show that the bond length for the fresh sample (A) is similar to CrO_3 (i.e., Cr=O double bond species), whereas after ethane treatment (B), a longer bond length, similar to the Cr–O single bond in Cr_2O_3 is observed. On the other hand, the latter sample after CO_2 treatment (C) exhibits similar bond length with the fresh sample. Hence, the authors have correlated such changes to the regeneration of the Cr^{6+}=O species to the initial state, which are consistent with FTIR results [56]. However, the activation of alkanes on Cr=O is quite similar to that on vanadia catalysts, whereby isolated tetrahedral vanadium oxide species containing terminal V=O double bond species have been suggested to be the active centers for the selective formation of alkenes [84].

Hence, based on such findings, a reaction scheme (Scheme 7.3) was suggested by Mimura et al. [56] for the CO_2-ODH of ethane over the active Cr/HZSM-5 catalysts.

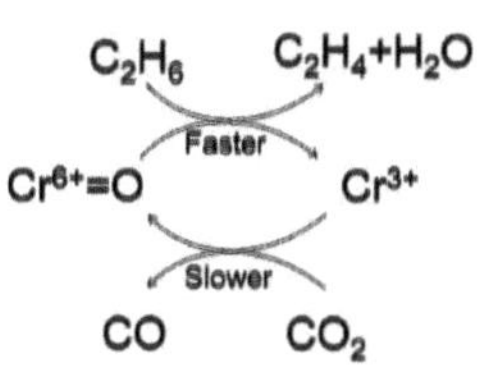

Scheme 7.3 Redox cycle of active Cr species on zeolite surface [56]

The oxidation of Cr^{3+} to Cr^{6+} during the catalyst regeneration with CO_2 can lead to redispersion of chromia crystallites, which is a particularly important step, because without the presence of a mild oxidant such as CO_2, the sintering of chromia under dehydrogenation conditions has been reported to be a contributing factor to the reversible deactivation-activation of chromium-based catalysts [80]. The important role of the Cr redox cycle has also been highlighted for similar reactions over monolithic Cr/SBA-15/Al_2O_3/FeCrAl catalysts [61]. The difference in configuration of monolithic catalysts (known for their high heat and mass transfer rates) compared to conventional fixed-bed reactor setups is that the Cr-based monolithic metal catalysts are prepared with FeCrAl foil with honeycomb structure which acts as the support and SBA-15/Al_2O_3 as washcoat [61]. Similarly, for Cr supported on mesoporous silica such as MCM-41 [53], CO_2 was observed to participate as the oxidizing agent whereby the redox process via the consumption and re-filling of the defect sites formed from lattice oxygen uptake from gas phase CO_2 was deemed to be the key factor in the reaction process.

A fine example demonstrating the role of CO_2 in the redox mechanism is given by the classical redox type of catalysts such as vanadia and ceria. As such, an interesting example for this is represented by the investigation by Park et al. [37] who studied the CO_2-ODH of butane on $V_2O_5/CeO_2–ZrO_2$ catalyst, of which both the vanadia and the ceria support are known for their redox nature. The abundance of lattice oxygen in the catalyst/support results in participation of the lattice oxygen in the dehydrogenation process, although the oxidation state of the metal is reduced to a lower state (7.12).

$$C_4H_{10} + MO_x \rightarrow C_4H_8 + MO_{x-1} + H_2O \text{ (ODH by lattice oxygen)} \quad (7.12)$$

$$MO_{x-1} + CO_2 \rightarrow CO + MO_x \text{ (reoxidation with } CO_2\text{)} \quad (7.13)$$

In this process, which proceeds via a Mars–van Krevelen (redox) type mechanism [85], the role of CO_2 as an oxidant is critical in re-oxidizing the lower oxidation state metal oxide (7.13) and as shown in Scheme 7.4 [37, 47].

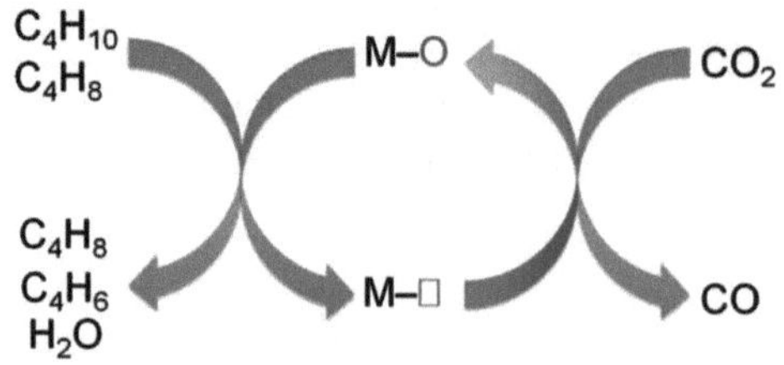

Scheme 7.4 Proposed reaction pathway (O, lattice oxygen, and anion vacancy) for butane-ODH with CO_2 [37]

7.2.2.2.2 Surface Reactions for Removal of Hydrogen via RWGS

The presence of CO_2 in the ODH reaction has been found to promote desorption of olefin from the catalyst surface and aid removal of hydrogen produced by alkane dehydrogenation via the RWGS reaction. For example, Xu et al. [86] studied the CO_2-ODH of propane over gallium-based catalysts with various supports such as alumina, zirconia, titania, and silica. Ga_2O_3/Al_2O_3 exhibited the best performance in promoting the dehydrogenation of propane because of the surface acidity. Their results match the findings of Chen et al. [87] who attributed the good catalytic performance of analogous catalysts to the formation of gallia-alumina solid solutions, which allow the favourable generation of numerous coordinatively unsaturated surface Ga^{3+} sites with weak Lewis acidity [88]. Unlike conventional vanadia or chromium redox-based catalysts, gallium is known to be quite stable as Ga^{3+} [89]. Therefore, instead of a redox mechanism, the authors have proposed that propane is adsorbed on gallium oxide where it is heterolytically dissociated, forming gallium hydrido-alkoxo species as shown in (7.14).

$$Ga^{x+}\text{–}O^{2-}\text{–}M^{y+} + C_3H_8 \rightarrow \overset{H^-}{\overset{|}{Ga^{x+}}}\text{–}\overset{C_3H_7^+}{\overset{|}{O^{2-}}}\text{–}M^{y+} \tag{7.14}$$

The alkoxo species further decompose to form olefin (7.15) via β-H shift. Hydrogen can be formed from the coupling of the hydride and proton (7.16) or removed by CO_2 (via the RWGS reaction route) (7.17).

$$\overset{H^-}{\overset{|}{Ga^{x+}}}\text{–}\overset{C_3H_7^+}{\overset{|}{O^{2-}}}\text{–}M^{y+} \rightarrow \overset{H^-}{\overset{|}{Ga^{x+}}}\text{–}\overset{H^+}{\overset{|}{O^{2-}}}\text{–}M^{y+} + C_3H_6 \tag{7.15}$$

$$\overset{H^-}{\overset{|}{Ga^{x+}}}\text{–}\overset{H^+}{\overset{|}{O^{2-}}}\text{–}M^{y+} \rightarrow Ga^{x+}\text{–}O^{2-}\text{–}M^{y+} + H_2 \tag{7.16}$$

$$\overset{H^-}{\overset{|}{Ga^{x+}}}\text{–}\overset{H^+}{\overset{|}{O^{2-}}}\text{–}M^{y+} + CO_2 \rightarrow Ga^{x+}\text{–}O^{2-}\text{–}M^{y+} + CO + H_2O \tag{7.17}$$

The rate-limiting step in this process is the propene formation via β-H shift (7.15). Moreover, as Ga_2O_3 and H^+ are on the catalyst surface, the propyl carbenium ion on Ga_2O_3 might readily exchange with a proton via surface migration reaction (7.18) [90]:

$$\overset{H^-}{\overset{|}{Ga^{x+}}}\text{–}\overset{C_3H_7^+}{\overset{|}{O^{2-}}}\text{–}M^{y+} + H^+S \rightleftharpoons \overset{H^-}{\overset{|}{Ga^{x+}}}\text{–}\overset{H^+}{\overset{|}{O^{2-}}}\text{–}M^{y+} + C_3H_7^+S \tag{7.18}$$

where S is the catalyst surface site. The formation of propene is again the result of a β-H shift with desorption from the active site (7.19):

$$-S-\underset{\alpha}{CH_2}-\underset{\beta}{CH_2}-CH_3 \rightleftharpoons -S-H + CH_3CH{=}CH_2 \tag{7.19}$$

Equations (7.18) and (7.19) are considered to be fast with respect to the kinetically slow reaction (7.15) because they benefit from the conjugated effect of gallium oxide and proton migration, which increases the dehydrogenation of propane [86]. Hence, the reduction of the hydrogen partial pressure shifts the dehydrogenation equilibrium towards production of olefin. In addition to removing hydrogen (7.17), the role of CO_2 is also that of removing any carbon formed on the surface in this process via the reverse Boudouard reaction [86, 91].

So far in the literature, the kinetics of alkane dehydrogenation has mostly been studied on direct dehydrogenation reactions or by using oxygen as the oxidant. Recently, Liu et al. [42] have studied the kinetics of ODH of propane in the presence of CO_2 over Cr/MSU-1 catalyst. The sequence of mechanistic steps shown in Scheme 7.5 was modeled based on the Langmuir–Hinshelwood mechanism and the surface reaction between adsorbed propane and adsorbed CO_2 (7.22) was found to be the rate-determining step. Such data correlate well with other studies on ODH of propane with O_2 as oxidant, affirming that the H-abstraction from the secondary carbon atom in adsorbed propane is the irreversible rate-determining step [85].

$$C_3H_8 + * \leftrightarrows C_3H_8* \tag{7.20}$$

$$CO_2 + * \leftrightarrows CO_2* \tag{7.21}$$

$$C_3H_8* + CO_2* \leftrightarrows C_3H_6* + CO* + H_2O \quad \text{(Rate Determining Step)} \tag{7.22}$$

$$C_3H_6* \leftrightarrows C_3H_6 + * \tag{7.23}$$

$$CO* \leftrightarrows CO + * \tag{7.24}$$

Scheme 7.5 Langmuir–Hinshelwood mechanistic steps for the CO_2-ODH of propane [42]

For the CO_2-ODH of alkanes, the proposed pathway advocated by several groups [92–94] was found to be in agreement with the experimental results of some studies in which the formation of notable amounts of CO and H_2O during the reaction was observed [43, 44]. According to the proposed general pathways (Scheme 7.6), hydrogen from the alkane dehydrogenation is oxidized by CO_2 via RWGS reaction (7.26), producing CO and H_2O. In addition, the CO_2 reforming of ethane (more favorable at higher temperatures) was proposed to be one of the side reactions (7.28) [50]. As mentioned earlier, CO_2 also aids in oxidizing the deposited carbon, forming more CO via the reverse Boudouard reaction (7.29) [44].

$$C_nH_{n+2} \leftrightarrows C_nH_n + H_2 \tag{7.25}$$

$$H_2 + CO_2 \rightarrow CO + H_2O \ \textbf{(RWGS)} \tag{7.26}$$

$$C_nH_{n+2} + CO_2 \rightarrow C_nH_n + CO + H_2O \tag{7.27}$$

$$C_nH_{n+2} + 2CO_2 \rightarrow 4CO + 3H_2 \tag{7.28}$$

$$CO_2 + C \rightarrow CO \tag{7.29}$$

Scheme 7.6 General pathway proposed for CO_2-ODH of alkanes via RWGS reaction

In recent investigations, Peng et al. [95] have studied the effect of co-feeding methane in the CO_2-ODH of ethane at high reaction temperatures (1,073 K) using a Ni–La/SiO_2 catalyst. They noted that methane affected the equilibrium of CO_2 reforming of ethane (7.30) and the CO_2-ODH of ethane.

$$C_2H_6 + 2CO_2 \rightarrow 4CO + 3H_2 \tag{7.30}$$

In fact, the presence of methane induced the CO_2 (dry) reforming of methane to take place, hence retarding the CO_2 reforming of ethane. More ethane can, thus, be selectively converted into ethene through the CO_2-ODHE process (7.30) instead of undergoing dry reforming. It was also found that the type of carbon deposited on the catalyst bed was more reactive (i.e., filamentous type) compared to the case in which no methane was co-fed. Hence, Peng et al. correlated the improvement in the ethene selectivity to the easier removal of filamentous carbon with CO_2 [95].

Solymosi et al. [46] found that, despite the slow reaction of CO_2 with Mo_2C at 873 K, the appearance of new peaks at 233.3 and 236.2 eV in the XPS spectra indicates the occurrence of a fast reaction between Mo-O species and propane. In fact, based on the catalytic activity data and XPS measurements, they found that propane and ethane were effective carburizing compounds. Hence, they proposed that CO_2 oxidizes Mo_2C forming Mo oxycarbide, whereby propane has the possibility of forming a surface complex with the active oxygen of the carbide as shown in (7.31) [46].

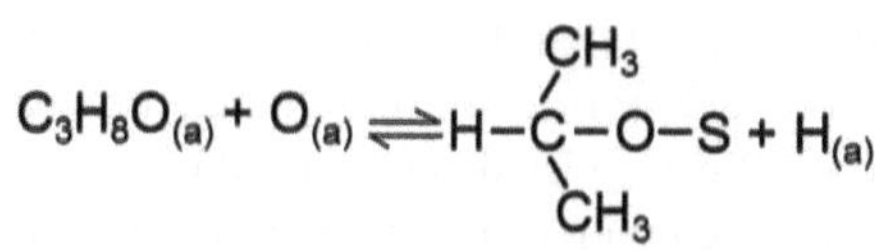

(7.31)

The C–H cleavage is assumed most likely to involve the methylenic C–H bonds, which are weaker in energy than the methyl C–H bonds (7.32) [46] .

$$C_3H_8O_{(a)} \rightarrow C_3H_7O_{(a)} + H_{(a)} \tag{7.32}$$

Michorczyk et al. [47, 96, 97] studied the CO_2 dehydrogenation of propane on chromium oxide supported on various types of mesoporous siliceous supports such as SBA-1 [47], MCM-41 [96], and SBA-15 [97]. In general, they found that the RWGS reaction can proceed over redox and non-redox Cr sites [47, 96]. This inadvertently enhances the alkane conversion as hydrogen is eliminated from the dehydrogenation process. Based on their experimental evidence, it was found that in the RWGS reaction the CO formation rate grows together with the concentration of Cr redox species in CrOx/SBA-1 catalysts, a scenario similar to the dehydrogenation of propane in the presence of CO_2. Their conclusion tallies with previous reported results [98, 99], whereby the RWGS reaction is postulated to occur via the redox mechanism in which Cr^{3+} species are successively reduced to Cr^{2+} by H_2 and then reoxidized by CO_2 (Scheme 7.7). The reduction of Cr^{3+} to Cr^{2+} is promoted over a silica support which facilitates the electron fluctuations necessary for the reduction process [96, 100], prior to reoxidation with CO_2. In Scheme 7.7, the CrO_x and CrO_{x-1} represent the oxidized and reduced redox Cr sites, respectively, whereas the CrO_y represent non-redox Cr sites [47].

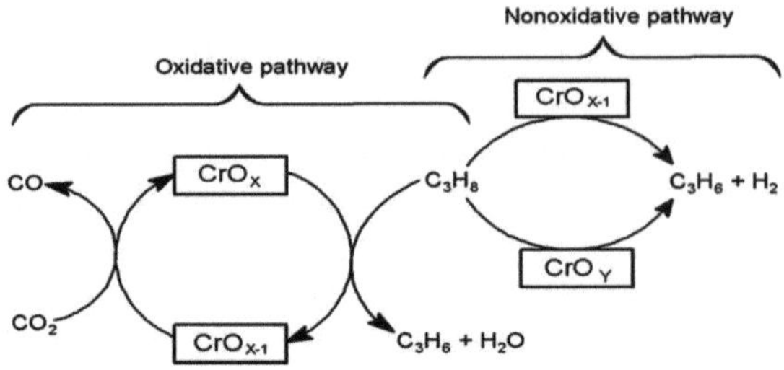

Scheme 7.7 Postulated reaction pathways for propene formation via CO_2-ODH reaction

Based on Scheme 7.7, the oxidative pathway relates to the presence of reducible Cr sites forming a redox cycle CrO_x (oxidized sites) and CrO_{x-1} (reduced sites). The reduced forms identified as Cr^{3+} and Cr^{2+} sites are active in the non-oxidative dehydrogenation of propane. It was also presumed that the non-redox Cr^{3+} species (CrO_y) forming chromium(III) oxide agglomerates may participate in the non-oxidative pathway to a lesser extent because of the lower number of active sites [47].

Similarly, using an iron promoted Fe–Cr/ZrO_2 catalyst for the CO_2-ODH of ethane, Deng et al. [99] observed that the dehydrogenation of ethane is activated by the Cr^{3+} species whereas Fe_3O_4 formed during the reaction can promote the RWGS reaction. The formation of Fe_3O_4 was detected on the basis of XPS spectra: it is an important finding because the inverse spinel structure of Fe_3O_4 can produce a rapid electron exchange between Fe^{3+} and Fe^{2+} which can facilitate the progress of the reaction [99, 101].

7.2.3 *Oxidative Dehydrogenation of Ethylbenzene to Styrene*

The oxidative dehydrogenation of ethylbenzene (ODE) to styrene is a promising alternative or complement to the current industrial technology based on steam, which may contribute to meeting the growing demand for styrene and the strong market incentive [93, 102]. Styrene is one of the most widely used feedstocks in the production of high-value commercial products such as synthetic rubbers, polystyrene, and styrene–acrylonitrile copolymers [93, 103, 297]. The global styrene annual production is 14.6 Mt, with the US producing 4 Mt and Europe 3.4 Mt, respectively [104]. Figure 7.3 shows the global styrene consumption growth over 2008–2015 [105], following the very similar trend observed in the previous decade. The growing rate is higher for developing regions than for mature markets in America and Europe. Significant downstream expansion and emerging plants in

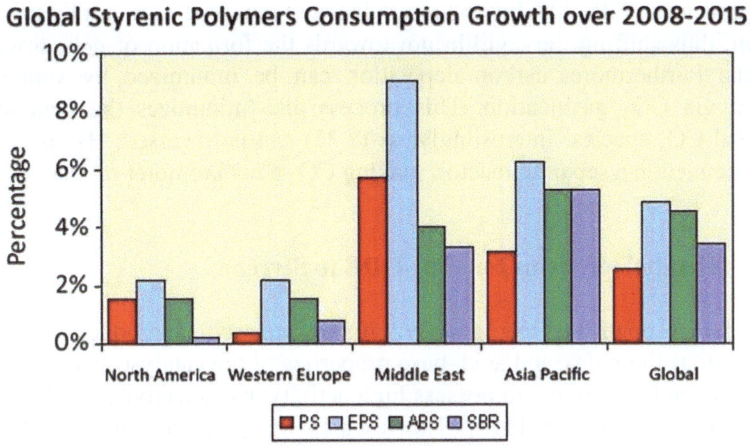

Fig. 7.3 Global styrene consumption growth over 2008–2015 [105]

the Middle East mean that the Middle East is the dominant world exporter of styrene [105].

The well-established industrial dehydrogenation of ethylbenzene to styrene is carried out with a large excess of steam to overcome carbon deposition and the thermodynamic barriers of the reaction [49, 106]. However, the high endothermicity of the process causes a low conversion rarely exceeding 50 %. The occurrence of some side reactions such as production of toluene, benzene, and coke also affects the process yield and selectivity, and leads to catalyst deactivation [41]. A large excess of steam, a growing energetic and environmental issue, is often required to improve the catalytic selectivity to above 90 % and overcome the loss in selectivity and yield caused by irreversible coke formation [49, 107].

Oxygen can be used as dehydrogenating agent, but this raises problems of radical formation. A promising alternative is to use CO_2 as dehydrogenating agent, which presents several advantages such as enhanced styrene selectivity (97 %), inhibition of the total oxidation, and prevention of hot spots [41, 106, 108, 109]. In fact, compared to other types of oxidants such as N_2O and SO_2, there is no loss of latent heat for CO_2 [110]. In addition, compared to steam, the energy consumption is about $1.5–1.9 \times 10^5$ instead of 1.5×10^6 kcal/mol of styrene produced [41]. The overall process, based on a single step pathway is shown in (7.33), whereas the two-step pathway is defined by (7.34) and (7.35) as shown in Scheme 7.8 [111].

$$C_6H_5CH_2CH_3 + CO_2 \rightarrow C_6H_5CH{=}CH_2 + H_2O + CO \quad (7.33)$$

$$C_6H_5CH_2CH_3 \rightarrow C_6H_5CH{=}CH_2 + H_2 \quad (7.34)$$

$$H_2 + CO_2 \rightarrow CO + H_2O \quad (7.35)$$

Scheme 7.8 Scheme of pathways for CO_2-ODE to styrene

As shown in (7.35), the hydrogen produced is removed as steam by the RWGS reaction, thus shifting the equilibrium towards the formation of dehydrogenation products. Furthermore, carbon deposition can be minimized by simultaneous removal via CO_2 gasification. This process also minimizes the generation of undesired CO_x species. Interestingly, as (7.35) can be reversed, H_2 and CO_2 can be regenerated in a separate reactor, making CO_2 a net promoter of reaction (7.34).

7.2.3.1 Catalytic Systems for CO_2-ODE to Styrene

Several factors contribute to good selectivity towards styrene production from CO_2-ODE, such as the redox and acid–base properties of the catalyst. Iron oxide-based catalysts have been shown to possess high activity and selectivity towards styrene production via CO_2-ODE [112]. The research focus has been on improving the viability of Fe-based catalysts for this process. Some early work has shown that iron oxide in the presence of CO_2 induces an increased selectivity towards styrene. Different supports such as alumina and activated carbon have been used [111, 113, 114]. Nevertheless, the iron-based catalysts show several drawbacks and efforts have been made for identifying promoters or alternative catalysts. Some of the drawbacks include low surface area, lack of stability, and vulnerability towards deactivation of the catalytically active Fe^{3+} sites compared to metal oxides such as titania [102, 108]. Castro et al. [102] synthesized porous iron-based catalysts by adding Ti, Al, or Zr, and found a higher catalytic performance in the case of Ti modification. The in situ formed stable mixed phases such as $FeTiO_3$, Fe_2TiO_5, and $FeTi_2O_5$ resulted in highly active species resistant to phase transformation and sintering, thus, assuring high catalytic performance, although coke formation increased because of ethylbenzene cracking. In addition, styrene oligomerization took place [102].

Mg-Fe/Al-based catalysts were developed via impregnation and sol gel methods. Spinel $MgFe_{0.1}Al_{0.9}O_4$ showed high catalytic activity and stability. The good performance was attributed to the weak acid sites, as shown by NH_3-temperature programmed desorption studies [115]. In addition, the Fe^{3+} active site species are well distributed in the spinel lattice, hence demonstrating properties such as resistance to reduction and avoidance of sintering. Fe^{3+} being the active sites was suggested by Zhu et al. [116] from results from transient kinetic experiments.

Ceria-based iron oxide catalysts can form solid solutions ($Ce_{1-x}Fe_xO_2$) and were found to perform better than ceria-based Zr, Ti, Pr, and Y mixed oxides [117]. In order to stabilize the lattice structure of CeO_2 and the oxygen diffusion capacity, the $Ce_{1-x}Fe_xO_2$ catalyst was prepared via the hydrothermal method and a highly dispersed Fe_2O_3 on the surface of the $Ce_{1-x}Fe_xO_2$ solid solution was obtained [117]. The formation of solid solution with Fe^{3+} in the CeO_2 lattice is confirmed via Raman analysis (Fig. 7.4) as demonstrated by the shift and decrease of the intensity of the diffraction peak at 462 cm^{-1}, corresponding to CeO_2. The Fe^{3+} doping has the potential to improve the reaction kinetics [118], but can lead to strong structural distortions, with a lower energy for the formation of oxygen

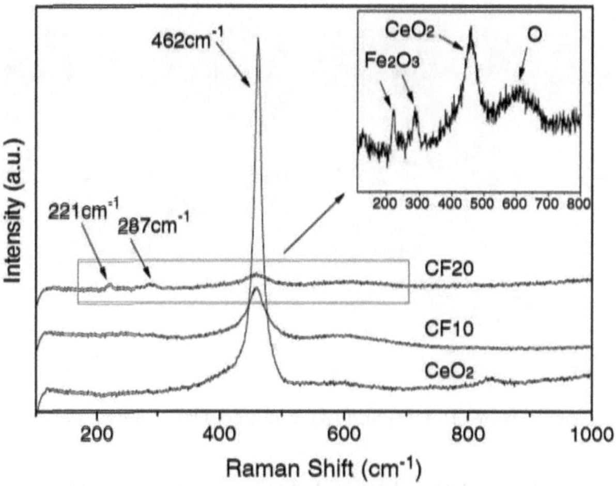

Fig. 7.4 Raman spectra of the fresh $Ce_{1-x}Fe_xO_2$ catalysts compared to CeO_2 [117]

vacancies, compared to the undoped system. As a result, the authors found that this structure distortion results in the creation of more oxygen vacancies promoting the adsorption and activation of CO_2, leading to a higher conversion by 13 %. They attributed the higher catalytic performance to the rapid oxygen exchange between Ce^{4+} and Ce^{3+} redox sites promoted by Fe^{3+} and improvement in oxygen storage capacity of ceria, which enabled iron to maintain its high shift activity [119].

Vanadium-based catalysts have been widely explored. For example, Sakurai et al. [120] observed high EB conversion (67.1 %) and selectivity to styrene (80 %) for vanadium supported on activated carbon in the presence of CO_2 at 823 K (EB conversion was 14 % higher than in the presence of argon). Vanadium-substitution in Mg–Al hydrotalcite-like catalysts resulted in Mg–V–Al mixed oxides with high surface area and mesoporous characteristics, which are favourable for ethylbenzene dehydrogenation to styrene under CO_2 flow [121]. In addition, V^{5+} was identified as the active site for the dehydrogenation process [121]. The key properties of this catalyst, besides its mesoporous characteristics, are the weak Lewis acid sites contributed by aluminum, which reduce the catalyst deactivation.

Titanium oxides doped with vanadium were prepared using the acid-catalyzed sol gel technique, shown to be an effective method for preparing a catalyst with a well determined ratio of V to Ti [122]. Based on TEM images, Fig. 7.5, the authors showed that loading V^{5+} up to 6 mol% (the optimal amount for best catalytic activity) caused the diminution of the particle size because of incorporation of vanadia into TiO_2. Such a preparation method is capable of producing a good dispersion of vanadia, hence favoring the formation of small crystallites.

Likewise, Park et al. [123, 124] have studied the effect of ceria on the structure and catalytic activity of $V_2O_5/TiO_2–ZrO_2$ catalyst for the CO_2-ODE reaction. In their previous research, the authors had found that titania–zirconia mixed oxides exhibited excellent catalytic activity and selectivity for the CO_2-ODE to styrene because of the high specific surface area, good acid–base and redox properties,

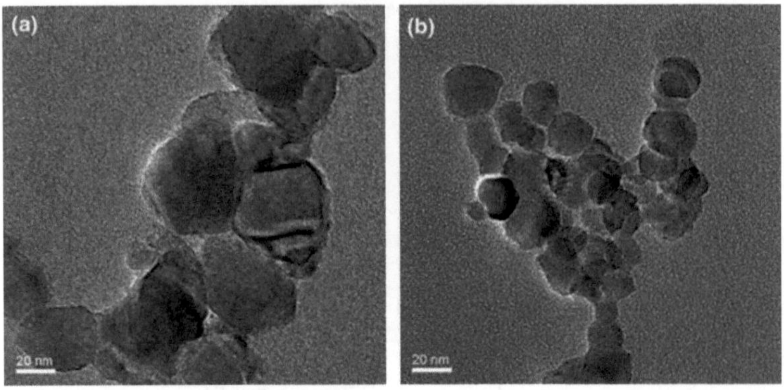

Fig. 7.5 TEM micrographs of (**a**) TiO_2 and (**b**) V-doped TiO_2 catalyst [122]

allied with the high thermal stability and strong mechanical strength of TiO_2–ZrO_2 mixed oxides [125, 126]. Because of the preparation method, the formation of crystalline $ZrTiO_4$ spinel phase was observed. ZrV_2O_7 and $CeVO_4$ phases were also produced. A high EB conversion of 56 % and product selectivity of 98 % was obtained over the ceria doped catalyst. In addition, Raman analysis revealed that the catalyst particles were nanosized (<7 nm) with a narrow particle size distribution. Furthermore, without ceria, the catalyst suffered a faster rate of deactivation. Adding ceria produces a catalyst, the redox properties of which suppress carbon deposition. Park et al. [127] have explored the use of redox-based CeO_2–ZrO_2 mixed oxide supported on a high surface area mesoporous silica such as SBA-15. Positive results were obtained thereby, with an increase of the reaction temperature from 723 to 923 K, and a remarkable increase in EB conversion from 21 to 65 % was observed. In fact, the redox nature of the catalyst enables its oxidation-reduction activity to be more pronounced at the higher temperatures because of the surface and sub-surface rearrangements involved during the reaction, which enables the increase in availability of surface oxygen [127].

Another type of mesoporous silica, i.e., MCM-41, has also been employed as a high surface area support for vanadia catalysts [128]. Upon comparison with the conventional silica support, it was evident that better dispersion could be obtained on the mesoporous material, as demonstrated by Raman spectra showing the presence of strong bands of isolated tetrahedral VO_4 [128].

7.2.3.2 Reaction Mechanism for CO_2-ODE to Styrene

Several possible mechanisms have been identified by Sun et al. [129] taking into consideration that EB is activated by acid–base catalysts or lattice oxygen (via redox mechanism) in the presence of CO_2. Based on the CO_2-ODE reaction over V- and Fe-supported catalysts, Sun et al. [130] found that a synergistic effect exists between EB dehydrogenation and the RWGS reaction. In fact, CO_2 can be either reduced to CO via interaction with redox sites (one-step mechanism) or by reaction with hydrogen formed in EB dehydrogenation (two-step pathway) as shown in Scheme 7.9 [130].

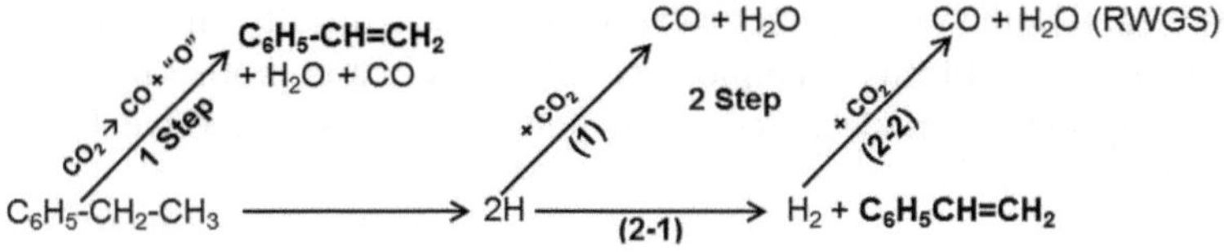

Scheme 7.9 Role of CO_2 in the CO_2-ODE process

Based on these two pathways, three vital mechanistic steps are proposed and are discussed with the relevant examples. Mechanism 1 (one-step pathway) is redox-based: vanadia or ceria type catalysts play an important role in the activation of both CO_2 and EB. The C–H bond breaking proceeds via a Mars-van-Krevelen (redox) type mechanism and lattice oxygen of the catalyst is consumed during the dehydrogenation process, producing styrene and water (7.36)–(7.40) in Scheme 7.10 [131]. The lattice oxygen, however, is replenished by CO_2 dissociation to CO and surface oxygen species. Different types of V and Ce species act as oxygen carrier through their oxidized-reduced pairs [132].

Mechanism 1 – One step pathway

$$CO_2\,(g) + \bullet \leftrightarrows CO\,(g) + O\bullet \qquad (7.36)$$

$$C_8H_{10} + {}^{\circ} \leftrightarrows C_8H_{10}{}^{\circ} \qquad (7.37)$$

$$O\bullet + C_8H_{10}{}^{\circ} \rightarrow C_8H_8\bullet + H_2O^{\circ} \text{ (Redox type C-H bond breaking of EB to Styrene)} \qquad (7.38)$$

$$C_8H_8\bullet \rightarrow C_8H_8 + \bullet \qquad (7.39)$$

$$H_2O^{\circ} \rightarrow H_2O\,(g) + {}^{\circ} \qquad (7.40)$$

Scheme 7.10 Mechanism 1 via one-step pathway for CO_2-ODE process (*open circles* – oxidative site; *closed circles* – reductive site)

Sakurai et al. [133] have investigated the possible reaction mechanism of the ODE to styrene over conventional ODH catalysts such as VO_x supported on MgO. In the presence of CO_2, a high styrene yield of 73.8 % was obtained with a selectivity of 90.1 % at a temperature of 873 K. Based on the data collected by using several characterization techniques such as UV-vis, FTIR, and XRD, it was ascertained that the active phase of vanadium was the V^{5+} species in vanadia or $Mg_3V_2O_8$ on highly dispersed MgO. The lower reduction states of vanadia were deemed to be less reactive sites for the dehydrogenation. As shown in Scheme 7.11, the lattice oxygen from the magnesium vanadate structure abstracts hydrogen from EB, producing water and styrene and, as a result, the vanadium species is simultaneously reduced. Another recent study by Chen et al. [134] on CO_2-ODE over silica-supported vanadium (VO_x/SiO_2) also suggests the redox mechanism and highlights the active site as the V^{5+} species. The role of CO_2 is found to be effective to maintain or regain the high oxidation state of vanadium, enhancing its activity for EB dehydrogenation. They also found that both the coke deposition and deep reduction of active vanadium species contribute to the catalyst deactivation. However, because of the chemically inert nature of the silica support, CO_2 in this example was only effective in protecting the oxidized vanadium species from deep reduction, but was not so effective in inhibiting coke formation.

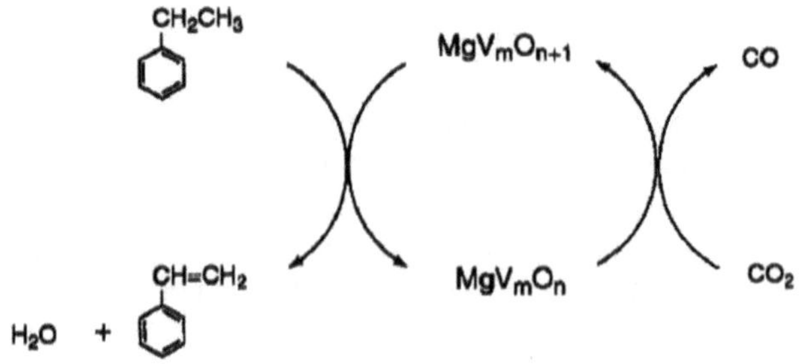

Scheme 7.11 Proposed redox mechanism over Mg loaded vanadia catalyst [133]

Similar to CO_2-ODH, Liu et al. [135] have proposed a redox mechanism (as shown in Scheme 7.12) for highly dispersed vanadium oxide supported on $Ce_{0.6}Zr_{0.4}O_2$–Al_2O_3. Zirconia stabilized with ceria is well known to be an excellent redox material, especially when ceria is enriched in defect sites. At the optimal vanadium oxide loading in Ce-Zr oxides, high styrene selectivity of up to 95 % can be obtained.

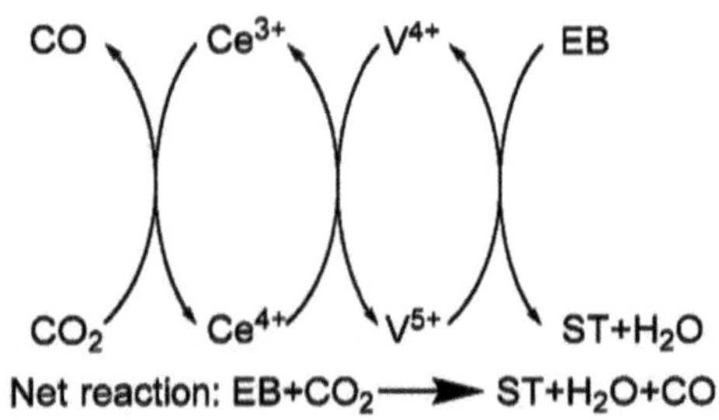

Scheme 7.12 Proposed redox mechanism for CO_2-ODE over $V_2O_5/Ce_{0.6}Zr_{0.4}O_2$–$Al_2O_3$ [135]

As explained by Liu et al. [135], the simultaneous oxidation of Ce^{3+} to Ce^{4+} by CO_2 and the reduction of V^{5+} to lower oxidation states (very likely V^{4+}) by ethylbenzene are the main pathways involved in the CO_2-ODE. Subsequently, with the reduction of Ce^{4+} to Ce^{3+}, V^{4+} is oxidized to V^{5+} to complete the full cycle. Figure 7.6 shows the Raman spectra of the support (1) and at various V_2O_5

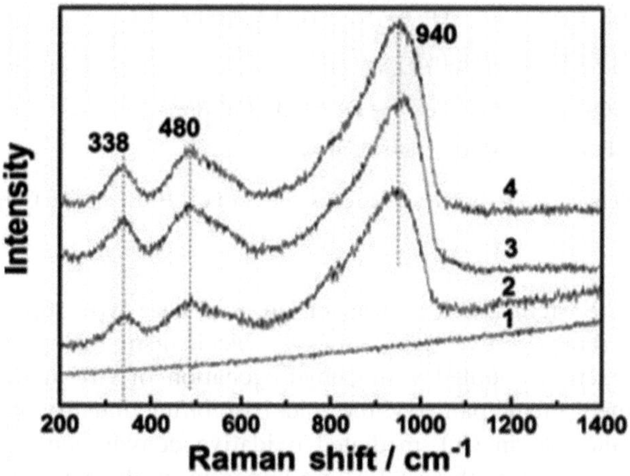

Fig. 7.6 UV-Raman spectra of 0, 6, 9, and 15 wt % V_2O_5/CZS (traces 1, 2, 3, and 4, respectively) [135]

loadings (6 wt% vanadium oxide is the optimal loading) on CZS. The broad Raman shift centered at about 940 cm^{-1} can be assigned to the interface mode of bridging V–O–S (where S refers to metal cations from the support, i.e., Al, Ce, and Zr) which is found to be the main catalytic active site for the CO_2-ODE. The other two peaks relate to V–O–V and V≡O bonds, which increase when the vanadium oxide loading is increased, leading to poorer dispersion and lower catalytic activity. A high dispersion of vanadium oxide on the support is critical for the high catalyst stability. The stabilization of V^{5+} with the reduction of Ce^{4+} to Ce^{3+} has been discovered by density functional theory (DFT) calculations [136] which support the proposed Scheme 7.12 [135].

The presence of ceria is influential in promoting the redox pathway, even in the presence of iron oxide. Studies by Wang et al. [117] have shown that the incorporation of iron oxide in the ceria lattice results in the formation of a solid solution. Hence, during the reaction, the surface Fe^{3+}/Ce^{4+} absorbs the EB molecule, and the surface oxygen of the catalyst abstracts H from the ethyl moiety, generating H_2O and forming styrene, whilst the reduction of the surface active centers to Fe^{2+}/Ce^{3+} takes place, generating oxygen vacancies on the surface (7.41). CO_2 is adsorbed on the oxygen vacancy (7.42), and dissociates to CO and surface oxygen by accepting electrons from the transformation of $Ce^{3+} \leftrightarrow Ce^{4+}$ ($Fe^{3+} \leftrightarrow Fe^{2+}$) (7.43) [298]. In this mechanism, the defective structure containing oxygen vacancies is important to accelerate the adsorption of CO_2 as well as the diffusion rate of oxygen. A similar mechanism is prevalent on perovskite-type oxides such as $Cr_{0.75}Fe_{1.25}O_3$. It should be noted that the low surface area of the perovskite structures results in lower EB conversion, despite having good selectivity to styrene [108] (Scheme 7.13).

$$EB + Fe^{3+}/Ce^{4+} + O^{2-} \rightarrow ST + H_2O + [] + Fe^{2+}/Ce^{3+} \quad (7.41)$$

$$[] + CO_2 \rightarrow CO_2\ (ads) \quad (7.42)$$

$$CO_2\ (ads) + Ce^{3+}/Fe^{3+} \rightarrow O^{2-} + CO + Ce^{4+}/Fe^{3+} \quad (7.43)$$

where [] is a O vacancy; ads: adsorbed.

Scheme 7.13 Influence of redox sites of Fe^{3+}/Ce^{3+} on the CO_2-ODE reaction

Systems such as iron on an acid–base-type support follow Mechanism 2 or 3 (both are two-step pathways). Mechanism 2 proposes the occurrence of RWGS reaction coupled with dehydrogenation of EB in the presence of CO_2. In this mechanistic pathway, the acid–base bifunctionality of the catalyst plays an important role in CO_2-mediated oxidative dehydrogenation reactions [49]. The acid properties of the catalyst are important in the activation of EB, whereas the basic sites are for CO_2 adsorption to form carbonates for hydrogen blockage. The cooperative action of acid sites and basic sites is fundamental [49].

Mechanism 2: Two-step pathway to EB dehydrogenation

$$C_8H_{10} + {}^{\circ} \leftrightarrows C_8H_{10}{}^{\circ} \quad (7.44)$$

$$C_8H_{10}{}^{\circ} + {}^{\circ} + \bullet \rightarrow C_8H_8\bullet + 2H^{\circ} \quad \text{(Dehydrogenation step)} \quad (7.45)$$

$$H^{\circ} + CO_2\ (g) + * \rightarrow CO^{\circ} + OH\oplus \quad \text{(RWGS step)} \quad (7.46)$$

$$CO^{\circ} \leftrightarrows CO\ (g) + {}^{\circ} \quad (7.47)$$

$$OH\oplus + H^{\circ} \rightarrow H_2O\ (g) + {}^{\circ} + * \quad (7.48)$$

$$C_8H_8\bullet \rightarrow C_8H_8 + \bullet \quad (7.49)$$

Scheme 7.14 Two-step pathway for the CO_2-ODE to styrene via RWGS reaction (*crossed circle*/open circle – acid/oxidative site; *asterisk*/*closed circle* – basic/reductive site)

Acccording to Nederlof et al. [137], the reaction of CO_2 with hydrogen produced in EB dehydrogenation (RWGS) can shift the equilibrium towards higher conversions. The difference between the two-step Mechanisms 2 and 3 are represented in Schemes 7.14 and 7.15, respectively. In Mechanism 2, the H atoms formed upon EB dehydrogenation attack one of the oxygen atoms of CO_2 with the formation of CO and OH moieties (step of RWGS). The latter species generates H_2O. In Mechanism 3, the direct dehydrogenation of EB takes place on a vacant oxygen site producing hydrogen atoms [111]. The latter associate to form molecular hydrogen, which reacts with the O atoms generated on the catalyst by CO_2 conversion to CO [137].

Mechanism-3: Two-step pathway for EB dehydrogenation

$$C_8H_{10} + ^{\circ} \leftrightarrows C_8H_{10}^{\circ} \tag{7.50}$$

$$C_8H_{10}^{\circ} + ^{\circ} + \bullet \rightarrow C_8H_8\bullet + 2H^{\circ} \quad \text{(Dehydrogenation Step)} \tag{7.51}$$

$$H^{\circ} + H^{\circ} \rightarrow H_2\,(g) + 2^{\circ} \tag{7.52}$$

$$CO_2\,(g) + \bullet \leftrightarrows CO\,(g) + O\bullet \tag{7.53}$$

$$H_2\,(g) + O\bullet \leftrightarrows H_2O\,(g) + \bullet \tag{7.54}$$

$$C_8H_8\bullet \rightarrow C_8H_8 + \bullet \tag{7.55}$$

Scheme 7.15 Two-step pathway for the CO_2-ODE to styrene (*open circles* – oxidative site; *closed circles* – reductive site)

Mechanism 3 (Scheme 7.15) is accepted to be more thermodynamically favorable [135]. Nederlof [137] and Badstube [299] have investigated the CO_2-ODE reaction over iron oxide catalyst supported on activated carbon. Their experimental data correlate with this postulated two-step mechanism, whereby EB dehydrogenation is followed by redox-type RWGS reactions (Fig. 7.7).

He et al. [138] have carried out studies on the role of CO_2 in EB dehydrogenation over Fe_2O_3 using DFT calculations. They have combined the two general mechanisms (one-step, two-step) on iron catalysts, i.e., the redox cycle and the dehydrogenation (Scheme 7.16). The two Mechanisms related to the RWGS reaction are combined together and labeled as the "coupling mechanism," as shown in Scheme 7.16.

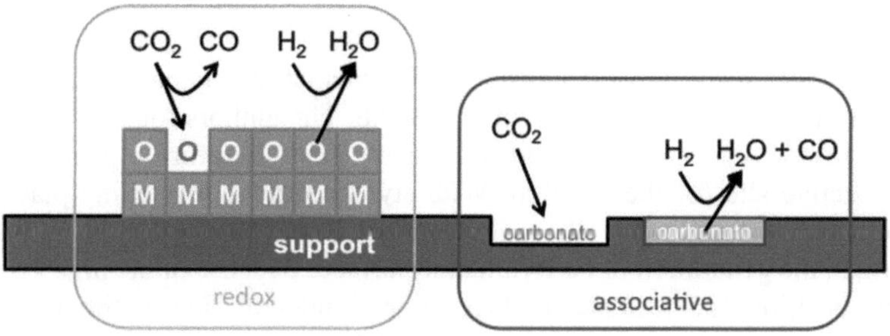

Fig. 7.7 Different RWGS mechanisms which accompany the CO_2-ODE reaction [137]

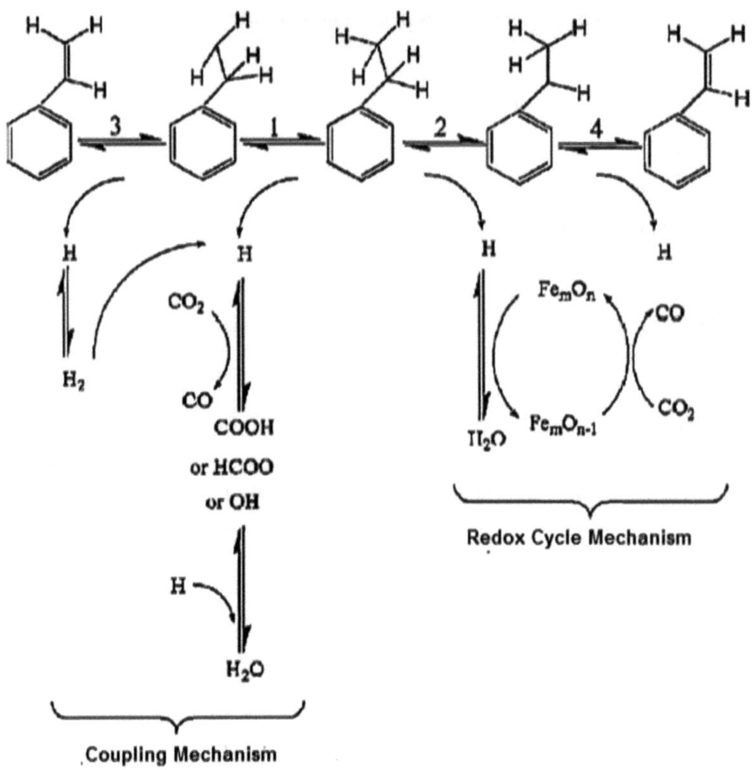

Scheme 7.16 General mechanism pathway over Fe-based catalysts [138]

Three model surfaces were considered in calculations, namely the Fe-, ferryl-, and O- terminated surfaces as shown in Fig. 7.8. The authors found that the lowest energy barriers were on O-terminated surfaces. They therefore suggested that this was the active site for the reaction. With styrene production taking place on the O-terminated surface, some lattice O-atoms are removed through H_2O formation, resulting in the generation of Fe-terminated surface. Because of the high TS energy, CO_2 dissociation and formate production are hindered. Hence, the authors concluded that a synergistic effect between EB dehydrogenation and RWGS reaction takes place, producing H_2O and/or H_2 via carboxyl species (CO_2H) or formate (HCO_2) intermediates (see Scheme 7.16) following the "coupling mechanism." Such theoretical results correlate with Mechanisms 2 and 3 proposed by Sun et al. [129] centered on the RWGS reaction, which may take place on the iron-based catalysts.

Badstube et al. [114] have investigated the CO_2-ODE over iron oxide catalyst supported on activated carbon in the presence of CO_2. In addition to styrene, benzene, and toluene, they detected CO and H_2O. Comparing the experimental data with the postulated mechanisms, they concluded that both the RWGS reaction and the redox cycle mechanism contribute to ethylbenzene dehydrogenation in the

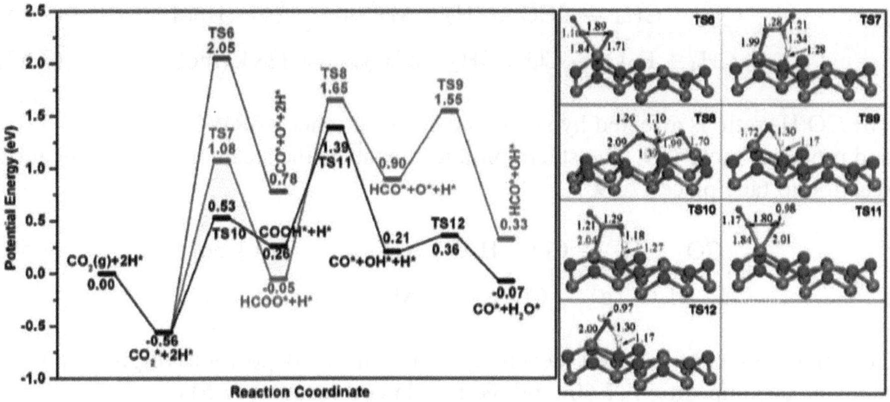

Fig. 7.8 Potential energy diagrams and transition states for three reaction pathways of CO_2 consumption on the Fe-terminated surface [138]

presence of CO_2. Likewise, Chang et al. [139] scrutinized the role of CO_2 on the dehydrogenation of ethylbenzene to styrene over zeolite-supported iron oxide catalyst. Production of styrene, H_2O, CO, and a small amount of H_2 were observed. This makes evident that CO and H_2O are obtained from CO_2 reduction coupled to H abstraction from EB by surface oxygen or via the involvement of RWGS reaction. Therefore, the authors proposed that both the coupling and redox cycle mechanisms are possible for CO_2-ODE over their catalytic system. However, CO_2 plays a dual role in the mechanisms discovered with experiments and calculations: either forms an active site on the catalyst by transferring O atoms to a reduced site, or acts as scavenger of H_2 formed upon EB dehydrogenation shifting the equilibrium to the right.

7.2.4 CO_2 (Dry) Reforming of Methane

In industrial practice, steam reforming of methane (SMR) is widely used for hydrogen/Syngas production. Steam reforming of methane is known to release a large amount of CO_2 which is eventually vented to the atmosphere [140]. However, in view of global climate change concerns, which have been increasingly attributed to GHG emissions, CO_2 (dry) reforming of methane (DRM) is becoming an attractive alternative. DRM reaction produces Syngas (1:1 mixture of CO and H_2 in the specific case (7.57)), which can be used as feedstock for industrial processes such as Fischer–Tropsch, methanol synthesis, or hydrocarbonylations [141]. DRM is a highly endothermic reaction (7.56), which has similar thermodynamics and equilibrium characteristics with steam reforming of methane (7.57) except that the former reaction produces synthesis gas with a lower H_2/CO ratio [142].

$$CO_2 + CH_4 \rightarrow 2CO + 2H_2 \quad \Delta H^\circ_{298K} = 247\ \text{kJ mol}^{-1} \tag{7.56}$$

$$CH_4 + H_2O \rightarrow CO + 3H_2 \quad \Delta H^\circ_{298K} = 228\ \text{kJ mol}^{-1} \tag{7.57}$$

The CO:H_2 ratio is affected by simultaneous occurrence of RWGS reactions (7.58) and produces more CO. The steam produced enables the steam/carbon gasification reaction to take place (7.59).

$$CO_2 + H_2 \rightarrow CO + H_2O \quad \Delta H^\circ_{298K} = 41\ \text{kJ mol}^{-1} \tag{7.58}$$

$$C + H_2O\ (g) \rightarrow CO + H_2 \quad \Delta H^\circ_{298K} = 131\ \text{kJ mol}^{-1} \tag{7.59}$$

The main issue of catalyst deactivation is carbon deposition originated from methane decomposition (7.60) and the Boudouard reaction (7.61).

$$CH_4 \rightarrow C_{ads} + 2H_2 \quad \Delta H^\circ_{298K} = 75\ \text{kJ mol}^{-1} \tag{7.60}$$

$$2CO \rightarrow C_{ads} + CO_2 \quad \Delta H^\circ_{298K} = -173\ \text{kJ mol}^{-1} \tag{7.61}$$

The two side reactions (7.60) and (7.61), which generate coke deposits, are responsible for catalyst deactivation and may damage the reactor [143]. Based on Boudouard equilibrium [144], coke formation by CO disproportionation is less favored at high temperatures (973 K and above). The highest carbon deposition usually occurs on transition metal catalysts such as Ni, an otherwise interesting catalyst, the cost of which is up to three orders of magnitude lower than noble metals used in such a process (see below). In general, the dissociation of methane occurs on a Ni surface to generate highly reactive C_α, which can be further gasified by reactions with H_2O, CO_2, or H_2 [144]. The encapsulation of Ni by carbon is the major step for rapid growth of carbon nanotubes, which can block the catalyst and plug the reactor system [145]. It is therefore important to design rationally a highly active, stable, and economically viable catalyst system, which both kinetically inhibits the coke formation and improves the conversion in the DRM reaction.

Throughout the past decades, researchers have tried to integrate DRM, SRM, and partial oxidation of methane (POM) to mitigate the negative thermodynamics of DRM and SRM. The concept of 'tri-reforming" was introduced by Song et al. [146]. It involves simultaneous reaction of methane with CO_2, H_2O, and O_2, and it is of practical interest because of the advantages it produces. A variable H_2/CO ratio can be obtained by adjusting the amounts of CO_2, H_2O, and O_2 in the feedstock. Carbon formation on the catalyst, which is a major issue in DRM, can also be suppressed by the addition of H_2O and O_2 [147]. Finally, the exothermic nature of POM reaction provides energy for the reforming process and hence increases the energy efficiency. This is an interesting reaction concept, which may help the industrial exploitation of DRM.

7.2.4.1 Catalytic Systems for the DRM Process

The catalysts used for the DRM process generally consist of noble metals (Rh, Pt, Rh) and some transition metal catalysts such as Ru and Ni. Noble metals are used for their resistance to corrosion and oxidation, high activity, and good carbon resistance [148].

Rh, Pd, Pt, and Ru are known to reduce carbon deposition. Despite the high activity and resistance against carbon formation, noble metals are also known for their high cost and low availability. As such, low metal loadings have been investigated in the DRM reaction. For example, Lercher et al. [149, 150] studied the deactivation and coke accumulation during DRM reaction with a loading as low as ca. 0.5–1 wt% Pt on zirconia. The size of the Pt particles and the stability of the carbonates formed on the support (from CO_2 activation) were found to be the crucial factors in determining the rate of carbon formation. Smaller Pt particles and supports such as zirconia with high affinity for carbonate stability were found to favor catalyst stability during the DRM. On the other hand, with Rh, the concentration of accessible surface atoms seems to be the key factor, with an apparently less important influence of the catalyst support [147]. In order to enhance the catalytic performance, it is inevitable that higher loadings of active metals are required. Most of the Group 8 transition metals (which are of lower cost compared to noble metals) have shown promising catalytic performance for this reaction. Nickel, among non-noble metals, is the one that has comparable activity to noble metals [151]. Nevertheless, as mentioned earlier, nickel catalysts are prone to catalyst deactivation caused by carbon formation. As such, there is a need for suitable catalyst supports or promoters to be used to counteract extensive carbon formation on Ni-based catalysts.

Several studies have been carried out aimed at synthesizing nickel-based catalysts in the most suitable form for providing high activity and better carbon resistance. Some of the nickel-based catalysts synthesized include bimetallic catalysts, oxide-supported Ni catalysts, Ni-based perovskites, and Ni core-shell structure catalysts, which are discussed in this section.

As reported by Ruckenstein and Hu [152], DRM activity is enhanced by addition of various promoters onto the catalyst surface. They have been found to be useful in CO_2 activation and for decreasing coke formation [153], and include oxides of Group 2 elements or La, e.g., La_2O_3, MgO, BaO, and CaO, which also improve metal dispersion [154, 155]. In addition, they contribute to a high catalytic activity and stability, strong metal-support interaction, and lattice oxygen mobility, factors of paramount importance for this reaction. Pan et al. [156] found that Ga_2O_3 on SiO_2 support activates adsorbed CO_2 to form surface carbonate and hydrogen-carbonate species. This leads to a more facile reaction with the deposited carbon than the η^1-O-linked CO_2 or physically adsorbed CO_2. Consequently, higher stability and coke resistance was achieved with Ga_2O_3-promoted SiO_2-supported Ni catalyst. A promotional effect by rare earth element over the catalytic DRM performance of Ni–La_2O_3 catalysts was recently explored [157]. The Ni–La_2O_3

catalyst doped with a small amount of Sr significantly improves the DRM process, even at temperatures as low as 873 K, with low carbon deposition. This was attributed to the highly mobile surface oxygen species, which enable C–H bond activation. The surface oxygen species can adsorb CO_2 molecules, thus forming bidentate carbonate species, which can facilitate carbon removal and result in higher CO_2 conversion. Similarly, the positive effect of alkaline metals was observed on borated-alumina-supported Ni catalysts [158]. In this case, instead of surface oxygen species, surface hydroxyl (OH) groups were found to play a dominant role in oxidizing surface carbon species derived from methane decomposition.

Bimetallic structures have been proposed as an alternative to single nickel metal catalysts. Alloying with a second metal (either a noble metal or another transition metal) results in the modification of the catalyst properties caused by a "synergistic" effect between the two metals [159]. For example, Oemar et al. [160] studied the effect of Pd on Ni-based catalysts supported on yttria for the oxidative DRM reaction. The formation of bimetallic Pd–Ni species enabled the catalyst to have a good metal support interaction, and the presence of α-oxygen on Y_2O_3 surface was beneficial in dissociating methane. Because the amount of α-oxygen is very little, a small amount of oxygen in the feed was required to regenerate the α-oxygen to maintain the stability of the catalyst. Zhang et al. [161, 162] prepared Ni–Me (Me=Co, Fe, Cu, or Mn) bimetallic catalysts with the general formula Ni–Co–Al–Mg–O using the coprecipitation method. Among bimetallic couples, Ni–Co provided the best results with respect to catalyst stability and activity: it underwent a 2 000-h stability test for DRM and no significant catalyst deactivation was observed. Perovskite-type catalysts represented by the general formula ABO_3 (A = lanthanide, B = transition metal) have been widely investigated in high temperatures applications because of their high thermal stability [65, 163]. In the perovskite structure, the B-cation is surrounded octahedrally and the A-cation is located in the cavity among the octahedral sites. A variation is the spinel oxide Ni-based catalysts of formula ABO_4 or A_2BO_4 [164]. The substoichiometric spinel structures are active catalysts caused by (1) high dispersion of NiO into the spinel, (2) stability of the structure at low temperature, or (3) formation of Ni phase in the same temperature region than the catalytic reaction and presence of Ni oxide. The latter is able to limit the growing of particles, probably by formation of interaction between nickel metal out of the structure and Ni oxide of the structure [65, 165]. Gallego et al. [166] have used reduced La_2NiO_4 as catalyst precursor for DRM at 973 K and have observed a conversion rate of 85 and 93 % for methane and CO_2, respectively, with no carbon deposition after 160 h of experiment. According to the authors, the inhibition in carbon deposition was because of the enhanced reaction between $La_2O_2CO_3$ and the Ni–CH_x species to produce CO and H_2. In addition, the high activity of the catalyst was attributed to the formation of the small nickel particles (average diameter 7 nm) via reduction of the perovskite La_2NiO_4. A recent investigation by Sutthiumporn et al. [167] on the performance of partially-substituted $LaSrNiO_3$ perovskite catalysts in DRM revealed that lattice oxygen species in these perovskite catalysts are crucial in activating the C–H bond of the

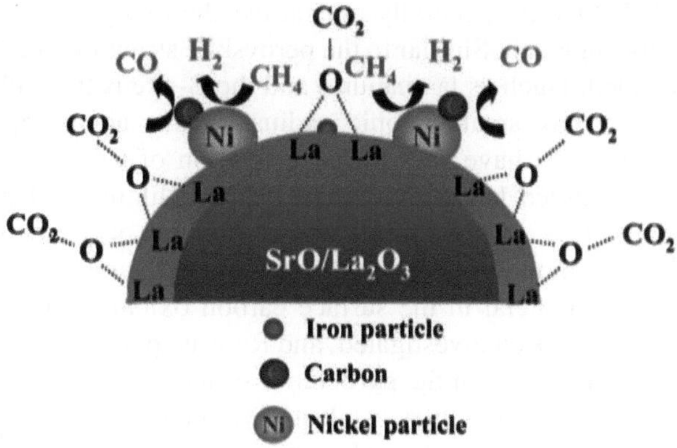

Fig. 7.9 Carbon removal on reduced Ni-supported SrO/La_2O_3 during DRM reaction [167]

methane molecule. In addition, carbon formation is reduced because of the reaction with CO_2 to form $La_2O_2CO_3$. The Cu-substituted $LaSrNiO_3$ catalyst was found to display initially high methane conversion; because of the weak metal–support interaction, the Ni particles agglomerate to form larger Ni particles, causing lower catalytic stability. In contrast, although the Fe-substituted $LaSrNiO_3$ catalyst possesses lower initial DRM activity, a high activity and stability could be achieved at prolonged reaction time as the Fe species begin to be reduced and the lattice oxygen species become increasingly mobile as the reaction proceeds in the reducing environment. In summary, key issues of the reaction mechanism on this catalyst are shown in Fig. 7.9 and are caused by (1) strong metal-support interaction, hindering thermal agglomeration of Ni particles, and (2) the presence of the abundant lattice oxygen species, which are not so active for C–H bond activation but active enough to form $La_2O_2CO_3$.

Spinel structures such as $NiAl_2O_4$ formed from the Ni/Al_2O_3 mixed oxide matrix as a result of high temperature calcinations result in strong metal support interaction [168]: the Ni–O bond in $NiAl_2O_4$ is stronger than in NiO crystals. However, the bulk structure is difficult to reduce and possesses lower activity at low reaction temperatures. On the other hand, surface spinel structures are more desirable as they are more active. Wang and Lu [169] have reported a higher activity and stability of Ni/Al_2O_3 catalyst compared to Ni/CeO_2 catalyst produced by strong metal support interactions. Nevertheless, it has been reported by Bhattacharya and Chang [170] that γ-alumina in the $NiAl_2O_4$ spinel structure formed via solid state reaction is prone to transform into the α-phase, especially over prolonged reaction time at high temperatures, resulting in gradual catalyst deactivation caused by sintering of the alumina support. In a recent study, $LaAlO_3$ perovskite oxide phase was pre-formed on alumina-based support and calcined [171]. Upon further impregnation with Ni and calcination steps, this structure was found to form a predominantly "inverse" surface nickel aluminate. This catalyst was shown to be beneficial in hindering the phase transformation of the γ-alumina to α-alumina phase [171].

Spivey et al. [172] have recently reported some interesting works on structurally stable lanthanum-zirconate pyrochlores ($La_2Zr_2O_7$). The pyrochlore structures

($A_2B_2O_7$) are generally crystalline thermally stable ternary metal oxides with a cubic unit cell. Similar to the perovskite structure, the A-site is typically a rare-earth element (such as lanthanum) and the B-site is typically a transition metal element, which has smaller ionic radius. Using an isotopic pulse technique, Spivey et al. [153] have shown that activation of CO_2 takes place on the La site of the Rh-modified lanthanum-zirconate pyrochlore via formation of three polymorphs (types I, Ia, and II) of $La_2O_2CO_3$, of which only type II is reactive. Promotion with higher amounts of Rh resulted in a greater number of Rh–La interfacial sites, which were beneficial in the surface carbon oxidation. Other metals such as Pt and Ru have also been investigated, and Ru was found to exhibit a more interesting activity compared to Pt in the pyrochlore structure [173, 174].

More recently, nanocatalysts have emerged as potential candidates for the DRM reaction. The formation of nano-sized Ni particles facilitates high dispersion which, in turn, has the tendency to minimize surface energy by increasing the interaction with the support to prevent agglomeration [175–177]. For example, Corma et al. [178, 179] have prepared delaminated Ni-ITQ-6 catalysts from silica, which was tested for DRM reaction. The catalytic performance was compared with conventional mesoporous silica-supported catalysts such as Ni/Silicalite-1 and Ni/MCM-41. The presence of defect sites in delaminated ITQ-6 material was found to permit high dispersion of Ni particles and prevent metal sintering, contributing to its high activity and stability during DRM reactions.

Core-shell type of catalysts has emerged as a widely studied topic for various catalytic processes. The metal@metal oxide nanoreactor core shell structure has the potential to resist sintering under high temperature operations because of partial blocking by the peripheral silica shells. However, in the past few years some attempts have been made to investigate such catalysts for the DRM reaction. Li et al. [180] developed Ni@SiO_2 core shell catalysts and studied the effect of shell thickness modifications for the DRM reaction. The high catalytic stability at 1 073 K was attributed to the formation of nickel phyllosilicate species with strong interaction between the Ni core and silica shell. Upon reduction, the species formed small satellite Ni particles within the silica shell, resulting in good catalytic DRM performance at the optimum shell thickness of 11.2 nm. The schematic representation of this structure is shown in Fig. 7.10 [180].

Porous silica supports are gaining prominence for the preparation of highly dispersed Ni particles. For instance, Copéret et al. [181] successfully prepared nano-ensembles of Ni on porous silica support for DRM reaction in a low temperature range (below 823 K). This preparation method could effectively eliminate particle size factors in investigating the competitive coverage of reaction intermediates, mostly $*CH_x$, $^*O$, and $^*CO$ on the nickel surface. Copéret et al. [182] also recently described a versatile method for the preparation of small Ni nanoparticles on oxide supports via a colloidal approach. In this approach, Ni-silicide (Ni_3–Si_2) colloids were prepared and dispersed on supports such as silica and ceria. In particular, by dispersing these narrowly distributed supported particles on ceria, the high oxygen storage capacity of ceria was found to improve significantly the performance of the reforming catalyst with 60 % higher Syngas productivity [182].

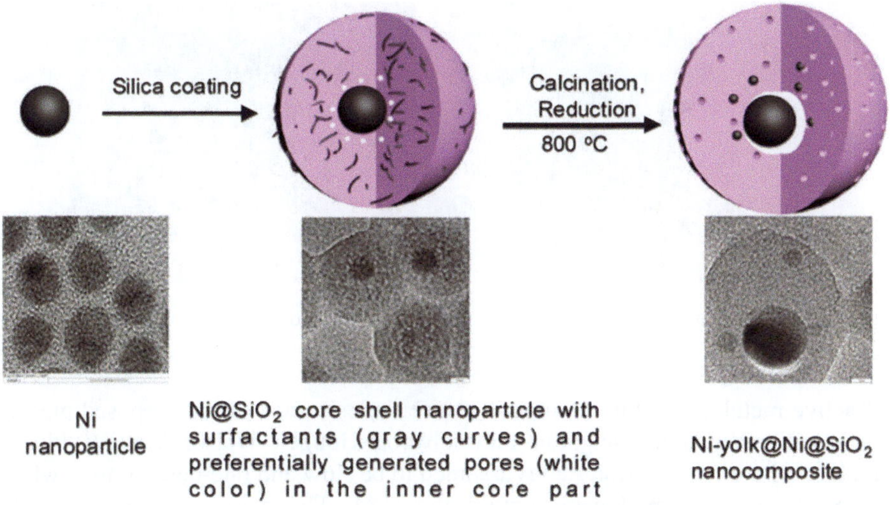

Fig. 7.10 Illustration for the formation process of Ni-yolk@Ni@SiO_2 [180]

Recently Mo et al. [183] have developed a facile and economical method to prepare highly dispersed Ni on silica via the core-shell precursor route. In this method the catalyst was promoted with oleic acid (OA) ($d_{Ni} < 3.0$ nm) to form core-shell (nickel nitrate species as core and nickel oleate as shell) precursor in the impregnated solution. Further promotion with only 1 % La_2O_3 resulted in the NiLaSi–OA catalyst achieving near equilibrium CO_2 and methane conversions (above 80 %) at 973 K with negligible carbon deposition even after 100 h on stream. Figure 7.11 shows the TEM images of the NiLaSi catalyst with (A) and without (B) OA promotion. It is evident that good dispersion can be observed upon OA promotion on the catalyst, whereas agglomerated species are evident on the unpromoted catalyst.

Therefore, the substantial progress achieved in catalyst development for DRM clearly says that the key to optimize the carbon resistance and promotion of activity lies in good catalyst dispersion and the promotion with adequate basic elements to enhance CO_2 adsorption in the DRM reaction. In addition, design and synthesis of novel nanostructures pave the way for future co-implementation of DRM processes in existing industrial systems.

7.2.4.2 Reaction Mechanism for the DRM Process

Numerous publications propose several reaction mechanisms for the DRM reaction, and a variety of differing opinions because the mechanism depends on the type

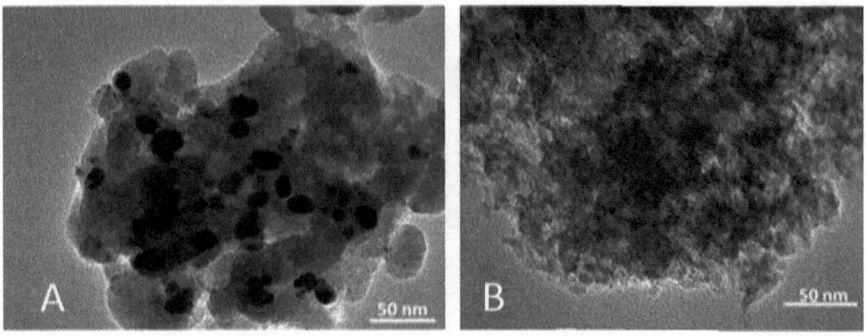

Fig. 7.11 TEM images of NiLaSi catalysts (**a**) without OA promotion; (**b**) with OA promotion [183]

of active metal and nature of support. Nevertheless, most of the proposed mechanisms are based on the Langmuir–Hinshelwood–Hougen–Watson (LHHW) kinetic models, where one reaction step is assumed to be slow and rate-determining, whilst the others are at thermodynamic equilibrium [184]. The major steps involved in the DRM reaction are dissociation of methane and CO_2 followed by adsorption of intermediates on the active sites leading to formation of products, i.e., CO, H_2, and H_2O, that are eventually desorbed. Among these steps, activation of both methane and CO_2 play the dominant role in the DRM reaction.

7.2.4.2.1 Methane Activation

The most kinetically significant step in the DRM reaction is the adsorption and subsequent activation of methane which dissociates directly on the active metal, forming carbon, CH_x radicals or formyl intermediates, and hydrogen adatoms as shown in Scheme 7.17 [154, 185, 186].

$$CH_{4(g)} \rightarrow CH_{3(a)} + H_{(a)} \quad (7.62)$$

$$CH_{3(a)} \rightarrow CH_{2(a)} + H_{(a)} \quad (7.63)$$

$$CH_{2(a)} \rightarrow CH_{(a)} + H_{(a)} \quad (7.64)$$

$$CH_{(a)} \rightarrow C_{(a)} + H_{(a)} \quad (7.65)$$

Scheme 7.17 Stepwise decomposition of methane on active metal sites

Even though the activation of methane is generally proposed to be taking place on the metal sites [187], Zhang et al. [188] found that when Ni was supported on more acid sites such as Al_2O_3 compared to the basic La_2O_3, the conversion of methane was higher because the acid sites of the alumina support promoted cracking of methane to carbon species. Nevertheless, a significant drop in activity was observed with time because of the greater amount of carbon deposition on the more acid alumina.

According to Rostrup Nielsen and Bak Hansen [148], during the reforming process there exists the possibility for the formation of three types of carbon species, namely:

1. Whisker-type carbon (formation temperature above 723 K)
2. Encapsulating amorphous carbon formed via polymerization at temperatures less than 773 K
3. Pyrolytic carbon nanotubes formed by methane cracking at temperatures above 873 K

The type of carbon formed depends on the reaction conditions, whereby higher reaction temperatures favour the formation of the more inert-graphitic carbon nanotubes. Based on kinetic and isotopic investigations on the mechanism of DRM reaction over Ni/MgO catalysts, Wei and Iglesia [189] observed a similarity in turnover rates and first-order rate constants with methane decomposition. Hence, they concluded that the dissociation is the kinetically relevant step for the DRM reaction and that the Ni behavior resembles that of supported noble metal catalysts (Rh, Pt, Ir, Ru).

According to Ferreira-Aparicio et al. [190], the supply of surface oxygen species from the hydroxyls of the acidic supports can aid the formation of methoxo (CH_xO) species. Based on FTIR spectroscopy analysis of methane adsorption on alumina, Li et al. [191] observed the presence of two hydroxyl signals at 3 750 and 3 665 cm^{-1} which shifted to 3 707 and 3 640 cm^{-1} upon adsorption of methane. Their results indicate the possibility of weak interaction between methane and surface hydroxyls, a phenomenon also observed with Ir catalysts during methane decomposition [192]. Similarly, on perovskite- and pyrochlore-type catalysts, the lattice oxygen species on the surface were found to assist the methane activation [167, 174, 193].

The same authors have studied the effect of various metal substitutions, and have found that Cu-modified catalysts possess the highest initial activity, whereas the Fe-modified catalysts begin to improve in catalytic activity over time, reaching results analogous to the Cu-substituted catalysts while showing a higher stability. Based on XPS analysis of the various transition LSNM bimetallic perovskites (Fig. 7.12a), it can be observed that lattice oxygen species (O^{2-}) at ca. 528–529 eV exist in all the samples, and present a lower binding energy in LSN(Cu)O compared to the LSN(Fe)O. This shows that the species is more mobile in the LSN (Cu)O sample. On the other hand, the peak area of the deconvoluted O^{2-} species in the LSN(Fe)O is much higher than that of the LSN(Cu)O sample, indicating a greater amount of lattice oxygen species. This phenomenon is generated by the greater difference in electronegativity of Fe (Fe = 1.83, and oxygen = 3.44) compared to Cu (Cu = 1.91, oxygen = 3.44). The authors confirmed their analysis by conducting methane -TPR experiments (Fig. 7.12b) which showed that methane was activated at a higher temperature using the LSN(Fe)O sample compared to the LSN(Cu)O sample. This difference in methane activation temperature correlates with the lattice oxygen mobility of the respective perovskites, as shown by XPS analysis, which eventually correlates with the catalytic stability [167].

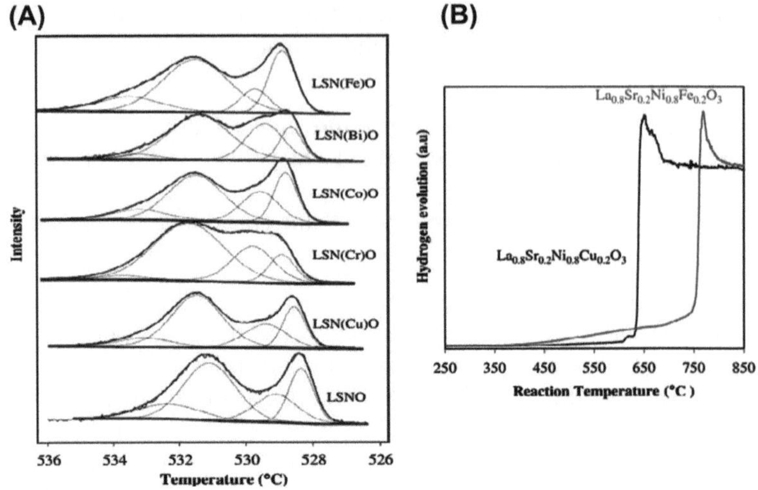

Fig. 7.12 (**a**) XPS profiles of O 1 s of $La_{0.8}Sr_{0.2}Ni_{0.8}M_{0.2}O_3$ perovskites (where M = Bi, Fe, Cr, Cu, and Co). (**b**) CH-TPR profiles of reduced $La_{0.8}Sr_{0.2}Ni_{0.8}Cu_{0.2}O_3$ and $La_{0.8}Sr_{0.2}Ni_{0.8}Fe_{0.2}O_3$ perovskite catalyst precursors [167]

7.2.4.2.2 CO_2 Activation

The dissociation of CO_2 depends on the nature of the catalyst support. On an acidic or inert support such as SiO_2, the CO_2 chemisorption and dissociation occurs on a transition metal surface and is dominated by electron transfer, requiring the formation of an anionic CO_2^- precursor [142]:

$$CO_{2(g)} \leftrightarrows CO_{2(a)} \tag{7.66}$$

$$CO_{2(a)} \leftrightarrows CO_{(a)} + O_{(a)} \tag{7.67}$$

The adsorbed O atoms then proceed to react with either the methyl radicals or adsorbed H atoms as in Scheme 7.18.

$$CH_{x(a)} + O_{(a)} \rightarrow CO_{(a)} + xH_{(a)} \tag{7.68}$$

$$H_{(a)} + O_{(a)} \leftrightarrows OH_{(a)} \tag{7.69}$$

$$H_{(a)} + OH_{(a)} \leftrightarrows H_2O_{(g)} \tag{7.70}$$

$$2H_{(a)} \rightarrow H_{2(g)} \tag{7.71}$$

$$CO_{(a)} \rightarrow CO_{(g)} \tag{7.72}$$

Scheme 7.18 Reaction of adsorbed O atoms (from CO_2 dissociation) with methyl radicals

However, the limitation in CO adsorption, based on the availability of the active metal sites, results in severe accumulation of carbon deposits, which eventually deactivate the catalyst. In the presence of basic catalytic sites or basic supports such as La_2O_3, CO_2 can easily be activated on basic sites, forming lanthanum

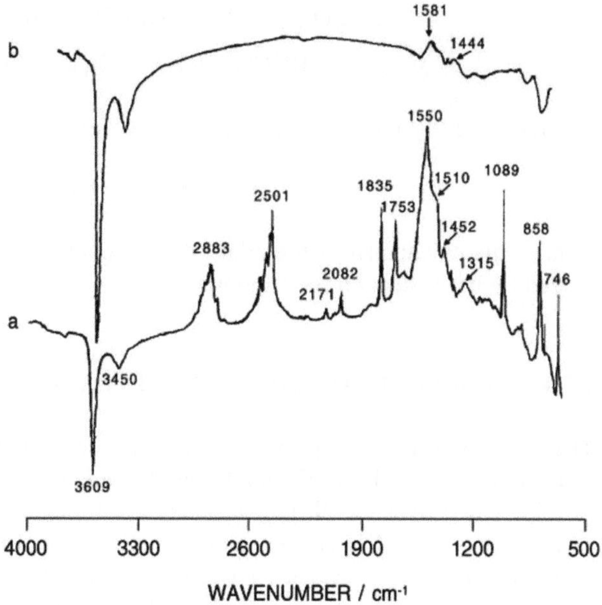

Fig. 7.13 Lanthanum oxycarbonate formation upon exposure of La_2O_3 to (**a**) CO_2 for 2 h; (**b**) methane/CO_2 mixture for 5 h at 1 023 K [295]

oxycarbonate species. For example, on a Ni/La_2O_3 catalyst, CO_2 forms the $La_2O_2CO_3$ phase at the interface with Ni surface and participates in the formation of CO. Under high temperature conditions the oxygen species in the lanthanum oxycarbonate species acts as a scavenger of C and removes deposited carbon. This is demonstrated by FTIR spectra as shown in Fig. 7.13.

Zhang et al. [194] proposed the following mechanistic steps, involving a Ni/La_2O_3 catalyst, which has been found to be applicable to various other lanthanum-based supports [157, 195]:

$$CH_4 + Ni_n \rightarrow C - Ni_n + 2\,H_2 \tag{7.73}$$

$$CO_2 + La_2O_3 \rightarrow La_2O_2CO_3 \tag{7.74}$$

Under the high temperature conditions of DRM reaction (above 873 K), the lanthanum oxycarbonate species formed at the interfacial sites becomes mobile and able to gasify the carbon species formed on the metallic nickel active sites. The La_2O_3 acts as an intermediate that stores CO_2 from the gaseous phase and releases it to the active sites:

$$C - Ni_n + La_2O_2CO_3 \rightarrow 2\,CO + Ni_n + La_2O_3 \tag{7.75}$$

Based on their FTIR study, Zhang et al. [194] observed significant amounts of formate species and considered that they may be formed by hydrogen spillover from the Ni crystallites. Eventually, formates are thermally decomposed into H_2O and CO as follows:

$$CO_{2(a)} + H^+ \rightarrow HCCO^- \quad (support) \tag{7.76}$$

$$HCCO^- + H^+ \rightarrow H_2O + CO \tag{7.77}$$

The resulting H_2O may further react with C forming more Syngas as follows:

$$H_2O + C \rightarrow CO + H_2 \tag{7.78}$$

Hence, this mechanism can explain the removal of carbon and the important role played by basic sites on the supports.

7.2.4.3 Kinetic Modeling for the DRM Process

The rate-determining steps were found to vary depending on the nature of the catalyst. Zhang et al. [194] found that methane activation is the rate-determining step for DRM over the Ni/La_2O_3 catalyst, although the reaction between surface carbon species via CH species and the oxidant (including CO_2 activation) is the rate-determining step over the $Ni/\gamma\text{-}Al_2O_3$ catalyst. On the other hand, Wang and Lu [196] did a comprehensive study on DRM over $Ni/\gamma\text{-}Al_2O_3$ catalyst within the temperature range 773–973 K and showed that methane and CO_2 both chemisorbed on metallic Ni sites with CO formation as the rate-determining step. They proposed the mechanism shown in Scheme 7.19, where [*] denotes the catalyst site.

$$CH_4 + * \rightarrow CH_4* \tag{7.79}$$

$$CO_2 + * \rightarrow CO_2* \tag{7.80}$$

$$CH_4* \rightarrow CH_3* + H \tag{7.81}$$

$$CH_3* \rightarrow CH_2* + H \tag{7.82}$$

$$CH_2* \rightarrow CH* + H \tag{7.83}$$

$$CH* \rightarrow C* + H \tag{7.84}$$

$$C* + CO_2* \rightarrow 2CO + 2* \qquad \textbf{<RDS>} \tag{7.85}$$

$$CO \rightarrow C + O \tag{7.86}$$

$$2H + O \rightarrow H_2O \tag{7.87}$$

$$H + H \rightarrow H_2 \tag{7.88}$$

Scheme 7.19 Proposed LHHW mechanism for DRM reaction over $Ni/\gamma\text{-}Al_2O_3$ catalyst

Wang and Lu [196] observed that the rate is first order in methane, and it is first order in CO_2 at lower partial pressures and zero order at high pressures. This observation shows that, at higher partial pressures, CO_2 may not be the rate-determining factor.

With regards to the role of CO_2 activation on support, Tsipouriari et al. [185] proposed a mechanism of DRM over Ni/La_2O_3 catalyst. They observed that the reversible adsorption of methane on the surface of nickel leads to the decomposition of methane and production of carbon species and hydrogen. $La_2O_2CO_3$ demonstrated the direct affinity of CO_2 towards the basic support. In their proposed steps, the adsorption of methane on the surface of Ni site was considered to be a reversible reaction although the methane decomposition was the rate-limiting step. The interaction between CO_2 and La_2O_3 in the formation of $La_2O_2CO_3$ was considered to be an equilibrium reaction. Finally, the reaction of oxycarbonate species with carbon deposited onto Ni particles was also considered to be a slow step. The overall mechanism is shown in Scheme 7.20, where [*] denotes the catalyst site:

$$CH_4 + * \leftrightarrows CH_4^* \quad (7.89)$$

$$CH_4^* \rightarrow C^* + 2H_2 \quad (7.90)$$ <RDS>

$$CO_2 + La_2O_3 \leftrightarrows La_2O_2CO_3 \quad (7.91)$$

$$La_2O_2CO_3 + C^* \rightarrow La_2O_3 + 2CO + * \quad (7.92)$$ <RDS>

$$H_2 + 2^* \leftrightarrows 2H^* \quad (7.93)$$

$$La_2O_2CO_3 + H^* \leftrightarrows La_2O_3 + CO + OH\text{-}^* \quad (7.94)$$

$$OH\text{-}^* + C^* \leftrightarrows CO^* + H^* \quad (7.95)$$

$$OH^* + H^* \leftrightarrows H_2O + 2^* \quad (7.96)$$

$$CO^* \leftrightarrows CO + * \quad (7.97)$$

$$CO_2 + * \leftrightarrows CO_2^* \quad (7.98)$$

$$CO_2^* + H^* \leftrightarrows CO^* + OH^* \quad (7.99)$$

$$OH^* + H^{*}\bullet \leftrightarrows H_2O + 2^* \quad (7.100)$$

Scheme 7.20 Overall mechanism of DRM reaction over Ni/La_2O_3 with 2 RDS, i.e., methane decomposition to carbon and lanthanum oxycarbonate decomposition to La_2O_3 and CO

7.3 Hydrogenation of CO_2

The thermodynamic stability of CO_2 requires high energy substances or electroreductive processes for its transformation into valuable chemicals in which the carbon atom has a lower oxidation state than 4 [1, 3, 4, 197]. Catalytic hydrogenation of CO_2 has been acknowledged as one of the major potential steps for CO_2 valorization to fuels, or other products (e.g., HCOOH, methanol, H_2CO, and C1+) which are considered to be potential hydrogen carriers or useful chemicals or fuels [1, 198].

Catalytic conversion of CO_2 to high energy chemicals such as light olefins and liquid hydrocarbons can generally proceed via two routes: (1) conversion of CO_2 into liquid hydrocarbons via CO formation (F-T route), which is a combination of CO_2 hydrogenation by RWGS and further hydrogenation of CO to hydrocarbons, and (2) direct (without CO formation) hydrogenation of CO_2 to methanol, DME, higher alcohols, methane, and higher hydrocarbons. A common issue for such processes is the production of hydrogen from renewable sources if large volumes of CO_2 should be converted. Key issues are: (1) the stability of the CO_2 molecule, (2) the amount of H_2 necessary to hydrogenate CO_2, and (3) the kinetics of the reaction [199]. Hence, it is highly crucial to employ suitable catalysts which can decrease the energy barriers, thus resulting in lower energy states of the system. Exploring the mechanistic pathways of such reactions may help to gain a deeper insight into the feasibility of CO_2 conversion at lower temperatures.

7.3.1 *Reverse Water Gas Shift Reaction (RWGS)*

As already stated, Syngas is widely used to produce methanol or other long-chain hydrocarbons. CO, which is considered to be the initial step of CO_2 hydrogenation on metal catalysts [200], can be formed in the RWGS reaction. Such a reaction is crucial for the production of CO, which is an important precursor for various other chemical productions [201].

This endothermic reaction, which occurs at high temperatures (>973 K), is shown in (7.101).

$$CO_2 + H_2 \leftrightarrows CO + H_2O \quad \Delta H^\circ_{298K} = 41.2\ \mathrm{kJ\,mol^{-1}} \tag{7.101}$$

However, several undesired parallel and side reactions tend to occur as well:

$$\text{Sabatier reaction}: CO_2 + 4H_2 \leftrightarrows CH_4 + 2H_2O \quad \Delta H^\circ_{298K} = -165\ \mathrm{kJ\,mol^{-1}} \tag{7.102}$$

$$\text{Boudouard reaction}: 2CO \leftrightarrows C + CO_2 \quad \Delta H^\circ_{298K} = -172.6\ \mathrm{kJ\,mol^{-1}} \tag{7.103}$$

$$\text{Methanation reaction}: CO + 3H_2 \leftrightarrows CH_4 + H_2O \quad \Delta H^\circ_{298K} = -206\ \mathrm{kJ\,mol^{-1}} \tag{7.104}$$

All such reactions have been encountered in the processes discussed above and occur simultaneously any time H_2O, CO, CO_2, H_2, C are present in the reaction medium.

7.3.2 *CO_2 Hydrogenation to Methanol and DME*

7.3.2.1 Methanol

Methanol is a common feedstock considered as a building block in the chemical industry because of its versatility in the synthesis of several important chemicals such as chloromethane, acetic acid, methyl *tert*-butyl ether (MTBE), alkyl halides, and formaldehyde [202, 203]. Methanol can be converted into DME which is attracting much attention as it can be used in fuel cells and diesel engines [204]. In addition, methanol is used for the production of higher hydrocarbons (through the methanol-to-gasoline route [205–207]) or of unsaturated hydrocarbons (through the methanol-to-olefins (MTO) or methanol-to-propene (MTP) routes).

Methanol is currently produced from Syngas as feedstock. The first industrial plant for methanol production was commissioned by BASF in 1923 using Syngas and operated at a high pressure of 20 MPa and 573 K using zinc oxide/chromium oxide catalysts [208].

Increasing concerns of the environmental impact of CO_2 emissions have recently led to increased research interest in the potential use of CO_2 to replace CO in methanol synthesis [209], although it requires one mol of hydrogen in excess. In general, CO_2 is gaining prominence as a substitute for toxic CO, derived from fossil carbon [210], in several processes. Olah et al. [211] have promoted the use of methanol as an alternative way to store, transport and use energy, and have proposed the "Methanol Economy." The hydrogen required for the chemical recycling of CO_2 must be sourced from water (by electrolysis or other cleavage technologies using perennial energy sources) if the reduction of CO_2 emission is targeted. Currently, methane is also an interesting source for methanol as it is quite abundant and can be found in many different environments, such as natural gas deposits, shale gas, coal-bed methane, methane hydrate, and biomethane from agricultural, domestic, and industrial waste [211].

Even though hydrogenation of CO_2 to methanol has favorable thermodynamics, high activation energy barriers must be overcome and this requires the use of appropriate catalysts. Formation of other by-products during CO_2 hydrogenation, such as CO, hydrocarbons, and higher alcohols [4], requires highly selective heterogeneous catalysts with the important function of improving the process sustainability. The reaction for methanol production via CO_2 hydrogenation is

$$CO_2 + 3H_2 \rightarrow CH_3OH + H_2O \quad \Delta H^{\circ}_{298K} = -49.5\ \mathrm{kJ\,mol^{-1}} \tag{7.105}$$

Thermodynamically, a decrease in reaction temperature or an increase in reaction pressure can favor the synthesis of methanol. A reaction temperature higher than 513 K facilitates CO_2 activation and, consequently, methanol formation [4, 5, 197].

7.3.2.2 DME

DME (CH_3OCH_3) is generally formed upon methanol dehydration but can be directly produced by CO_2 hydrogenation. DME is an efficient alternative to petroleum-based transportation fuels (because of its high cetane number) and liquefied natural gas, and has been found to have a low environmental impact because it does not generate sulfur oxides or soot [211, 212]. In the chemical industry, DME is a useful intermediate for the synthesis of chemicals such as methyl acetate, dimethyl acetate, and light olefins [213, 214]. Even though methanol has been used in diesel engines, its low cetane number limits its application as a replacement for diesel fuel. In contrast, DME has about 40 % higher cetane number compared to diesel fuel and is, thus, far superior compared to methanol for such application. Two main pathways bring to DME from CO_2: (1) a two-step process, i.e., methanol is synthesized (7.105) on a suitable metal catalyst, which is then dehydrated on an acid:

$$2CH_3OH \rightarrow CH_3OCH_3 + H_2O \quad \Delta H^\circ_{298K} = -23.4\,kJ\,mol^{-1} \tag{7.106}$$

catalyst (7.106) and, (2) a single-step process, using a bifunctional catalyst producing methanol and causing its dehydration in the same reactor, forming DME [215]. Scheme 7.21 summarizes the thermodynamic properties of the conventional (CO-based) synthesis of DME via the two-step process [213, 216, 217] as compared to the one-step synthesis based on CO_2.

Conventional 2-step synthesis based on CO

(i) Methanol synthesis (eq. 7.107)

$$CO + 2H_2 \rightarrow CH_3OH \qquad \Delta H^\circ_{298K} = -90.6\ kJ.mol^{-1} \tag{7.107}$$

(ii) Methanol dehydration (in the presence of solid acid catalyst)

$$2CH_3OH \rightarrow CH_3OCH_3 + H_2O \qquad \Delta H^\circ_{298K} = -23.4\ kJ.mol^{-1} \tag{7.106}$$

One-step reaction for the synthesis of DME from CO_2

$$2CO_2 + 6H_2 \rightarrow CH_3OCH_3 + 3H_2O \qquad \Delta H^\circ_{298K} = -122.2\ kJ.mol^{-1} \tag{7.108}$$

Scheme 7.21 Conventional synthesis of DME via methanol production and dehydration (two-step) and direct synthesis via CO_2 hydrogenation (one-step)

The one-step process, which integrates the methanol synthesis and its dehydration to DME, reduces the investment and operational cost because only one reactor is required [218]. A typical process flow diagram is shown in Fig. 7.14. However, for the one-step process, catalyst selection is extremely crucial to minimize CO formation via RWGS reaction [217].

7.3.3 Catalytic Systems for CO_2 Hydrogenation

Heterogeneous catalysts are generally used for most CO_2 hydrogenation reactions. A number of investigations have addressed the effects of active components, supports, promoters, preparation methods, and surface morphology on CO_2 hydrogenation. The ternary $CuO/ZnO/Al_2O_3$ mixed oxides are highly effective for the RWGS and methanol/DME synthesis reactions [219–221]. In fact, Cu- and Cu/ZnO-based catalysts, which are active for water gas shift (WGS) reactions, are also active for the RWGS based on the principle of microscopic reversibility [2]. Cu has

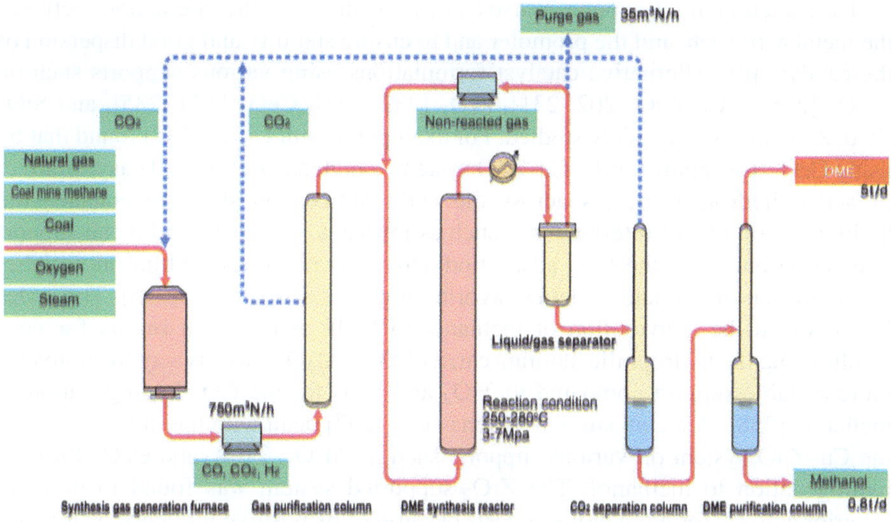

Fig. 7.14 Process flow diagram for single-step DME synthesis from coal/natural gas [296]

good properties for catalyzing the RWGS reaction; hence it aids the preliminary hydrogenation of the CO_2 molecule. The combination of Cu with Zn further accelerates the formation of methanol (because ZnO plays the role of inhibitor of RWGS reaction) [205–207]. In addition, ZnO has a key role in the dispersion and stabilization of the Cu active metal centers. Besides, the presence of lattice oxygen vacancies in ZnO proves to be important for methanol synthesis [4, 202]. Other modifiers such as Zr, Ce, Si, Al, V, Ti, Ga, B, Cr, La, and Fe have also been explored [222].

Cu–ZrO_2-based mixed oxide catalysts were reported to give better performances compared to Cu–ZnO in CO_2 hydrogenation [223–225]. The nanocrystalline form of ZrO_2 leads to higher catalyst surface areas compared to the conventional catalysts prepared via the coprecipitation method and this leads to greater CO_2 conversion and higher selectivity to methanol [226]. Sintering of Cu can lead to poor thermal stability of the catalyst and Cu-based catalysts are deactivated at temperatures higher than 573 K [200]. Cu/ZnO catalysts modified with Ga_2O_3 were found to show superior stability and high catalytic activity and selectivity towards methanol because of the ability of Ga_2O_3 to promote dispersion of Cu^+ species on the surface [227]. By supporting the combination of Cu/ZnO/Ga_2O_3 on hydrophobic silica, high activity and selectivity (>99 %) towards methanol in the temperatures range 523–543 K was obtained [228]. Furthermore, addition of Ga_2O_3 and Cr_2O_3 to Cu/ZnO catalysts can increase the selectivity towards DME production. DME synthesis is usually accompanied by side reactions such as F-T synthesis and WGS [229], and hence high selectivity, high activity, and high stability are extremely important for this process.

The selection of the support is also a key issue in tuning the interactions between the metal active site and the promoter and to ensure stability and good dispersion of the catalyst [4]. Alternative catalyst formulations using various supports such on Al_2O_3 [228, 230], ZnO [202, 231], ZrO_2 [232, 233], CeO_2 [234, 235], and SiO_2 [200, 236] have been widely studied. For example, Bonura et al. [235] found that by using CeO_2 as support for the Cu–ZnO system, a high adsorption and easy spillover of active hydrogen species across the ceria lattice contribute to enhance the hydrogenation rate of intermediates such as formates. The increased metal support interaction enhances the CO_2 adsorption-activation process, resulting in a higher concentration of formate moieties favoring higher reaction rates [235]. The water formation in the dehydration of methanol to DME is the main reason for poor results using the hydrophilic alumina carrier [237]. Al_2O_3 was anyway found to be a more suitable support compared to TiO_2 and SiO_2 for the CO_2 hydrogenation to methanol [238]. A recent study by Arena et al. [237] dealt with the investigation of the Cu–ZnO system on various supports such as Al_2O_3, ZrO_2, and CeO_2 for CO_2 hydrogenation to methanol. The ZrO_2-supported system was found to promote superior activity for methanol synthesis because of its positive influence on textural and chemical effects, in addition to its acid/base features and ability to adsorb CO_2. The addition of M^{3+} ions such as Al^{3+} to Cu–ZnO-based catalysts has the beneficial effect of enhancing the Cu dispersion and the catalyst surface area [239]. With respect to the methanol dehydration pathway for DME production, solid acid catalysts such as HZSM-5, HMCM-49, HMCM-22, SAPOs, and ferrierite are generally used [213]. Several studies have reported that the use of HZSM-5 favors CO_2 hydrogenation and methanol dehydration [240–242]. In the one-step process, the bifunctional catalyst for DME production can create synergetic effects by combining a hydrogenation component with a dehydration component [243]. For example, An et al. [244] reported that CuO–ZnO–Al_2O_3–ZrO_2/HSZM-5 catalyst had a higher activity compared to the CuO–ZnO–Al_2O_3/γ-Al_2O_3 catalyst used for CO_2 hydrogenation to DME under high water concentrations. Addition of ZrO_2 provides some variations in the porosity and microstructural features and can positively affect the surface area, pore size distribution, and the reducibility of the active centers. Besides, the CO_2 hydrogenation capacity of the catalyst can be further improved [235, 245, 246]. The activation of H_2 and CO_2 and the formation-hydrogenation of the formate intermediate are greatly influenced by the neighboring metal and oxide sites reactivity on the Cu–ZnO/ZrO_2 system [235, 247, 248]. The addition of chromium oxides to the Cu/ZrO_2 metal-support combination can positively suppress the sintering of Cu particles and enhance the catalyst thermal stability by shifting the ZrO_2 crystallization towards higher temperatures [249]. The optimum distribution of basic sites on the catalyst is also important in the selectivity towards methanol formation. Basic rare earths such as La_2O_3 can actually increase the Cu surface area which, in turn, effectively enhances the CO_2 conversion [250, 251].

Recently, multi-walled carbon nanotubes (CNT) have drawn increasing attention as catalyst support because of their unique features such as nanometer-sized channels and sp^2-C bonding, possessing high electron- and thermal-conductivity and large surface area. Zha et al. [229] have deposited CuO–ZnO–Al_2O_3 and

HZSM-5 composite nanoparticles outside and inside the wall of carbon nanotubes by ultrasonic assisted coprecipitation and have obtained higher DME yields when the surface of the catalyst was functionalized with H_2SO_4/HNO_3 mixture to enhance the acidity. This system was found to perform better than conventional $CuO–ZnO–Al_2O_3$/HZSM-5 bifunctional catalysts in terms of improvement in CO_2 conversion to DME. Zhang et al. [252] used Pd-decorated-CNT coupled to a Cu–ZrO_2 catalyst admixed with the HZSM-5 zeolite component. The Pd-decorated CNT was found to have excellent capability to adsorb H_2 and CO_2, facilitating the rate of CO_2 hydrogenation, which resulted in it being 1.22 times that of the catalyst without the Pd-CNT promotion.

Silica is commonly used as support because of some desirable properties such as porosity, acidity, and thermal stability [211]. To enhance the strength of catalysts, and especially for lower temperature operations, binders are sometimes added. At higher temperatures, these binders are chemically reactive. For example, adding an alumina binder to an Fe–K/Al_2O_3 catalyst has been shown to improve the activity and selectivity towards higher hydrocarbons (C5+), whereas when using silica as binder, the activity and selectivity dramatically decreased. This was attributed to the change of catalyst acidity and of the structure and metal-support interaction, which influence the activity and selectivity of the catalyst [253].

The most common methods used to enhance the thermal stability and selectivity of catalysts include alloying with other metals. Iron is known to prevent active centers from sintering as well as to favor the dispersion of the active centers on the catalyst. Several studies have found that iron can potentially have a significant role in enhancing the catalyst stability, because reactions producing DME generate more water and thus the catalyst can be affected by the hydrophilic properties of ZnO [228]. The introduction of Fe into a Cu-based catalyst can result in a strong bimetallic interaction between Cu and Fe, which facilitates strong CO_2 and H_2 adsorption, thus promoting the catalytic hydrogenation activity. Iron can act as a textural promoter to prevent Cu sintering at higher temperatures, as well as to protect Cu particles from being oxidized during the reaction [254]. There is an increasing interest to understand better the catalyst structure-function relationships so as to be able to design the optimal catalyst system [255]. Addition of Zr into the $CuO–Fe_2O_3$/HZSM-5 bifunctional catalyst has been found to improve DME selectivity [243]. As mentioned earlier, the catalysts for the one-step CO_2 hydrogenation to DME should be able to catalyze efficiently both the methanol synthesis and the methanol dehydration, keeping the yield of CO formed via RWGS reaction to the minimum level [218]. After alloying with noble metals such as Pd, the Cu–Zn–Al–Zr/HZSM-5 catalyst has been found to enhance remarkably DME production, whilst retarding CO formation. This was attributed to the spillover effect of hydrogen from Pd to the neighboring phase [256].

Cobalt catalysts are widely used for the traditional Fisher–Tropsch synthesis based on CO and H_2. Despite being active for CO_2 hydrogenation, Co catalysts have been found to promote the formation of undesired methane [257]. By replacing the Syngas feed-stream with CO_2 and H_2, it was found that cobalt is prone to be more selective towards methanation rather than methanol or other alcohol production [4, 258, 259]. However, it is important to ensure that the catalyst composition is such to minimize the methanation reaction for an efficient and economically feasible liquid-fuel synthesis based on CO_2 hydrogenation. Methane is considered a "low value" molecule, which unnecessarily consumes 4 mol of hydrogen per CO_2 [238]. Mixed Fe/Co catalysts have also been shown to give low selectivity towards the desired hydrocarbons [4]. Based on the study carried out by Akin et al. [260] on Co/Al_2O_3 catalyst, considering the higher amount of methane in the product stream, the CO_2 hydrogenation pathway was proposed to imply surface bound intermediates of H–C–O and O–H. Likewise, nickel is also found to favor methanation reactions [261–264]. For example, Chang et al. [262] have reported that Ni, when highly dispersed on amorphous silica, can afford a high turnover frequency for methane production. On suitable supports, such as ceria, it has been reported that high activity and stability for RWGS reaction can be achieved. According to Wang et al. [265], the main active sites for the RWGS reaction are the vacancies present in the CeO_2 as well as the highly dispersed Ni species. The incorporation of Ni in the CeO_2 lattice can also modulate the properties of the catalyst [266]. Recently, Zonetti et al. [266] have synthesized the $Ni_xCe_{0.75}Zr_{0.25-x}O_2$ solid solution, where Ni was dissolved in the lattice structure by coprecipitation or impregnation techniques. The high activity and stability for 60 h was attributed to the active sites being the interface of Ni metal with the NiCeZr interface. Based on temperature programmed desorption studies, the authors postulate that the catalysts seem to promote both associative (related to carboxyl and formate intermediates) as well as redox mechanisms.

Several innovative catalyst preparation routes have also been adopted by different research groups to improve the activity, selectivity, and stability of the catalytic systems for CO_2 hydrogenation reactions. The recent trend is to produce and use nanoparticles, which have higher surface area compared to conventional catalysts. The superior activity of Cu nanoparticles compared to conventionally prepared Cu (111) for methanol and other liquid hydrocarbon synthesis relate to the presence of more active corner sites as well as structural flexibilities, which can reduce the activation energy of the rate-determining steps. For example, Chen et al. [200] have used the atomic layer epitaxy (ALE) method to improve the thermal stability of Cu/SiO_2 catalysts for RWGS reaction. This ALE preparation method proved to be effective in the formation of small Cu nanoparticles, which could strongly bind CO, resulting in high activity for RWGS reaction [200]. Another method is the modification by functionalization of Cu on SiO_2 using organic precursors such as oleic acid and butyric acid. This facile and cheap preparation method was reported by Mo et al. [267] who tested the catalytic performance of the catalysts in the WGS reaction. Coteron and Hayhurst [268] found that the "spark erosion" preparation method resulted in high methanol selectivity. The sol gel preparation technique was found to boost the total and metal surface areas [223, 269], whereas the "combustion" route could result in enhanced turnover frequency [231]. On the other hand, the oxalate coprecipitation method has been found to exhibit fine and uniformly distributed catalyst particles, resulting in high catalytic activity for CO_2 hydrogenation to methanol [270, 271]. It has also been established that the citrate precursor method can result in solid catalyst formation with high surface area [272, 273]. Via this method, a homogeneous gel (containing the metal precursors) is formed by complexing the metal nitrate salts with citric acid. Hence, such a coprecipitation technique can ensure that the various precursors are adequately mixed during the solid formation step [224, 274]. Using simple solid state reaction techniques, Guo et al. [275] have prepared Cu/ZnO/ZrO_2 catalysts from hydrated metal salts using citric acid as ligand, and have found that the calcination temperature greatly affects the catalytic activity in the reaction of CO_2 hydrogenation to methanol. The calcination temperature of 673 K gave the optimum performance, whereas above 873 K an undesirable phase transformation of ZnO from tetragonal to monoclinic takes place, which decreases the dispersion of the Cu species.

7.3.4 *Reaction Mechanism for CO_2 Hydrogenation*

In the CO_2 hydrogenation, three reactions are main competitors, namely RWGS, methanol synthesis, and DME production [244, 276, 277]. The methanol and DME production reactions have a slower rate in reaching thermodynamic equilibrium compared to RWGS reaction.

7.3.4.1 Reaction Mechanism for RWGS

It has generally been assumed that most of the reaction mechanisms that occur for WGS are also applicable for RWGS. The two main mechanisms proposed for WGS reactions are (1) surface redox mechanism and (2) dissociative mechanism of water to afford OH and H species, with subsequent formation of formate species (HCO_2–M, obtained by an attack of OH groups on M–CO moieties) or metalla carboxyl species (MO_2H, which might result from the reaction of M–H moieties with CO_2 or, even better, by protonation of metal-bound CO_2; see Chaps. 2 and 4) as intermediates [200].

7.3.4.1.1 Surface Redox Mechanism

The main concept of surface redox mechanism is represented by the oxidation and reduction cycle of the active sites. For example, with Cu-based catalysts, Cu^0 atoms provide active sites for the dissociation of CO_2 (7.109). The reduction of the oxidized Cu catalyst (7.110) should be faster than the oxidation process [200, 278] according to Scheme 7.22 [219].

$$CO_{2(g)} + 2Cu^{\circ}_{(s)} \rightarrow CO_{(g)} + Cu_2O_{(s)} \tag{7.109}$$

$$H_{2(g)} + Cu_2O_{(s)} \rightarrow H_2O_{(g)} + 2Cu^{\circ}_{(s)} \tag{7.110}$$

Scheme 7.22 Oxidation of Cu metal by CO_2 and subsequent hydrogenation to form water

Gines et al. [219] have studied the kinetics and mechanism involved in RWGS over commercial and laboratory $CuO/ZnO/Al_2O_3$-based catalysts. Their results show that the composition of the H_2/CO_2 mixture has a tremendous role, as the reaction shifts from first order in H_2 to first order in CO_2 depending on the pressure of the two gases [219]. Hence, they surmised that the surface redox mechanism pathway is formed by the presence of subsurface hydrogen trapped on the reconstructed copper surface caused by oxygen overlayer as also postulated by Waugh [279]. This serves in making the surface more reactive to CO_2 adsorption. The changes of the surface reactivity and of the rate-limiting reagents were taken into consideration in the formulation of the redox mechanism reported in Scheme 7.23.

$$CO_{2(g)} + L \rightleftharpoons CO_{(g)} + O.L \qquad r = k_1 P_{CO_2} C_L - k_1' P_{CO} C_{OL} \tag{7.111}$$

$$H_{2(g)} + 2L \rightleftharpoons 2H.L \qquad K_2 = \frac{C_{HL}^2}{C_L^2 P_{H_2}} \tag{7.112}$$

$$2H.L + O.L \rightleftharpoons H_2O_{(g)} + 3L \qquad K_3 = \frac{P_{H_2O} C_L^3}{C_{HL}^2 C_{OL}} \tag{7.113}$$

Where L = **reaction site**

Scheme 7.23 Simplified redox mechanism proposed for RWGS reaction [219]

7.3.4.1.2 Dissociative Mechanism

The dissociative mechanism involving formate intermediates has been widely considered as the major route for CO production in the RWGS reaction [280–282]. With Cu as the catalyst active site, the formate species derived from association of H_2 and CO_2 is mainly proposed to be the key intermediate for CO production [281, 282]. Several studies have been carried out to improve the formate species decomposition to CO [282–284].

For example, Chen et al. [283] have added potassium to Cu/SiO_2 and have found that the enhancement of CO_2 adsorption on K_2O is the important factor, which results in improved catalytic activity for RWGS reaction. By using temperature programmed desorption (TPD) of H_2/CO_2, they suggested the mechanism reported in Scheme 7.24. According to them, H atoms are produced on copper and spill over to associate with CO_2 adsorbed on K_2O, forming formates which can be decomposed to CO and H_2O. Yoshihara and Campbell [282] investigated a similar system and proposed the possibility that formate species are formed on potassium, migrate to Cu, and then decompose to CO:

$$K - CO_2 + Cu - H \rightarrow K - (HCOO) + Cu \rightarrow K + Cu - (HCOO) \tag{7.114}$$

It was also postulated that CO_2 adsorbed on K_2O may also migrate to Cu and then associate with H atom forming formate species [282].

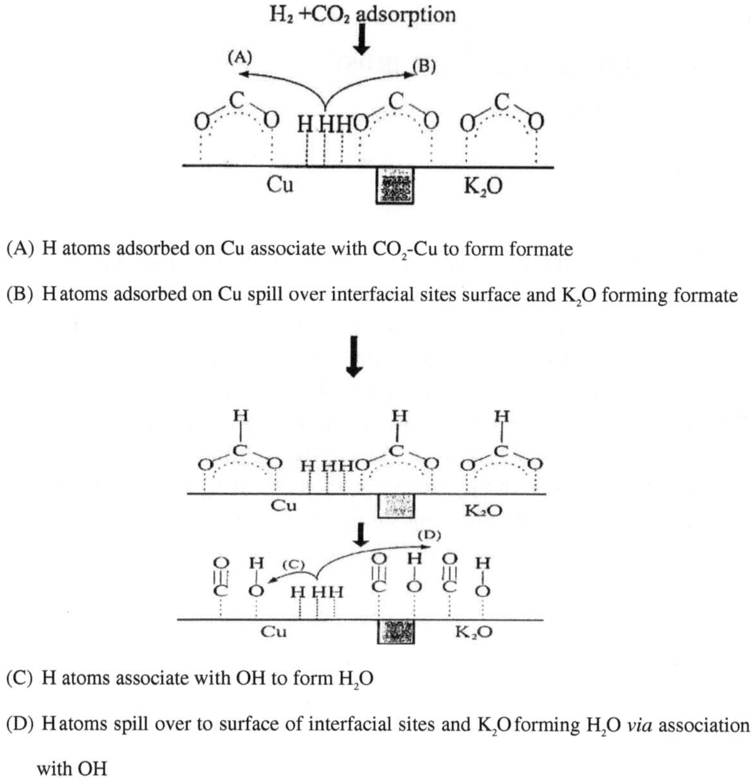

Scheme 7.24 Role of K_2O in RWGS reaction over Cu/SiO_2 catalyst [283]

7.3.4.2 Reaction Mechanism for Methanol Synthesis

Despite the numerous theories on the mechanistic pathways for the methanol synthesis, and the extensive investigation, there is still uncertainty about the definition of the role of Cu°, Cu^+, Cu–Zn alloy, and the carrier sites [222]. In the past, it has been assumed that CO hydrogenation is the main pathway for methanol synthesis. However, isotopic labeling experiments conducted by Chinchen et al. [285] have revealed that methanol is formed directly from CO_2, whilst CO scavenges the oxygen atoms which hinder the active metal sites [237, 286].

According to earlier proposed mechanisms, formate hydrogenation on Cu/ZnO catalysts is the rate-limiting steps. This was based on data reported by Fujitani et al. [287] who found that the formate coverage on Cu/ZnO catalysts is proportional to the turnover frequency for methanol formation. The proposed reaction pathway is shown in Scheme 7.25.

$$CO_2 + \frac{1}{2} H_2 \leftrightarrows HCOO_{(a)} \quad (7.115)$$

$$HCOO_{(a)} + 2H_{(a)} \leftrightarrows CH_3O_{(a)} + O_{(a)} \qquad \textbf{(RDS)} \quad (7.116)$$

$$CH_3O_{(a)} + H_{(a)} \leftrightarrows CH_3OH \quad (7.117)$$

Scheme 7.25 Proposed reaction pathway with formate hydrogenation as rate-limiting step on Cu/ZnO catalyst system

Arena et al. [288] have conducted a detailed study to probe the solid state interactions, adsorption sites, and functionality of ZnO-promoted Cu/ZrO_2 systems for CO_2 hydrogenation to methanol. They found that ZnO promotes the Cu dispersion because of strong Cu–ZnO interaction, whereas presence of ZnO on the ZrO_2 support markedly enhances the surface for CO_2 adsorption. Based on various FTIR and TPD studies, they have proposed that the hydrogenation of the reactive formate intermediate on ZnO and ZrO_2 surface sites in the vicinity of Cu is the rate-limiting step. The metallic Cu sites, the hydroxyl groups of ZrO_2, and the basic Lewis sites of ZnO were proposed to play vital roles in the main CO_2 hydrogenation pathway as shown in the proposed mechanism in Scheme 7.26 [288].

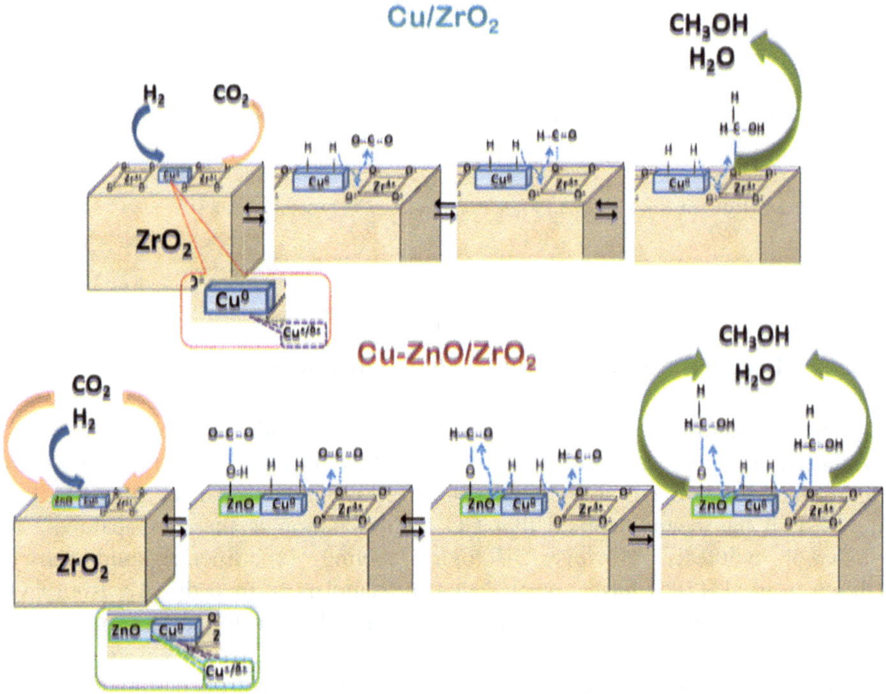

Scheme 7.26 Sketch of catalysts surface and functionality of Cu/ZrO_2-based sites in the CO_2 hydrogenation to methanol [288]

Arena et al. [237] have recently further extended their previous work on Cu–Zn/ ZrO_2 and carried out some kinetic studies. They have demonstrated that methanol is the main reaction product at pressure > 0.1 MPa and low temperature (≤473 K), whereas CO is formed via both parallel RWGS and consecutive methanol decomposition paths, mainly at higher temperatures (>473 K) (Scheme 7.27). The authors have emphasized the essential role of the metal/oxide interface for the CO_2 hydrogenation, enabling the adsorption of CO_2 on basic sites of ZnO and ZrO_2 in the vicinity of the hydrogenation sites provided by Cu^0. The recent kinetic studies correlate to the originally proposed mechanism in which the rate-limiting step was suggested to be the formation and hydrogenation of the dioxomethylene reactive intermediate on ZnO and ZrO_2 surface sites in the interface with Cu. It should be noted that the dioxomethylene intermediate, which plays such a key role according to the authors, is only rarely mentioned by other authors and has not been identified, neither via experimental nor theoretical studies. It seems certain that the support may play a key role in driving the mechanism in different directions.

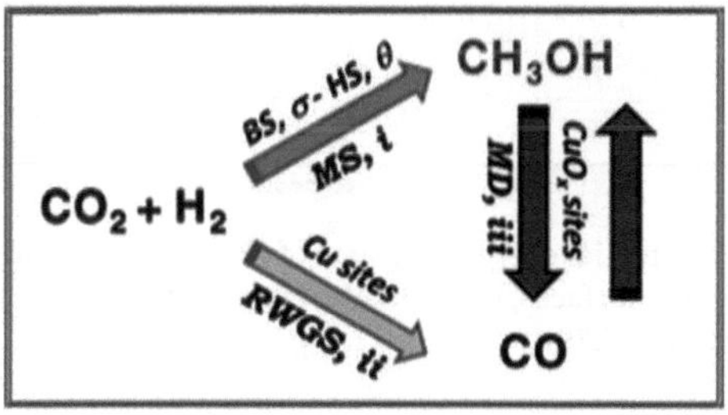

Scheme 7.27 CO_2 hydrogenation reactions network of supported Cu–ZnO catalysts [237]

Zhang et al. [289] have used DFT studies to gain insight on the effect of surface hydroxyls on CO_2 hydrogenation over Pd/γ-Al_2O_3 and have deciphered the elementary steps leading to the formation of HCOO and CO, which are important intermediates for methanol and methane formation. The results are summarized in Scheme 7.28 as a simplified potential energy diagram considering different sites: i.e., Pd4 cluster on dry-γAl_2O_3 support (D(Pd4)), Pd4 cluster on hydroxylated-γ Al_2O_3 support (H(Pd4)), and ideal Pd(111) surface. The highest barrier for CO_2 conversion to HCOO and CO is found on four-atom Pd (Pd4) clusters. Such data show that CO is the initial product of CO_2 hydrogenation and therefore the Pd cluster on a dry alumina surface is likely to lead to dominant formation of methane. However, for the Pd cluster on hydroxylated alumina, HCOO is the initial product and hence can lead to dominant formation of methanol. On the other hand, for the Pd(111) surface, both CO and HCOO are dominant products of the initial CO_2 hydrogenation, meaning there is no obvious selectivity towards either methane or methanol production. Their results also show that the surface hydroxyls, despite an initial greater selectivity towards methanol, result in weakened Pd-support interaction, thus destabilizing the catalyst. It was, thus, proposed that the Pd/γ Al_2O_3 catalyst system requires modification and promoters in order to improve the Pd dispersion or to control the hydroxylation prevalent on alumina-based supports.

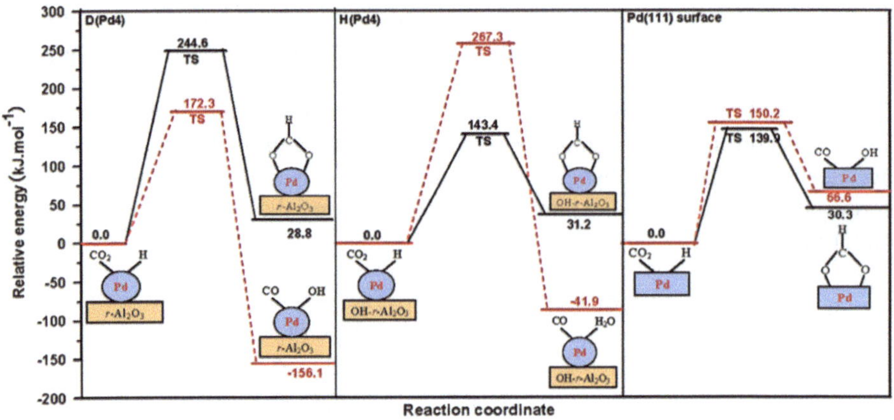

Scheme 7.28 Schematic potential energy diagrams of CO_2 hydrogenation to HCO_2 and to CO on D(Pd4), H(Pd4), and Pd(111) surfaces, respectively. Only the transition state (TS) with the highest barrier for each path is shown [289]

Using Pd/Ga_2O_3 systems [290–292], it was established that the reaction intermediates were chemisorbed onto gallia forming (bi)carbonates, which were successively hydrogenated to mono-, bidentate- and bridged-formate, methylenebisoxy and methoxy species as shown in Scheme 7.29 [292]. It was revealed that the bimetallic Pd-Ga particles were responsible for the CO_2 hydrogenation to methanol and it was postulated that the methoxo/methanol formation from H_2/CO_2 took place as a consequence of monodentate formate hydrogenation, whereas the bridged formate behaved more as a spectator or laggard reaction intermediate [291]. These authors also invoke the methylenebisoxy species as an intermediate in the hydrogenation of formate to the methoxo moiety.

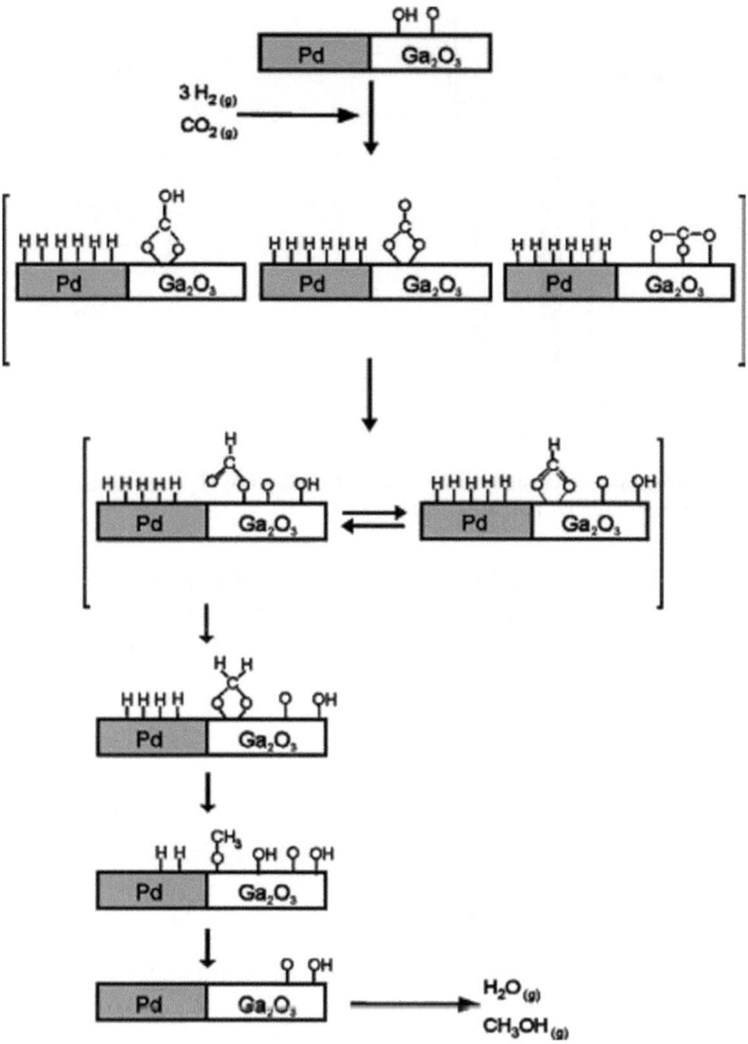

Scheme 7.29 Proposed reaction pathway for methanol synthesis from CO_2 hydrogenation over Pd/Ga catalyst [292]

Initial research work generally focused on the industrial Cu–ZnO/Al_2O_3 systems because of favourable economics of using alumina as catalyst carrier for industrial applications. Hence, based on such a configuration, Sun et al. [293] utilized the DRIFTS spectroscopy method (Fig. 7.15) and observed that methanol was directly formed from CO_2 hydrogenation. Based on their proposed mechanism shown in Scheme 7.30, in the absence of CO, dissociative adsorption of CO_2 takes place on the catalyst surface, with a greater tendency to go through pathway (I) in Scheme 7.30. Even if the reaction proceeds through pathway (II) (Scheme 7.30), the surface $(O{=}C{-}OH)_s$ species could decompose to CO_s and OH_s because of the absence of CO gas phase. However, in the presence of CO in the CO_2/H_2 system, the reaction proceeds towards pathway (II) to form $(O{=}C{-}OH)_s$, which then

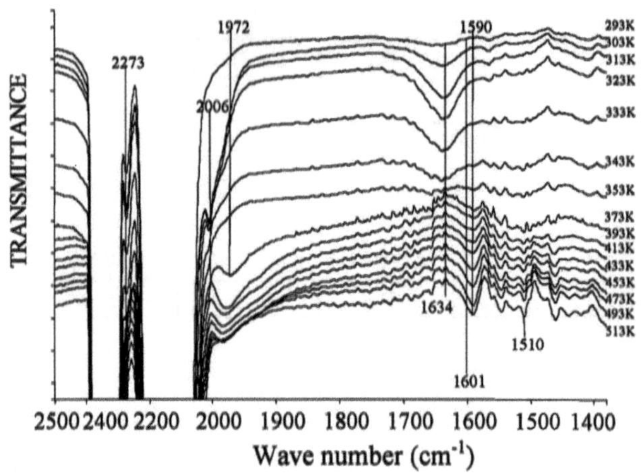

Fig. 7.15 IR spectra for methanol synthesis from CO/CO_2 and H_2 obtained during temperature reaction programmed studies at 2 MPa [293]

isomerizes to form bidentate formate species that cannot easily decompose to CO_s and OH_s, thus enhancing the methanol production rate. These authors too mention the methylenebisoxy species as key intermediate towards the formation of methoxo species from formate.

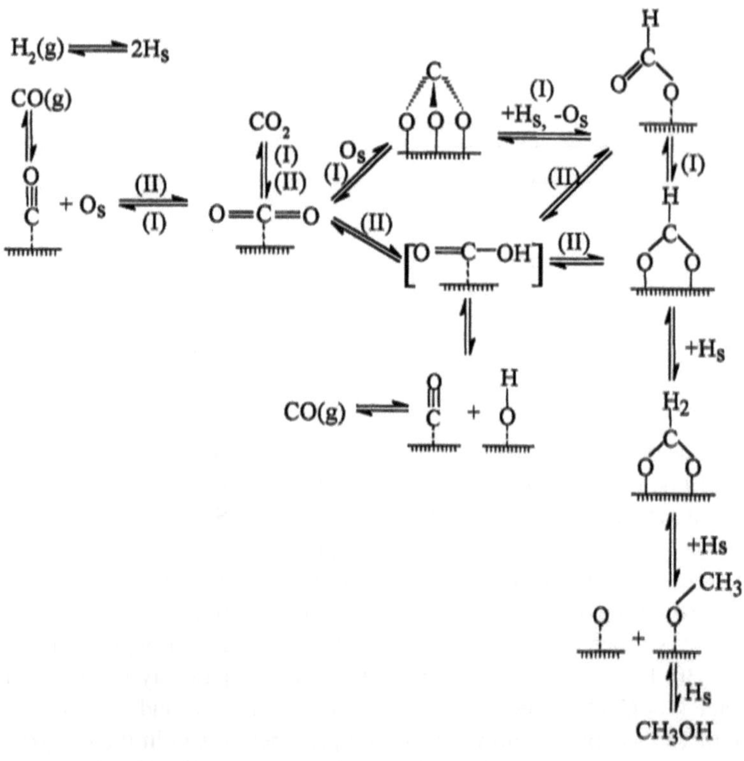

Scheme 7.30 Mechanism of methanol synthesis from CO_2/H_2 and $CO/CO_2/H_2$ over ultrafine $Cu/ZnO/Al_2O_3$ catalyst [293]

On the other hand, with incorporation of ZrO_2 in the catalyst system, Lim et al. [246] have developed a kinetic model for methanol synthesis for Cu/ZnO/Al_2O_3/ZrO_2 based on the rate-determining steps for the various complex reaction networks. Such a model was tested and applied for various CO_2 partial pressures. On the basis of their experimental data (as shown in Scheme 7.31), the rate-determining steps were found to be the surface reaction of a methoxo species (for CO hydrogenation reaction), the hydrogenation of a formate (HCO_2) intermediate (for CO_2 hydrogenation reaction), and the formation of a formate intermediate (for the WGS reaction), respectively. However, a certain optimum threshold of CO_2 as a reactant was found to be required in order to ensure maximum methanol synthesis.

Surface Reaction	Elementary Steps	
(A) CO hydrogenation reaction	Step 1: $CO \cdot s_1 + H \cdot s_2 \rightleftharpoons HCO \cdot s_1 + s_2$	
	Step 2: $HCO \cdot s_1 + H \cdot s_2 \rightleftharpoons H_2CO \cdot s_1 + s_2$	
	Step 3: $H_2CO \cdot s_1 + H \cdot s_2 \rightleftharpoons H_3CO \cdot s_1 + s_2$	
	Step 4: $H_3CO \cdot s_1 + H \cdot s_2 \rightleftharpoons CH_3OH + s_1 + s_2$	(RDS)
(B) water–gas shift reaction	Step 1: $CO_2 \cdot s_3 + H \cdot s_2 \rightleftharpoons HCO_2 \cdot s_3 + s_2$	(RDS)
	Step 2: $HCO_2 \cdot s_3 + H \cdot s_2 \rightleftharpoons CO \cdot s_3 + H_2O \cdot s_2$	
(C) CO_2 hydrogenation reaction	Step 1: $CO_2 \cdot s_3 + H \cdot s_2 \rightleftharpoons HCO_2 \cdot s_3 + s_2$	
	Step 2: $HCO_2 \cdot s_3 + H \cdot s_2 \rightleftharpoons H_2CO_2 \cdot s_3 + s_2$	(RDS)
	Step 3: $H_2CO_2 \cdot s_3 + H \cdot s_2 \rightleftharpoons H_3CO_2 \cdot s_3 + s_2$	
	Step 4: $H_3CO_2 \cdot s_3 + H \cdot s_2 \rightleftharpoons H_2CO \cdot s_3 + H_2O \cdot s_2$	
	Step 5: $H_2CO \cdot s_3 + H \cdot s_2 \rightleftharpoons H_3CO \cdot s_3 + s_2$	
	Step 6: $H_3CO \cdot s_3 + H \cdot s_2 \rightleftharpoons CH_3OH + s_3 + s_2$	

Scheme 7.31 Surface reactions involved in the methanol synthesis from CO_2 and the corresponding elementary and rate-determining steps (RDS)

7.3.4.3 Reaction Mechanism for DME Synthesis

Study of the mechanism for the one-step CO_2 hydrogenation to DME is still in its infancy, as most studies are devoted to the optimization of the bifunctional catalyst. Chen et al. [213] have conducted a kinetic and thermodynamic study on the DME production via the direct synthesis route over a bifunctional Cu–ZnO–Al_2O_3/HZSM-5 catalyst at various pressures. Their studies reveal that DME production is governed by two different factors above or below 498 K. At the lower temperatures (<498 K), the reaction is dominated by chemical kinetics, whereas for the higher temperatures (>498 K), the thermodynamic equilibrium dominates the reaction.

Based on a recent study on the Cu–ZnO–ZrO_2/H-ZSM-5 system, the reaction pathway is proposed to be a multi-site reaction, taking place with the primary formation of methanol and the consecutive production of DME on the acid sites of the zeolite framework from methanol dehydration [218]. According to Frusteri et al. [218], on the Cu–ZnO–ZrO_2/H-ZSM-5 system, H_2 is initially adsorbed and activated (H_2^*) on the Cu° sites, thus resulting in reduction of the Cu species, whereas CO_2 is bridge adsorbed on the basic surface sites of ZnO and ZrO_2 forming activated CO_2 (CO_2^*). Thereafter, the H_2^* spillover to CO_2^* leads to formation of intermediate species (such as formate/dioxomethylene/methoxy) stabilized at the metal-oxide interface prior to evolving as methanol from the hydrogenation reaction [228, 294].

1) $CO_2 + 3H_2 \leftrightarrows CH_3OH + H_2O$

2) $CO_2 + H_2 \leftrightarrows CO + H_2O$

3) $2CH_3OH \leftrightarrows CH_3OCH_3 + H_2O$

4) $CO + 2H_2 \leftrightarrows CH_3OH$

Scheme 7.32 Reaction scheme for the direct CO_2 hydrogenation to DME [217]

Based on the thermodynamic evaluation by Bonura et al. [217], the conversion of methanol to DME on the acidic zeolite sites rapidly approaches the equilibrium level at all the investigated temperatures because it is a fast reaction. Moreover, they also observed a promoting effect of the reaction temperature on the relative rate of methanol synthesis from CO_2 rather than CO, on the basis of thermodynamics. Hence, based on their findings, the authors have proposed the reaction in Scheme 7.32. It turns out that, apart from the methanol dehydration reaction, 3 which takes place very fast at any temperature, reactions 1–2–4 affect the methanol formation depending on their relative reaction rates. The thermodynamic analysis of the main reaction paths involved in the synthesis of DME by CO_2 hydrogenation reveals that, under kinetic conditions, the CO concentration dramatically increases with temperature; therefore, low reaction temperature and recycling unreacted CO_x/H_2 mixtures could be a solution to improving the methanol/DME productivity.

7.4 Conclusions

CO_2 can act either as a "soft oxidant" for DH for olefin production or can be hydrogenated itself to afford CO, methanol (or methane), and DME. Such reactions make a complex network, inter-related not only in terms of reaction pathways (Scheme 7.33) but also in terms of the catalysts used.

Most of the high temperature reactions associated with the use of CO_2 as oxidant involve the role of CO_2 as mediator in saturated hydrocarbon conversion into olefin with the release of hydrogen. The reaction pathways are mainly determined by the nature and properties of the catalyst, i.e., redox properties or acid–base properties, and supports. For example, a redox-based catalyst such as ceria or vanadia would promote the redox-type mechanism to take place. Prevalent issues for the production of olefins are the low selectivity and yield. Several studies have shown that selectivity can be enhanced using CO_2 as "soft-oxidant." Further improvement of conversion yield and selectivity is a focal research point. On the other hand, the role of CO_2 in hydrocarbon reforming (albeit at higher temperatures than DH for olefin production) is an interesting route for Syngas production. The catalysts used are mainly nickel-based (which are more active in reforming than in olefin production). The main issue in this reaction, which produces Syngas with a H_2:CO ratio equal to 1, is the catalyst deactivation caused by carbon formation. Recent advances in the synthesis of highly dispersed nanocatalysts appear to be promising in terms of carbon resistance. Co-reforming the hydrocarbon with CO_2, oxygen and steam (tri-reforming process) is a strategy for increasing the H_2/CO ratio of the Syngas in less energy-requiring conditions, because the POM contributes heat to the reacting system.

Today, Syngas is used for downstream production of methanol (which is a building block for the production of other chemicals). The simplest CO_2 hydrogenation reaction is the RWGS reaction, which converts CO_2 into CO, that can then be used for methanol production. Instead of going through the Syngas route, methanol and DME can be produced directly via CO_2 hydrogenation. The most widely used catalysts for CO_2 hydrogenation are Cu-based. This route seems to be more efficient. More and more evidence of its existence has recently been gained, changing the belief that CO_2 is first converted into CO and then the Syngas route is followed. Catalysts that are good for RWGS can also be modified and utilized for direct methanol synthesis from CO_2. Methanol molecules can undergo further dehydration to produce DME in a separate process, but the direct conversion of CO_2 to DME in a single reactor is now being considered. Such an approach requires modification of the methanol synthesis catalysts so as to have acid functions to promote DME formation.

As a conclusion, the use of CO_2 as a carbon source for the synthesis of various energy rich molecules via the high temperature reactions discussed in this chapter shows great potential for industrial exploitation on a large scale. Rational designs of active catalysts play a key role in terms of improvement towards reaction activity and selectivity. The use of solar energy via solar power concentrators would be highly beneficial in this field which deals with endoergonic reactions carried out at high temperature.

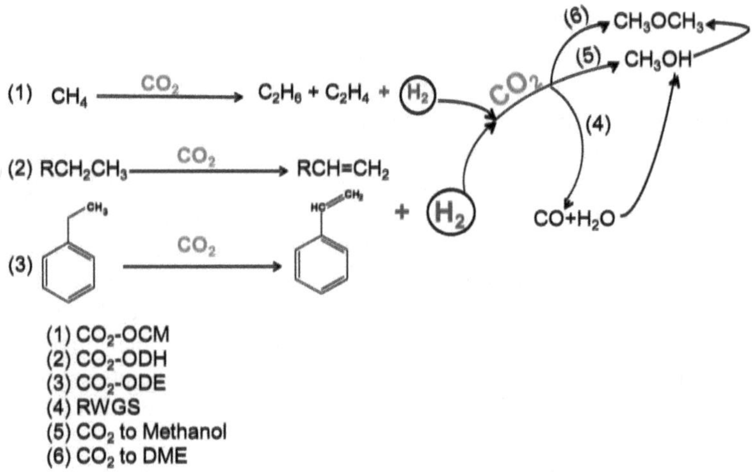

Scheme 7.33 Network of reactions showing the role of CO_2 in dehydrogenation or hydrogenation reactions

References

1. Aresta M, Dibenedetto A, Angelini A (2014) Catalysis for the valorization of exhaust carbon: from CO_2 to chemicals, materials and fuels. Technological use of CO_2. Chem Rev 114:1709–1742
2. Styring P, Armstrong K (2011) Catalytic carbon dioxide conversions to value-added chemicals. Chem Today 29:28–31
3. Chueh WC, Falter C, Abbott M, Scipio D, Furler P, Haile SM, Steinfeld A (2010) High-flux solar-driven thermochemical dissociation of CO_2 and H_2O using nonstoichiometric ceria. Science 330:1797–1801
4. Wang W, Wang S, Ma X, Gong J (2011) Recent advances in catalytic hydrogenation of carbon dioxide. Chem Soc Rev 40:3703–3727
5. Zou JJ, Liu CJ (2010) Utilization of carbon dioxide through nonthermal plasma approaches. In: Aresta M (ed) Carbon dioxide as chemical feedstock. Wiley, New York, pp 267–289
6. Aresta M, Dibenedetto A (2004) The contribution of the utilization option to reducing the CO_2 atmospheric loading: research needed to overcome existing barriers for a full exploitation of the potential of the CO_2 use. Catal Today 98:455–462
7. Colmenares JC (2010) Novel trends in the utilization of CO_2 as a reagent and mild oxidant in the C-C coupling reactions. Curr Org Synth 7:533–542
8. Tang P, Zhu Q, Wu Z, Ma D (2014) Methane activation: the past and future. Energy Environ Sci 7:2580–2591
9. Zaman J (1999) Oxidative processes in natural gas conversion. Fuel Process Technol 58:61–81
10. Corey EJ, Cheng XM (1989) The logic of chemical synthesis. Wiley, New York, pp P1–P456
11. Seshan K (2010) Oxidative conversion of lower alkanes to olefins. Catalysis 22:119–143
12. Wolf EE (ed) (1992) Methane conversion by oxidative processes: fundamental and engineering aspects. Van Nostrand Reinhold, New York, pp P1–P548

13. Ansari MB, Park SE (2012) Carbon dioxide utilization as a soft oxidant and promoter in catalysis. Energy Environ Sci 5:9419–9437
14. Dury F, Gaigneaux EM, Ruiz P (2003) The active role of CO_2 at low temperature in oxidation processes: the case of the oxidative dehydrogenation of propane on $NiMoO_4$ catalysts. Appl Catal Gen 242:187–203
15. Lunsford JH (1995) The catalytic oxidative coupling of methane. Angew Chem Int Ed 34:970–980
16. Dubois JL, Cameron CJ (1990) Common features of oxidative coupling of methane cofeed catalysts. Appl Catal 67:49–71
17. Choudhary VR, Mulla SAR, Uphade BS (1999) Oxidative coupling of methane over alkaline earth oxides deposited on commercial support precoated with rare earth oxides. Fuel 78:427–437
18. Palermo A, Vazquez JPH, Lee AF, Tikhov MS, Lambert RM (1998) Critical influence of the amorphous silica-to-cristobalite phase transition on the performance of $Mn/Na_2WO_4/SiO_2$ catalysts for the oxidative coupling of methane. J Catal 177:259–266
19. Pak S, Lunsford JH (1998) Thermal effects during the oxidative coupling of methane over $Mn/Na_2WO_4/SiO_2$ and $Mn/Na_2WO_4/MgO$ catalysts. Appl Catal Gen 168:131–137
20. Wang D, Rosynek MP, Lunsford JH (1995) Oxidative coupling of methane over oxide-supported sodium-manganese catalysts. J Catal 155:390–402
21. Yu L, Hou R, Liu X, Xue J, Li S (1998) Performance of Na_2WO_4-Mn/SiO_2 catalyst for conversion of CH_4 with CO_2 into C_2 hydrocarbons and its mechanism. Stud Surf Sci Catal 119:307–311
22. Suzuki T, Wada K, Watanabe Y (1990) Effects of carbon dioxide and catalyst preparation on the oxidative dimerization of methane. Appl Catal 59:213–225
23. Tong Y, Rosynek MP, Lunsford JH (1990) The role of sodium carbonate and oxides supported on lanthanide oxides in the oxidative dimerization of methane. J Catal 126:291–298
24. Wang D, Xu M, Shi C, Lunsford JH (1993) Effect of carbon dioxide on the selectivities obtained during the partial oxidation of methane and ethane over Li^+/MgO catalyst. Catal Lett 18:323–328
25. Wang Y, Ohtsuka Y (2001) Mn-based binary oxides as catalysts for the conversion of methane to C_2 hydrocarbons with carbon dioxide as oxidant. Appl Catal Gen 219:183–193
26. Chen C, Xu Y, Li G, Guo X (1996) Oxidative coupling of methane by carbon dioxide: a highly C_2 selective La_2O_3/ZnO catalyst. Catal Lett 42:149–153
27. Xu Y, Yu L, Guo X (1997) Effect of basicity and adding CO_2 in the feed on the oxidative coupling of methane over K_2O and SrO promoted La_2O_3/ZnO catalysts. Appl Catal Gen 164:47–57
28. Oshima K, Tanaka K, Yabe T, Kikuchi E, Sekine Y (2013) Oxidative coupling of methane using carbon dioxide in an electric field over La–ZrO_2 catalyst at low external temperature. Fuel 107:879–881
29. Oumghar A, Legrand JC, Diamy AM, Turillon N (1995) Methane conversion by an air microwave plasma. Plasma Chem Plasma Process 15:87–107
30. Wang Y, Takahashi Y, Ohtsuka Y (1998) Carbon dioxide-induced selective conversion of methane to C_2 hydrocarbons on CeO_2 modified with CaO. Appl Catal Gen 172:L203–L206
31. Wang Y, Takahashi Y, Ohtsuka Y (1999) Carbon dioxide as oxidant for the conversion of ethane to ethane and ethylene using modified CeO_2 catalysts. J Catal 186:160–168
32. Istadi I, Amin NAS (2006) Synergistic effect of catalyst basicity and reducibility on performance of ternary CeO_2-based catalyst for CO_2-OCM to C_2 hydrocarbons. J Mol Catal A Chem 259:61–66
33. Sanderson RT (1960) Chemical periodicity. Physical and inorganic chemistry textbook series. Chapman & Hall, Reinhold
34. He Y, Yang B, Cheng G (2004) On the oxidative coupling of methane with carbon dioxide over CeO_2/ZnO nanocatalysts. Catal Today 98:595–600

35. Aika K, Nishiyama T (1988) Utilisation of CO_2 in the oxidative coupling of methane over PbO–MgO and PbO–CaO. J Chem Soc Chem Commun 1:70–71
36. Asami K, Kusakabe K, Ashi N, Ohtsuka Y (1997) Synthesis of ethane and ethylene from methane and carbon dioxide over praseodymium oxide catalysts. Appl Catal Gen 156:43–56
37. Raju G, Reddy BM, Park SE (2014) CO_2 promoted oxidative dehydrogenation of *n*-butane over $VO_x/MO_2–ZrO_2$ (M = Ce or Ti) catalysts. J CO_2 Util 5:41–46
38. Urlan F, Marcu IC, Sandulescu I (2008) Oxidative dehydrogenation of *n*-butane over titanium pyrophosphate catalysts in the presence of carbon dioxide. Catal Commun 9:2403–2406
39. Qiao A, Kalevaru VN, Radnik J, Düvel A, Heitjans P, Hari Kumar AS, Sai Prasad PS (2014) Oxidative dehydrogenation of ethane to ethylene over V_2O_5/Al_2O_3 catalysts: effect of source of alumina on the catalytic performance. Ind Eng Chem Res. doi:10.1021/ie5008344
40. Cavaliere VN, Crestani MG, Pinter B, Pink M, Chen C, Baik M, Mindiola DJ (2011) Room temperature dehydrogenation of ethane to ethylene. J Am Chem Soc 133:10700–10703
41. Rangel MC, Monteiro APM, Oportus M, Reyes P, Ramos MS, Lima SB (2012) Ethylbenzene dehydrogenation in the presence of carbon dioxide over metal oxides. In: Liu G (ed) Greenhouse gases-capturing, utilization and reduction. In Tech, Rijeka, pp 117–136
42. Liu H, Zhang Z, Li H, Huang Q (2011) Intrinsic kinetics of oxidative dehydrogenation of propane in the presence of CO_2 over Cr/MSU-1 catalyst. J Nat Gas Chem 20:311–317
43. Nakagawa K, Kajita C, Ikenaga NO, Gamo MN, Ando T, Suzuki T (2003) Dehydrogenation of light alkanes over oxidized diamond-supported catalysts in the presence of carbon dioxide. Catal Today 84:149–157
44. Raju G, Reddy BM, Abhishek B, Mo YH, Park SE (2012) Synthesis of C4 olefins from n-butane over a novel $VO_x/SnO_2–ZrO_2$ catalyst using CO_2 as soft oxidant. Appl Catal Gen 423–424:168–175
45. Grzybowska B (1997) Active centres on vanadia-based catalysts for selective oxidation of hydrocarbons. Appl Catal Gen 157:409–420
46. Solymosi F, Nemeth R, Oszko A (2001) The oxidative dehydrogenation of propane with CO_2 over supported Mo_2C catalyst. Stud Surf Sci Catal 136:339–344
47. Michorczyk P, Pietrzyk P, Ogonowski J (2012) Preparation and characterization of SBA-1–supported chromium oxide catalysts for CO_2 assisted dehydrogenation of propane. Microporous Mesoporous Mater 161:56–66
48. Geske M, Korup O, Horn R (2013) Resolving kinetics and dynamics of a catalytic reaction inside a fixed bed reactor by combined kinetic and spectroscopic profiling. Catal Sci Technol 3:169–175
49. Sato S, Ohhara M, Sodesawa T, Nozaki F (1988) Combination of ethylbenzene dehydrogenation and carbon dioxide shift-reaction over a sodium oxide/alumina catalyst. Appl Catal 37:207–215
50. Deng S, Li H, Li S, Zhang Y (2007) Activity and characterization of modified Cr_2O_3/ZrO_2 nano-composite catalysts for oxidative dehydrogenation of ethane to ethylene with CO_2. J Mol Catal A Chem 268:169–175
51. Evans OR, Bell AT, Tilley TD (2004) Oxidative dehydrogenation of propane over vanadia-based catalysts supported on high-surface-area mesoporous $MgAl_2O_4$. J Catal 226:292–300
52. Collins SE, Baltanas MA, Bonivardi AL (2006) Infrared spectroscopic study of the carbon dioxide adsorption on the surface of Ga_2O_3 polymorphs. J Phys Chem B 110:5498–5507
53. Takehira K, Ohishi Y, Shishido T, Kawabata T, Takaki K, Zhang Q, Wang Y (2004) Behavior of active sites on Cr-MCM-41 catalysts during the dehydrogenation of propane with CO_2. J Catal 224:404–416
54. Zhang X, Yue Y, Gao Z (2002) Chromium oxide supported on mesoporous SBA-15 as propane dehydrogenation and oxidative dehydrogenation catalysts. Catal Lett 83:19–25
55. Zhao X, Wang X (2006) Oxidative dehydrogenation of ethane to ethylene by carbon dioxide over Cr/TS-1 catalysts. Catal Commun 7:633–638

56. Mimura N, Okamoto M, Yamashita H, Oyama ST, Murata K (2006) Oxidative dehydrogenation of ethane over Cr/ZSM-5 catalysts using CO_2 as an oxidant. J Phys Chem B 110:21764–21770
57. Liu L, Li H, Zhang Y (2007) Mesoporous silica-supported chromium catalyst: characterization and excellent performance in dehydrogenation of propane to propylene with carbon dioxide. Catal Commun 8:565–570
58. Weckhuysen BM, Schoonheydt RA (1999) Alkane dehydrogenation over supported chromium oxide catalysts. Catal Today 51:223–232
59. Cavani F, Koutyrev M, Trifiro F, Bartolini A, Ghisletti D, Iezzi R, Santucci A, Del Piero G (1996) Chemical and physical characterization of alumina-supported chromia-based catalysts and their activity in dehydrogenation of isobutene. J Catal 158:236–250
60. Botavina MA, Martra G, Agafonov YA, Gaidai NA, Nekrasov NV, Trushin DV, Coluccia S, Lapidus AL (2008) Oxidative dehydrogenation of C_3–C_4 paraffins in the presence of CO_2 over CrO_x/SiO_2 catalysts. Appl Catal Gen 347:126–132
61. Shi X, Ji S, Wang K, Li C (2008) Oxidative dehydrogenation of ethane with CO_2 over novel Cr/SBA-15/Al_2O_3/FeCrAl monolithic catalysts. Energy Fuel 22:3631–3638
62. Liu L, Li H (2006) A comparative study on catalytic performances of chromium incorporated and supported mesoporous MSU-x catalysts for the oxidehydrogenation of ethane to ethylene with carbon dioxide. Catal Today 115:235–241
63. Baek J, Yun HJ, Yun D, Choi Y, Yi J (2012) Preparation of highly dispersed chromium oxide catalysts supported on mesoporous silica for the oxidative dehydrogenation of propane using CO_2: insight into the nature of catalytically active chromium sites. ACS Catal 2:1893–1903
64. Zhang W, Glomski B, Pauly TR, Pinnavaia TJ (1999) A new nonionic surfactant pathway to mesoporous molecular sieve silicas with long range framework order. Chem Commun 18:1803–1804
65. Peña MA, Fierro JLG (2001) Chemical structures and performance of perovskite oxides. Chem Rev 101:1981–2018
66. Pak S, Qiu P, Lunsford JH (1998) Elementary reactions in the oxidative coupling of methane over Mn/Na_2WO_4/$SiO_2$and Mn/Na_2WO_4/MgO catalysts. J Catal 179:222–230
67. Zhu J, Qin S, Ren S, Peng X, Tong D, Hu C (2009) Na_2WO_4/Mn/SiO_2 catalyst for oxidative dehydrogenation of ethane using CO_2 as oxidant. Catal Today 148:310–315
68. Blasco T, Nieto JML, Dejoz A, Vazaquez MI (1995) Influence of the acid–base character of supported vanadium catalysts on their catalytic properties for the oxidative dehydrogenation of *n*-butane. J Catal 157:271–282
69. Liu YM, Cao Y, Yi N, Feng WL, Dai WL, Yan SR, He HY, Fan KN (2004) Vanadium oxide supported on mesoporous SBA-15 as highly selective catalysts in the oxidative dehydrogenation of propane. J Catal 224:417–428
70. Madeira LM, Aranda RMM, Hodar FJM, Fierro JLG, Portela MF (1997) Oxidative dehydrogenation of *n*-butane over alkali and alkaline earth-promoted α-$NiMoO_4$ catalysts. J Catal 169:469–479
71. Postole G, Chowdhury B, Pinki K, Banerji J, Auroux A (2010) Knoevenagel condensation reaction over acid–base bifunctional nanocrystalline $Ce_xZr_{1-x}O_2$ solid solutions. J Catal 269:110–121
72. de Lima SM, da Cruz IO, Jacobs G, Davis BH, Mattos LV, Noronha FB (2008) Steam reforming, partial oxidation, and oxidative steam reforming of ethanol over Pt/$CeZrO_2$ catalyst. J Catal 257:356–368
73. Karamullaoglu G, Dogu T (2007) Oxidative dehydrogenation of ethane over chromium – vanadium mixed oxide and chromium oxide catalysts. Ind Eng Chem Res 46:7079–7086
74. Solsona B, Blasco T, Nieto JML, Peña ML, Rey F, Vidal-Moya A (2001) Vanadium oxide supported on mesoporous MCM-41 as selective catalysts in the oxidative dehydrogenation of alkanes. J Catal 203:443–452

75. Michorczyk P, Kuśtrowski P, Kolak A, Zimowska M (2013) Ordered mesoporous Ga_2O_3 and Ga_2O_3–Al_2O_3 prepared by nanocasting as effective catalysts for propane dehydrogenation in the presence of CO_2. Catal Commun 35:95–100
76. Michorczyk P, Ogonowski J (2003) Dehydrogenation of propane to propene over gallium oxide in the presence of CO_2. Appl Catal Gen 251:425–433
77. Aresta M, Fragale C, Quaranta E, Tommasi I (1992) Carbon dioxide as modulator of the oxidative properties of dioxygen in the presence of transition metal systems. J Chem Soc Chem Commun 4:315–317
78. Chang JS, Vislovskiy VP, Park MS, Hong DY, Yoo JS, Park SE (2003) Utilization of carbon dioxide as soft oxidant in the dehydrogenation of ethylbenzene over supported vanadium–antimony oxide catalysts. Green Chem 5:587–590
79. Ge S, Liu C, Zhang S, Li Z (2003) Effect of carbon dioxide on the reaction performance of oxidative dehydrogenation of *n*-butane over V-Mg-O catalyst. Chem Eng J 94:121–126
80. Sattler JJHB, Ruiz-Martinez J, Santillan-Jimenez E, Weckhuysen BM (2014) Catalytic dehydrogenation of light alkanes on metals and metal oxides. Chem Rev 114 (20):10613–10653
81. Solymosi F, Nemeth R (1999) The oxidative dehydrogenation of ethane with CO_2 over Mo_2C/SiO_2 catalyst. Catal Lett 62:197–200
82. Lezla O, Bordes E, Courtine P, Hecquet G (1997) Synergetic effects in the Ni-Mo-O system: influence of preparation on catalytic performance in the oxidative dehydrogenation of propane. J Catal 170:346–356
83. Dury F, Centeno MA, Gaigneaux EM, Ruiz P (2003) An attempt to explain the role of CO_2 and N_2O as gas dopes in the feed in the oxidative dehydrogenation of propane. Catal Today 81:95–105
84. Blasco T, Nieto JML (1997) Oxidative dehydrogenation of short chain alkanes on supported vanadium oxide catalysts. Appl Catal Gen 157:117–142
85. Chen KD, Khodakov A, Yang J, Bell AT, Iglesia E (1999) Isotopic tracer and kinetic studies of oxidative dehydrogenation pathways on vanadium oxide catalysts. J Catal 186:325–333
86. Xu B, Zheng B, Hua W, Yue Y, Gao Z (2006) Support effect in dehydrogenation of propane in the presence of CO_2 over supported gallium oxide catalysts. J Catal 239:470–477
87. Chen M, Xu J, Su FZ, Liu YM, Cao Y, He HY, Fan KN (2008) Dehydrogenation of propane over spinel-type gallia–alumina solid solution catalysts. J Catal 256:293–300
88. Chen M, Xu J, Cao Y, He HY, Fan KN (2010) Dehydrogenation of propane over In_2O_3–Al_2O_3 mixed oxide in the presence of carbon dioxide. J Catal 272:101–108
89. Kathiraser Y, Wang Z, Yang NT, Zahid S, Kawi S (2013) Oxygen permeation and stability study of $La_{0.6}Sr_{0.4}Co_{0.8}Ga_{0.2}O_{3-\delta}$ (LSCG) hollow fiber membrane with exposure to CO_2, CH_4 and He. J Membr Sci 427:240–249
90. Meriaudeau P, Naccache C (1990) The role of Ga_2O_3 and proton acidity on the dehydrogenating activity of Ga_2O_3-HZSM-5 catalysts: evidence of a bifunctional mechanism. J Mol Catal A Chem 59:L31–L36
91. Nakagawa K, Kajita C, Ikenaga N, Suzuki T, Kobayashi T, Gamo NN, Ando T (2003) The role of chemisorbed oxygen on diamond surfaces for the dehydrogenation of ethane in the presence of carbon dioxide. J Phys Chem B 107:4048–4056
92. Zheng B, Hua WM, Yue YH, Gao Z (2005) Dehydrogenation of propane to propene over different polymorphs of gallium oxide. J Catal 232:143–151
93. Cavani F, Trifiro F (1995) Alternative processes for the production of styrene. Appl Catal Gen 133:219–239
94. Wang S, Murata K, Hayakawa T, Suzuki K (2000) Dehydrogenation of ethane with carbon dioxide over supported chromium oxide catalysts. Appl Catal Gen 196:1–8
95. Peng X, Zhu J, Yao L, Hu C (2013) Effect of methane co-feeding on the selectivity of ethylene produced from oxidative dehydrogenation of ethane with CO_2 over a Ni-La/SiO_2 catalyst. J Energy Chem 22:653–658

96. Michorczyk P, Ogonowski J, Kustrowski P, Chmielarz L (2008) Chromium oxide supported on MCM-41 as a highly active and selective catalyst for dehydrogenation of propane with CO_2. Appl Catal Gen 349:62–69
97. Michorczyk P, Ogonowski J, Zenczak K (2011) Activity of chromium oxide deposited on different silica supports in the dehydrogenation of propane with CO_2—a comparative study. J Mol Catal A Chem 349:1–12
98. Ohishi Y, Kawabata T, Shishido T, Takaki K, Zhang Q, Wang Y, Takehira K (2005) Dehydrogenation of ethylbenzene with CO_2 over Cr-MCM-41 catalyst. J Mol Catal A Chem 230:49–58
99. Deng S, Li S, Li H, Zhang Y (2009) Oxidative dehydrogenation of ethane to ethylene with CO_2 over Fe-Cr/ZrO_2 catalysts. Ind Eng Chem Res 48:7561–7566
100. Weckhuysen BM, Schoonheydt RA, Jehna JM, Wachs CSJ, Ryoo R, Kijlstra S, Poels E (1995) Combined DRS–RS–EXAFS–XANES–TPR study of supported chromium catalysts. J Chem Soc Faraday Trans 91:3245–3253
101. Lund CRF, Kubsh JE, Dumesic JA (1985) Water gas shift over magnetite-based catalysts: nature of active sites for adsorption and catalysis. In: Solid state chemistry in catalysis, Chap 19. American Chemical Society, Washington, DC, pp. 313–318
102. Castro AJR, Soares JM, Filho JM, Oliveira AC, Edwin AC, Milet RC (2013) Oxidative dehydrogenation of ethylbenzene with CO_2 for styrene production over porous iron-based catalysts. Fuel 108:740–748
103. Liu BS, Rui G, Chang RZ, Au CT (2008) Dehydrogenation of ethylbenzene to styrene over $LaVO_x$/SBA-15 catalysts in the presence of carbon dioxide. Appl Catal Gen 335:88–94
104. http://www.essentialchemicalindustry.org/polymers/polyphenylethene.html
105. http://www.plastemart.com/upload/literature/styrenics-styrene-business-further-restructuring-consolidation-merger.asp
106. Jiang N, Han DS, Park SE (2009) Direct synthesis of mesoporous silicalite-1 supported TiO_2–ZrO_2 for the dehydrogenation of EB to styrene with CO_2. Catal Today 141:344–348
107. Devoldere KR, Froment GF (1999) Coke formation and gasification in the catalytic dehydrogenation of ethylbenzene. Ind Eng Chem Res 38:2626–2633
108. De Araujo JCS, Sousa CBA, Oliveira AC, Freire FNA, Ayala AP, Oliveira AC (2010) Dehydrogenation of ethylbenzene with CO_2 to produce styrene over Fe-containing ceramic composites. Appl Catal Gen 377:55–63
109. Irún O, Sadosche SA, Lasobras J, Soler J, Francés E, Herguido J, Menéndez M (2013) Catalysts for the production of styrene from ethylbenzene: redox and deactivation study. Catal Today 203:53–59
110. Adams CR, Jennings TJ (1970) Catalytic oxidations with sulfur dioxide: II. Alkylaromatics. J Catal 17:157–177
111. Mimura N, Saito M (2000) Dehydrogenation of ethylbenzene to styrene over Fe2O3/Al2O3 catalysts in the presence of carbon dioxide. Catal Today 55:173–178
112. Liao SJ, Chen T, Miao CX, Yang WM, Xie ZK, Chen QL (2008) Effect of TiO_2 on the structure and catalytic behavior of iron–potassium oxide catalyst for dehydrogenation of ethylbenzene to styrene. Catal Commun 9:1817–1821
113. Mimura N, Saito M (1999) Dehydrogenation of ethylbenzene to styrene over Fe_2O_3/Al_2O_3 catalysts in the presence of carbon dioxide. Catal Lett 58:59–62
114. Badstube T, Papp H, Kustrowski P, Dziembaj R (1998) Oxidative dehydrogenation of ethylbenzene with carbon dioxide on alkali-promoted Fe/active carbon catalysts. Catal Lett 55:169–172
115. Ji M, Chen G, Wang J, Wang X, Zhang T (2010) Dehydrogenation of ethylbenzene to styrene with CO_2 over iron oxide-based catalysts. Catal Today 158:464–469
116. Zhu XM, Schon M, Bartmann U, van Veen AC, Muhler A (2004) The dehydrogenation of ethylbenzene to styrene over a potassium-promoted iron oxide-based catalyst: a transient kinetic study. Appl Catal Gen 266:99–108

117. Wang Q, Li X, Li W, Feng J (2014) Promoting effect of Fe in oxidative dehydrogenation of ethylbenzene to styrene with CO_2 (I) preparation and performance of $Ce_{1-x}Fe_xO_2$ catalyst. Catal Commun 50:21–24
118. Li KZ, Wang H, Wei YG, Yan DX (2009) Selective oxidation of carbon using iron-modified cerium oxide. J Phys Chem C 113:15288–15297
119. Reddy GK, Boolchand P, Smirniotis PG (2012) Unexpected behavior of copper in modified ferrites during high temperature WGS reaction—aspects of $Fe^{3+} \leftrightarrow Fe^{2+}$ redox chemistry from Mössbauer and XPS studies. J Phys Chem C 116:11019–11031
120. Sakurai Y, Suzaki T, Ikenaga N, Suzuki T (2000) Dehydrogenation of ethylbenzene with an activated carbon-supported vanadium catalyst. Appl Catal Gen 192:281–288
121. Carja G, Nakamura R, Aida T, Niiyama H (2003) Mg–V–Al mixed oxides with mesoporous properties using layered double hydroxides as precursors: catalytic behavior for the process of ethylbenzene dehydrogenation to styrene under a carbon dioxide flow. J Catal 218:104–110
122. Li W, Li X, Feng J (2009) Behaviors of V-doped titanium mixed oxides in the catalytic dehydrogenation of ethylbenzene. Catal Lett 130:575–582
123. Rao KN, Reddy BM, Abishek B, Seo YH, Jiang N, Park SE (2009) Effect of ceria on the structure and catalytic activity of $V_2O_5/TiO_2–ZrO_2$ for oxidehydrogenation of ethylbenzene to styrene utilizing CO_2 as soft oxidant. Appl Catal Environ 91:649–656
124. Burri A, Jiang N, Ji M, Park SE, Khalid Y (2013) Oxidative dehydrogenation of ethylbenzene to styrene with CO_2 over $V_2O_5–Sb_2O_5–CeO_2/TiO_2–ZrO_2$ catalysts. Top Catal 56:1724–1730
125. Reddy BM, Jin H, Han DS, Park SE (2008) Oxidative dehydrogenation of ethylbenzene to styrene with carbon dioxide over $Fe_2O_3/TiO_2–ZrO_2$ catalyst: influence of chloride. Catal Lett 124:357–363
126. Burri DR, Choi KM, Han SC, Burri A, Park SE (2007) Selective conversion of ethylbenzene into styrene over K_2O/TiO_2-ZrO_2 catalysts: unified effects of K_2O and CO_2. J Mol Catal A Chem 269:58–63
127. Burri DR, Choi KM, Lee JH, Han DS, Park SE (2007) Influence of SBA-15 support on CeO_2–ZrO_2 catalyst for the dehydrogenation of ethylbenzene to styrene with CO_2. Catal Commun 8:43–48
128. Qiao Y, Miao C, Yue Y, Xie Z, Yang W, Hua W, Gao Z (2009) Vanadium oxide supported on mesoporous MCM-41 as new catalysts for dehydrogenation of ethylbenzene with CO_2. Microporous Mesoporous Mater 119:150–157
129. Sun A, Qin Z, Wang J (2002) Reaction coupling of ethylbenzene dehydrogenation with water-gas shift. Appl Catal Gen 234:179–189
130. Sun A, Qin Z, Chen S, Wang J (2004) Role of carbon dioxide in the ethylbenzene dehydrogenation coupled with reverse water–gas shift. J Mol Catal A Chem 210:189–195
131. Saito K, Okuda K, Ikenaga N, Miyake T, Suzuki T (2010) Role of lattice oxygen of metal oxides in the dehydrogenation of ethylbenzene under a carbon dioxide atmosphere. J Phys Chem A 114:3845–3854
132. Satterfield CN (1980) Heterogeneous catalysis in practice. McGraw-Hill, New York, pp P1–P460
133. Sakurai Y, Suzaki T, Nakagawa K, Ikenaga N, Aota H, Suzuki T (2002) Dehydrogenation of ethylbenzene over vanadium oxide-loaded MgO catalyst: promoting effect of carbon dioxide. J Catal 209:16–24
134. Chen S, Qin Z, Wang G, Dong M, Wang J (2013) Promoting effect of carbon dioxide on the dehydrogenation of ethylbenzene over silica-supported vanadium catalysts. Fuel 109:43–48
135. Liu ZW, Wang C, Fan WB, Liu ZT, Hao QQ, Long X, Lu J, Wang JG, Qin ZF, Su DS (2011) V_2O_5/ $Ce_{0.6}Zr_{0.4}O_2–Al_2O_3$ as an efficient catalyst for the oxidative dehydrogenation of ethylbenzene with carbon dioxide. ChemSusChem 4:341–345
136. Baron M, Abbott H, Bondarchuk O, Stacchiola D, Uhl A, Shaikhutdinov S, Freund HJ, Popa C, Ganduglia-Pirovano MV, Sauer J (2009) Resolving the atomic structure of vanadia

monolayer catalysts: monomers, trimers, and oligomers on ceria. Angew Chem Int Ed 121:8150–8153
137. Nederlof N, Talay G, Kapteijn F, Makkee M (2012) The role of RWGS in the dehydrogenation of ethylbenzene to styrene in CO_2. Appl Catal Gen 423–424:59–68
138. He XX, Fan C, Gu XY, Zhou XG, Chen D, Zhu YA (2011) Role of CO_2 in ethylbenzene dehydrogenation over Fe_2O_3(0 0 0 1) from first principles. J Mol Catal A Chem 344:53–61
139. Chang JS, Park SE, Park MS (1997) Beneficial effect of carbon dioxide in dehydrogenation of ethylbenzene to styrene over zeolite-supported iron oxide catalyst. Chem Lett 26:1123–1124
140. Broda M, Kierzkowska AM, Baudouin D, Imtiaz Q, Copéret C, Müller CR (2012) Sorbent-enhanced methane reforming over a Ni–Ca-based, bifunctional catalyst sorbent. ACS Catal 2:1635–1646
141. Choudhary VR, Mondal KC (2006) CO_2 reforming of methane combined with steam reforming or partial oxidation of methane to syngas over $NdCoO_3$ perovskite-type mixed metal-oxide catalyst. Appl Energy 83:1024–1032
142. Bradford MCJ, Vannice MA (1999) CO_2 reforming of CH_4. Catal Rev Sci Eng 41:1–42
143. Zhu X, Huo P, Zhang Y, Cheng D, Liu C (2008) Structure and reactivity of plasma treated Ni/Al_2O_3 catalyst for CO_2 reforming of methane. Appl Catal Environ 81:132–140
144. Liu CJ, Ye J, Jiang J, Pan Y (2011) Progresses in the preparation of coke resistant Ni-based catalyst for steam and CO_2 reforming of methane. ChemCatChem 3:529–541
145. Rostrup-Nielsen J, Trimm DL (1977) Mechanisms of carbon formation on nickel-containing catalysts. J Catal 48:155–165
146. Song C, Pan W (2004) Tri-reforming of methane: a novel concept for catalytic production of industrially useful synthesis gas with desired H_2/CO ratios. Catal Today 98:463–484
147. Jiang HT, Li HQ, Zhang Y (2007) Tri-reforming of methane to syngas over Ni/Al_2O_3–thermal distribution in the catalyst bed. J Fuel Chem Technol 35:72–78
148. Rostrup-Nielsen J, Bak-Hansen JH (1993) CO_2-reforming of methane over transition metals. J Catal 144:38–49
149. Bitter JH, Seshan K, Lercher JA (1999) Deactivation and coke accumulation during CO_2/CH_4 reforming over Pt catalysts. J Catal 183:336–343
150. Nagaoka K, Seshan K, Aika K, Lercher JA (2001) Carbon deposition during carbon dioxide reforming of methane—comparison between Pt/Al_2O_3 and Pt/ZrO_2. J Catal 197:34–42
151. Gao J, Hou Z, Lou H, Zheng X (2011) Dry (CO_2) reforming. In: Fuel cells: technologies for fuel processing, Chap. 7. Elsevier, Oxford, pp 191–221
152. Ruckenstein E, Hu YH (1996) Role of support in CO_2 reforming of CH_4 to syngas over Ni catalysts. J Catal 162:230–238
153. Pakhare D, Schwartz V, Abdelsayed V, Haynes D, Shekhawat D, Poston J, Spivey J (2014) Kinetic and mechanistic study of dry (CO_2) reforming of methane over Rh-substituted $La_2Zr_2O_7$ pyrochlores. J Catal 316:78–92
154. Pakhare D, Spivey J (2014) A review of dry (CO_2) reforming of methane over noble metal catalysts. Chem Soc Rev. doi:10.1039/C3CS60395D
155. Hu YY, Ruckenstein E (2004) Catalytic conversion of methane to synthesis gas by partial oxidation and CO_2 reforming. Adv Catal 48:297–345
156. Pan YX, Kuai P, Liu Y, Ge Q, Liu CJ (2010) Promotion effects of Ga_2O_3 on CO_2 adsorption and conversion over a SiO_2-supported Ni catalyst. Energy Environ Sci 3:1322–1325
157. Sutthiumporn K, Kawi S (2011) Promotional effect of alkaline earth over Ni–La_2O_3 catalyst for CO_2 reforming of CH_4: role of surface oxygen species on H_2 production and carbon suppression. Int J Hydrogen Energy 36:14435–14446
158. Ni J, Chen L, Lin J, Kawi S (2012) Carbon deposition on borated alumina supported nano-sized Ni catalysts for dry reforming of CH_4. Nano Energy 1:674–686
159. Sankar M, Dimitratos N, Miedziak PJ, Wells PP, Kiely CJ, Hutchings GJ (2012) Designing bimetallic catalysts for a green and sustainable future. Chem Soc Rev 41:8099–8139

160. Oemar U, Hidajat K, Kawi S (2011) Role of catalyst support over PdO–NiO catalysts on catalyst activity and stability for oxy-CO_2 reforming of methane. Appl Catal Gen 402:176–187
161. Zhang J, Wang H, Dalai AK (2007) Development of stable bimetallic catalysts for carbon dioxide reforming of methane. J Catal 249:300–310
162. Zhang J, Wang H, Dalai AK (2008) Effects of metal content on activity and stability of Ni-Co bimetallic catalysts for CO_2 reforming of CH_4. Appl Catal Gen 339:121–129
163. Tejuca LG, Fierro JLG, Tascón JMD, Tejuca LG (1989) Structure and reactivity of perovskite-type oxides. Adv Catal 36:237–328
164. Corthals S, Van Nederkassel J, Geboers J, De Winne H, Van Noyen J, Moens B, Sels B, Jacobs P (2008) Influence of composition of $MgAl_2O_4$ supported $NiCeO_2ZrO_2$ catalysts on coke formation and catalyst stability for dry reforming of methane. Catal Today 138:28–32
165. Sahli N, Petit C, Roger AC, Kiennemann A, Libs S, Bettahar MM (2006) Ni catalysts from $NiAl_2O_4$ spinel for CO_2 reforming of methane. Catal Today 113:187–193
166. Gallego GS, Mondragón F, Tatibouët JM, Barrault J, Batiot-Dupeyrat C (2008) Carbon dioxide reforming of methane over La_2NiO_4 as catalyst precursor—characterization of carbon deposition. Catal Today 133–135:200–209
167. Sutthiumporn K, Maneerung T, Kathiraser Y, Kawi S (2012) CO_2 dry-reforming of methane over $La_{0.8}Sr_{0.2}Ni_{0.8}M_{0.2}O_3$ perovskite (M = Bi, Co, Cr, Cu, Fe): roles of lattice oxygen on C–H activation and carbon suppression. Int J Hydrogen Energy 37:11195–11207
168. Chen Y, Ren J (1994) Conversion of methane and carbon dioxide into synthesis gas over alumina-supported nickel catalysts, effect of Ni-Al_2O_3 interactions. Catal Lett 29:39–48
169. Wang S, Lu GQ (1998) Role of CeO_2 in Ni/CeO_2–Al_2O_3 catalysts for carbon dioxide reforming of methane. Appl Catal Environ 19:267–277
170. Bhattacharyya A, Chang VW (1994) CO_2 reforming of methane to syngas: deactivation behavior of nickel aluminate spinel catalysts. Stud Surf Sci Catal 88:207–213
171. Kathiraser Y, Thitsartarn W, Sutthiumporn K, Kawi S (2013) Inverse $NiAl_2O_4$ on $LaAlO_3$–Al_2O_3: unique catalytic structure for stable CO_2 reforming of methane. J Phys Chem C 117:8120–8130
172. Pakhare D, Wu H, Narendra S, Abdelsayed V, Haynes D, Shekhawat D, Berry D, Spivey J (2013) Characterization and activity study of the Rh-substituted pyrochlores for CO_2(dry) reforming of CH_4. Appl Petrochem Res 3(3–4):117–129
173. Gaur S, Pakhare D, Wu H, Haynes DJ, Spivey JJ (2012) CO_2 reforming of CH_4 over Ru-substituted pyrochlore catalysts: effects of temperature and reactant feed ratio. Energy Fuels 26:1989–1998
174. Pakhare D, Shaw C, Haynes D, Shekhawat D, Spivey J (2013) Effect of reaction temperature on activity of Pt- and Ru-substituted lanthanum zirconate pyrochlores ($La_2Zr_2O_7$) for dry (CO_2) reforming of methane (DRM). J CO_2 Util 1:37–42
175. Shinde VM, Madras G (2014) Catalytic performance of highly dispersed Ni/TiO_2 for dry and steam reforming of methane. RSC Adv 4:4817–4826
176. Ligthart DAJM, Van Santen RA, Hensen EJM (2011) Influence of particle size on the activity and stability in steam methane reforming of supported Rh nanoparticles. J Catal 280:206–220
177. Schulz PG, Gonzales MG, Quincoces CE, Gigola CE (2005) Methane reforming with carbon dioxide. The behavior of Pd/γ-Al2O3 and Pd-CeO_x/γ-Al_2O_3 catalysts. Ind Eng Chem Res 44:9020–9029
178. Corma A, Diaz U, Domine ME, Fornes V (2000) AlITQ-6 and TiITQ-6: synthesis, characterization, and catalytic activity. Angew Chem Int Ed 39:1499–1500
179. Corma A, Fornes V, Guil JM, Pergher S, Maesen Th LM, Buglass JG (2000) Preparation, characterisation and catalytic activity of ITQ-2, a delaminated zeolite. Microporous Mesoporous Mater 38:301–309
180. Li Z, Mo L, Kathiraser Y, Kawi S (2014) Yolk–satellite–shell structured Ni–yolk@Ni@SiO_2 nanocomposite: superb catalyst toward methane CO_2 reforming reaction. ACS Catal 4:1526–1536

181. Baudouin D, Rodemerck U, Krumeich F, Mallmann AD, Szeto KC, Menard H, Veyre L, Candy JP, Webb PB, Thieuleux C, Copéret C (2013) Particle size effect in the low temperature reforming of methane by carbon dioxide on silica-supported Ni nanoparticles. J Catal 297:27–34
182. Baudouin D, Szeto KC, Laurent P, Mallmann AD, Fenet B, Veyre L, Rodemerck U, Copéret C (2012) Nickel–silicide colloid prepared under mild conditions as a versatile Ni precursor for more efficient CO_2 reforming of CH_4 catalysts. J Am Chem Soc 134:20624–20627
183. Mo L, Leong KKM, Kawi S (2014) A highly dispersed and anti-coking Ni–La_2O_3/SiO_2 catalyst for syngas production from dry carbon dioxide reforming of methane. Catal Sci Technol 4:2107.2114
184. Mark MF, Maier WF, Mark F (1997) Reaction kinetics of the CO_2 reforming of methane. Chem Eng Technol 20:361–370
185. Tsipouriari VA, Verykios XE (2001) Kinetic study of the catalytic reforming of methane with carbon dioxide to synthesis gas over Ni/La_2O_3 catalyst. Catal Today 64:83–90
186. Ferreira-Aparicio P, Rodriguez-Ramos I, Anderson JA, Guerrero-Ruiz A (2000) Mechanistic aspects of the dry reforming of methane over ruthenium catalysts. Appl Catal Gen 202:183–196
187. Bitter JH, Seshan K, Lercher JA (2000) On the contribution of X-ray absorption spectroscopy to explore structure and activity relations of Pt/ZrO_2 catalysts for CO_2/CH_4 reforming. Top Catal 10:295–305
188. Zhang Z, Verykios XE (1996) Mechanistic aspects of carbon dioxide reforming of methane to synthesis gas over Ni catalysts. Catal Lett 38:175–179
189. Wei J, Iglesia E (2004) Isotopic and kinetic assessment of the mechanism of reactions of CH_4 with CO_2 or H_2O to form synthesis gas and carbon on nickel catalysts. J Catal 24:370–383
190. Ferreira-Aparicio P, Fernandez-Garcia M, Guerrero-Ruiz A, Rodriguez-Ramos I (2000) Evaluation of the role of the metal–support interfacial centers in the dry reforming of methane on alumina-supported rhodium catalysts. J Catal 190:296–308
191. Li C, Yan W, Xin Q (1994) Interaction of methane with surface of alumina studied by FT-IR spectroscopy. Catal Lett 24:249–256
192. Solymosi F, Cserenyi J (1994) Decomposition of CH_4 over supported Ir catalysts. Catal Today 21:561–569
193. Lisi L, Bagnasco G, Ciambelli P, De Rossi S, Porta P, Russo G, Turco M (1999) Perovskite-type oxides: II. Redox properties of $LaMn_{1-x}Cu_xO_3$ and $LaCo_{1-x}Cu_xO_3$ and methane catalytic combustion. J Solid State Chem 146:176–183
194. Zhang Z, Verykios XE, Macdonald SM, Affrossman S (1996) Comparative study of carbon dioxide reforming of methane to synthesis gas over Ni/La_2O_3 and conventional nickel-based catalysts. J Phys Chem 100:744–754
195. Munera JF, Irusta S, Cornaglia LM, Lombardo EA, Cesar DV, Schmal M (2007) Kinetics and reaction pathway of the CO_2 reforming of methane on Rh supported on lanthanum-based solid. J Catal 245:25–34
196. Wang S, (Max) Lu GQ (1999) A comprehensive study on carbon dioxide reforming of methane over Ni/γ-Al_2O_3 catalysts. Ind Eng Chem Res 38:2615–2625
197. Ma J, Sun NN, Zhang XL, Zhao N, Mao FK, Wei W, Sun YH (2009) A short review of catalysis for CO_2 conversion. Catal Today 148:221–231
198. Inui T (1996) Highly effective conversion of carbon dioxide to valuable compounds on composite catalysts. Catal Today 29:329–337
199. Cheng D, Negreiros FR, Apra E, Fortunelli A (2013) Computational approaches to the chemical conversion of carbon dioxide. ChemSusChem 6:944–965
200. Chen CS, Wu JH, Lai TW (2010) Carbon dioxide hydrogenation on Cu nanoparticles. J Phys Chem C 114:15021–15028
201. Pekridis G, Kalimeri K, Kaklidis N, Vakouftsi E, Iliopoulou EF, Athanasiou C, Marnellos GE (2007) Study of the reverse water gas shift (RWGS) reaction over Pt in a solid oxide fuel cell (SOFC) operating under open and closed-circuit conditions. Catal Today 127:337–346

202. Liu XM, Lu GQ, Yan ZF, Beltramini J (2003) Recent advances in catalysts for methanol synthesis *via* hydrogenation of CO and CO_2. Ind Eng Chem Res 42:6518–6530
203. Mei DH, Xu LJ, Henkelman G (2008) Dimer saddle point searches to determine the reactivity of formate on Cu(111). J Catal 258:44–51
204. Olah GA (2013) Towards oil independence through renewable methanol chemistry. Angew Chem Int Ed 52:104–107
205. Fujiwara M, Kieffer R, Ando H, Souma Y (1995) Development of composite catalysts made of Cu-Zn-Cr oxide/zeolite for the hydrogenation of carbon dioxide. Appl Catal Gen 121:113–124
206. Fujiwara M, Kieffer R, Ando H, Xu Q, Souma Y (1997) Change of catalytic properties of **Fe–ZnO/zeolite composite catalyst in the hydrogenation of carbon dioxide. Appl Catal Gen 154:87–101
207. Lunev NK, Shmyrko YI, Pavlenko NV, Norton B (2001) Synthesis of iso-hydrocarbons mixture from CO_2 and H_2 on hybrid catalysts. Appl Organomet Chem 15:99–104
208. Raudaskoski R, Turpeinen E, Lenkkeri R, Pongrácz E, Keiski RL (2009) Catalytic activation of CO_2: use of secondary CO_2 for the production of synthesis gas and for methanol synthesis over copper-based zirconia-containing catalysts. Catal Today 144:318–323
209. Yang C, Ma Z, Zhao N, Wei W, Hu T, Sun Y (2006) Methanol synthesis from CO_2-rich syngas over a ZrO_2 doped CuZnO catalyst. Catal Today 115:222–227
210. Zhang R, Wang B, Liu H, Ling L (2011) Effect of surface hydroxyls on CO_2 hydrogenation over Cu/γ-Al_2O_3 catalyst: a theoretical study. J Phys Chem C 115:19811–19818
211. Olah GA, Goeppert A, Surya Prakash GK (2009) Chemical recycling of carbon dioxide to methanol and dimethyl ether: from greenhouse gas to renewable, environmentally carbon neutral fuels and synthetic hydrocarbons. J Org Chem 74:487–498
212. Kim IH, Kim S, Cho W, Yoon ES (2010) Simulation of commercial dimethyl ether production plant. Comput Aided Chem Eng 28:799–804
213. Chen WH, Lin BJ, Lee HW, Huang MN (2012) One-step synthesis of dimethyl ether from the gas mixture containing CO_2 with high space velocity. Appl Energy 98:92–101
214. Kang SW, Bae JW, Jun KW, Potdar HS (2008) Dimethyl ether synthesis from syngas over the composite catalysts of Cu–ZnO–Al_2O_3/Zr-modified zeolites. Catal Commun 9:2035–2039
215. Wang S, Mao D, Guo X, Wu G, Lu G (2009) Dimethyl ether synthesis *via* CO_2 hydrogenation over CuO–TiO_2–ZrO_2/HZSM-5 bifunctional catalysts. Catal Commun 10:1367–1370
216. Pellegrini LA, Soave G, Gamba S, Lange S (2011) Economic analysis of a combined energy–methanol production plant. Appl Energy 88:4891–4897
217. Bonura G, Cordaro M, Spadaro L, Cannilla C, Arena F, Frusteri F (2013) Hybrid Cu–ZnO–ZrO_2/H-ZSM5 system for the direct synthesis of DME by CO_2 hydrogenation. Appl Catal Environ 140–141:16–24
218. Frusteri F, Cordaro M, Cannilla C, Bonura G (2015) Multifunctionality of Cu–ZnO–ZrO_2/H-ZSM5 catalysts for the one-step CO_2-to-DME hydrogenation reaction. Appl Catal Environ 162:57–65
219. Gines MJL, Marchi AJ, Apesteguia CR (1997) Kinetic study of the reverse water-gas shift reaction over CuO/ZnO/Al_2O_3 catalysts. Appl Catal Gen 154:155–171
220. Behrens M, Kisner S, Girsgdies F, Kasatkin I, Hermerschmidt F, Mette K, Ruland H, Muhler M, Schlögl R (2011) Knowledge-based development of a nitrate-free synthesis route for Cu/ZnO methanol synthesis catalysts *via* formate precursors. Chem Commun 47:1701–1703
221. Kaluza S, Behrens M, Schiefenhövel N, Kniep B, Fischer R, Schlögl R, Muhler M (2011) A novel synthesis route for Cu/ZnO/Al_2O_3 catalysts used in methanol synthesis: combining continuous consecutive precipitation with continuous aging of the precipitate. ChemCatChem 3:189–199
222. Gnanamani MK, Jacobs G, Pendyala VRR, Ma W, Davis BH (2014) Hydrogenation of carbon dioxide to liquid fuels. In: Centi G, Perathoner S (eds) Green carbon dioxide: advances in CO_2 utilization, 1st edn. Wiley, New York, pp 99–118

223. Köppel RA, Stöcker C, Baiker A (1998) Copper- and silver–zirconia aerogels: preparation, structural properties and catalytic behavior in methanol synthesis from carbon dioxide. J Catal 179:515–527
224. Arena F, Barbera K, Italiano G, Bonura G, Spadaro L, Frusteri F (2007) Synthesis, characterization and activity pattern of Cu–ZnO/ZrO_2 catalysts in the hydrogenation of carbon dioxide to methanol. J Catal 249:185–194
225. Rhodes MD, Pokrovski KA, Bell AT (2005) The effects of zirconia morphology on methanol synthesis from CO and H_2 over Cu/ZrO_2 catalysts: part II. Transient-response infrared studies. J Catal 233:210–220
226. Schilke TC, Fisher IA, Bell AT (1999) *In situ* infrared study of methanol synthesis from CO_2/H_2 on titania and zirconia promoted Cu/SiO_2. J Catal 184:144–156
227. Jung KT, Bell AT (2002) Effects of zirconia phase on the synthesis of methanol over zirconia-supported copper. Catal Lett 80:63–68
228. Arena F, Italiano G, Barbera K, Bonura G, Spadaro L, Frusteri F (2009) Basic evidences for methanol-synthesis catalyst design. Catal Today 143:80–85
229. Zha F, Tian H, Yan J, Chang Y (2013) Multi-walled carbon nanotubes as catalyst promoter for dimethyl ether synthesis from CO_2 hydrogenation. Appl Surf Sci 285B:945–951
230. Zhang Y, Fei J, Yu Y, Zheng X (2006) Methanol synthesis from CO_2 hydrogenation over Cu based catalyst supported on zirconia modified γ-Al_2O_3. Energy Convers Manag 47:3360–3367
231. Arena F, Spadaro L, Blasi OD, Bonura G, Frusteri F (2004) Integrated synthesis of dimethylether *via* CO_2 hydrogenation. Stud Surf Sci Catal 147:385–390
232. Guo X, Mao D, Lu G, Wang S, Wu G (2010) Glycine–nitrate combustion synthesis of CuO–ZnO–ZrO_2 catalysts for methanol synthesis from CO_2 hydrogenation. J Catal 271:178–185
233. Pokrovski KA, Rhodes MD, Bell AT (2005) Effects of cerium incorporation into zirconia on the activity of Cu/ZrO_2 for methanol synthesis *via* CO hydrogenation. J Catal 235:368–377
234. Pokrovski KA, Bell AT (2006) An investigation of the factors influencing the activity of Cu/$Ce_xZr_{1-x}O_2$ for methanol synthesis *via* CO hydrogenation. J Catal 241:276–286
235. Bonura G, Arena F, Mezzatesta G, Cannilla C, Spadaro L, Frusteri F (2011) Role of the ceria promoter and carrier on the functionality of Cu-based catalysts in the CO_2-to-methanol hydrogenation reaction. Catal Today 171:251–256
236. Mo L, Kawi S (2014) An *in situ* self-assembled core–shell precursor route to prepare ultrasmall copper nanoparticles on silica catalysts. J Mater Chem A 2:7837–7844
237. Arena F, Mezzatesta G, Zafarana G, Trunfio G, Frusteri F, Spadaro L (2013) Effects of oxide carriers on surface functionality and process performance of the Cu–ZnO system in the synthesis of methanol *via* CO_2 hydrogenation. J Catal 300:141–151
238. Rodemerck U, Holena M, Wagner E, Smejkal Q, Barkschat A, Baerns M (2013) Catalyst development for CO_2 hydrogenation to fuels. ChemCatChem 5:1948–1955
239. Baltes C, Vukojevic S, Schuth F (2008) Correlations between synthesis, precursor, and catalyst structure and activity of a large set of CuO/ZnO/Al_2O_3 catalysts for methanol synthesis. J Catal 258:334–344
240. Yang G, Thongkam M, Vitidsant T, Yoneyama Y, Tan Y, Tsubaki N (2011) A double shell capsule catalyst with core-shell-like structure for one-step exactly controlled synthesis of dimethyl ether from CO_2 containing syngas. Catal Today 171:229–235
241. Aguayo AT, Ereña J, Sierra I, Olazar M, Bilbao J (2005) Deactivation and regeneration of hybrid catalysts in the single-step synthesis of dimethyl ether from syngas and CO_2. Catal Today 106:265–270
242. Mao D, Xia J, Zhang B, Lu G (2010) Highly efficient synthesis of dimethyl ether from syngas over the admixed catalyst of CuO–ZnO–Al_2O_3 and antimony oxide modified HZSM-5 zeolite. Energy Convers Manag 51:1134–1139
243. Liu RW, Qin ZZ, Ji HB, Su TM (2013) Synthesis of dimethyl ether from CO_2 and H_2 using a Cu–Fe–Zr/HZSM-5 catalyst system. Ind Eng Chem Res 52:16648–16655

244. An X, Zuo YZ, Zhang Q, Wang DZ, Wang JF (2008) Dimethyl ether synthesis from CO_2 hydrogenation on a CuO–ZnO–Al_2O_3–ZrO_2/HZSM-5 bifunctional catalyst. Ind Eng Chem Res 47:6547–6554
245. Zhang Q, Zuo YZ, Han MH, Wang JF, Jin Y, Wei F (2010) Long carbon nanotubes intercrossed Cu/Zn/Al/Zr catalyst for CO/CO_2 hydrogenation to methanol/dimethyl ether. Catal Today 150:55–60
246. Lim HW, Park MJ, Kang SH, Chae HJ, Bae JW, Jun KW (2009) Modeling of the kinetics for methanol synthesis using Cu/ZnO/Al_2O_3/ZrO_2 catalyst: influence of carbon dioxide during hydrogenation. Ind Eng Chem Res 48:10448–10455
247. Fisher IA, Bell AT (1997) *In-situ* infrared study of methanol synthesis from H_2/CO_2 over Cu/SiO_2 and Cu/ZrO_2/SiO_2. J Catal 172:222–237
248. Kakumoto T, Watanabe T (1997) A theoretical study for methanol synthesis by CO_2 hydrogenation. Catal Today 36:39–44
249. Jansen WPA, Beckers J, Van Der Heuvel JC, Van der Gon AWD, Bliek A, Brongersma HH (2002) Dynamic behavior of the surface structure of Cu/ZnO/SiO_2 catalysts. J Catal 210:229–236
250. Saito M, Fujitani T, Takeuchi M, Watanabe T (1996) Development of copper/zinc oxide-based multicomponent catalysts for methanol synthesis from carbon dioxide and hydrogen. Appl Catal Gen 138:311–318
251. Gao W, Wang H, Wang Y, Guo W, Jia M (2013) Dimethyl ether synthesis from CO_2 hydrogenation on La-modified CuO-ZnO-Al_2O_3/HZSM-5 bifunctional catalysts. J Rare Earths 31:470–476
252. Zhang MH, Liu ZM, Lin GD, Zhang HB (2013) Pd/CNT-promoted Cu_ZrO_2/HZSM-5 hybrid catalysts for direct synthesis of DME from CO_2/H_2. Appl Catal Gen 451:28–35
253. Lee SC, Jang JH, Lee BY, Kang MC, Kang M, Choung SJ (2003) The effect of binders on structure and chemical properties of Fe-K/γ-Al_2O_3 catalysts for CO_2 hydrogenation. Appl Catal Gen 253:293–304
254. Chen CS, Cheng WH, Lin SS (2004) Study of iron-promoted Cu/SiO_2 catalyst on high temperature reverse water gas shift reaction. Appl Catal Gen 257:97–106
255. Ribeiro MC, Jacobs G, Davis BH, Cronauer DC, Kropf AJ, Marshall CL (2010) Fischer – Tropsch synthesis: an in-situ TPR-EXAFS/XANES investigation of the influence of Group I alkali promoters on the local atomic and electronic structure of carburized iron/silica catalysts. J Phys Chem C 114:7895–7903
256. Sun K, Lu W, Wang M, Xu X (2004) Low-temperature synthesis of DME from CO_2/H_2 over Pd-modified CuO-ZnO-Al_2O_3-ZrO_2/HZSM-5 catalysts. Catal Commun 5:367–370
257. Riedel T, Schaub G, Jun KW, Lee KW (2001) Kinetics of CO_2 hydrogenation on a K-promoted Fe catalyst. Ind Eng Chem Res 40:1355–1363
258. Dorner RW, Hardy DR, Williams FW, Davis BH, Willauer HD (2009) Influence of gas feed composition and pressure on the catalytic conversion of CO_2 to hydrocarbons using a traditional cobalt-based Fischer – Tropsch catalyst. Energy Fuel 23:4190–4195
259. Riedel T, Claeys M, Schulz H, Schaub G, Nam SS, Jun KW, Choi MJ, Kishan G, Lee KW (1999) Comparative study of Fischer–Tropsch synthesis with H_2/CO and H_2/CO_2 syngas using Fe- and Co-based catalysts. Appl Catal Gen 186:201–213
260. Akin AN, Ataman M, Aksoylu AE, Onsan ZI (2002) CO_2 fixation by hydrogenation over coprecipitated Co/Al_2O_3. React Kinet Catal Lett 76:265–270
261. Chang FW, Kuo MS, Tsay MT, Hsieh MC (2003) Hydrogenation of CO_2 over nickel catalysts on rice husk ash-alumina prepared by incipient wetness impregnation. Appl Catal Gen 247:309–320
262. Chang FW, Tsay MT, Liang SP (2001) Hydrogenation of CO_2 over nickel catalysts supported on rice husk ash prepared by ion exchange. Appl Catal Gen 209:217–227
263. Perkas N, Amirian G, Zhong ZY, Teo J, Gofer Y, Gedanken A (2009) Methanation of carbon dioxide on Ni catalysts on mesoporous ZrO_2 doped with rare earth oxides. Catal Lett 130:455–462

264. Ocampo F, Louis B, Roger AC (2009) Methanation of carbon dioxide over nickel-based $Ce_{0.72}Zr_{0.28}O_2$ mixed oxide catalysts prepared by sol–gel method. Appl Catal Gen 369:90–96
265. Wang L, Zhang S, Liu Y (2008) Reverse water gas shift reaction over Co-precipitated $Ni-CeO_2$ catalysts. J Rare Earths 26:66–70
266. Zonetti PC, Letichevsky S, Gaspar AB, Sousa-Aguiar EF, Appel LG (2014) The $Ni_xCe_{0.75}Zr_{0.25-x}O_2$ solid solution and the RWGS. Appl Catal Gen 475:48–54
267. Mo L, Kawi S (2014) US Patent Application No: PCT/SG2014/000108
268. Coteron A, Hayhurst AN (1993) Methanol synthesis by amorphous copper-based catalysts prepared by spark-erosion. Appl Catal Gen 101:151–165
269. Sun Y, Sermon PA (1994) Evidence of a metal-support interaction in sol–gel derived $Cu-ZrO_2$ catalysts for CO hydrogenation. Catal Lett 29:361–369
270. Ma Y, Sun Q, Wu D, Fan WH, Zhang YL, Deng JF (1998) A practical approach for the preparation of high activity $Cu/ZnO/ZrO_2$ catalyst for methanol synthesis from CO_2 hydrogenation. Appl Catal Gen 171:45–55
271. Deng J, Sun Q, Zhang Y, Chen S, Wu D (1996) A novel process for preparation of a $Cu/ZnO/Al_2O_3$ ultrafine catalyst for methanol synthesis from CO_2+H_2: comparison of various preparation methods. Appl Catal Gen 139:75–85
272. Słoczynsky J, Grabowski R, Kozłowska A, Olszewski P, Stoch J, Skrzypek J, Lachowska M (2004) Catalytic activity of the $M/(3ZnO \cdot ZrO_2)$ system (M = Cu, Ag, Au) in the hydrogenation of CO_2 to methanol. Appl Catal Gen 278:11–23
273. Słoczynsky J, Grabowski R, Kozłowska A, Olszewski P, Lachowska M, Skrzypek J, Stoch J (2003) Effect of Mg and Mn oxide additions on structural and adsorptive properties of $Cu/ZnO/ZrO_2$ catalysts for the methanol synthesis from CO_2. Appl Catal Gen 249:129–138
274. Raudaskoski R, Niemela MV, Keiski RL (2007) The effect of ageing time on co-precipitated $Cu/ZnO/ZrO_2$ catalysts used in methanol synthesis from CO_2 and H_2. Top Catal 45:57–60
275. Guo X, Mao D, Lu G, Wang S, Wu G (2011) CO_2 hydrogenation to methanol over $Cu/ZnO/ZrO_2$ catalysts prepared *via* a route of solid-state reaction. Catal Commun 12:1095–1098
276. Kenji U, Kozo M, Takeshi K, Taiki W, Masahiro S (2000) Methanol synthesis from CO_2 and H_2 in a bench-scale test plant. Appl Organomet Chem 14:819–825
277. Takeshi K, Itaru H, Hirotaka M, Kozo M, Kenji U, Taiki W, Masahiro S (2001) Kinetic study of methanol synthesis from carbon dioxide and hydrogen. Appl Organomet Chem 15:121–126
278. Fujita SI, Usui M, Takezawa N (1992) Mechanism of the reverse water gas shift reaction over Cu/ZnO catalyst. J Catal 134:220–225
279. Waugh KC (1992) Methanol synthesis. Catal Today 15:51–75
280. Shido T, Iwasawa Y (1993) The effect of coadsorbates in reverse water-gas shift reaction on ZnO, in relation to reactant-promoted reaction mechanism. J Catal 140:575–584
281. Herwijnen TV, Guczalski RT, Jong WA (1980) Kinetics and mechanism of the CO shift on CuZnO: II. Kinetics of the decomposition of formic acid. J Catal 63:94–101
282. Yoshihara J, Campbell CT (1996) Methanol synthesis and reverse water–gas shift kinetics over Cu(110) model catalysts: structural sensitivity. J Catal 161:776–782
283. Chen CS, Cheng WH, Lin SS (2003) Study of reverse water gas shift reaction by TPD, TPR and CO_2 hydrogenation over potassium-promoted Cu/SiO_2 catalyst. Appl Catal Gen 238:55–67
284. Jung KD, Bell AT (2000) Role of hydrogen spillover in methanol synthesis over Cu/ZrO_2. J Catal 193:207–223
285. Chinchen GC, Waugh KC, Whan DA (1986) The activity and state of the copper surface in methanol synthesis catalysts. Appl Catal 25:101–107
286. Melian-Cabrera I, Granados ML, Fierro JLG (2002) Reverse topotactic transformation of a Cu-Zn-Al catalyst during wet Pd impregnation: relevance for the performance in methanol synthesis from CO_2/H_2 mixtures. J Catal 210:273–284

287. Fujitani T, Nakamura I, Uchijima T, Nakamura J (1997) The kinetics and mechanism of methanol synthesis by hydrogenation of CO_2 over a Zn-deposited Cu(111) Surface. Surf Sci 383:285–298
288. Arena F, Italiano G, Barbera K, Bordiga S, Bonura G, Spadaro L, Frusteri F (2008) Solid-state interactions, adsorption sites and functionality of Cu-ZnO/ZrO_2 catalysts in the CO_2 hydrogenation to CH_3OH. Appl Catal Gen 350:16–23
289. Zhang R, Liu H, Wang B, Ling L (2012) Insights into the effect of surface hydroxyls on CO_2 hydrogenation over Pd/-Al_2O_3 catalyst: a computational study. Appl Catal Environ 126:108–120
290. Collins SE, Baltanás MA, Bonivardi AL (2005) Mechanism of the decomposition of adsorbed methanol over a Pd/α, β-Ga2O3. Appl Catal Gen 295:126–133
291. Collins SE, Delgado JJ, Mira C, Calvino JJ, Bernal S, Chiavassa DL, Baltanás MA, Bonivardi AL (2012) The role of Pd–Ga bimetallic particles in the bifunctional mechanism of selective methanol synthesis *via* CO_2 hydrogenation on a Pd/Ga_2O_3 catalyst. J Catal 292:90–98
292. Collins SE, Baltanás MA, Bonivardi AL (2004) An infrared study of the intermediates of methanol synthesis from carbon dioxide over Pd/β-Ga_2O_3. J Catal 226:410–421
293. Sun Q, Liu CW, Pan W, Zhu QM, Deng JF (1998) In situ IR studies on the mechanism of methanol synthesis over an ultrafine Cu/ZnO/Al_2O_3 catalyst. Appl Catal Gen 171:301–308
294. Bonura G, Cordaro M, Cannilla C, Aresta F, Frusteri F (2014) The changing nature of the active site of Cu-Zn-Zr catalysts for the CO_2 hydrogenation reaction to methanol. Appl Catal Environ 152–153:152–161
295. Verykios XE (2003) Catalytic dry reforming of natural gas for the production of chemicals and hydrogen. Int J Hydrogen Energy 28:1045–1063
296. Dimethyl ether production technology. http://www.jcoal.or.jp/eng/cctinjapan/2_4A4.pdf
297. Kaeding WW (1974) Styrene monomer. Catal Rev 8:307–316
298. Ji M, Zhang X, Wang J, Park SE (2013) Ethylbenzene dehydrogenation with CO_2 over Fe-doped $MgAl_2O_4$ spinel catalysts: synergy effect between Fe^{2+} and Fe^{3+}. J Mol Catal A Chem 371:36–41
299. Badstube T, Papp H, Dziembaj R, Kustrowski P (2000) Screening of catalysts in the oxidative dehydrogenation of ethylbenzene with carbon dioxide. Appl Catal Gen 204:153–165

Chapter 8
One- and Multi-electron Pathways for the Reduction of CO_2 into C1 and C1+ Energy-Richer Molecules: Some Thermodynamic and Kinetic Facts

Abstract This chapter deals with the mechanism of reduction of "free" and coordinated CO_2 by electron transfer. One-e^- and multi-e^- transfer pathways are compared energetically and their role in the conversion of CO_2 into higher energy C1 or C1+-species is highlighted. The state of the knowledge is presented through the analysis of reference cases.

8.1 Introduction

Electron transfer to CO_2 experiences both thermodynamic and kinetic hindrance. The Gibbs free energy change (ΔG) of formation of several C1 molecules is represented in Fig. 8.1. It says that if CO_2 has to be converted into species in which the C-oxidation state is lower than +4, energy is needed – if not also hydrogen. With the exception of CO, which can be produced from CO_2 through routes that may or may not involve hydrogen (CO_2 splitting to afford CO and "O" does not require hydrogen), all other C1 molecules have an H-to-C ratio higher than the value in CO_2, where it is zero. This means that the reduction process requires "H" or "e^- plus H^+" transfer. The source of "H, H^+" must be water if the reduction of the heterocumulene is to be performed large scale. Using water as a source of hydrogen, in principle, the reduction of CO_2 can be performed following two quite different routes: (1) electrolysis of water to afford H_2 and use of the latter in thermal catalytic reactions with CO_2 bearing to energy-rich C1 or C1+ molecules; (2) electro-co-processing of CO_2 and water to afford C-reduced species. The two routes are quite different; the former requires production and possibly storage of H_2, whereas the latter directly converts CO_2 into its reduced forms without any need for a preliminary production of H_2. In this chapter, the latter option is discussed.

In Chap. 1, it was said that CO_2 can be reduced through a 1-e^- transfer to the radical anion $CO_2^{\cdot-}$ Figure 8.1 shows that any reduction of CO_2 to other C1 species is an "n × $2e^-$" reduction. A process that may be born from the transfer of "$1e^-$" to CO_2 is the synthesis of C–C bonds. For example, the C2 species oxalic acid (HO_2C–CO_2H), in which the oxidation state of C is +3, can be formed by coupling two $CO_2^{\cdot-}$ units followed by protonation; the direct carboxylation of hydrocarbons

M. Aresta et al., *Reaction Mechanisms in Carbon Dioxide Conversion*,
DOI 10.1007/978-3-662-46831-9_8

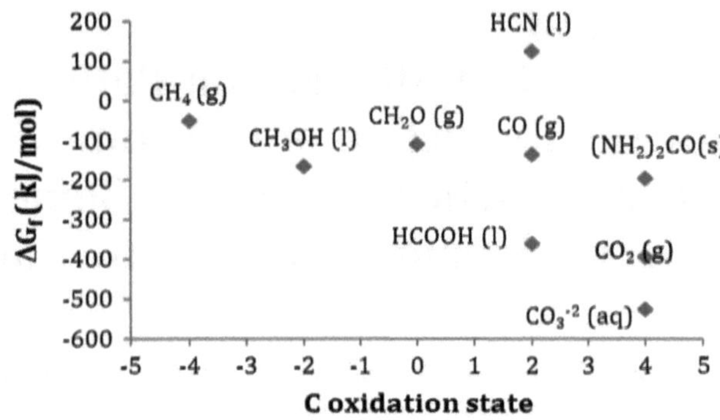

Fig. 8.1 Free Gibbs energy of formation for CO_2 and other C1 molecules

forming carboxylic acids ($R\text{–}H + CO_2 \rightarrow R\text{–}CO_2H$), in which again the formal oxidation state of C in CO_2 is +3, is another interesting reaction. As observed in Chaps. 1 and 2, the $1e^-$ transfer to CO_2 can also originate a second species that is the carbonyl-carbonate, $^-OC(=O)\text{–}OC(=O)^-$, obtained upon coupling of CO_2 with $CO_2^{\cdot -}$ and subsequent transfer of a second electron to the product of coupling. The fate of such a product was discussed in Chap. 2.

However, several questions may arise in the mind of the reader, such as, Are radical $CO_2^{\cdot -}$ anions to be likely intermediates in the synthesis of other C1 molecules? Is such one-electron transfer thermodynamically more or less feasible than a multielectron process? Is the multielectron transfer a process occurring at different potentials for each e^--transfer? Or are iso-potential (iso-energetic) synchronous multiple e^--transfers possible? It has also been shown in Chaps. 1 and 2 that when a metal interacts with the C atom of CO_2, or electrons are transferred to the same atom, the molecular geometry of the latter is dramatically changed and the change requires the reorganization of the energy levels of the molecular orbitals. Does such a geometry change represent an influential barrier to the reduction of CO_2? Therefore, do the facility of coordination and the mode influence the $1e^-$–$2e^-$ transfer? The electroreduction of CO_2 is not a trivial case. In this chapter, an answer to such questions is attempted, and the state of our knowledge is presented.

8.2 Key Steps and Aspects in CO_2 Reduction

The potentials of the various reactions of reduction of CO_2 to other C1 molecules are shown in Table 8.1. It is quite evident that the one-electron transfer is a process that requires higher energy than the multi-electron transfers. If the reductions are considered to take place in water, then it is useful to recall the Frost diagram built by correlating the number of electrons transferred in a process and the free energy change (or electrochemical potential; Table 8.1) of that process (Fig. 8.2).

An important message from such a diagram is that any species that lies above a line joining two other different species to which it is correlated is not stable with

Table 8.1 Potential of reduction of CO_2 to other C1 molecules (da 1) vs the standard hydrogen electrode (SHE). Reprinted with permission from [1]. Copyright (2009) Royal Society of Chemistry

Entry	Reaction		E° V (SHE)
1	$CO_2 + e^- = CO_2^{\cdot -}$	−2.21 in aprotic solvents	−1.90 in water
2	$CO_2 + 2H^+ + 2e^- \rightarrow CO + H_2O$		−0.53
3	$CO_2 + 2H^+ + 2e^- \rightarrow HCO_2H$		−0.61
4	$CO_2 + 4H^+ + 4e^- \rightarrow CH_2O + H_2O$		−0.48
5	$CO_2 + 6H^+ + 6e^- \rightarrow CH_3OH + H_2O$		−0.38
6	$CO_2 + 8H^+ + 8e^- \rightarrow CH_4 + 2H_2O$		−0.24

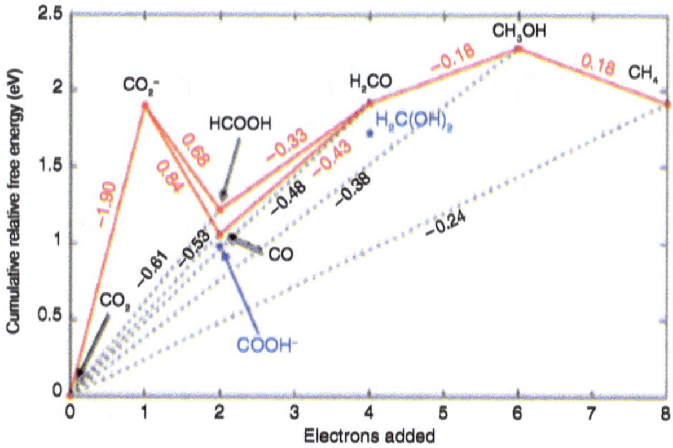

Fig. 8.2 Frost diagram applied to the reduction of CO_2 through $n(e^- + H^+)$ transfer processes Reprinted with permission from [2]. Copyright (2012) Royal Society of Chemistry

respect to the latter two and converts into them. So, for example, formaldehyde, which lies above the line joining carbon monoxide and methanol, converts into the latter two species. If coordinated CO_2, more than free CO_2, has to be reduced, then a key step in the entire reduction process is the bonding of CO_2 to the metal centre. This part was discussed in Chap. 2, where it was shown that ligands may influence the mode of bonding of CO_2 to the metal centre. Moreover, carrying out the electroreduction in water requires that the metal centre is stabilized by ligands with a good water affinity, proper donor properties and correct steric hindrance.

Consequently, N- or O-ligands are more frequently used than phosphanes, unless the latter are conveniently functionalized (for example with sulphonic groups or other water soluble moieties) for enhancing their water solubility or mixed organic-water solvent media are used. Figure 8.3 presents a list of N-ligands, from the historical ones such as *cyclam* used in the first studies in this field and Fig. 8.4 lists some of the P-ligands used in electrochemical reductions. Table 8.2 highlights the electroreduction of CO_2 and showcases the electrodes and the catalyst used, the products obtained, the operative conditions, the yield, efficiency and selectivity.

Another important factor that influences the reduction process is the solvent. Table 8.3 shows the solubility of CO_2 in various organic solvents [31]. It should be

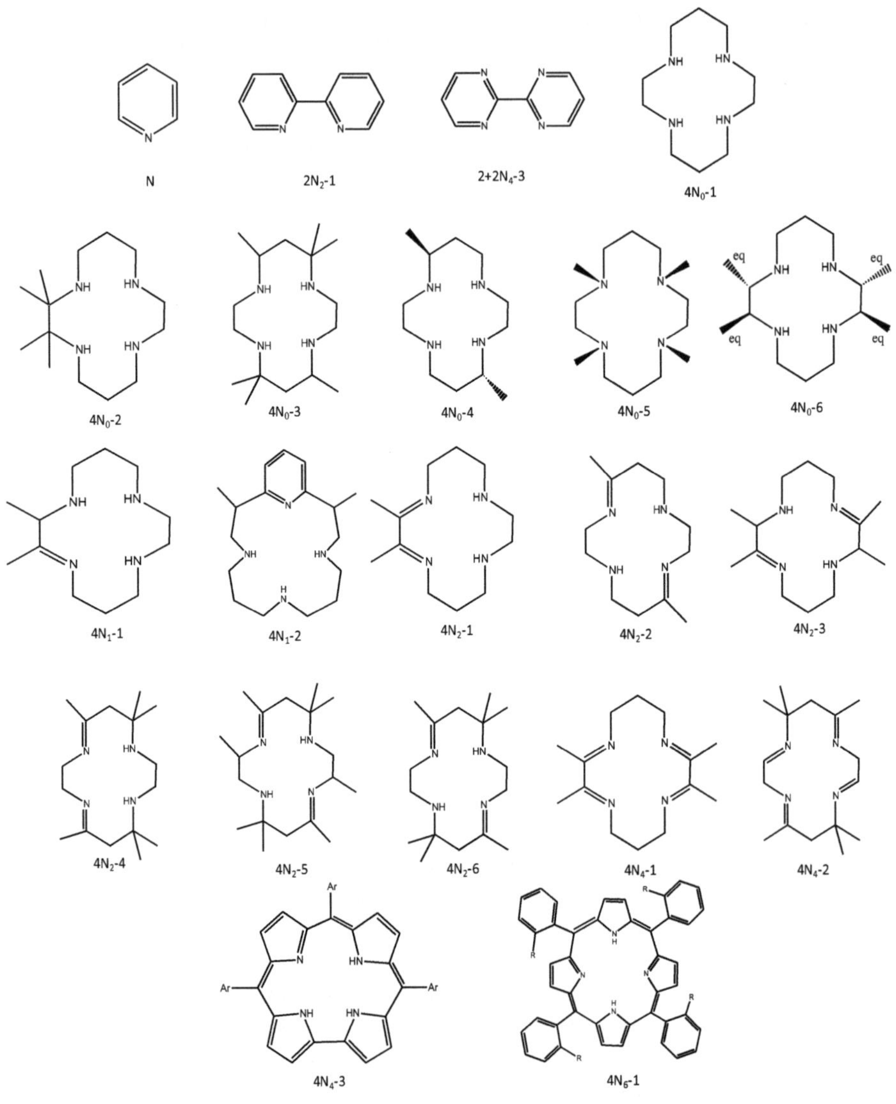

Fig. 8.3 N-ligands used as catalysts or for stabilizing metal centres in the electrochemical reduction of CO_2. The *first figure* indicates the number of N-donor atoms in the ligand, the *second* gives the number of unsaturated N=C bonds in the ligand and the *third* is an ordering number of ligands in the same class. Adapted with permission from [2]. Copyright (2012) Royal Society of Chemistry

noted that water is not one of the best solvents for CO_2 (Fig. 8.5). At 293 K its value (0.034 mol/L) is very close to one order of magnitude lower than in acetone, and lower than that in most organic solvents. The low solubility of CO_2 in water sets the limit to operative parameters such as the current density (usually below 10 mA m^{-2}),

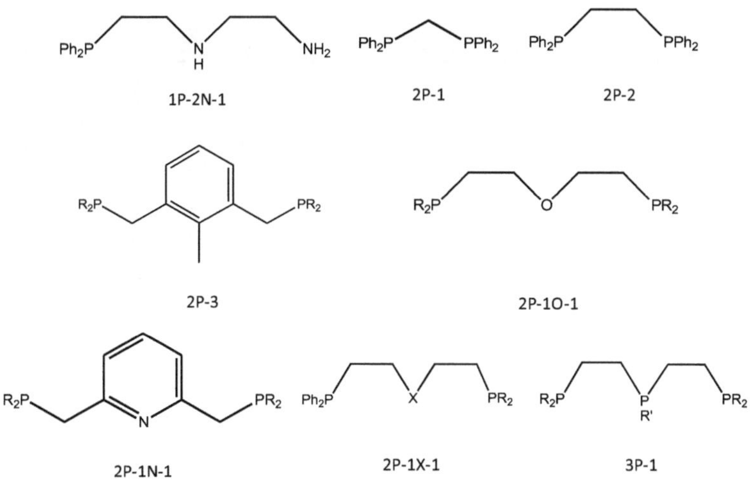

Fig. 8.4 Selected P-ligands used in the electrochemical reduction of CO_2. X can be As, S, O, N or more sophisticated centres

and the diffusion limitation influences the kinetics of a process. Water is often added to organic solvents as a source of protons when the latter are necessary. As discussed above, multielectron transfer coupled to proton transfer produces better energetic conditions for CO_2 reduction. In such a case, up to ca. 1 vol.% water can be added without changing the solubility of CO_2 in the organic solvent. The use of devices which operate under pressure [32] is now attracting much attention in order to eliminate barriers caused by the low concentration of CO_2 in solution; gas diffusion electrodes, which concentrate CO_2, mimicking natural systems, are also used to this end [33]. The equilibria which occur in water ((8.1)–(8.4)) make the situation much more complex as CO_2 is present in water in many different forms which have completely different behaviours and reactivities. The equilibria are influenced by the pH of the solution and Fig. 8.6 [32] gives the abundance of the various species as a function of the pH of the solution. The concentration of $[CO_2]$ in a given pH condition represents its solubility whereas the sum of the concentrations of all the species is the "apparent" solubility of CO_2 under the selected conditions. This is an important distinction as the reactivity of $[CO_2]$ is different from that of hydrogen carbonate or carbonate and different conditions may be required. However, it is important to refer to $[CO_2]$ and to know what is its real value at a given temperature and pH in a solvent.

A paradox is that in a basic solution (pH = 10) a much greater amount of CO_2 is captured in water than at pH = 7, but the available $[CO_2]$ at pH = 10 is almost equal to zero, although it is appreciable at pH = 7 (Fig. 8.6). Consequently, basic conditions are not suitable for studying the reactivity of CO_2 itself: other species such as HCO_3^- or CO_3^{2-} react instead. This is a key point when dealing with reactions in

Table 8.2 Selected examples of CO_2 reduction to C1 and C1+ species, electrodes, electrocatalysts, ligands, operative conditions

E	Product of reduction of CO_2	Structure of the ligand	Metal centre	Potential V vs. SCE	Current[a] efficiency, $\%\eta_c$	Co-product %	Solvent	Process: EC, PEC, PC	TON (TOF)	Notes	Refs.
1	HCO_2	2N – 1	Rh	−1.55	64 (12 % for H_2)	H_2 (12), from support electrolytes	H_2O	EC	6.8–12.3		[3]
2		2N – 1, CO	Os, Ru	−1.55	25	CO	Water	EC			[4]
3		2P – 1	Rh	−1.55	42 (down to 22)	Cyanoacetate	CH_3CN–H_2O	EC			[5]
4		S, -SR	Fe_4S_4	−1.7			DMF	EC			[6]
5		2N – 1, CO, H	Ru sensitizer Re catalyst		0.12 (Φ)			PC	170 (80.5)	DYAD better performance than single systems	[7]
6		None	Mg_{doped} $CuFeO_2$	−0.9	10	H_2		PEC		Blue LED , 470 nm, 2.1 mW/cm^2	[8]
7	CO		Co, Ni	−1.3 to −1.6	98 (Ni, 4 N_0 – 2)	H_2	H_2O	EC	(2–9)/h	High overpotentials	[9]
8		$4N_0$ – 1	Ni	−0.86	96	H_2	H_2O	EC		Catalyst absorbed on Hg, Sensitive to pH. Support electrolytes influence selectivity. KNO_3 and $KClO_4$ are the best	[10]
9		$4N_6$ – 1	Fe	−1.5	>99	$H_2 < 1$	DMF	EC	(350/h)	Need weak BrØnsted acids. Slow decay (1 %) per run	[11]

10		Corroles $4N_1-3$	Ni, Fe	−1.7				PEC		Decay of the catalyst after 10 h. NEt_3 sacrificial Terphenyl as sensitizer	[12]
11		2N − 1, CO	Re	−1.49	98	H_2, <1	9:1 DMF: H_2O	EC	(21.4)/h	High selectivity	[13]
12		2N − 1, CO	Ru	−1.4	>50	H_2, HCO_2^-	No water	EC			[14]
13		3P − 1, 2P − 2, 2P1N − 1, 2PX − 1, 2PO − 1	Pd	−0.63 to −0.83	>90	H_2	Acidic CH_3CN	EC	10–300	Hydroxycarbene intermediate.	[5, 15–17]
14		2P − 1, CNR, MeCN	Ni_2	−0.85				EC		CO is bound to the metal	[18]
15		2P − 1. MeCN, I	Ni_3	−1.08 to −1.18		CO_3^{2-}	CH_3CN dry	EC		1e⁻- transfer, no oxalates	[19–21]
16		1P2N − 1, 1N − 1	Cu_2	−1.35, −1.53		CO_3^{2-}	CH_3CN	EC	(2)/h	Two waves	[22]
17		2N − 1, CO, CH_3CN	Re		0.59 (Φ)		DMF	PC		Φ = rate/light intensity	[23]
18		2N (4,4′) Bu^t, CO	Re	−0.9				PEC		p-type H-Si electrode	[24]
19		2N–(4,4′) Bu^t,CO	Mn	−1.5	100			EC	(340)/s	Weak acids are needed	[25]
20	CH_2O		$1N–H^+$								[26]

(continued)

Table 8.2 (continued)

E	Product of reduction of CO_2	Structure of the ligand	Metal centre	Potential V vs. SCE	Current[a] efficiency, $\%\eta_c$	Co-product %	Solvent	Process: EC, PEC, PC	TON (TOF)	Notes	Refs.
21	CH_3OH		1N–H$^+$		18–28	HCO_2H					[26]
		None	n-GaAs							Selective towards CH_3OH	[27]
22	CH_4	2N – 1	2N – 1, Ru or Os colloid		10^{-4} (Φ_{CH4}) 10^{-3} (Φ_{H2})	H_2		PC		Triethanolamine as donor	[28]
23		None	Cu poly	−2.3	12 %. 85 %	CO(8) C2 (7)	H_2O			Phosphate buffer	[29]
24		None	Cu	−1.44 (SHE)	103.5	C_2H_4 (25.5), H_2 (20.5), EtOH (5.7), PrOH (3)		EC		Several other metal electrodes used. A classification attempted	[30]

[a]The faradaic efficiency is defined as: (mol product/mol e^-)ne^- , where ne^- is the variation of the oxidation umber of C from +4 in CO_2 to that of the product

Table 8.3 Solubility of CO_2 in organic solvents at 293 K

Acetone	0.29 (mol/L)
Dimethylformamide	0.20
Methanol	0.18
Pyridine	0.16
Chloroform	0.15
Ethanol (97 %)	0.12
Benzene	0.11
Toluene	0.10

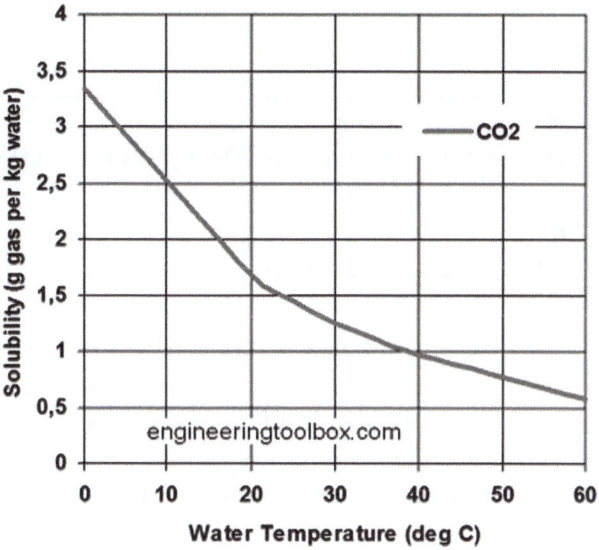

Fig. 8.5 Solubility of $CO_{2(g)}$ in water

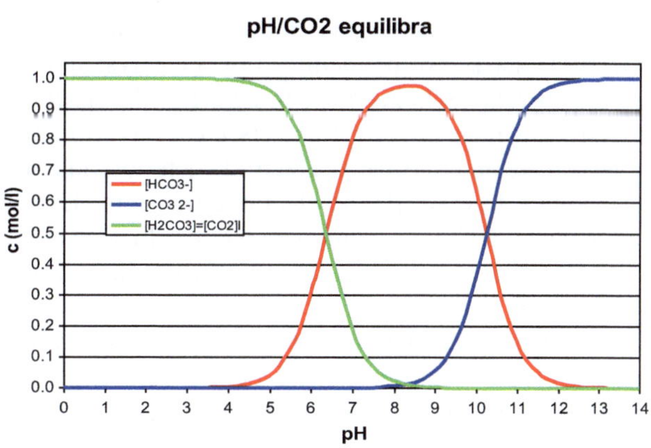

Fig. 8.6 Abundance of CO_2 species as function of pH of a water solution

water. The risk is high that when one discusses the reactivity of CO_2, it is the corresponding properties of hydrogen carbonate or carbonate which have been measured.

$$CO_{2(g)} + H_2O_{(l)} \rightarrow [CO_2 \cdot H_2O]_{(aq)} \tag{8.1}$$

$$[CO_2 \cdot H_2O]_{(aq)} \rightarrow H_2CO_{3(aq)} \tag{8.2}$$

$$H_2CO_{3(aq)} + H_2O \rightarrow H^+{}_{(aq)} + HCO_3{}^-{}_{(aq)} \tag{8.3}$$

$$HCO_3{}^-{}_{(aq)} + H_2O \rightarrow H^+{}_{(aq)} + CO_3{}^{2-}{}_{(aq)} \tag{8.4}$$

(A)

Surface + CO_2 + ne^- → Surface-CO_2($^{n-}$) → Surface + CO_2-reduction products + ...

(B)

Surface + Catalyst + ne^- → Surface + Catalyst($^{ne-}$)

substrate

Catalyst ($^{ne-}$) + CO_2 → Catalyst-CO_2($^{ne-}$) → Catalyst + CO_2-reduction products +...

CO_2

Catalyst ($^{ne-}$) + substrate → Catalyst-substrate($^{ne-}$) → Catalyst + CO_2-reduction products +...

Scheme 8.1 Direct (**A**) and catalyst-mediated (**B**) electroreduction of CO_2 in solution

Another key fact is that the electroreduction of CO_2 may take place at the electrode surface or be mediated by a catalyst attached to the electrode or soluble in the solvent (Scheme 8.1). In case (**A**) the reaction occurs at the surface of the electrode which directly transfers electrons to CO_2: this generally requires high overvoltage and may cause electrode deterioration or even consumption and scarce selectivity. In case (**B**) the electrode transfers electrons to the catalyst which then interacts with CO_2 causing its reduction. In such a case, CO_2 is not in contact with the electrode, and this results in better stability of the electrode and its longer life, which may be important in the overall costing determination of a process.

Another advantage of electrocatalysis is that overvoltages are reduced almost to zero and the electron transfer occurs near the thermodynamic potential of the reaction, e.g. E°(substrates/products), which one wishes to perform.

Solar to Electricity	>>>	**Electricity to Hydrogen**	>>>	**Hydrogen (and CO_2) to Fuels**
PV		(water as H-source)		(CH_3OH or CH_4)
StE		**EtH**		$\mathbf{HtF}_{l\ or\ g}$
η=20%		η=70%		70-80%

Overall "Solar to Fuel-StF" maximum efficiency = (20x70x80)%= 11.2%

Scheme 8.2 Efficiency in the conversion of solar energy, water and CO_2 into fuels

In addition to electrochemical reduction (EC), photoelectrochemical (PEC) reduction is playing a more and more important role. In PEC reduction, a photoactive material is attached to the electrode surface and the excitons produced upon irradiation generate electrons used for the reduction process. Alternatively, photochemical (PC) processes can be used in which semiconductors produce excitons upon irradiation and generate electrons and holes used in the reduction of CO_2 and oxidation of an electron donor (water should be used). All such methodologies have in common the "*electron plus proton*" transfer to CO_2. Recently the use of solar energy for powering the reduction of CO_2 is much under consideration as this would represent a way to convert low-value CO_2 into useful chemicals and fuels, mimicking nature and natural photosynthesis. The use of PV as source of electrons in reduction reactions is receiving much attention today as it may represent an effective way to the conversion of large volumes of CO_2. As a matter of fact, the use of PV for CO_2 reduction using water as electron donor is already an attractive technology, although the cost may appear high. From the solar to chemicals/fuels (StC/F) energy conversion point of view, it must be pointed out that the most ready technology is based on the sequence shown in Scheme 8.2.

Having said that, we can now consider in detail the electron-transfer processes and catalysts in pathways for CO_2 (and water) conversion into fuels.

8.3 One-Electron Transfer to CO_2 vs Multi-electron Transfer

The one-electron transfer to free CO_2 affords $CO_2^{\cdot -}$, as seen in Chap. 1. Table 8.1 shows that it requires much more energy than multielectron transfers. Moreover, the potential is dependent on the medium in which the electron transfer occurs: polar media (−1.9 V) or apolar media (−2.21 V). The one-electron transfer generates the $CO_2^{\cdot -}$ species which can evolve according to two different routes (Scheme 8.3).

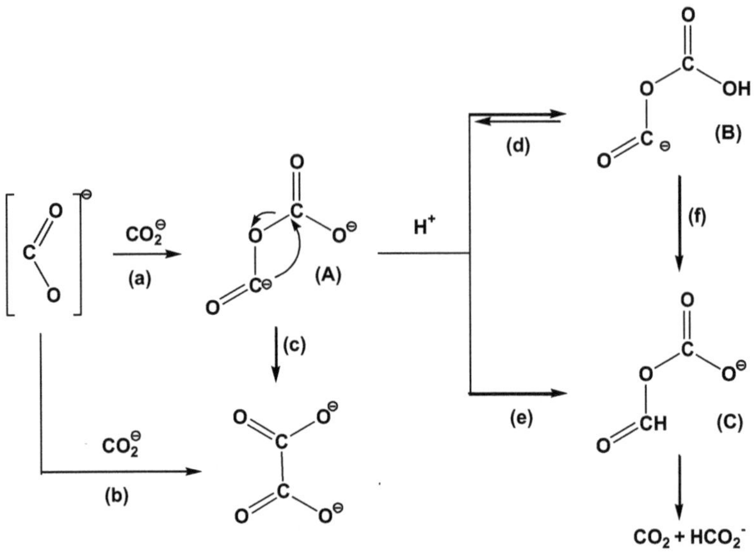

Scheme 8.3 Reaction mechanisms for the conversion of the radical anion $CO_2^{\cdot -}$

The radical anion $CO_2^{\cdot -}$ has a life as short as microseconds, as discussed in Chap. 1. It can be stabilized by interaction with the solvent or radical traps. It can attack a second CO_2 molecule via a *hetero* O–C coupling and generate the dianion [A], in Scheme 8.3 which can be the source of CO and carbonate, as already seen. However, by protonation, the following couple of products can be obtained: either CO and hydrogencarbonate HCO_3^- or CO_2 and formate HCO_2^-.

Vice versa, the *homo* C–C coupling of two radical anions with antiparallel spin of the electron localized on C produces the oxalate anion [B]. The e^--transfer to "free" CO_2 has been attempted for the synthesis of oxalates, with alternate fortune [34]. Whether the homo-coupling or the hetero-coupling is favoured depends on several parameters such as the solvent, the current density, and the electrode.

All anions need a cation for neutralization, which is provided by the electrode, in general, which is consumed. Such technology is scarcely exploited at the industrial level.

8.4 Competitive Coordination of CO_2 and H^+ to a Catalytic Centre and Their Reduction

When the electron transfer involves CO_2 coordinated to a metal centre, then several factors determine both the kinetics and thermodynamics of the reaction. Such factors can be categorized as follows: the metal, the nature and structure of the ligand, the geometry of the metal-complex, the solvent: they may affect the electron transfer process in many ways. Moreover, the addition of a proton to the metal centre and its consequent reduction to H_2 is a side-by-side process with CO_2

coordination-reduction. However, as shown in Table 8.2, the reduction of CO_2 in water is not a simple process and the selectivity towards a target C1 or C1+ molecule containing CO_2 in a reduced form is quite variable with the system used, having the formation of either hydrogen or other C-species as a competing process. Some of the competitive reactions are shown in Scheme 8.4.

$$L_nM^{n+} + H^+ \rightarrow M\text{-}H^{(n+1)+} \quad (8.5)$$

$$L_nM^{n+} + CO_2 \rightarrow L_nM^{n+}(CO_2) \quad (8.6)$$

$$L_nM\text{-}H^{(n+1)+} + e^- \rightarrow L_nM^{n+} + \text{“H”} \rightarrow L_nM^{n+} + \mathbf{1/2H_2} \quad (8.7)$$

$$L_nM\text{-}H^{(n+1)+} + CO_2 \rightarrow L_nM\text{-}OC(O)H^{(n+1)+} \quad (8.8)$$

$$L_nM\text{-}OC(O)H^{(n+1)+} + e^- + H^+ \rightarrow L_nM^{n+} + \mathbf{HCO_2H} \quad (8.9)$$

$$L_nM^{n+}(CO_2) + e^- \rightarrow L_nM^{n+}(CO_2)^- \quad (8.10)$$

$$L_nM^{n+}(CO_2)^- + H^+ \rightarrow L_nM^{n+}(CO_2H) \quad (8.11)$$

$$L_nM^{n+}(CO_2H) + e^- + H^+ \rightarrow L_nM(\mathbf{CO})^{n+} + \mathbf{H_2O} \quad (8.12)$$

Scheme 8.4 Competitive coordination to a metal centre and metal-catalysed reduction reactions implying the proton and CO_2. The formation of H_2, CO and HCO_2H are highlighted

A variety of metal systems have been used as catalysts in the electroreduction of CO_2, the most studied being Ni, Co, Pd and Ru. In the photochemical or photoelectrochemical processes, Ru and Re are mainly used. Table 8.2 presents a list of metals, ligands, conditions and products of the electrocatalysed reaction of CO_2.

A general route is the sequential “$e^- + H^+$” transfer to CO_2 coordinated to the metal centre.

In electrocatalytic processes, electrons are provided by the electrode and protons come from the solution. In photochemical processes, electrons are provided by a semiconductor irradiated with the suitable radiation (UV or visible) and used for the reduction of CO_2, whereas the cogenerated “hole” is used for the oxidation of water or a sacrificial molecule. In photoelectrochemistry the two effects are coupled. A key issue is the “overpotential” necessary for e^--transfer. This means that the reduction process does not occur at the “thermodynamic” potential but at more negative values. The difference plays a key role in determining the cost and exploitability of the process. The overpotential can be reduced by using electrocatalysts which may present an easier adjustability of the energy of the molecular orbitals so that the electrons can be transferred close to the thermodynamic potential. Table 8.2 shows that the same reduction process can occur under quite different (column 4) conditions, depending on the catalyst used.

The first step in the reduction process is the binding of CO_2 to the metal centre ((8.6) in Scheme 8.4). The reversible binding to a tetracoordinated $(4N_2-6)$Co (I) was first electrochemically measured [35] to range around $7 \times 10^4\ M^{-1}$ in dry dimethylsulphoxide (dmso). Spectroscopic methods were also used, including XANES (X-ray Adsorption Near Edge Structure) [36–42]. The bound-CO_2 experiences a large charge-shift from Co(I) as shown by the IR shift of the $v_{asym}(C{=}O)$ located at 1 710 cm^{-1}. Very interestingly, upon simple cooling of the solution a further shift down to 1 544 cm^{-1} is observed. XANES measurements have shown that at room temperature (298 K) one full unit of charge is transferred from Co to CO_2 which results in a CO_2^- state with Co in a +2 oxidation state. Conversely, when the XANES measurements were carried out at low temperature (198 K), two electron-transfer to CO_2 is demonstrated so that the adduct results in a $Co^{III}–CO_2^{2-}$ state with CO_2 bonded in a carbene-like mode $Co^{III}{=}CO_2^{2-}$. Such a two-electron transfer simplifies the reduction of CO_2 with respect to the one-electron transfer which is energetically hindered (Table 8.1). Interestingly, it has been shown via FTIR study [36–42] that the CO_2^- species bound to Co is stabilized by interactions between the O atoms and the amine hydrogens of the $4N_2-6$ ligand, as represented in Fig. 8.7. Moreover, when co-ordinating solvents are used, a solvent molecule (such as CH_3CN) can occupy the sixth coordination position on Co, which results in different properties as shown by the changes in the UV–vis spectrum [36].

DFT calculations (B3LYP/6-31 + G(d,p) set basis and CPCM for solvent effect) have been used [2] to investigate the $Co \rightarrow CO_2$ e^--transfer in both $(4N_2-6)$ $Co^I(CO_2)$ and its solvated (CH_3CN) form (Fig. 8.8).

The CoL–CO_2 system above has been largely investigated [36, 38, 39, 43] for the determination of the CO_2-binding constants and rate constants found to depend largely on the ligand used and on the solvent which may effectively solvate CO_2 [37]. The binding-constants have been found to vary in the interval <1 to $10^6\ M^{-1}$, although the rate constants range around 10^6 to $10^8\ M^{-1}\ s^{-1}$. A correlation has been found between the two values. These studies have been coupled with the determination of the $E_{1/2}$ shown to vary in the interval -0.1 to -1.4 V (vs SHE).

All such complexes show a limited stability as they convert CO_2 into CO (bound or free).

The same complexes can also coordinate "H" (transferred as a hydrogen atom or a proton [39, 44, 45]) as shown in (8.5). Interestingly, no insertion of CO_2 into the Co–H bond was observed although the Co–H system showed quite high pK_a of the order of 11–14.

Extended studies have also been carried out on Ni systems: $Ni(NH_3)_4F_2$ [2, 46] and Ni–N–macrocyclic ligand [47–50] complexes have been studied and modelled. The ligands and oxidation state of Ni play a key role in the stabilization of Ni–CO_2 adducts. The $E_{1/2}$ value for Ni(II)-complexes with a number of the ligands shown in Fig. 8.3 varies over the range -0.2 to -1.20 V (vs SHE), depending on the ligand and the solvent used.

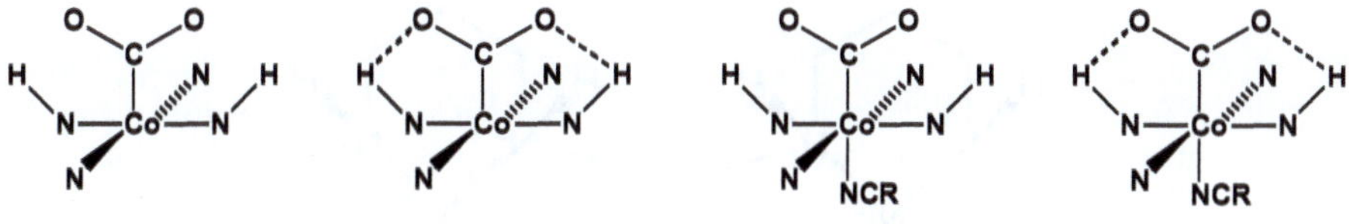

Fig. 8.7 Bonding of CO_2 to $(4N_2-6)Co^+$: both pentacoordinated Co-species and hexacoordinated complexes bearing a solvent molecule co-ordinated in the sixth position are shown. Intramolecular hydrogen-bonding contributes to stabilizing the bound CO_2. Reprinted with permission from [40]. Copyright (1993) American Chemical Society

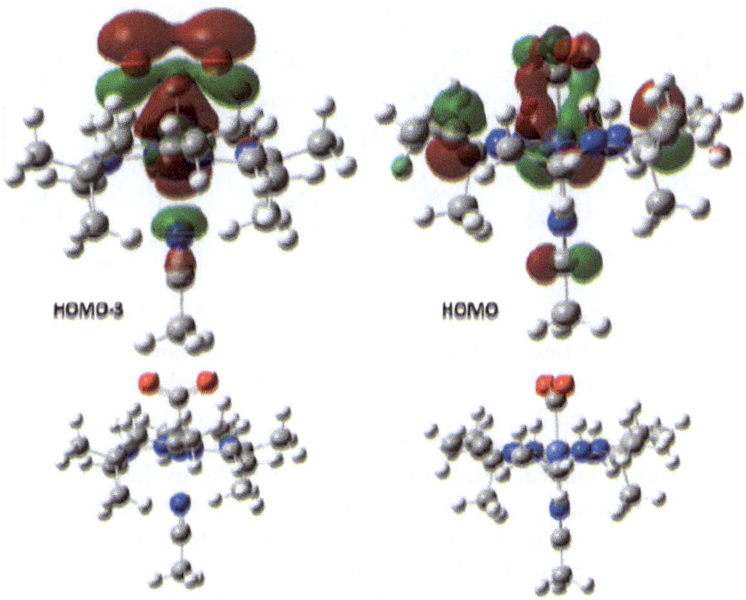

Fig. 8.8 Calculated molecular orbitals for the species $(4N_2-6)(CH_3CN)Co^I(CO_2)^+$. On the *left* the formation of a single bond between Co and C of the heterocumulene is shown. On the *right* the formation of the carbene-like species $Co{=}CO_2^{2-}$ is shown. Reprinted with permission from [2]. Copyright (2012) Royal Society of Chemistry

Using cyclam-related ligands characterized by various conformational configurations (Fig. 8.9), it was demonstrated that the ligand RSRS in Fig. 8.9 favours the binding of CO_2 as the four hydrogens on the same side interact with O atoms of the heterocumulene, reinforcing the bonding to Ni. The mode of bonding Ni–CO_2 is shown to be quite different from that encountered in Co–CO_2 adducts because of the nature of the orbitals involved. In the case of Ni, a doubly occupied molecular orbital shares its electrons with the C atom of CO_2 and the electron in a singly occupied orbital also contributes to the bonding.

The adducts of CO_2 with Co(I) and Ni(I) complexes show different properties and binding free energies [51].

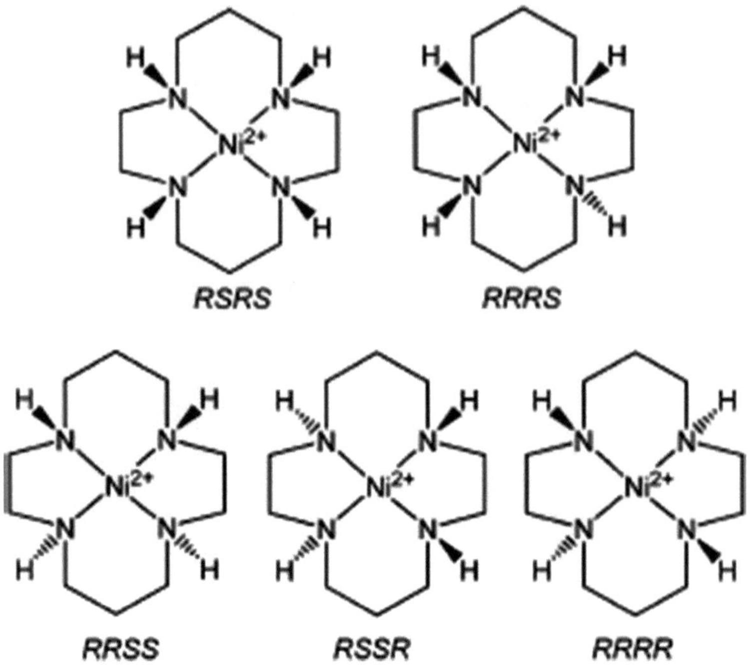

Fig. 8.9 Conformational isomers of Ni(II)-complexes with cyclam-related ligands. Reprinted with permission from [2]. Copyright (2012) Royal Society of Chemistry

All the parameters discussed above influence the catalytic activity of the metal systems in electrochemical reactions. In particular, the cyclam-related Ni complexes show an interesting selectivity towards CO_2 reduction to CO, also repressing the proton reduction in acidic water solutions. This distinctive behaviour with respect to Co has been ascribed to the different pK_a of the "LnNiH" systems with respect to the analogous Co complexes. For example, the pK_a of (cyclam)NiH^{2+} is <2, making the interaction of "cyclamNi^+" with CO_2 much easier than with H^+ in a water solution saturated with CO_2 in moderately acidic conditions (the addition of H^+ to "cyclamNi^+" only occurs at very low pH).

8.5 Sequential "*One-Electron Plus One-Proton*" Pathways in Multi-electron Reduction of Bound CO_2

The sequential transfer of "*one electron plus one proton*" mentioned above has been proposed to operate in several catalytic pathways. The reaction mechanism of CO_2 reduction has been investigated using several transition metal complexes as catalysts (see, for example, [2, 14, 15, 24]). For some of them a reaction mechanism has been demonstrated. A separate mention should be given to the alternative "organic" electrocatalysts, which do not use any metal (see below).

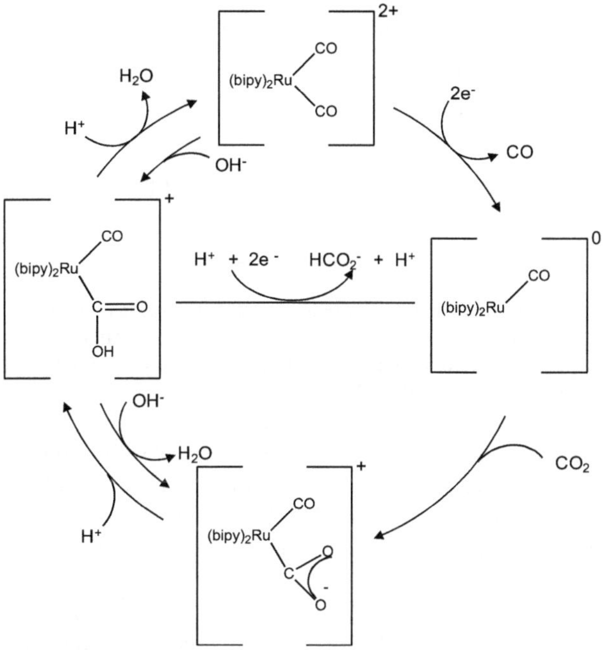

Scheme 8.5 Ru-catalysed reduction of CO_2 to CO, HCO_2^- and H_2 (external cycle). The key intermediate, the "$Ru(CO_2H)$" species, was also formed by addition of a hydroxyl group to co-ordinated CO (internal cycle). Reprinted with permission from [14]. Copyright (1987) American Chemical Society

Early studies carried out by the J.-M. Lehn group [13] with $Re(2N_2 - 1)(CO)_2Cl$ have clearly demonstrated the competition proton-CO_2 towards electrons. The reduction in anhydrous solvents selectively (>98 %, TOF 21.4 h^{-1}, -1.49 V vs SCE) produces CO, whereas a growing amount of added water produces increasing amounts of H_2. Such behaviour is not a negative fact as one may decide to produce Syngas by electroreduction of CO_2 in water, supposing that robust electrocatalysts with high TOF and TON are developed. In 1987, Tanaka et al. [14] proposed the catalytic cycle in Scheme 8.5 for the Ru-catalysed reduction of CO_2 to CO, H_2 and HCO_2^- at -1.40 V (SCE).

The protonation of "$RuCO_2H$" to afford "$RuC(O)OH_2$" causes water release and the formation of CO. The oxidation state of Ru shuttles between +2 and 0. The $2e^-$ transfer to Ru(II) produces the Ru(0) which is responsible for CO_2 coordination followed by a double protonation. In such a mechanism *two electrons* are transferred to Ru in one step and two protons are then added to CO_2

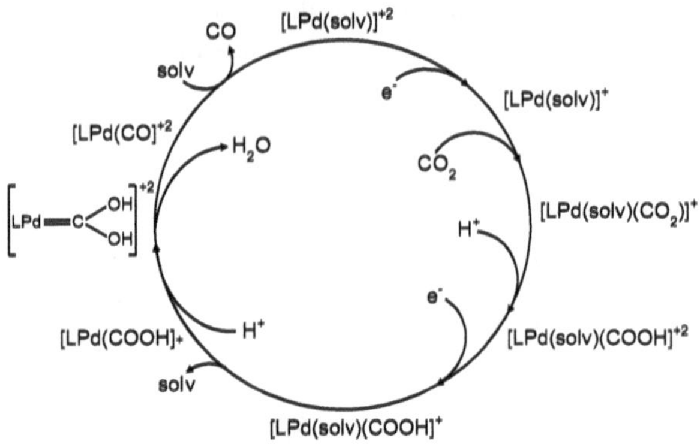

Scheme 8.6 Proposed mechanism for CO_2 reduction: the formation of a dihydroxo-carbene (*left*) as intermediate and the two-($1-e^- + H^+$) consecutive transfers are distinctive features with respect to the Tanaka mechanism shown in Scheme 8.5. Reprinted with permission from [17]. Copyright (1997) Comments Inorg Chem 19:307–325

under slightly acidic conditions (pH = 6) with increase of the oxidation state of the metal back to +2. Under more basic conditions, formate is obtained instead of CO. Interestingly, changing Rh for Ru (E1, Table 8.2) with the same ligand produces formate more than CO, also with average TOF and selectivity.

Other transition metal complexes (Co, Fe, Ni, Pd) with variously decorated phosphane ligands have been used, showing different behaviours and catalytic activities depending on the ligand (Fig. 8.4), the metal, and the conditions (solvent and water presence). Pd-systems have been by far the most studied by DuBois and coworkers [15–17]. The authors have shown that on using the dicationic complex bearing 3P – 1 (Fig. 8.4) and a tertiary phosphane as ligands to Pd (namely, $[(3P-1)(PR_3)Pd(BF_4)_2]$), the latter is first converted into the active form through the substitution of the phosphane with a solvent molecule (CH_3CN). Such form is able to promote *two-(1-e^- + H^+)* consecutive transfers as shown in Scheme 8.6.

The Pd oxidation number is reduced from +2 to +1 by accepting a single electron. The Pd(I) is the active form which picks-up CO_2. Further protonation and electron transfer form the dihydroxocarbene which eliminates water to afford the dicationic labile-carbonyl complex, which eliminates CO under solvent co-ordination. In an attempt to improve the catalytic properties (TOF and TON) the polydentate P-ligand has been modified in many different ways by changing the central P-donor atom as shown in Fig. 8.4 without any apparent gain in efficiency.

A different mechanism always based on a double consecutive ($e^- + H^+$) transfer has been proposed by the same authors (Scheme 8.7), in which the formation of the "Pd-C(CO)OH_2" intermediate more than the hydroxyl-carbene is postulated.

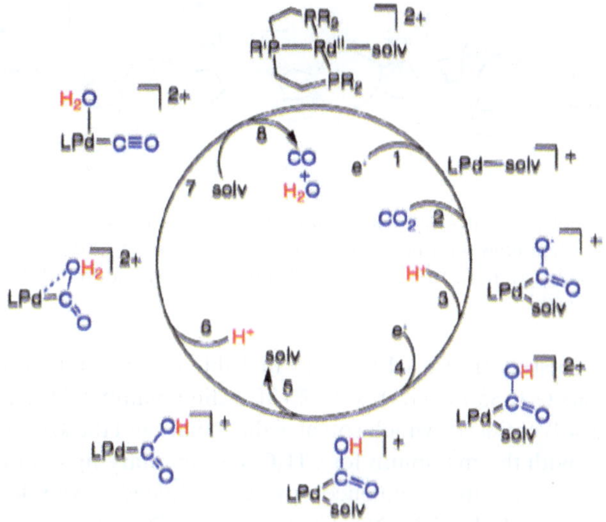

Scheme 8.7 Alternative mechanism for the Pd-electrocatalysed reduction of CO_2. Reprinted with permission from [52]. Copyright (2013) American Chemical Society

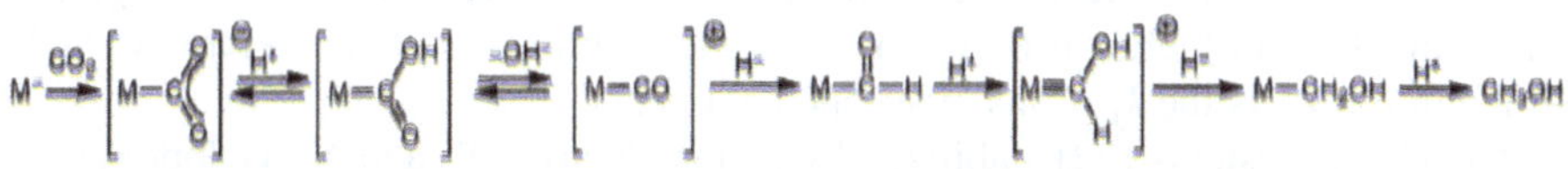

Scheme 8.8 Multiple "$e^- + H^+$" transfer to CO_2 with formation of formate, CO, formyl, hydroxyl carbene and hydroxymethylene intermediates and methanol as end product. Reprinted with permission from [53]. Copyright (2010) Elsevier

Very recently [54] a $2e^-$ reduction of *N*-heterocyclic carbene-carbonyl Mn(I) to Mn(−I) has been reported which occurs at the same potential (−1.35 V vs SCE) instead of two consecutive potentials separated by a few hundred mV as shown for analogous $MnBr(CO)_x(2N-1)$. Such Mn complexes catalyse very selectively the CO_2 to CO reduction with a faradaic efficiency of 34.6 %.

All the above catalytic systems afford mainly CO, an interesting molecule as it may release energy in combustion with air, but more useful C1 or C1+ molecules can also be produced.

Scheme 8.8 shows the multiple "$e^- + H^+$" transfer to afford methanol. Such a sequence is very similar to the enzymatic reduction of CO_2 to methanol which is discussed in Chap. 9.

Cluster complexes consisting of di- and tri- homo- or hetero-nuclear species have also been used (Scheme 8.9), the latter with the aim of taking advantage of the possible co-operative effect of two different metal centres each able to perform a different activity.

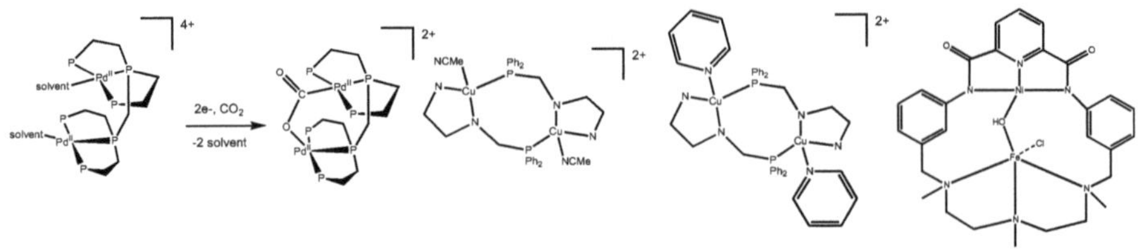

Scheme 8.9 Di-nuclear homo- (*left* and *centre*) and hetero-nuclear (*right*) complexes for the reduction of CO_2. Reprinted with permission from [47] Copyright (2013) American Chemical Society, from [55] Copyright (2010) American Chemical Society and from [1] Copyright (2009) Royal Society of Chemistry

Ni_2 and Ni_3 clusters (Entries 14 and 15 in Table 8.2) are able to reduce CO_2 at various potentials (−0.85 to −1.18 V SCE) affording mainly CO. The kinetics [56] depends on the solvent used (which solvates the metal) and the k_{CO2} range from 0.1 to 1.6 $M^{-1}\,s^{-1}$, with the maximum for CH_3CN as ancillary ligand in a Ni_3 cluster. The k_{CO2} trend in the sequence of investigated cyanides [56] was: $CH_3CN{=}C_6H_{11}CN > C_6H_5CH_2CN > CO > (t\text{-}C_4H_9)CN$, demonstrating that both electronic and steric hindrance factors may play a role in the kinetics.

The metal-catalysed [57] reduction is favoured with respect to the participation of the cyanide in the reduction of CO_2 and shows two peculiar features: the nature of the ancillary ligand (various cyanides as reported above or CO) does not greatly affect the $E_{1/2}$ (which varies in the restricted range −1.08 to −1.18 V SCE) although it affects the k_{CO2}, as mentioned above.

The Cu_2(I) systems [22] stabilized by ligands bearing P- and N- as donor atoms (Scheme 8.9, middle) reduce CO_2 to CO and the CV shows two consecutive e^--transfers at −1.35 and −1.53 V (vs SCE). As reported in Table 8.2, carbonates are also formed, inferring a mechanism such as that shown in Scheme 8.3.

The use of pyridine as an organic catalyst [26, 27, 58] has been proposed by Bocarsly and coworkers. The interesting aspect of such electrocatalysts is that they can promote several "$1e^- + H^+$" transfer steps, affording several reduced species of CO_2, such as HCO_2H, CH_2O and CH_3OH, but also C2 and C3 species. Quite recently, the authors have critically reviewed [59] the proposed mechanisms and report the possible role of H-bonded pyridinium-to-pyridine dimers in the electrocatalytic reduction of CO_2 (Scheme 8.10).

A possible role is also attributed to pyridine-*N*-carboxylic acid (Scheme 8.11) for the generation of the $^{\bullet}CO_2H$ moiety which is then further reduced to other C1 species. Scheme 8.12 provides a mechanism for the formation of several C1 species.

This chemistry is relevant to biochemical systems discussed in Chap. 9.

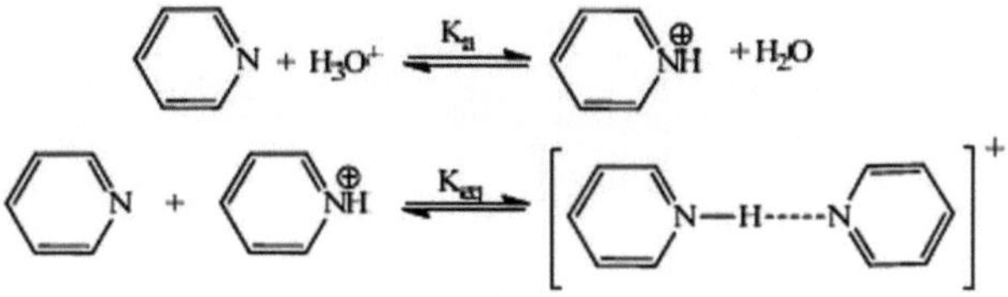

Scheme 8.10 H-bonded pyridine-pyridinium ion dimers as possible actors in the e^--transfer. Reprinted with permission from [59]

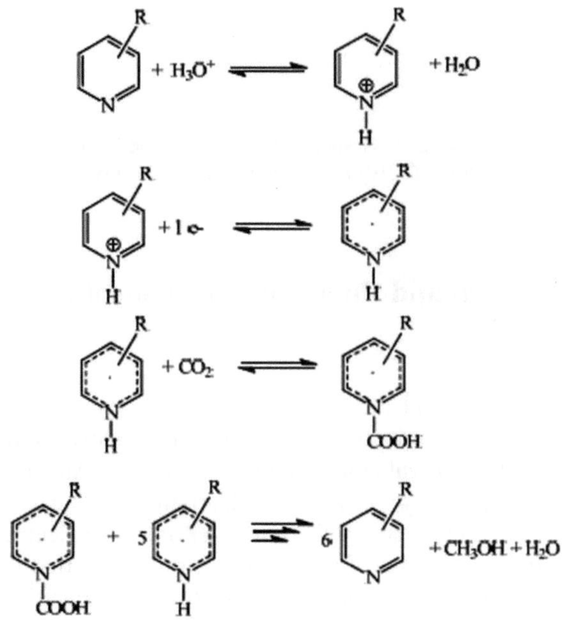

Scheme 8.11 Formation of pyridine *N*-carboxylic acid as intermediate in the reduction of CO_2 to other energy-rich C1 species such as HCO_2H, CH_2O and CH_3OH. Reprinted with permission from [59]

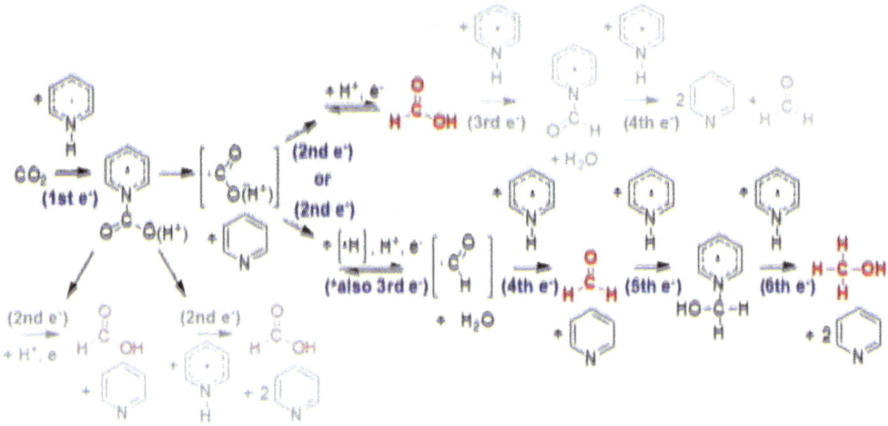

Scheme 8.12 Formation of several C1 species mediated by the pyridinium ion. Reprinted with permission from [26]. Copyright (2010) American Chemical Society

8.6 Photochemical and Photoelectrochemical Reduction of CO_2

The use of solar energy and other perennial primary sources of energy (wind, geothermal, hydro) for the reduction of CO_2 has great relevance to the conversion of large volumes of CO_2 to valuable chemicals and fuels. All such sources can be used in a indirect way, as they can generate electricity which can be used either, as indicated above, for the reduction of CO_2 in water or for the electrolysis of water to afford hydrogen that can be used for CO_2 reduction as reported in Chap. 7. In the very short term the use of excess electric energy (off-peak electricity), especially if produced from perennial sources, is a suitable strategy; in the medium term the use of PV can results in a viable approach [60]. Today, the cost of PV is too high (Table 8.4) for practical application.

This implies that the price of chemicals produced by using the PV-generated H_2 is much higher than that of products obtained using H_2 produced by reforming methane.

Table 8.4 Comparison of the cost of production of H_2 (1 kg) using different technologies

Technology	Gas reforming	Electrolysis with nuclear	Electrolysis with oil as primary energy source	Electrolysis with PV	Electrolysis with wind	Electrolysis with SPC
Cost €/kg	1.10 1.15	1.45 1.50	1.50 2.20	2.8 4.0	3.8 5.20	3.50

Anyway, if one considers the perspective increase of the efficiency of PV (from 20 % to 40 %) and the reduced cost of PV materials, one can foresee that the price of PV-generated H_2 decreases and gets close to that of H_2 by reforming. So, in the short term, H_2 generated via PV or wind- or geothermal- or hydro-electricity may compete with H_2 produced by methane reforming.

The use of solar energy assumes a particular role as solar radiation can be used directly in photochemical (PC) and photoelectrochemical (PEC) processes. The former use either soluble (homogeneous) or insoluble (heterogeneous) compounds which may capture solar radiation which generates excitons (e^- + hole) used for the direct reduction of CO_2 and oxidation of the donor (water or sacrificial reagents) whereas in PEC processes, electrons generated in photoreactions are used in conjunction with an electrode.

In a homogeneous photochemical process (Scheme 8.13) solar radiation is captured by a soluble sensitizer (P) in the presence of a donor (D) whereas a solid photomaterial is used in heterogeneous photochemical processes (Scheme 8.14). The photosensitizer is excited to P* with production of an exciton (e^- + hole). The hole oxidises the donor (water or a sacrificial reagent) as the negative charge is transferred to a metal centre (M_{red}) where the CO_2 is reduced to C1 or C1+ species with reoxidation of the metal centre.

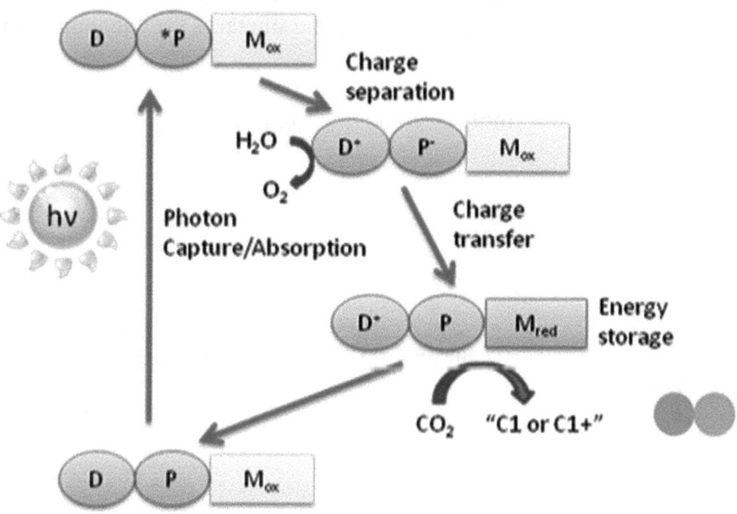

Scheme 8.13 Photochemical reduction of CO_2. Adapted with permission from [61]. Copyright (2013) Elsevier

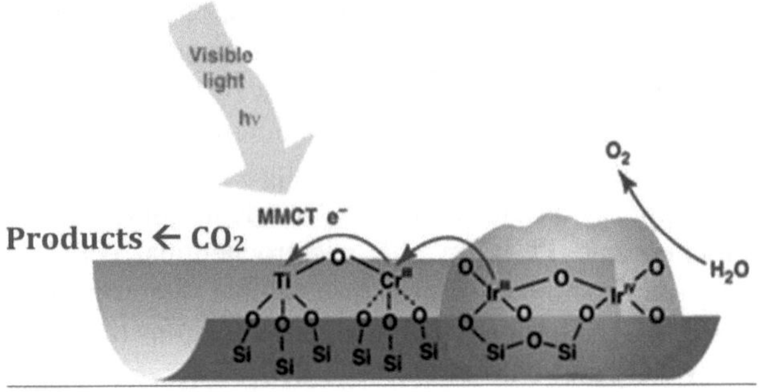

Scheme 8.14 Inorganic photomaterials for CO_2 reduction in water. Adapted with permission from [61]. Copyright (2013) Elsevier

Several sensitizers have been used, such as metal complexes (Ru, Re) with the 2N − 1 ligand and its derivatives or organic complex molecules. Solid inorganic materials (http://www2.lbl.gov/mfea/assets/docs/posters/14_Artificial_Photosynthesis) can also be used as heterogeneous photocatalysts.

Another approach to the use of solar energy is the hybrid system [62–66] formed by enzymes for the selective reduction of CO_2 and solar energy and photomaterials for the regeneration of cofactors (Scheme 8.15). This topic is discussed in Chap. 9.

In all the above-mentioned systems, a key parameter is the "quantum efficiency" expressed as

$$\Phi = \left(\text{mol}_{\text{products}}/n_{\text{photons absorbed}}\right) \times \text{ne}^- \tag{8.14}$$

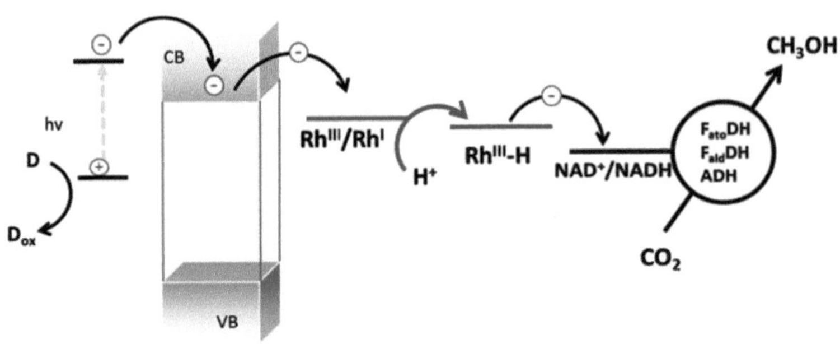

Scheme 8.15 Hybrid systems for the reduction of CO_2 to methanol in water at room temperature

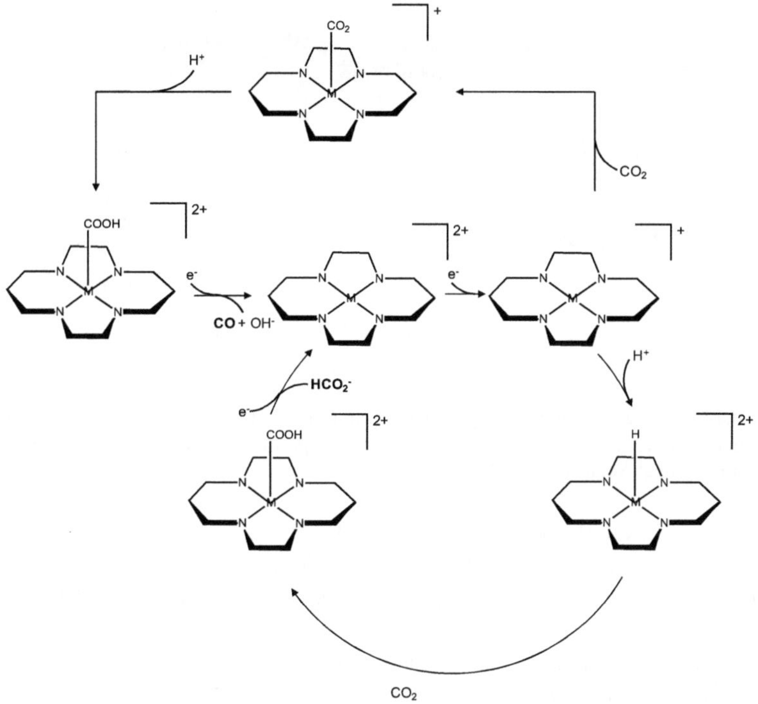

Scheme 8.16 Reaction mechanism for the formation of CO and HCO_2^- in the photochemical reduction of CO_2. Two consecutive e^- ($+H^+$) transfers of co-ordinated CO_2 to the metal in a low oxidation state were proposed with an unusual OH^- elimination in the formation of CO. Adapted from [67]

where ne^- is again the number of electrons transferred to CO_2. The quantum yield is a parameter somewhat similar to the "faradaic efficiency" defined in Table 8.2.

Among homogeneous photocatalysts, it is worth citing metal complexes with ligands having an extended system of conjugated double-bonds able to absorb solar light. phthalocyanines [68], corroles [12], corrins [69] or similar macrocyclic systems are typical ligands. The metal centres are Co or Fe or other metals. The negative point is that a low selectivity is observed with a large production of hydrogen, and this is also true for those systems, which under electrochemical conditions are quite effective for CO_2 reduction to C1 species (Scheme 8.16).

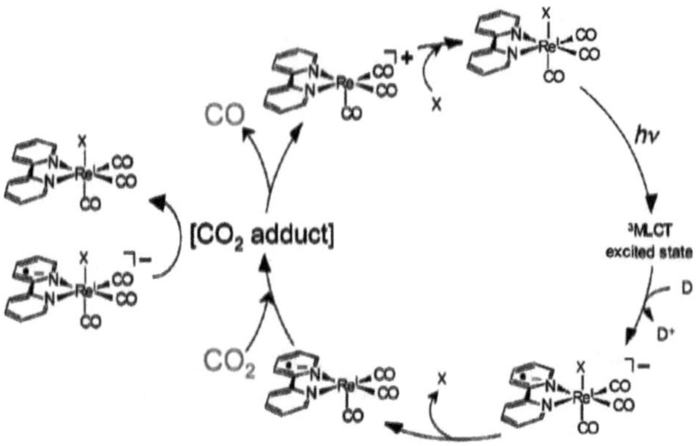

Scheme 8.17 Efficient Re-mediated photochemical reduction of CO_2 to CO (^{3}MLCT = metal-to-ligand charge transfer excited state). Reprinted with permission from [23]. Copyright (2008) American Chemical Society

In early studies in the 1980s [56, 70, 71], $Ru(2N-1)_3^{n+}$ was used as an efficient photosensitizer and organic molecules as e^--donors. The drawback was that the sacrificial reagents were more expensive than the products formed from CO_2, making such systems non-viable economically. An interesting case in which a unique complex acts as sensitizer and catalyst is represented by $Re(2N-1)(CO)_3X$ (X = Cl, Br) [70, 72] which is able to produce CO selectively.

Re complexes bearing phosphite ligands are the ones which show the best performance as photosensitizers for what concerns the quantum efficiency (0.38–0.59) [23, 73] using $Re(2N-1)(CO)_3(CH_3CN)$ as catalyst and $N(CH_2CH_2OH)_3$ as sacrificial reagent. The mechanism is shown in Scheme 8.17.

The use of solid photomaterials has been investigated since the 1980s using semiconductors as light absorbers and catalysts in several solvent media. The key issue in this case is that the band gap of the semiconductor used as photomaterials must match the potential of CO_2 reduction. Figure 8.10 shows the band gap expressed in V of several semiconductors compared with the reduction potential of CO_2 in one- and multi-electron transfer processes.

Although quite a few semiconductors are good photomaterials for CO_2 multielectron reduction under proton transfer regime, the one-electron transfer is not easily feasible. Moreover, in this case too, as in the electrochemical reduction of CO_2, kinetic issues cause an overvoltage increase which makes more energy required for the reduction processes.

Figure 8.10 shows that the most popular photomaterial (TiO_2) is not the best suited for such reduction processes. Moreover, TiO_2 absorbs mainly UV radiation, although the main interest is to use the visible part of the solar spectrum.

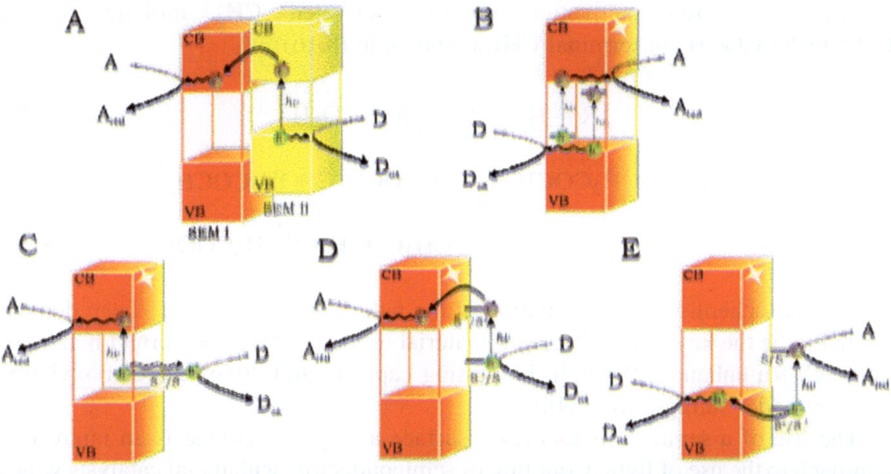

Scheme 8.18 Mechanism of wide band gap semiconductor photosensitization: (**a**) formation of composite semiconductors; (**b**) bulk doping resulting in formation of acceptor or donor levels; (**c**) direct photosensitization (optical charge transfer); (**d**) dye-photosensitization involving an electron injection to the conduction band from the excited photosensitizer; (**e**) dye-photosensitization involving a hole injection to the valence band from the excited photosensitizer

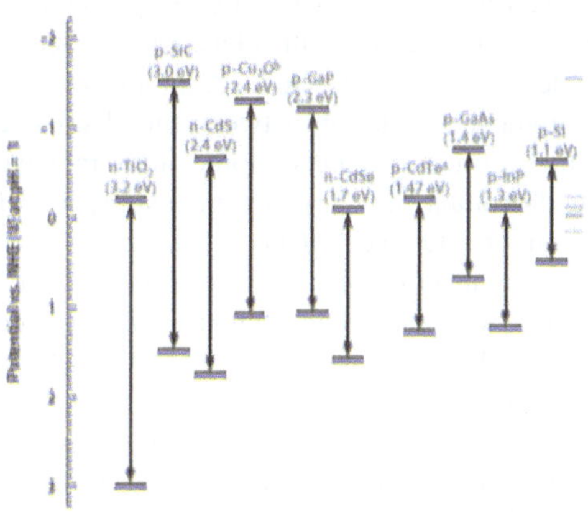

$CO_2 + e^- = CO_2^{\bullet-}$	-1.90 in water -2.21 in aprotic solvents;
$CO_2 + 2H^+ + 2e^- \rightarrow CO + H_2O$	-0.53
$CO_2 + 2H^+ + 2e^- \rightarrow HCO_2H$	-0.61
$CO_2 + 4H^+ + 4e^- \rightarrow CH_2O + H_2O$	-0.48
$CO_2 + 6H^+ + 6e^- \rightarrow CH_3OH + H_2O$	-0.38
$CO_2 + 8H^+ + 8e^- \rightarrow CH_4 + 2H_2O$	-0.24

Fig. 8.10 Comparison of the band gap of some semiconductors with the reduction potential of CO_2 in acidic water (pH = 1) in several one- and multi-electron transfer processes. Adapted from [71]

This problem can be solved by modifying the TiO_2 properties by depositing on its surface materials which can influence the band gap by shifting it more towards the visible region. This is possible in several modes as shown in Scheme 8.18.

p-Si has received attention for the reduction of CO_2 under quasi-neutral conditions (pH = 6–7). Another approach to photochemically induced CO_2 conversion is its formal $1e^-$ reduction to afford C–C bond formation in carboxylates – see (8.15). ZnS decorated with Ru [63] has been used for the carboxylation of C–H bonds having a different activation energy (activated CH_2 moieties as in $CH_3COCH_2COCH_3$ or terminal CH_3 as shown in (8.16)).

$$R - H + CO_2 \xrightarrow[cat]{hv} R - CO_2H \tag{8.15}$$

$$CH_3COCH_2COCH_3 \xrightarrow[cat]{hv} CH_3COCH(COOH)COCH_3 + CH_3COCH_2COCH_2COOH \tag{8.16}$$

In this photochemistry, the formation of the radical anion $CO_2^{\cdot -}$ has been demonstrated to be the key issue. The photomaterial used (ZnS) has the correct band gap, although Ru enhances the light harvesting capacity and drives the carboxylation reaction under sunlight irradiation.

The use of a semiconductors level surface as a photocathode is an interesting approach to the use of light. Coupling of semiconductors with metal catalysts which may drive the electron transfer and the selectivity of the reaction is an interesting option. The integration of the two properties in a PEC system should not change the FE (see Table 8.2) and selectivity towards products, eventually showing only a positive photovoltage shift of the onset voltage. Therefore, the interaction with surfaces and metal centres are analogous to those discussed in the previous chapters and the reaction mechanisms remain those discussed above. An open problem in some cases is whether the multielectron reduction occurs at the electrode surface or in solution, either in proximity to the electrode or in the bulk. Another point of discussion is whether there exists any univocal correlation between the electronic structure of the metal electrode and the formed products. Some authors have suggested that metals with incomplete d-orbitals may lead to an insurgence of π-bonding and thus favour the formation of CO over HCO_2^-, whereas d^{10} or Mg-like metal systems may favour the formation of oxalates and formates [30, 74]. An interesting case is represented by Cu, which has long been known [30] to be able to produce C2 and C3 species. The preliminary formation of CO has been postulated as the necessary step for C1+ formation.

PEC based on the semiconductor–molecular catalysts junction is an attractive case, as the molecular catalyst may better drive the sequential "$e^- + H^+$ *transfer*" than the flat electrode surface. Several pairings of semiconductor–metal complex are reported in the literature, including p-Si [75, 76], p-GaAs [77], p-InP [78] and p-GaP [79] coupled to several Co-, Ni- or Re-macrocyclic N-ligands or to other metal couples such as V(III)–V(II) [80] or TiO_2 [81]. In general, the reduction product of CO_2 is CO. An interesting case is represented by the couple p-GaP: pyridinium ion (organic catalyst) [82], which is able to convert CO_2 into CH_3OH with a remarkable 63–100 % faradaic efficiency at $pH = 5.2$, and a conversion efficiency of 11 % at 365 nm. The energy conversion efficiency has for a long time [83] been known to be related to the so-called "Fermi level pinning" [84] in semiconductors–liquid junctions, which makes possible the photoreduction of species with redox potential more negative than the conduction band edge of p-type semiconductors, such as those mentioned above. Such biphasic solid–liquid systems are characterized by three critical steps, namely (1) exciton generation and charge separation in semiconductors, (2) e^--transfer from the solid to the metal system in the liquid phase and (3) CO_2 reduction on the metal centre. Step (1) can be affected by charge recombination which results in the conversion of light into thermal energy. Step (2), the charge transfer, is affected by several parameters such as the diffusion layer and properties of the semiconductor. Step (3) has already been discussed. The relation of photocurrent density to photon flux, diffusion and other factors has been theoretically investigated [85] and a correlation to several parameters found [86].

An alternative to the solid–liquid junction is represented by anchoring the catalyst on the level semiconductor surface. This practice is largely used in catalysis in the so-called heterogenization of homogeneous catalysts, and is useful for their easy recovery. The heterogenization can be carried out in several ways by deposition, inglobation and tethering of metal systems. The first two techniques may somehow affect the structural features of the metal centre, whereas the latter does not influence the geometry of the catalytic centre, which remains practically the same as in the homogeneous catalyst. In the decoration of semiconductors, the nature of the "tether" may play a key role in several ways. The use of organic polymers (polyaniline [87]) has been known for a long time. Alternative routes, such as coordination to the catalytic metal centre of unsaturated ligands, which may undergo polymerization (polyvinyl-M systems [78]) have also been attempted.

An attempt has been made to use such an approach for the conversion of solar light [88, 89] using enzymes supported on electrodes [90, 91]. The use of hybrid systems in which enzymes selectively drive the reduction and a photocatalyst regenerates the cofactor has been demonstrated to be a feasible approach [62, 64] and needs further improvement for exploitation. This area deserves attention and the growing interest in such research is fully justified by its potential of converting large volumes of CO_2 into bulk or specialty chemicals [67] by using solar energy and mimicking natural processes [52].

8.7 Perspective Electrochemical, Photochemical and Photoelectrochemical Reduction of CO_2

The interesting aspect of PEC is that it is possible to use such technology with a single dual-face electrode (Scheme 8.19) in a compact two-compartment single cell which allows simple product separation.

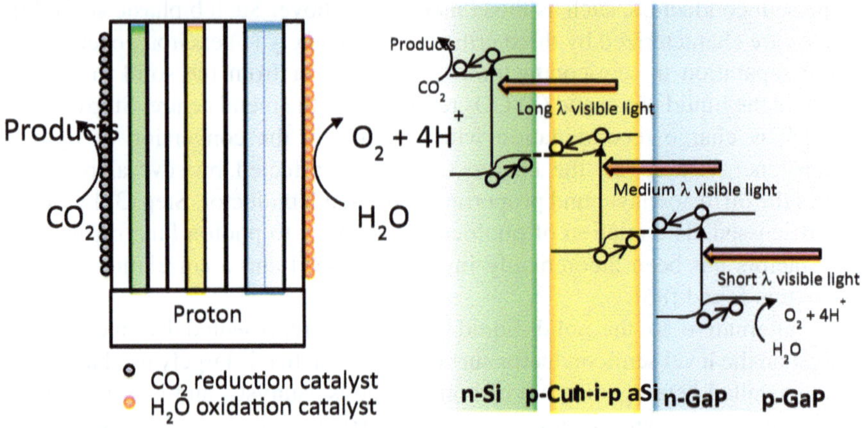

Scheme 8.19 Dual-face electrode for the co-processing of CO_2 and water under solar irradiation. Adapted with permission from [92]. Copyright (2009) American Chemical Society

Such a device needs a well-performing, fast and robust proton-exchange membrane. In principle, a cell designed for water splitting [92] under solar irradiation may also serve for CO_2-water co-processing. Therefore, work done on solar-driven water splitting [76] can be transferred to CO_2 conversion into chemicals/fuels under solar irradiation.

The interesting aspect is that, as discussed above, different electrocatalysts may afford different CO_2-reduction products. Therefore, such technology may result in a flexible tool for the production of selected chemicals or fuels from CO_2 and water, implementing a "*CO_2 economy*". Adaptation of the combinatorial approach to testing electrocatalysts [93] has been proposed as a route to shortening the time-to-market for such devices. Miniaturization [94] or development of new concept photobioreactors [95] can be a solution to the space issue, which is a predominant factor in the large scale development of CO_2 conversion. Coupling EC–PC–PEC with thermal processes and with enhanced biotechnological conversion of CO_2 [96] may shorten the time to exploitation of carbon-recycling, mimicking natural photosynthesis [97].

References

1. Benson EE, Kubiak CP, Sathrum AJ, Smieja JM (2009) Electrocatalytic and homogeneous approaches to conversion of CO_2 to liquid fuels. Chem Soc Rev 38:89–99
2. Schneider J, Jia H, Muckermana JT, Fujita E (2012) Thermodynamics and kinetics of CO_2, CO, and H^+ binding to the metal centre of CO_2 reduction catalysts. Chem Soc Rev 41:2036–2051
3. Bolinger CM, Story N, Sullivam BP, Meyer TJ (1988) Electrocatalytic reduction of carbon dioxide by 2,2'-bipyridine complexes of rhodium and iridium. Inorg Chem 27:4582–4587

4. Bruce MRM, Megehee E, Sullivan BP, Thorp H, O'Toole TR, Downard A, Meyer TJ (1988) Electrocatalytic reduction of carbon dioxide by associative activation. Organometallics 7:238–240
5. Slater S, Wagenknecht JH (1984) Electrochemical reduction of carbon dioxide catalyzed by Rh(diphos)$_2$Cl. J Am Chem Soc 106:5367–5368
6. Tezuka M, Yajima T, Tsuchiya A, Matsumoto Y, Uchida Y, Hidai M (1982) Electroreduction of carbon dioxide catalyzed by iron-sulfur cluster compounds $[Fe_4S_4(SR)_4]^{2-}$. J Am Chem Soc 104:6834–6836
7. Gholamkhass B, Mametsuka H, Koike K, Tanabe T, Furue M, Ishitani O (2005) Architecture of supramolecular metal complexes for photocatalytic CO_2 reduction: ruthenium-rhenium bi- and tetranuclear complexes. Inorg Chem 44:2326–2336
8. Gu J, Wuttig A, Krizan JW, Hu Y, Detweller ZM, Cava RJ, Bocarsly A (2013) Mg-doped $CuFeO_2$ photocatalysts for photoelectrochemical reduction of carbon dioxide. J Phys Chem 117:12415–12422
9. Fisher B, Eisenberg R (1980) Electrocatalytic reduction of carbon dioxide by using macrocycles of nickel and cobalt. J Am Chem Soc 102:7361–7363
10. Beley M, Collin JP, Ruppert R, Sauvage JP (1984) Nickel(II)-cyclam: an extremely selective electrocatalyst for reduction of CO_2 in water. J Chem Soc Chem Commun 1315–1316
11. Hammuche M, Lexa D, Momenteau M, Saveant JP (1991) Chemical catalysis of electrochemical reactions. Homogeneous catalysis of the electrochemical reduction of carbon dioxide by iron("0") porphyrins. Role of the addition of magnesium cations. J Am Chem Soc 113:8455–8466
12. Grodkowski J, Neta P, Fujita E, Mahammed A, Simkhovich L, Gross Z (2002) Reduction of cobalt and iron corroles and catalyzed reduction of CO_2. J Phys Chem A 106:4772–4778
13. Hawecker J, Lehn JM, Ziessel R (1984) Electrocatalytic reduction of carbon dioxide mediated by Re(bipy)$(CO)_3$Cl (bipy = 2,2-bipyridine). J Chem Soc Chem Comm 328–330
14. Ishida H, Tanaka K, Tanaka T (1987) Electrochemical CO_2 reduction catalyzed by ruthenium complexes $[Ru(bpy)_2(CO)_2]^{2+}$ and $[Ru(bpy)_2(CO)Cl]^+$. Effect of pH on the formation of CO and HCOO-. Organometallics 6:181–186
15. DuBois DL, Miedaner A, Haltiwanger RC (1991) Electrochemical reduction of carbon dioxide catalyzed by [Pd(triphosphine)(solvent)]$(BF_4)_2$ complexes: synthetic and mechanistic studies. J Am Chem Soc 113:8753–8764
16. Raebiger JW, Turner JW, Noll BC, Curtis CJ, Miedaner A, Cox B, DuBois DL (2006) Electrochemical reduction of CO_2 to CO catalyzed by a bimetallic palladium complex. Organometallics 25:3345–3351
17. Dubois DL (1997) Development of transition metal phosphine complexes as electrocatalysts for CO_2 and CO reduction. Comments Inorg Chem 19:307–325
18. DeLaet DL, Del Rosario R, Fanwick PE, Kubiak CP (1987) Carbon dioxide chemistry and electrochemistry of a binuclear cradle complex of nickel(0), Ni_2(.mu-CNMe) $(CNMe)_2(PPh_2CH_2PPh_2)_2$. J Am Chem Soc 109:754–758
19. Morgenstern DA, Ferrence GM, Washington J, Henderson JI, Rosenhein L, Heise JD, Fanwick PE, Kubiak CP (1996) A class of halide-supported trinuclear nickel clusters $[Ni_3(m_3\text{-}L)(m_3\text{-}X)(m_2\text{-}dppm)_3]n+$ (L = I-, Br-, CO, CNR; X = I-, Br-; n = 0,1; dppm = $Ph_2PCH2PPh_2$): novel physical properties and the fermi resonance of symmetric m3-h1 bound isocyanide ligands. J Am Chem Soc 118:2198–2207
20. Wittrig RE, Ferrence GM, Washington J, Kubiak CP (1998) Infrared spectroelectrochemical and electrochemical kinetics studies of the reaction of nickel cluster radicals $[Ni_3(\mu_2\text{-}dppm)_3(\mu_3\text{-}L)(\mu_3I)]$•(L = CNR, R = CH_3, i-C_3H_7, C_6H_{11}, $CH_2C_6H_5$, t-C_4H_9, 2,6-$Me_2C_6H_3$; L = CO) with carbon dioxide. Inorg Chim Acta 270:111–117
21. Ferrence GM, Fanwick PE, Kubiak CP (1996) A telluride capped trinuclear nickel cluster $[Ni_3(\mu_3\text{-}Te)_2(\mu\text{-}PPh_2CH_2PPh_2)_3]^{n+}$ with four accessible redox states (n=–1, 0, 1, 2). J Chem Soc Chem Commun 1575–1576

22. Haines RJ, Wittrig RE, Kubiak CP (1994) Electrocatalytic reduction of carbon dioxide by the binuclear copper complex $[Cu_2(6\text{-(diphenylphosphino-2,2'-bipyridyl})_2(MeCN)_2][PF_6]_2$. Inorg Chem 33:4723–4728
23. Takeda H, Koike K, Inoue H, Ishitani O (2008) Development of an efficient photocatalytic system for CO_2 reduction using rhenium(I) complexes based on mechanistic studies. J Am Chem Soc 130:2023–2031
24. Kumar B, Smieja JM, Kubiak CP (2010) Photoreduction of CO_2 on p-type silicon using Re (bipy-Bu^t)$(CO)_3$Cl: photovoltages exceeding 600 mV for the selective reduction of CO_2 to CO. J Phys Chem 114:14220–14223
25. Smieja JM, Sampson MD, Grice AF, Benson EE, Froehlich JD, Kubiak CP (2013) Manganese as substitute for rhenium in CO_2 reduction catalysts: the importance of acids. Inorg Chem 52:2484–2491
26. Barton CE, Lakkaraju PS, Rampulla DM, Morris AJ, Abelev E, Bocarsly AB (2010) Using a one-electron shuttle for the multielectron reduction of CO_2 to methanol: kinetic, mechanistic, and structural insights. J Am Chem Soc 132:11539–11551
27. Frese KW Jr, Canfield D (1984) Reduction of CO_2 on n-GaAs electrodes and selective methanol synthesis. J Electrochem Soc 131:2518–2522
28. Willner I, Maidan R, Mandler D, Durr H, Dorr G, Zengerle K (1987) Photosensitized reduction of CO_2 to CH_4 and H_2 evolution in the presence of ruthenium and osmium colloids: strategies to design selectivity of products distribution. J Am Chem Soc 109:6080–6086
29. Christophe J, Doneux T, Buess-Herman C (2012) Electroreduction of carbon dioxide on copper-based electrodes: activity of copper single crystals and copper-alloys. Electrocatalysis 3:139–146
30. Hori Y, Wakebe H, Tsukamoto T, Koga O (1994) Electrocatalytic process of CO selectivity in electrochemical reduction of CO_2 at metal electrodes in aqueous media. Electrochim Acta 39:1833–1839
31. Stephen H, Stephen T (eds) (1963) Solubilities of inorganic and organic compounds, vol 1, Binary systems, Part II. Pergamon, Oxford, pp 1057–1076
32. Hara K, Kudo A, Sakata T (1995) Electrochemical reduction of CO_2 under high pressure on various electrodes in aqueous electrolytes. J Electroanal Chem 391:141–147
33. Hara K, Kudo A, Sakata T, Watanabe M (1995) High efficiency electrochemical reduction of carbon dioxide under high pressure on a gas diffusion electrode containing Pt catalysts. J Electrochem Soc 142:L57–L59
34. Amatore C, Savéant J-M (1981) Mechanism and kinetic characteristics of the electrochemical reduction of carbon dioxide in media of low proton availability. J Am Chem Soc 103:5021–5023
35. Gangi DA, Durand RR Jr (1986) Binding of carbon dioxide to cobalt and nickel tetra-aza macrocycles. J Chem Soc Chem Commun 697–699
36. Fujita E, Creutz N, Sutin N, Szalda DJ (1991) Carbon dioxide activation by cobalt (I) macrocycles: factors affecting carbon dioxide and carbon monoxide binding. J Am Chem Soc 113:343–353
37. Fujita E, Szalda DJ, Creutz N, Sutin N (1988) Carbon dioxide activation: thermodynamics of carbon dioxide binding and the involvement of two cobalt centers in the reduction of carbon dioxide by a cobalt(I) macrocycle. J Am Chem Soc 110:4870–4871
38. Ogata T, Yanagida S, Brunshwig BS, Fujita E (1995) Mechanistic and kinetic studies of cobalt macrocycles in a photochemical CO_2 reduction system: evidence of Co-CO_2 adducts as intermediates. J Am Chem Soc 117:6708–6716
39. Creutz C, Schwarz HA, Wishart JF, Fujita E, Sutin N (1991) Thermodynamics and kinetics of carbon dioxide binding to two stereoisomers of a cobalt(I) macrocycle in aqueous solution. J Am Chem Soc 113:3361–3371
40. Fujita E, Creutz N, Sutin N, Brunschwig BS (1993) Carbon dioxide activation by cobalt macrocycles. Evidence of hydrogen bonding between bound CO_2 and the macrocycle in solution. Inorg Chem 32:2657–2662

41. Fujita E, Furenlid LR, Renner MW (1997) Direct XANES evidence for charge transfer in Co–CO_2 complexes. J Am Chem Soc 119:4549–4550
42. Fujita E, van Eldik R (1998) Effect of pressure on the reversible binding of acetonitrile to the "Co(I)-CO_2" adduct to form cobalt(III) carboxylate. Inorg Chem 37:360–362
43. Schmidt MH, Miskelly GM, Lewis NS (1990) Effects of redox potential, steric configuration, solvent, and alkali metal cations on the binding of carbon dioxide to cobalt(I) and nickel (I) macrocycles. J Am Chem Soc 112:3420–3426
44. Creutz C, Schwarz HA, Wishart JF, Fujita E, Sutin N (1989) A dissociative pathway for equilibration of a hydrido $CoL(H)^{2+}$ complex with CO_2 and CO: ligand binding constants in the macrocyclic [14]dienecobalt(I) system. J Am Chem Soc 111:1153–1154
45. Fujita E, Wishart JF, van Eldik R (2002) Mechanistic information from pressure acceleration of hydride formation via proton binding to a cobalt(I) macrocycle. Inorg Chem 41:1579–1583
46. Sasaki S (1992) An ab initio MO/SD-CI study of model complexes of intermediates in electrochemical reduction of CO_2 catalyzed by $NiCl_2$(cyclam). J Am Chem Soc 114:2055–2062
47. Fujita E, Haff J, Sanzenbacker R, Elias H (1994) High electrocatalytic activity of RRSS-[NiIIHTIM]$(ClO_4)_2$ and [NiIIDMC]$(ClO_4)_2$ for carbon dioxide reduction (HTIM = 2,3,9,10-tetramethyl-1,4,8,11-tetraazacyclotetradecane, DMC = C-meso-5,12-dimethyl-1,4,8,11-tetraazacyclotetradecane). Inorg Chem 33:4627–4628
48. Gagné RR, Ingle DM (1981) One-electron-reduced nickel(II)-macrocyclic ligand complexes. Four-coordinate nickel(I) species and nickel(II)-ligand radical species which form paramagnetic, five-coordinate nickel(I) adducts. Inorg Chem 20:420–425
49. Furenlid LR, Renner MW, Szalda DJ, Fujita E (1991) EXAFS studies of nickel(II), nickel(I), and Ni(I)-CO tetraazamacrocycles and the crystal structure of (5,7,7,12,14,14-hexamethyl-1,4,8,11-tetraazacyclotetradeca-4,11-diene)nickel(I) perchlorate. J Am Chem Soc 113:883–892
50. Kelly CA, Mulazzani QG, Blinn EL, Rodgers MAJ (1996) Kinetics of CO addition to Ni (cyclam) + in aqueous solution. Inorg Chem 35:5122–5126
51. Kelly CA, Mulazzani QG, Venturi M, Blinn EL, Rodgers MAJ (1995) The thermodynamics and kinetics of CO_2 and H+ binding to Ni(cyclam) + in aqueous solution. J Am Chem Soc 117:4911–4919
52. Appel AM, Bercaw JE, Bocarsly AB, Dobbek H, DuBois DL, Dupuis M, Ferry JG, Fujita E, Hille R, Kenis PJA, Kerfeld CA, Morris RH, Peden CHF, Portis AR, Ragsdale SW, Rauchfuss TB, Reek JNH, Seefeldt LC, Thauer RK, Waldrop GL (2013) Frontiers, opportunities, and challenges in biochemical and chemical catalysis of CO_2 fixation. Chem Rev 113:6621–6658
53. Doherty MD, Grills DC, Muckerman JT, Polyansky DE, Fujita E (2010) Toward more efficient photochemical CO_2 reduction: use of $scCO_2$ or photogenerated hydrides. Coord Chem Rev 254:2472–2482
54. Agarwal J, Shaw TW, Stanton CJ III, Majetich GF, Bocarsly AB, Schaefer HF III (2014) NHC-containing manganese(I) electrocatalysts for the two-electron reduction of CO_2. Angew Chem Int Ed 53:5152–5155
55. Huang D, Holm RH (2010) Reactions of the terminal Ni^{II}–OH group in substitution and electrophilic reactions with carbon dioxide and other substrates: structural definition of binding modes in an intramolecular $Ni^{II}\cdots Fe^{II}$ bridged site. J Am Chem Soc 132:4693–4701
56. Tinnemans AHA, Koster TPM, Thewissen DHMW, Mackor A (1984) Tetraaza-macrocyclic cobalt(II) and nickel(II) complexes as electron-transfer agents in the photo(electro)chemical and electrochemical reduction of carbon dioxide. Recl Trav Chim Pays-Bas 103:288–295
57. Kelly CP, Cramer CJ, Trulher DG (2006) Aqueous solvation free energies of ions and ion-water clusters based on an accurate value for the absolute aqueous solvation free energy of the proton. J Phys Chem B 110:16066–16081
58. Yan Y, Zeitler EL, Gu J, Hu Y, Bocarsly AB (2013) Electrochemistry of aqueous pyridinium: exploration of a key aspect of electrocatalytic reduction of CO_2 to methanol. J Am Chem Soc 135:14020–14023 and references to same authors

59. Yan Y, Gu J, Bocarsly AB (2014) Hydrogen bonded pyridine dimer: a possible intermediate in the electrocatalytic reduction of carbon dioxide to methanol. Aerosol Air Qual Res 14:515–521
60. Aresta M, Dibenedetto A, Angelini A (2013) The use of solar energy can enhance the conversion of carbon dioxide into energy-rich products: stepping towards artificial photosynthesis. Philos Trans A Math Phys Eng Sci 371:20120111
61. Aresta M, Dibenedetto A, Angelini A (2013) The changing paradigm in CO_2 utilization. J CO2 Utilization 3–4:65–73
62. Dibenedetto A, Stufano P, Macyk W, Baran T, Fragale C, Costa M, Aresta M (2012) Hybrid technologies for an enhanced carbon recycling based on the enzymatic reduction of CO_2 to methanol in water: chemical and photochemical NADH regeneration. ChemSusChem 5:373–378
63. Baran T, Dibenedetto A, Aresta M, Kruczała K, Macyk W (2014) Photocatalytic carboxylation of organic substrates with carbon dioxide at zinc sulfide with deposited ruthenium nanoparticles. ChemPlusChem 79:708–715
64. Aresta M, Dibenedetto A, Baran T, Angelini A, Łabuz P, Macyk W (2014) An integrated photocatalytic-enzymatic system for the reduction of CO_2 to methanol in bio-glycerol-water. Beilstein J Org Chem 10:2556–2565
65. Aresta M, Dibenedetto A, Macyk W (2015) Hybrid (enzymatic and photocatalytic) systems for CO_2-water co-processing to afford energy rich molecules. In: Rozhkova EA, Ariga K (eds) From molecules to materials-pathways to artificial photosynthesis. Springer, V, 400 p
66. Aresta M, Dibenedetto A, Macyk W, Baran T (2013) Fotocatalizzatori per la riduzione nel visibile di NAD^+ a NADH in un processo ibrido chemoenzimatico MI2013A001135
67. Kumar B, Llorente M, Froehlich J, Dang T, Satrum A, Kubiak CP (2012) Photochemical and photoelectrochemical reduction of CO_2. Annu Rev Phys Chem 63:541–569
68. Grodkowski J, Dhanasekaran T, Neta P, Hambright P, Brunschwig BS et al (2000) Reduction of cobalt and iron phthalocyanines and the role of the reduced species in catalyzed photoreduction of CO_2. J Phys Chem A 104:11332–11339
69. Grodkowski J, Neta P (2000) Cobalt corrin catalyzed photoreduction of CO_2. J Phys Chem A 104:1848–1853
70. Hawecker J, Lehn J-M, Ziessel R (1983) Efficient photochemical reduction of CO_2 to CO by visible light irradiation of systems containing Re(bipy)(CO)3X or $Ru(bipy)_3^{2+}$-Co^{2+} combinations as homogeneous catalysts. J Chem Soc Chem Commun 9:536–538
71. Ulman M, Tinnemans AHA, Mackor A, Aurian-Blajeni B, Halmann M (1982) Photoreduction of carbon dioxide to formic acid, formaldehyde, methanol, acetaldehyde and ethanol using aqueous suspensions of strontium titanate with transition metal additives. Int J Sol Energy 1 (3):213–222
72. Hawecker J, Lehn J-M, Ziessel R (1986) Photochemical and electrochemical reduction of carbon dioxide to carbon monoxide mediated by (2,2′-bipyridine) tricarbonyl-chloro-rhenium (I) and related complexes as homogeneous catalysts. Helv Chim Acta 69:1990–2012
73. Hori H, Johnson FPA, Koike K, Ishitani O, Ibusuki T (1996) Efficient photocatalytic CO_2 reduction using $[Re(bpy)\ (CO)_3\{P(OEt)_3\}]^+$. J Photochem Photobiol A Chem 96:171–174
74. Jitaru M, Lowy DA, Toma M, Toma BC, Oniciu L (1997) Electrochemical reduction of carbon dioxide on flat metallic cathodes. J Appl Electrochem 27:875–979
75. Bradley MG, Tysak T, Graves DJ, Viachiopoulos NA (1983) Electrocatalytic reduction of carbon dioxide at illuminated p-type silicon semiconducting electrodes. J Chem Soc Chem Commun 7:349–350
76. Walter MG, Warren EL, McKone JR, Boettcher SW, Mi Q et al (2010) Solar water splitting cells. Chem Rev 110:6446–6473
77. Zafrir M, Ulman M, Zuckerman Y, Halmann M (1983) Photoelectrochemical reduction of carbon dioxide to formic acid, formaldehyde and methanol on p-gallium arsenide in an aqueous V(II)-V(III) chloride redox system. J Electroanal Chem 159:373–389

78. Arai T, Sato S, Uemura K, Morikawa T, Kajino T, Motohiro T (2010) Photoelectrochemical reduction of CO_2 in water under visible-light irradiation by a p-type InP photocathode modified with an electropolymerized ruthenium complex. Chem Commun 46:6944–6946
79. Petit J-P, Chartier P, Beley M, Deville JP (1989) Molecular catalysts in photoelectrochemical cells: study of an efficient system for the selective photoelectroreduction of CO_2: p-GaP or p-GaAs/Ni(cyclam)$^{2+}$, aqueous medium. J Electroanal Chem 269:267–281
80. Flaisher H, Tenne R, Halmann M (1996) Photoelectrochemical reduction of carbon dioxide in aqueous solutions on p-GaP electrodes: an a.c. impedance study with phase-sensitive detection. J Electroanal Chem 402(1–2):97–105
81. Anfuso CL, Snoeberger RC, Ricks AM, Liu W, Xiao D et al (2011) Covalent attachment of a rhenium bipyridyl CO_2 reduction catalyst to rutile TiO_2. J Am Chem Soc 133:6922–6925
82. Barton EE, Rampulla DM, Bocarsly AB (2008) Selective solar-driven reduction of CO_2 to methanol using a catalyzed p-GaP based photoelectrochemical cell. J Am Chem Soc 130:6342–6344
83. Bard AJ, Bocarsly AB, Fan FRF, Walton EG, Wrighton MS (1980) The concept of Fermi level pinning at semiconductor/liquid junctions: consequences for energy conversion efficiency and selection of useful solution redox couples in solar devices. J Am Chem Soc 102:3671–3677
84. Bocarsly AB, Bookbinder DC, Dominey RN, Lewis NS, Wrighton MS (1980) Photoreduction at illuminated p-type semiconducting silicon photoelectrodes: evidence for Fermi level pinning. J Am Chem Soc 102:3683–3688
85. Soedergren S, Hagfeldt A, Olsson J, Lindquist S-E (1994) Theoretical models for the action spectrum and the current-voltage characteristics of microporous semiconductor films in photoelectrochemical cells. J Phys Chem 98:5552–5556
86. Saveant JM, Vianello E (1962) Potential-sweep chronoamperometry theory of kinetic currents in the case of a first order chemical reaction preceding the electron-transfer process. Electrochim Acta 8:905–923
87. Aurian-Blajeni B, Taniguchi I, Bockris JOM (1983) Photoelectrochemical reduction of carbon dioxide using polyaniline-coated silicon. J Electroanal Chem Interf Electrochem 149:291–293
88. Ogura K, Yoshida I (1987) Electrocatalytic reduction of carbon dioxide to methanol. VI. Use of a solar cell and comparison with that of carbon monoxide. Electrochim Acta 32:1191–1195
89. Ogura K, Yamada M, Nakayama M, Endo N (1998) Electrocatalytic reduction of CO_2 to worthier compounds on a functional dual-film electrode with a solar cell as the energy source. Stud Surf Sci Catal 114:207–212
90. Woolerton TW, Sheard S, Reisner E, Pierce E, Ragsdale SW, Armstrong FA (2010) Efficient and clean photoreduction of CO_2 to CO by enzyme-modified TiO_2 nanoparticles using visible light. J Am Chem Soc 132:2132–2133
91. Schlager S (2013) Electrochemical reduction of CO_2 with immobilized dehydrogenases enzymes, MRS Fall meeting and exhibition, Boston, MA
92. Yamane S, Kato N, Kojima S, Imanishi A, Ogawa S, Yoshida N, Nonomura S, Nakato Y (2009) Efficient solar water splitting with a composite n-Si/p.CuI/n-i-p a-Si/n-p GaP/RuO_2 semiconductor electrode. J Phys Chem C 113:14575–14581
93. Dang T, Ramsaran R, Roy S, Froehlich J, Wang J, Kubiac CP (2011) Design of a high-throughput 25-well parallel electrolyzer for the accelerated discovery of CO_2 reduction catalysts via a combinatorial approach. Electroanalysis 23:2335–2342
94. Yamada Y, Matsuki N, Ohmori T, Mametsuka H, Kondo M, Matsuda A, Suzuki E (2003) One chip photovoltaic water electrolysis device. Int J Hydrog Energy 28:1167–1169
95. Nguyen TV, Wu JCS (2008) Photoreduction of CO_2 in an optical fibres photo-bioreator: effects of metal addition and catalyst carrier. Appl Cat A 335:112–120
96. Aresta M, Dibenedetto A, Angelini A (2014) Catalysis for the valorization of exhaust carbon: from CO_2 to chemicals, materials, and fuels. Technological use of CO_2. Chem Rev 114:1709–1742
97. den Boef G (1977) Theoretische grondslagen van de analyse in waterige oplossingen, 4th edn. Elsevier, Amsterdam/Brussel

Chapter 9
Enzymatic Conversion of CO_2 (Carboxylation Reactions and Reduction to Energy-Rich C1 Molecules)

Abstract This chapter deals with the enzymatic conversion of CO_2. It covers the two aspects of the fixation of the entire CO_2 molecule into substrates (carboxylation) and the reduction of CO_2 to other C1 (or C2) energy-richer molecules. The known mechanisms are discussed and barriers to exploitation at the industrial level highlighted.

9.1 Introduction

Enzyme-catalyzed processes have been used for centuries to make wine, beer, bread, and cheese. In general, enzymatic processes are faster, cleaner, and easier to operate with respect to microbial processes, which need very strict culturing conditions and require careful separation and purification steps to obtain the products [1].

The efficient utilization of carbon dioxide (CO_2) has attracted considerable attention from fundamental research to industrial application in recent years. The efficient conversion of (renewable) CO_2 using enzymes could complement thermal, electrochemical, photochemical and photoelectrochemical processes for effective utilization of CO_2, eventually reducing the greenhouse gas effect while producing useful chemicals or fuels. Heterogeneous catalysis, electrocatalysis, and photocatalysis are, at present, the three predominant technologies for converting CO_2 into useful chemicals, such as methanol, formic acid, acrylic acid, and carbonates, among others [2–5]. However, such methods need high temperature and pressure or additional electric or luminous energy, with some selectivity and yield issues. In comparison, the enzymatic approach to convert CO_2 has several advantages, such as high yields and selectivity under mild reaction conditions, without any significant impact on the environment [6, 7].

M. Aresta et al., *Reaction Mechanisms in Carbon Dioxide Conversion*,
DOI 10.1007/978-3-662-46831-9_9

9.2 CO_2 Fixation in Biosynthesis

Autotrophic organisms are able to synthesize complex organic molecules by assimilation of CO_2 via reduction. Depending on the type of redox equivalents, two different main pathways exist, which are largely divided into organisms living in the dark or in the light (Scheme 9.1):

1. Chemoautotrophic organisms, such as bacteria or Archea living in the dark and in hostile environments such as the deep sea, which obtain their energy through the oxidation of inorganics such as H_2S, elemental sulfur, metal ions (Fe^{2+}, Mn^{2+}), ammonia, or nitrite;
2. Photoautotrophic organisms, such as plants or algae, protists, and cyano bacteria, which use water as an electron source and light as an energy source.

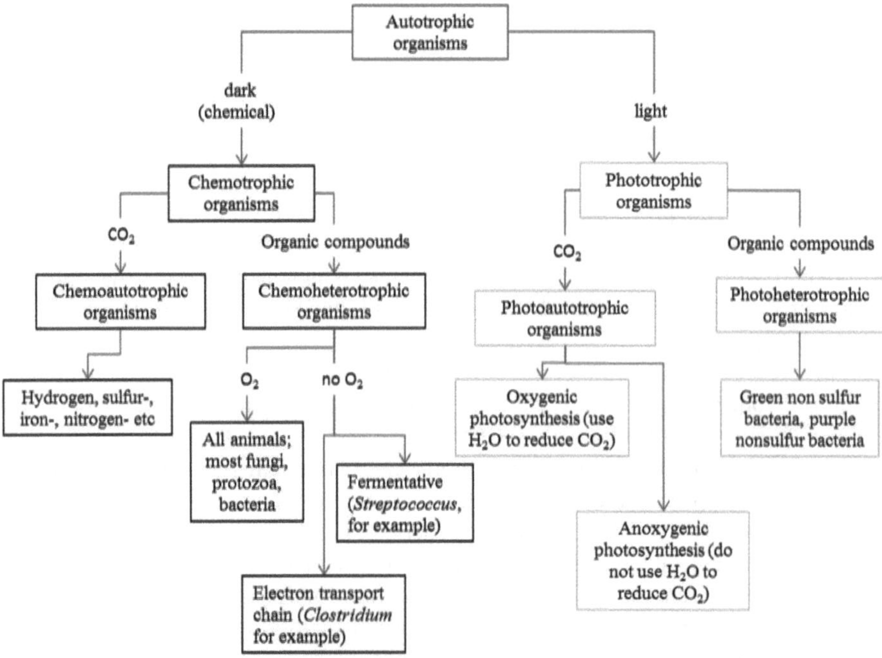

Scheme 9.1 CO_2 fixation in biosynthesis

It is estimated that more than 98 % of the CO_2 that enters the biological carbon cycle (~200 Gt_C/y) is fixed by carboxylases, which makes them crucial players in controlling atmospheric CO_2 concentrations [8, 9]. Biological CO_2 fixation can occur following four different pathways [10, 11]:

1. The Calvin–Benson–Bassham-cycle [12, 13].
2. The reductive tricarboxylic acid (TCA; Arnon–Buchanan) cycle [14].
3. The reductive acetyl-CoA (Wood–Ljungdahl) pathway [15–17].

4. Acyl-CoA carboxylase pathways: 3-hydroxypropionate/malonyl-CoA cycle [18, 19], 3-hydroxypropionate/4-hydroxybutyrate cycle [20], dicarboxylate/4-hydroxybutyrate pathway [21], and the ethylmalonyl-CoA pathway [22].

9.2.1 Calvin–Benson–Bassham-Cycle

Atmospheric CO_2 may be fixed through the Calvin–Benson–Bassham-cycle [13, 23–25]. The key carboxylating enzyme, D-ribulose-1,5-bisphosphate carboxylase/oxygenase (RubisCO) [26], binds CO_2 onto a pentose derivative, ribulose-1,5-bisphosphate. It has been estimated that about 4×10^{13} g (40 Mt) of RubisCO exists in the biosphere, which would correspond to an intangible 5 kg of RubisCO per person on Earth [27].

Scheme 9.2 illustrates the mechanism for RubisCO (http://chemwiki.ucdavis.edu/Organic_Chemistry/Organic_Chemistry_With_a_Biological_Emphasis/Chapter_13%3A_Reactions_with_stabilized_carbanion_intermediates_I/Section_13.5%3A_Carboxylation_and_decarboxylation_reactions). It consists of different steps. Magnesium ions play a key charge-stabilizing role throughout the reaction. During the first step, hydrogen is transferred from the alpha carbon to the keto group of ribulose and the enolic form is deprotonated. In Step 2 an intramolecular proton transfer occurs to afford an enolate at carbon 3, making C2 nucleophilic for an aldol-like attack on CO_2 (Step 3).

However, carbon dioxide is incorporated into the organic substrate and converted into a carboxylate group in the "six-carbon sugar" derivative. The hydrolysis of the carboxylate (Step 4) affords two molecules of 3-phosphoglycerate, which are incorporated into the 'gluconeogenesis' pathway of glucose synthesis.

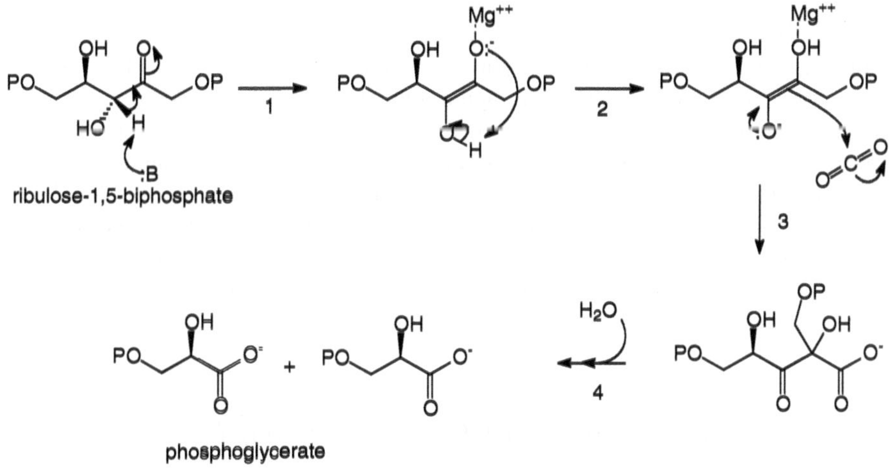

Scheme 9.2 Reaction mechanism of CO_2 fixation by RUBISCO. Adapted from http://chemwiki.ucdavis.edu/Organic_Chemistry/Organic_Chemistry_With_a_Biological_Emphasis/Chapter_13%3A_Reactions_with_stabilized_carbanion_intermediates_I/Section_13.5%3A_Carboxylation_and_decarboxylation_reactions

9.2.2 Reductive TCA (Arnon–Buchanan) Cycle

The reductive TCA cycle uses CO_2 for producing acetyl-CoA [28, 29]. Two CO_2 units are converted in each cycle into acetyl-CoA using one ATP and four NAD(P)H units. The full cycle was first reported to be found in a green sulfur photosynthetic bacterium (*Chlorobium limicola*) and was later also found to operate in *Aquificales*, *Archeal Crenarcheota*, and various types of proteobacteria.

The reductive citric acid cycle is generally considered to be the most energy-efficient CO_2 fixation pathway (~0.6 mol ATP/mol CO_2 for pyruvate) [20, 30, 31]. This is reflected by the activity of the pathway's key carboxylases, isocitrate dehydrogenase, α-ketoglutarate: ferredoxin oxidoreductase, and pyruvate: ferredoxin oxidoreductase. All three enzymes couple their carboxylation reaction to a subsequent reduction step. Whereas isocitrate dehydrogenase is an NAD(P)H-dependent enzyme, the latter two enzymes use ferredoxin as a reductant (Scheme 9.3) [32–34].

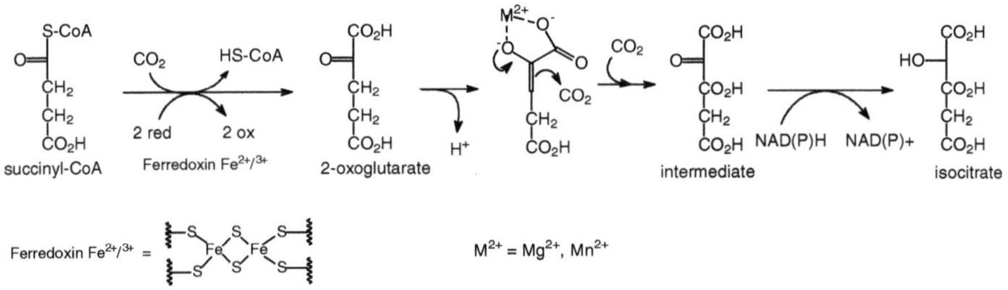

Scheme 9.3 Reaction mechanism in TCA cycle. Adapted from [29]. Copyright 2007 Springer

In a first step, succinyl-CoA is reductively carboxylated by 2-oxoglutarate synthase to form 2-oxoglutarate at the expense of 2 equiv. of reduced ferredoxin. Then, the oxoglutarate formed is again reductively carboxylated by isocitrate dehydrogenase. In the latter reaction, the (de)carboxylation step is assumed to proceed via the formation of an enolate moiety of an α-oxoglutarate, which is formed through the assistance of an essential divalent Mg^{2+} or Mn^{2+} ion (cf. RubisCO), leading to a highly unstable keto-tricarboxylic acid intermediate, which is immediately reduced to yield stable isocitrate. Both the enzymes active in the reductive TCA-cycle are highly specialized so it is supposed that they do not accept man-made substrates which occur in the Calvin cycle.

9.2.3 Reductive Acetyl-CoA (Wood–Ljungdahl) Pathway

If we compare the reductive acetyl CoA pathway with the Calvin cycle and the reductive TCA pathways, we see that it presents some differences: it is a linear unidirectional pathway, through which two molecules of CO_2 are reductively coupled to afford acetyl-CoA, and it proceeds through some rarely occurring organometallic enzyme–substrate intermediates, such as methyl-Co- and Ni-carbonyl-species [35]. This pathway contains some of the most sophisticated enzymatic catalysis known to date (Scheme 9.4) and it is not surprising that some of its components are extremely oxygen-sensitive.

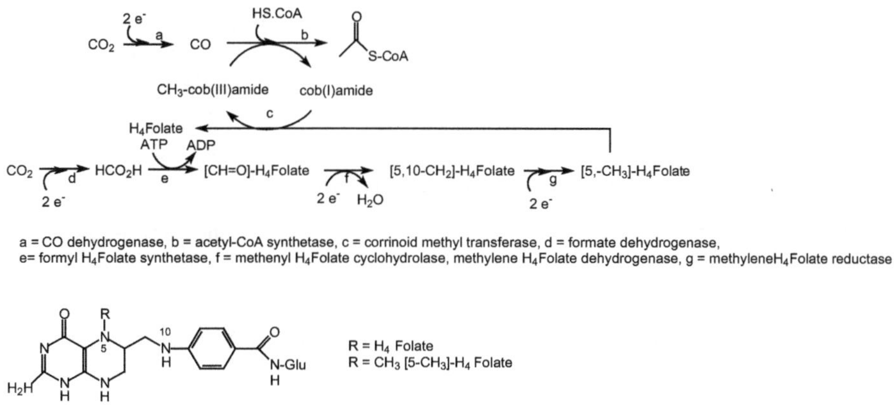

Scheme 9.4 Carbon dioxide fixation via the reductive acetyl-CoA pathway. Adapted from [11]. Copyright 2009 RSC

Overall, it represents the most important method by which anaerobic organisms, such as acetogenic bacteria and methanogenic *Archaea*, sequester carbon for biomass production and energy for living [36, 37].

9.2.4 *Acyl-CoA Carboxylation Pathways*

In contrast to the above-mentioned pathways for CO_2-fixation, which have been known for decades, several autotrophic CO_2 assimilation pathways occurring in anaerobic autotrophs have been elucidated only recently. If the 3-hydroxypropionate/4-hydroxybutyrate pathway can be listed as the 'fifth' pathway of CO_2-assimilation [11], it has been considered together with the other acyl-CoA carboxylation pathways because all the carboxylation steps show a common feature: they all take place on a (CoA-activated) carboxylic acid substrate. In the 3-hydroxypropionate/malyl-CoA cycle and the 3-hydroxypropionate/4-hydroxybutyrate pathway, one molecule of CO_2 is bound onto acetyl-CoA via an ATP-dependent step catalyzed by acetyl-CoA-carboxylase. After sequential reduction of the terminal carboxylate group, propionyl-CoA is carboxylated in the same way, yielding methylmalonyl-CoA (Scheme 9.5).

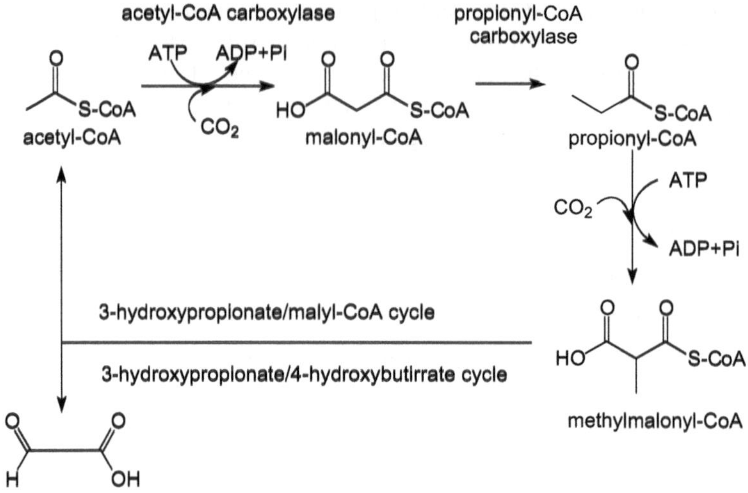

Scheme 9.5 Carboxylation of acetyl/propionyl-CoA. Adapted from [11]. Copyright 2009 RSC

For this reason (similar approach of conversion), it was assumed that both enzymes are identical [38]. In both cases a C2-unit is produced: the acetyl-CoA in the 3-hydroxypropionate/malyl-CoA cycle and the glyoxylate in the 3-hydroxypropionate/4-hydroxybutyrate pathway.

A variant of the above-mentioned carboxylation was found in the dicarboxylate/4-hydroxybutyrate pathway [39] (Scheme 9.6) where the acetyl-CoA is reductively carboxylated by pyruvate synthase at the expense of 2 equiv. of ferredoxin to afford pyruvate which is then phosphorylated.

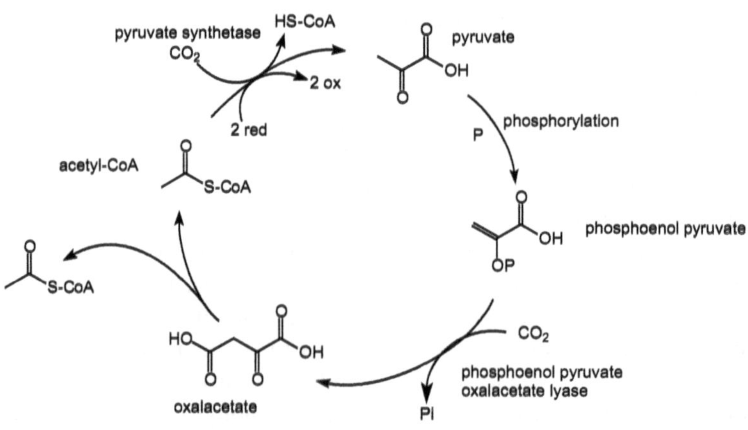

Scheme 9.6 Carboxylation of phosphoenol pyruvate. Adapted from [11]. Copyright 2009 RSC

The phosphoenol pyruvate so formed is carboxylated by phosphoenol pyruvate oxaloacetate lyase and oxaloacetate is obtained. The latter is transformed into 2 equiv. of acetyl-CoA, 1 equiv. of which re-enters the cycle.

Interesting is the carboxylation of crotonyl-CoA to afford ethylmalonyl-CoA.

The mechanism suggests that the β-carbon of crotonyl-CoA accepts a hydride from NADPH, then the α-carbon in the presence of CO_2 is carboxylated to yield ethylmalonyl-CoA (Scheme 9.7).

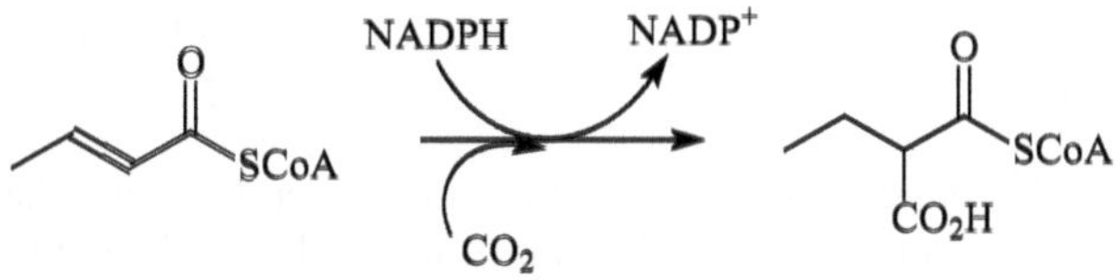

Scheme 9.7 Reductive carboxylation of an enoyl-CoA

In this case, an α,β-unsaturated acyl-CoA derivative is reductively carboxylated by crotonyl-CoA carboxylase/reductase at the expense of a hydride from nicotinamide.

9.3 Carboxylation Reactions

Carboxylation reactions occur widely in nature [40, 41] by the direct use of carbon dioxide or hydrogen carbonate and are mediated by enzymes, which may or may not have a metal as an active center. Such direct carboxylation reactions have found very few applications for synthetic purposes at industrial level [42–46]. An example of a carboxylation reaction is the assimilation of carbon into organic compounds which is the result of a complex series of enzymatically regulated chemical reactions.

9.3.1 Bio-Carboxylation of Aromatic and Hetero-Aromatic Compounds

Although the Kolbe–Schmitt reaction (9.1), known for more than a century, is used industrially for the synthesis of 2-OH-benzoic acid (**2**) and 4-OH-benzoic acid (**3**), the selectivity and yield of which are still not completely mastered, and the production of either pure isomer not easily achieved. For these reasons, purification processes are required after the synthesis, which consume energy and produce waste.

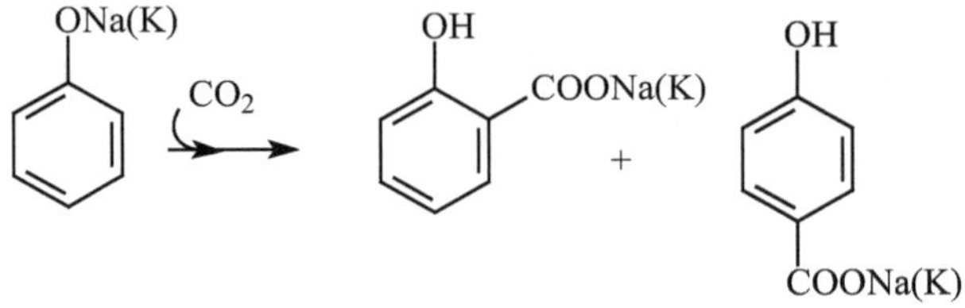

1 2 3 (9.1)

Aresta et al. [47] have found, during their studies on the degradation of phenol by *Thauera aromatica* under anaerobic conditions [48], that the pathway for phenol metabolism follows the route, quite common in nature, of 'increasing the complexity of the molecule for its easier degradation' [49, 50]. As depicted in Scheme 9.8, phenol is first phosphorylated (**7**), then carboxylated to 4-OH-benzoic acid (**8**), which is converted into the active form 4-OH-benzoyl-CoA (**9**), subsequently dehydroxylated to benzoyl-CoA (**10**), the active form of benzoic acid. The latter is then eventually converted into CO_2 and H_2O. 4-OH-benzoic acid is an interesting monomer for the production of polymers used in optics. It was therefore interesting to check whether it was possible to develop a biotechnological route to its production. Obviously, the entire bacteria were not suited for its production, as they would use it for their metabolism and growth. Therefore, the carboxylation enzymes were isolated and used. The bacterial cells were thus broken and the crude extract, obtained after ultracentrifugation, was partially purified by using membranes with a given molecular weight cut-off "MWCO" [6, 47]. The extracted-semipurified enzymes were used for the carboxylation of phenol in vitro. This was the first application of a phenol carboxylase enzyme for the functionalization of organic substrates. The process is clean and good results are obtained, with a turnover number of approximately 16,000.

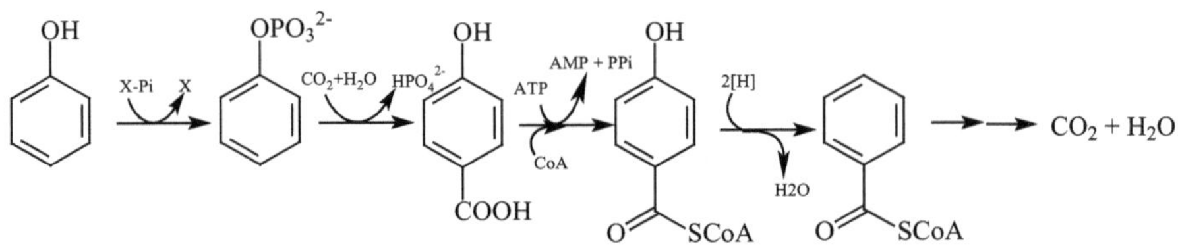

6 7 8 9 10

Scheme 9.8 Phenol degradation pathway. Adapted from [51]. Copyright 2002 Elsevier

Attempts to enhance the low stability by supporting the enzyme on low melting agar has been carried out successfully [51]. Supercritical carbon dioxide was also used as solvent/reagent for semi-purified phenylphosphate carboxylase extracted from *T. aromatica* [52].

Phenylphosphate carboxylase can also carboxylate catechol and *o*-cresol as it can phenol [53] to yield 3,4-dihydroxybenzoate and 4-hydroxy-3-methylbenzoate respectively. Similarly, *Sedimentibacter hydroxybenzoicus*, which catalyzes the decarboxylation of several benzoic acids, is also able to catalyze the reverse (carboxylation) reaction of phenol or catechol in the presence of hydrogencarbonate or CO_2 to yield 4-OH or 3,4-OH benzoic acids, respectively [54–56].

Acids such as pyruvic acid can be produced from acetaldehyde and carbon dioxide using the reverse reaction of pyruvate decarboxylase as shown in Scheme 9.9 [57].

The reaction, carried out in solventless conditions and using Brewer's yeast pyruvate decarboxylase at 500 mM $NaHCO_3$–Na_2CO_3 buffer and pH 11, gives a yield of 81 % in pyruvic acid [57].

L-Lactic acid may be produced from carbon dioxide and ethanol by using a multienzyme reaction system, with a designed internal cofactor regeneration loop [1].

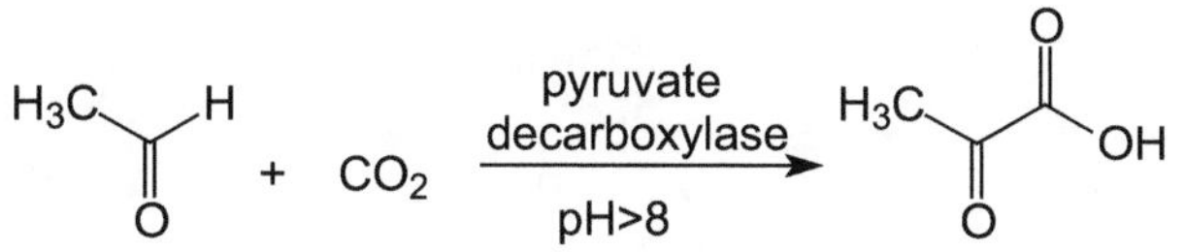

Scheme 9.9 Enzymatic conversion of carbon dioxide into pyruvic acid. Adapted from [57]. Copyright 2001 RSC

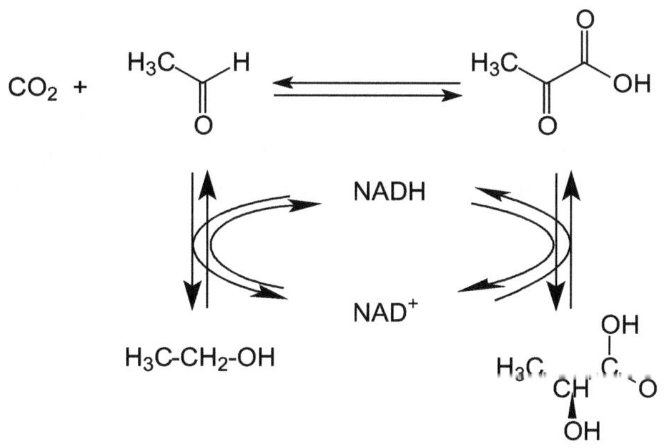

Scheme 9.10 Reaction mechanism for the enzymatic synthesis of L-lactic acid from CO_2 and ethanol. Adapted from [1]. Copyright 2010 Wiley

The process shown in Scheme 9.10 consists of three phases:

1. Synthesis of pyruvate from CO_2 and acetaldehyde using pyruvate decarboxylase (PyDC) as enzyme
2. Reduction of pyruvate to lactate by lactate dehydrogenase (LDH)
3. Regeneration of the cofactor NADH needed for reduction of pyruvate, achieved through the oxidation of ethanol catalyzed by alcohol dehydrogenase (ADH)

Also interesting is the synthesis of pyrrole-2-carboxylate from carbon dioxide and pyrrole using reverse reaction of pyrrole-2-carboxylate decarboxylase enzyme

[58], which was purified from *B. megaterium* PYR2910. The reverse reaction was performed at 298 K in a tightly closed reaction vessel containing potassium phosphate buffer (pH 5.5, necessary to counteract the pH increase by the HCO_3^- addition), ammonium acetate, pyrrole, dithiothreitol (useful to stabilize the enzyme), enzyme, and $KHCO_3$. The optimum temperature and pH for CO_2 fixation were 318 K and pH 7.0, respectively.

A possible mechanism (Scheme 9.11) of action of pyrrole-2-carboxylase decarboxylase has been proposed [58, 59], that requires further improvement, where the organic acid acts as "cofactor" and deprotonates the NH-moiety, causing a nucleophilic attack of Cα at the electrophile CO_2; then the intermediate regains aromaticity after tautomerisation.

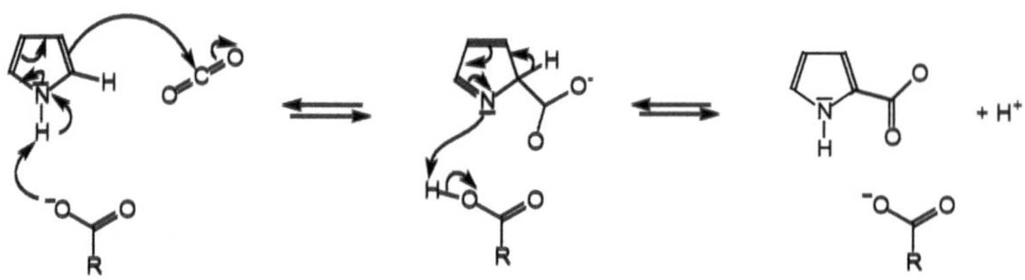

Scheme 9.11 Possible reaction mechanism for synthesis of pyrrole-2-carboxylate from carbon dioxide and pyrrole using pyrrole-2-carboxylate decarboxylase enzyme. Adapted from [58]. Copyright 2001 Elsevier

The yield after bioconversion was 80 %, limited by the equilibrium, and the overall yield after isolation was 52 % [58]. Similarly, indole was carboxylated at position 3 by indole-3-carboxylase (max. conversion 34 %) (Scheme 9.12) [61].

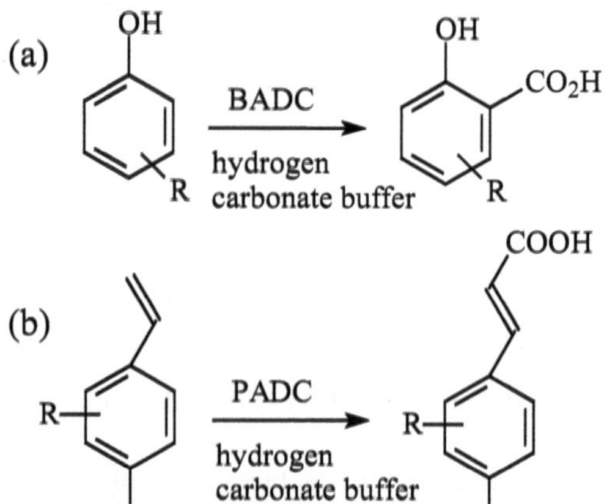

Scheme 9.12 Regio-complementary enzymatic carboxylation of (**a**) phenols and (**b**) hydroxystyrene derivatives. Adapted from [60]

The enzyme was found in *Arthrobacter nicotianae* FI1612 and several molds. Its activity was induced specifically by indole-3-carboxylate, but not by indole. The indole-3-carboxylate decarboxylase of *A. nicotianae* FI1612 catalyzed the nonoxidative decarboxylation of indole-3-carboxylate into indole, and efficiently carboxylated indole and 2-methylindole by the reverse reaction.

Some phenolic acid decarboxylases [60, 62–65] which catalyze the conversion of hydroxycinnamic acids into styrenes are also of interest. Considering the structural data available to date, both enzymes work with different mechanisms:

1. Benzoic acid decarboxylase (BADC) requires a metal active site such as Zn^{2+} [66].
2. Phenolic acid decarboxylase (PADC) is a metal-independent acid–base catalyst [67–69].

9.3.2 Bio-Carboxylation of Epoxides

Some bacteria can degrade epoxides, converting them into β-keto acids. For example, in *Xanthobacter* Py2, poly-β-hydroxyalkanoates are produced [70, 71]. The epoxide degradation seems to be regulated by the presence or absence of CO_2, as reported in Scheme 9.13.

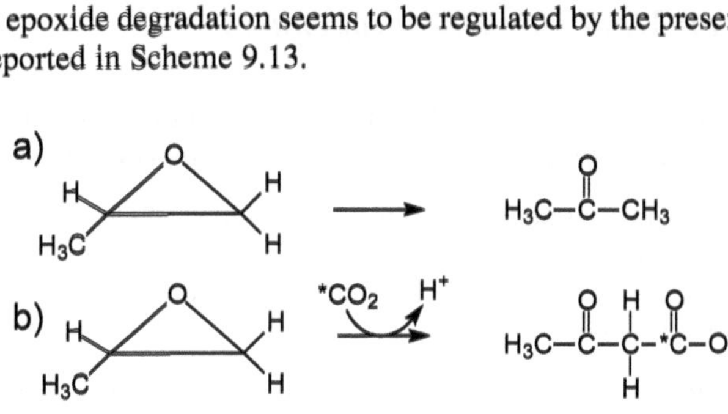

Scheme 9.13 Epoxide conversions in *Xanthobacter* Py2: (**a**) isomerization and (**b**) carboxylation. Adapted from [71]. Copyright 1996 ASM

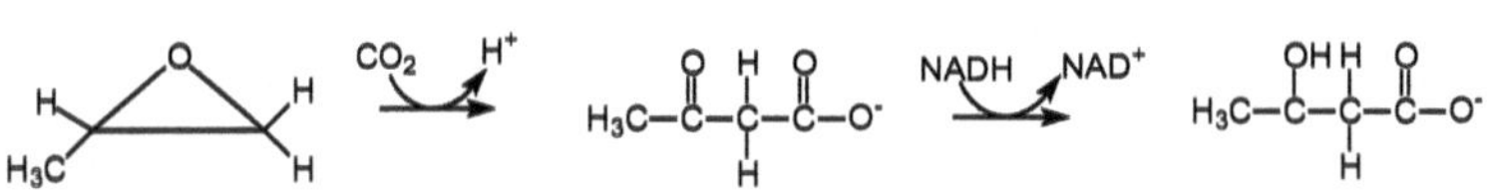

Scheme 9.14 Proposed mechanism of biodegradation of epoxypropane

As depicted in Scheme 9.13, in the absence of CO_2 the isomerization of aliphatic epoxides occurs with formation of a keto-compound (acetone) as a dead-end product, whereas in the presence of CO_2 the carboxylation reaction prevails with formation of acetoacetate. To explain the carboxylation of epoxypropane, the mechanism shown in Scheme 9.14 has been proposed where the epoxide is first carboxylated by the epoxide carboxylase, forming the acetoacetate as intermediate which undergoes NADH-dependent reduction catalyzed by β-hydroxybutyrate dehydrogenase to obtain β-hydroxybutyrate.

9.4 Reduction Reactions

The reduction of CO_2 in Nature occurs in several organisms, under anaerobic or aerobic conditions. The methanation of organic substrates involves, among others, enzymes which are able to reduce in a single step CO_2 to CO [carbon monoxide dehydrogenases (CODH)] or to formic acid [formate dehydrogenases ($F_{ate}DH$)]. CO_2 can be reduced to the methyl group $-CH_3$ through a tetrahydrofolate-THF-mediated process. CO and $-CH_3$ are coupled to give the acetyl moiety using an Fe_4S_4–Ni enzyme and vitamin B_{12}. Formate can be reduced to formaldehyde under the formaldehyde dehydrogenase ($F_{ald}DH$) enzyme, and formaldehyde can be converted into methanol by the alcohol dehydrogenase (ADH) enzyme.

9.4.1 *Carbon Monoxide Dehydrogenases*

Carbon monoxide dehydrogenase (CODH) enzymes catalyze the interconversion of CO and CO_2. These enzymes can be categorized into two classes:

1. Ni-CODH enzymes, found in anaerobic bacteria which have an $[Ni_4Fe_5S]$ cluster as the active site.
2. Mo-CODH enzymes, found in aerobic carboxydotrophic bacteria which have a CuMo-pterin active site.

There is also the related class of Ni-CODH/ACS (acetyl CoA synthetase), where the CODH enzyme is part of a larger protein in which CO_2 reduction is coupled to acetyl CoA synthesis.

Ni,Fe-containing CODHs play an important role in anaerobic bacteria and archea by allowing them to grow with CO or CO_2 as their sole carbon and/or energy source. The structures of CODHs are homodimers with ~130 kDa containing five metal clusters, called B, B′, C, C′, and D (Fig. 9.1) [72].

Each subunit contains the active site C-cluster and cubane-type $[Fe_4S_4]$ B-cluster. Another $[Fe_4S_4]$ D-cluster connects two subunits forming a covalent homodimer. The CODHs catalyze the reversible oxidation of CO to CO_2 at the active site C-cluster, which is composed of $[NiFe_4S_4OH_x]$ ($CO + H_2O \leftrightarrow CO_2 + 2e^- + 2H^+$). In addition to the reversible oxidation of CO, CODHs are able to catalyze further

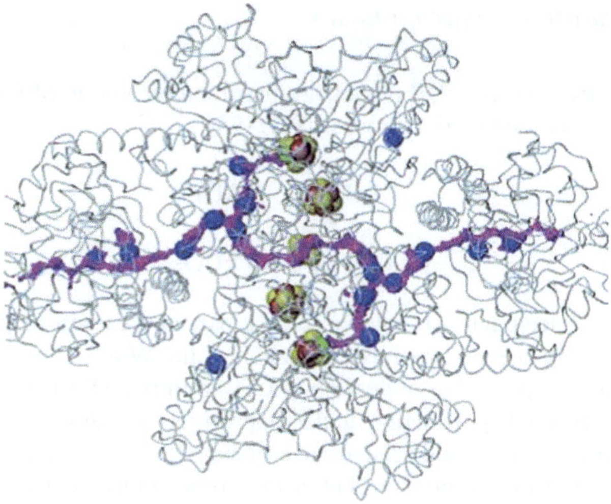

Fig. 9.1 Structure of Ni,Fe CODH. Adapted from [72]. Copyright 2005 RSC

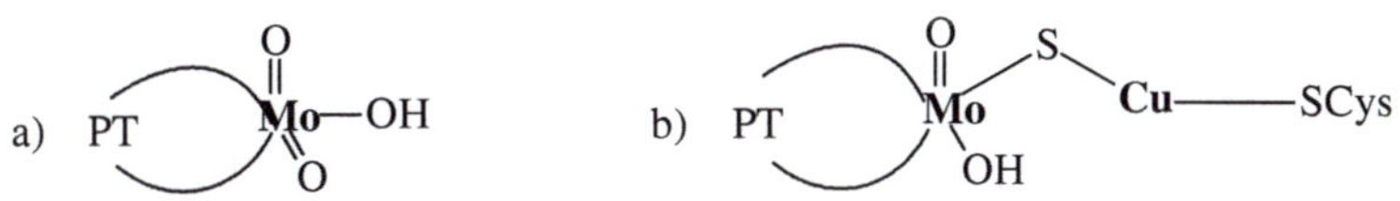

PT= pyranopterin cofactor

Fig. 9.2 (**a**) Mononuclear Mo active site. (**b**) Dinuclear Mo–Cu active site

reactions, such as the oxidation of H_2 and the reductions of protons, 2,4,6-trinitrotoluene (TNT), and hydroxylamine, and the oxidation of *n*-butylisocyanide (*n*-BIC). *n*-BIC is a slow-turnover substrate of CODHs, whose oxidation occurs at the C-cluster.

The Mo-CODH metalloprotein is known to have three subunits – L, M, and S. The molybdenum active site is found in the L subunit. The M subunit consists of a 2-flavoprotein, and the S subunit contains two [2Fe-2S] clusters [73].

There has been some controversy over the active site structure of the Mo-CODH. Dobbek et al. [74] determined the crystal structure to 2.2 Å, which revealed an active site where molybdenum is coordinated by two oxo ligands, a hydroxo group, and one pyranopterin cofactor (Fig. 9.2a) [74].

Later, upon further investigation, the crystal structure was resolved to 1.1 Å, which revealed different results. It was determined that the active site was heterobimetallic containing both molybdenum and copper (Fig. 9.2b).

9.4.2 Formate Dehydrogenases

Formate dehydrogenases (FDHs) catalyze the interconversion of CO_2 and formic acid through an oxidoreductive process (9.2) [75].

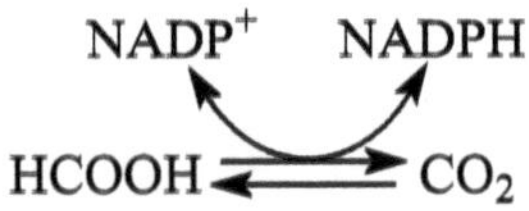

(9.2)

FDHs present in acetogens are known to take part in a carbon fixation metabolic pathway producing acetate (the Eastern branch of the Wood–Ljungdahl pathway), in which the first step involves reduction of CO_2 to formate [76]. Several FDHs are known to catalyze CO_2 reduction under appropriate conditions [77–81]. Those enzymes from acetogenic and related anaerobes, such as *Moorella thermoacetica* and *Clostridium pasteurianum*, are better than other FDHs as reduction catalysts but also show similar catalytic efficiency toward formate oxidation.

The structure of *Escherichia coli* FDH_H, as solved by multiple isomorphous replacement (MIR) and multiwavelength anomalous dispersion (MAD) methods, consists of four αβ domains [82]. It has been prepared with MOLSCRIPT and RASTER3D [83–85].

Formate dehydrogenase H from *E. coli* contains selenocysteine (SeCys), molybdenum, two molybdopterin guanine dinucleotide (MGD) cofactors, and an Fe4S4 cluster at the active site.

9.4.3 Formaldehyde Dehydrogenase

$F_{ald}DH$ was isolated and purified in an overall yield of 12 % from cell-free extract of *Pseudomonas putida C-83* by chromatography on a column of DEAE-cellulose, DEAE-Sephadex A-50, and hydroxyapatite. NAD and especially NADH stabilize the enzyme during storage and against denaturation at high temperatures.

The molecular mass of the enzyme was estimated to be 150,000 by gel filtration method, and analysis by SDS-polyacrylamide gel electrophoresis indicated that the enzyme was composed of two subunit monomers [86].

Kinetic analysis gave Km values of 67 μM for formaldehyde and 56 μM for NAD+, and suggested that the reaction proceeds by a "ping-pong" mechanism. The enzyme catalyzed the oxidation of formaldehyde accompanied by the stoichiometric reduction of NAD^+, but no reverse reaction was observed. The crystal structure of $F_{ald}DH$ from *Pseudomonas putida* (PFDH) has been solved by the multiwavelength anomalous diffraction method using intrinsic zinc ions and has been refined at a 1.65 Å resolution (Fig. 9.3) [87].

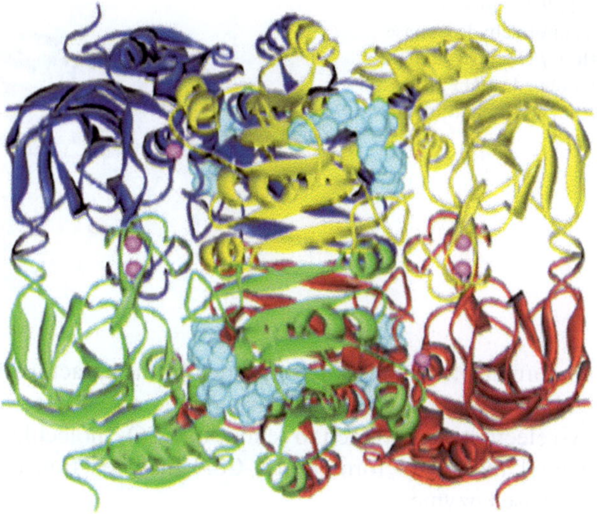

Fig. 9.3 Structure of formaldehyde dehydrogenase ($F_{ald}DH$) from *Pseudomonas putida*. (Reprinted with permission from [87]. Copyright 2003 Elsevier)

9.4.4 Alcohol Dehydrogenases

Alcohol dehydrogenases are a group of dehydrogenase enzymes which occur in many organisms and facilitate the interconversion between alcohols and aldehydes or ketones with the reduction of nicotinamide adenine dinucleotide (NAD^+ to NADH). The first-ever isolated alcohol dehydrogenase (ADH) was purified in 1937 from *Saccharomyces cerevisiae* (baker's yeast) [88]. Many aspects of the catalytic mechanism for the horse liver ADH enzyme were investigated by Hugo Theorell and coworkers [89]. ADH was also one of the first oligomeric enzymes that had its amino acid sequence and three-dimensional structure determined [90, 91]. Horse liver alcohol dehydrogenase, one of the most studied (along with the yeast enzyme), is a symmetrical dimer. Each chain is about 40 kDa, contains two Zn^{++} (orange), only one of which is catalytically active, and each binds a single NAD^+ (green) (Fig. 9.4)

9.4.5 Production of Acetic Acid

Anaerobic bacteria such as *Moorella thermoacetica* have the capacity to fix carbon dioxide with carbon monoxide and hydrogen for the production of ethanol, acetic acid, and other useful chemicals.

Acetogenic bacteria fix CO_2 via the Wood–Ljungdahl (W–L) metabolic pathway as shown in Scheme 9.15. The pathway comprises two branches, the methyl branch and the carbonyl branch, by which CO_2 and CO as carbon sources are converted into acetic acid and ethanol. In the methyl branch, one CO_2 molecule is converted,

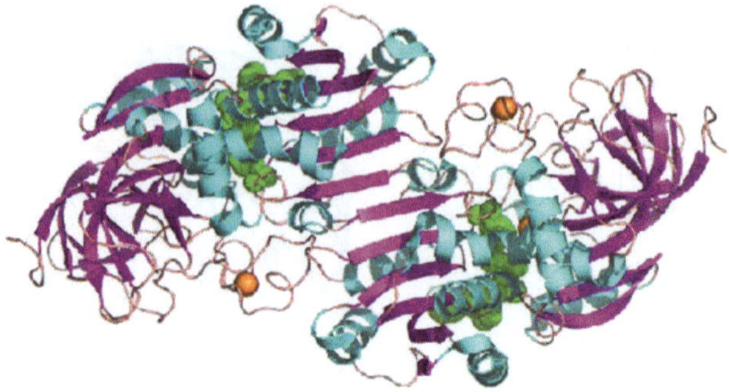

Fig. 9.4 Horse liver alcohol dehydrogenase (http://chemistry.umeche.maine.edu/CHY251/Enzym-Stereo.html)

via formate, to a methyl moiety aided by the cofactor tetrahydrofolate. Reduction of carbon dioxide to the methyl group requires six electrons. In the carbonyl branch, two electrons are utilized to reduce one CO_2 molecule to CO, which reacts with the methyl moiety to form acetyl-CoA in a reaction catalyzed by the acetyl-CoA synthase enzyme.

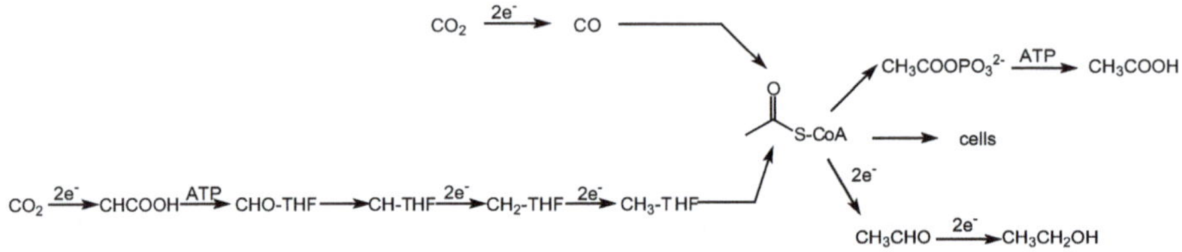

Scheme 9.15 Wood–Ljungdahl (W–L) metabolic pathway for syngas fermentation

CO can also be used directly to form acetyl-CoA, rather than being produced from CO_2 reduction. Acetic acid, ethanol, and cell mass are the main products produced from the intermediate acetyl-CoA [15].

9.4.6 Reduction of CO_2 to Carbon Monoxide or Formate

Classes of enzymes such as CODHs and FDHs are able to catalyze the reversible reduction of CO_2 to CO or $HCOO^-$, respectively. There are two types of CODHs: the first is an O_2-sensitive enzyme from obligate anaerobes such as *Moorella thermoacetica*, *Carboxydothermus hydrogenoformans*, *Methanosarcina barkerii* characterized by [Fe4S4Ni] active sites and the second are air stable [MoSCu]-containing enzymes (*Oligotropha carboxidovorans*). Because of their different active sites (different metals in particular) it is expected a different mode of actions.

In some prokaryotes the FDHs are complex enzymes which contain molybdenum or tungsten cofactors to transfer the electrons from formate oxidation to an independent active site, to reduce quinone, protons, or NAD(P) [82, 92]. These enzymes are suitable for adsorption onto an electrode, so that the electrode accepts the electrons from formate oxidation, and it may also donate electrons and drive

CO_2 reduction. Therefore, they are potential electrocatalysts for the reduction of CO_2 (see Scheme 9.16). Two electrons are transferred from the electrode to the active site (buried inside the insulating protein interior) by the iron–sulfur clusters, to reduce CO_2 to formate, forming a C–H bond. Conversely, when formate is oxidized, the two electrons are transferred from the active site to the electrode.

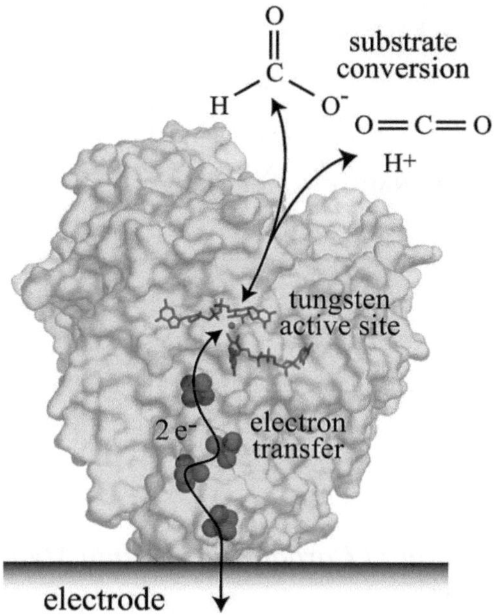

Scheme 9.16 Electrocatalytic interconversion of CO_2 and formate by a formate dehydrogenase from *Desulfovibrio gigas* adsorbed on an electrode surface. (Reprinted with permission from [80]. Copyright 2008)

Tungstoenzymes catalyze low-potential reactions [93] so tungsten-containing FDHs, in which the tungsten is coordinated by two pyranopterin guanosine dinucleotide cofactors and a selenocysteine [92], are those most associated with CO_2 reduction.

One of the tungsten-containing FDHs, namely FDH1, isolated by *Syntrophobacter fumaroxidans*, an anaerobic bacterium, oxidizes propionate to acetate, CO_2, and six reducing equivalents [94]. The reducing equivalents are used to reduce protons to hydrogen or to reduce CO_2 to formate.

The NADH-independent FDH enzymes from *S. fumaroxidans* are highly unstable and inactive in the presence of O_2, limiting their practical application. Conversely, the NADH-dependent FDH from *Candida boidinii* (CbsFDH) was sufficiently stable for commercial use and has been used to enzymatically regenerate NADH. CbsFDH requires NADH, protons, and electrons to convert CO_2 into formate, which can be supplied by the electrochemical system of the Cu electrode

(Scheme 9.17). The $[Cp^*Rh(bpy)Cl]^+$ complex was used as catalyst for the conversion of NAD^+ into the active NADH with almost 100 % selectivity [96].

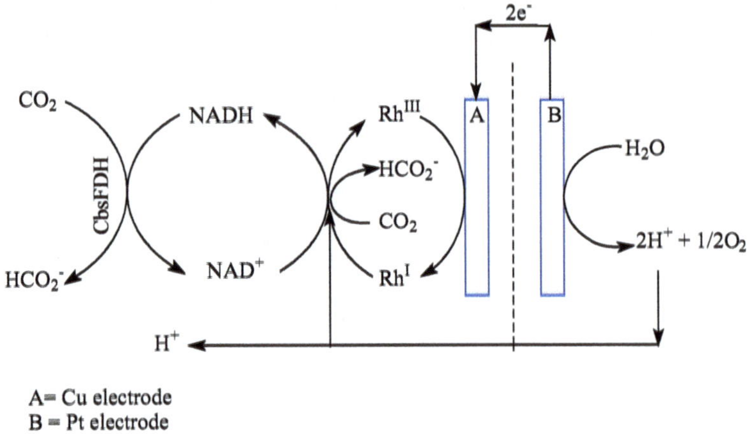

Scheme 9.17 Electro-enzymatic CO_2 reduction using CbsFDH. Adapted from [95] Copyright 2014 Elsevier

9.4.7 *Bioconversion of Carbon Dioxide into Methanol*

In an attempt to mimic Nature, a biotechnological approach to the conversion of CO_2 into methanol has been investigated based on the use of enzymes. In fact, it has been shown that the enzymes $F_{ate}DH$, $F_{ald}DH$, and ADH are able to reduce CO_2 in water at room temperature [7]. A minimum of 3 mol of NADH are consumed per mol of CH_3OH produced (Scheme 9.18).

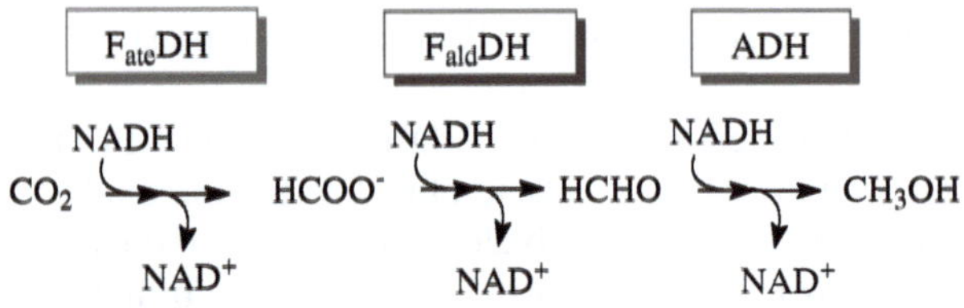

Scheme 9.18 CO_2 reduction to methanol in water promoted by $F_{ate}DH$, $F_{ald}DH$, and ADH

This approach, even if quite appealing, is not economically and energetically affordable unless NAD^+ formed upon oxidation of NADH is efficiently recycled, increasing thereby the ratio $CH_3OH/NADH$ to limits that can be acceptable for a biotechnological production of methanol.

More robust enzymes have been obtained [95] by encapsulation. TEOS is better used with Ca-alginate as agent for producing a Ca-silicate cage, as it avoids any interference with the production of methanol in the case of non-complete hydrolysis of TMOS [97, 98]. Co-encapsulated enzymes work better than singly encapsulated ones [99].

In order to increase the ratio CH_3OH/NADH, the regeneration of NAD^+ into NADH should be performed through a reduction reaction of NAD^+ to NADH. Different ways can be considered: (1) chemicals that act as reducing agents; (2) metal systems that may use dihydrogen for the reduction under thermal or irradiated conditions; (3) semiconductors, water, and light as sources of energy hydrogen for the reduction; (4) electrocatalysts that may reduce NAD^+ under electrochemical conditions.

Among the chemicals, hydrazine and sodium dithionite-SDT [100, 101] have been used. The latter has been slowly added up to a stoichiometric amount – see (9.3) – to NAD^+ after all NADH is oxidized to NAD^+, producing 5 mol of methanol per mol of NADH. Addition of excess of SDT causes a rapid deactivation of the enzymes.

$$NAD^+ + S_2O_4^{2-} + 2H_2O \rightarrow NADH + 3H^+ + 2SO_3^{2-} \tag{9.3}$$

In order to avoid such detrimental contact, the oxidation of NADH and the reduction of NAD^+ have been carried out in separate boxes to obtain a ratio of CH_3OH/NADH up to >35 [102].

Galarneau et al. [103] compared different regeneration systems: phosphite dehydrogenase (PTDH), one of the most efficient enzymatic recycling systems reported so far (together with formate dehydrogenase but already used here in the opposite direction) which converts phosphite into phosphate, a chemical used to buffer the solution, and glycerol dehydrogenase (GlyDH), which transforms glycerol into dihydroxyacetone (DHA) and a natural photosystem extracted from spinach leaves (chloroplasts). It has been shown that, under the catalytic conditions needed for the CO_2 reduction into methanol (pH 6.5, 310 K), PTDH is much more active than GlyDH.

For the chloroplast photosystem, only 0.25 mM NADH was obtained in 30 min, which is below the NADH regenerating activity of PTDH.

By using PTDH plus phosphite together with the polyenzymatic system encapsulated in silica nanocapsules (internal diameter 30 nm) nanostructured by phospholipids (NPS), an activity 55 times higher than the free enzymes in solution was observed.

A different biological system could be the acetogenic bacteria microbes [104] used as biocatalysts for the electrochemical reductive conversion of CO_2 with efficiency of 80–100 % in phosphate buffer solution (pH 7). The direct bacterial use for electrochemical CO_2 conversion could solve the problem related to the enzyme purification steps and widens the choice of catalysts to include the naturally developed and optimized microorganisms.

Also interesting is the hybrid bioinspired systems used for the enzymatic reduction of CO_2 to CH_3OH in water [105] that combines two processes: the enzymatic reduction of CO_2 to CH_3OH promoted by the reduced form of coenzyme NADH and the in situ photocatalytic reduction of NAD^+ to afford NADH, using semiconductors. The objective is to reach a highly efficient and selective reduction of NAD^+ upon visible-light irradiation to 1,4-NADH or other equally active isomers. To this end, a number of semiconducting materials which behave as photocatalysts and absorb visible-light, such as: Cu_2O, $InVO_4$, TiO_2 modified either with the organic compound rutin, or with the inorganic complex $[CrF_5(H_2O)]^{2-}$ [106] and the doped sulphide Fe/ZnS have been prepared and used for the reduction of NAD^+ to NADH. The expected mechanism of NADH regeneration is shown in Scheme 9.19.

Regeneration of the cofactor NADH from NAD^+ has been achieved by using visible light-active heterogeneous TiO_2-based photocatalysts. The efficiency of the regeneration process is enhanced by using a Rh^{III}-complex for facilitating the electron and hydride transfer from the H-donor (water or a water-glycerol solution) to NAD^+. In this way, 1 mol of NADH was used for producing 100–1000 mol of CH_3OH, opening the way to a practical application using as mediator a Rh-complex [105].

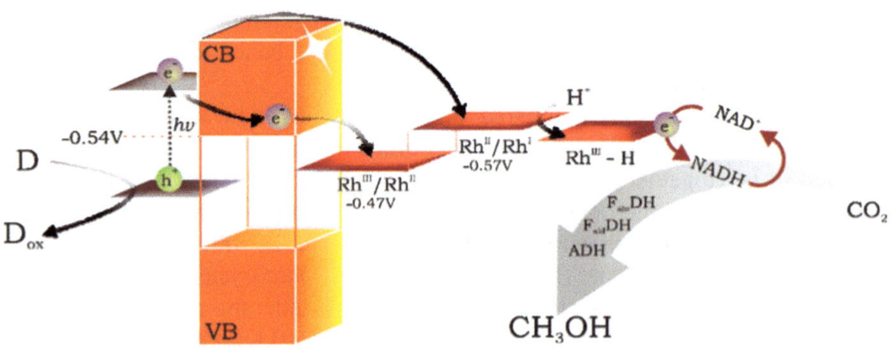

Scheme 9.19 Expected mechanism of NADH regeneration and further carbon dioxide reduction to methanol

References

1. Tong X, El-Zahab B, Zhao X, Liu Y, Wang P (2011) Enzymatic synthesis of L-lactic acid from carbon dioxide and ethanol with an inherent cofactor regeneration cycle. Biotechnol Bioeng 108(2):465–469
2. Park SW, Joo OS, Jung KD, Kim H, Han SH (2001) Development of ZnO/Al_2O_3 catalyst for reverse-water-gas-shift reaction of CAMERE (carbon dioxide hydrogenation to form methanol via a reverse-water-gas-shift reaction) process. Appl Catal A Gen 211:81–90

3. Azuma M, Hashimoto K, Hiromoto M (1990) Electrochemical reduction of carbon dioxide on various metal electrodes in low-temperature aqueous $KHCO_3$ media. J Electrochem Soc 137:1772–1778
4. Subrahmanyam M, Kaneco S, Alonso-Vante N (1999) A screening for the photo reduction of carbon dioxide supported on metal oxide catalysts for C1–C3 selectivity. Appl Catal B Environ 23(2–3):169–174
5. Kuwabata S, Nishida K, Tsuda R, Inoue H, Yoneyama H (1994) Photochemical reduction of carbon dioxide to methanol using ZnS microcrystallite as a photocatalyst in the presence of methanol dehydrogenase. J Electrochem Soc 141(6):1498–1503
6. Aresta M, Quaranta E, Liberio R, Dileo C, Tommasi I (1998) Enzymatic synthesis of 4-OH-benzoic acid from phenol and CO_2: the first example of a biotechnological application of a carboxylase enzyme. Tetrahedron 54(30):8841–8846
7. Dave BC, Obert R (1999) Enzymatic conversion of carbon dioxide to methanol: enhanced methanol production in silica sol-gel matrices. J Am Chem Soc 121:12192–12193
8. Field CB, Behrenfeld MJ, Randerson JT, Falkowski P (1998) Primary production of the biosphere: integrating terrestrial and oceanic components. Science 281:237–240
9. Thauer RK, Kaster A-K, Seedorf H, Buckel W, Hedderich R (2008) Methanogenic archaea: ecologically relevant differences in energy conservation. Nat Rev Microbiol 6:579–591
10. Hugler M, Huber H, Stetter KO, Fuchs G (2003) Autotrophic CO_2 fixation pathways in archaea (*Crenarchaeota*). Arch Microbiol 179:160–173
11. Glueck SM, Gumus S, Fabian WMF, Faber K (2010) Biocatalytic carboxylation. Chem Soc Rev 39:313–328
12. Calvin M (1961) Nobel prize for chemistry: Prof. M. Calvin, For. Mem. R.S. Nature 192:799
13. Hartman FC, Harpel MR (1994) Structure, function, regulation, and assembly of D-ribulose-1,5-bisphosphate carboxylase/oxygenase. Annu Rev Biochem 63:197–234
14. Evans MC, Buchanan BB, Arnon DI (1966) A new ferredoxin-dependent carbon reduction cycle in a photosynthetic bacterium. Proc Natl Acad Sci USA 55:928–934
15. Ljungdahl LG (1986) The autotrophic pathway of acetate synthesis in acetogenic bacteria. Annu Rev Microbiol 40:415–450
16. Drake HL, Goßner AS, Daniel SL (2008) Old acetogens, new light. In: Wiegel, J (ed.) Incredible anaerobes. Ann N Y Acad Sci, 1125:100–128
17. Ragsdale SW, Pierce E (2008) Acetogenesis and Wood-Ljungdahl pathway of CO_2 fixation. Biochim Biophys Acta 1784:1873–1898
18. Herter S, Fuchs G, Bacher A, Eisenreich W (2002) A bicyclic autotrophic CO_2 fixation pathway in *Chloroflexus aurantiacus*. J Biol Chem 277:20277–20283
19. Alber B, Olinger M, Rieder A, Kockelkorn D, Jobst B, Hugler M, Fuchs G (2006) Malonyl-coenzyme A reductase in the modified 3-hydroxypropionate cycle for autotrophic carbon fixation in *Archaeal Metallosphaera* and *Sulfolobus* spp. J Bacteriol 188:8551–8559
20. Berg IA, Kockelkorn D, Buckel W, Fuchs G (2007) A 3-hydroxypropionate/4-hydroxybutyrate autotrophic carbon dioxide assimilation pathway in *Archaea*. Science 318:1782–1786
21. Huber H, Gallenberger M, Jahn U, Eylert E, Berg IA, Kockelkorn D, Eisenreich W, Fuchs G (2008) A dicarboxylate/4-hydroxybutyrate autotrophic carbon assimilation cycle in the hyperthermophilic *Archaeum Ignococcus hospitalis*. Proc Natl Acad Sci USA 105:7851–7856
22. Erb TJ, Berg IA, Brecht V, Muller M, Fuchs G, Alber BE (2007) Synthesis of C5-dicarboxylic acids from C2-units involving crotonyl-CoA carboxylase/reductase: the ethylmalonyl-CoA pathway. Proc Natl Acad Sci USA 104:10631–10636
23. Andersson I, Backlund A (2008) Structure and function of Rubisco. Plant Physiol Biochem 46:275–291
24. Bowyer JR, Leegood RC (1997) Photosynthesis. In: Dey P, Harborne J (eds) Plant biochemistry. Academic, New York, pp 49–110
25. Ellis RJ (1979) The most abundant protein in the world. Trends Biochem Sci 4:241–244

26. Schneider G, Lindqvist Y, Branden C-I (1992) RUBISCO: structure and mechanism. Annu Rev Biophys Biomol Struct 21:119–143
27. Phillips R, Milo R (2009) A feeling for the numbers in biology. Proc Natl Acad Sci USA 106:21465–21471
28. Buchanan BB, Arnon DI (1990) A reverse KREBS cycle in photosynthesis: consensus at last. Photosynth Res 24:47–53
29. Aoshima M (2007) Novel enzyme reactions related to the tricarboxylic acid cycle: phylogenetic/functional implications and biotechnological applications. Appl Microbiol Biotechnol 75:249–255
30. Berg IA (2011) Ecological aspects of the distribution of different autotrophic CO_2 fixation pathways. Appl Environ Microbiol 77:1925–1936
31. Boyd JM, Ensign SA (2005) ATP-dependent enolization of acetone by acetone carboxylase from *Rhodobacter capsulatus*. Biochemistry 44:8543–8553
32. Mai XH, Adams MWW (1996) Characterization of a fourth type of 2-keto acid-oxidizing enzyme from a hyperthermophilic archaeon: 2-ketoglutarate ferredoxin oxidoreductase from *Thermococcus litoralis*. J Bacteriol 178:5890–5896
33. Ragsdale SW (2003) Pyruvate ferredoxin oxidoreductase and its radical intermediate. Chem Rev 103:2333–2346
34. Schut GJ, Menon AL, Adams MWW (2001) 2-Keto acid oxidoreductases from *Pyrococcus furiosus* and *Thermococcus litoralis*. Methods Enzymol 331:144–158
35. Ragsdale SW (2007) Nickel and the carbon cycle. J Inorg Biochem 101:1657–1666
36. Ragsdale SW (2004) Life with carbon monoxide. Crit Rev Biochem Mol Biol 39:165–195
37. Lindahl PA, Chang B (2001) The evolution of acetyl-CoA synthase. Orig Life Evol Biosph 31:403–434
38. Hugler M, Krieger RS, Jahn M, Fuchs G (2003) Characterization of acetyl-CoA/propionyl-CoA carboxylase in *Metallosphaera sedula*. Eur J Biochem 270:736–744
39. Jahn U, Huber H, Eisenreich W, Hugler M, Fuchs G (2007) Insights into the autotrophic CO_2 fixation pathway of the *Archaeon Ignicoccus hospitalis*: comprehensive analysis of the central carbon metabolism. J Bacteriol 189:4108–4119
40. Aresta M, Forti G (eds) (1987) Carbon dioxide as a source of carbon. Elsevier
41. Aresta M, Schloss JV (eds) (1990) Enzymatic and model reaction for carbon dioxide carboxylation and reduction reactions. Elsevier
42. Aresta M, Quaranta E, Tommasi I, Giannoccaro P, Ciccarese A (1995) Enzymatic versus chemical carbon dioxide utilisation. Part I. The role of metal centres in carboxylation reactions. Gazz Chim Ital 125:509–538
43. Kolbe H (1860) Ueber Synthese der Salicylsäure Justus Liebigs. Annalen der Chemie 113 (1):125–127
44. Ota K (1974) Conversion reaction of alkali 4-hydroxyisophthalates to hydroxybenzoic acids. Bull Chem Soc Jpn 47:2343–2344
45. Fumasoni S, Pochetti F, Roberti G (1974) Simultaneous manufacture of urea and glycol, Ger Offen 2,318,327, CA, 80, 14593j
46. Fromm D, Luetzow D (1979) Modern methods of industrial chemistry: urea. Chem Unserer Zeit 13:78–81
47. Aresta M, Tommasi I, Dileo C, Dibenedetto A, Narracci M (2001) Biotechnological synthesis of 4-OH benzoate mediated by a phenylphosphate-carboxylase enzyme *221st National Meeting, American Chemical Society*, San Diego, CA, April 1–5, Inorganic division, Abstract n° 581
48. Lack A, Fuchs G, Aresta M, Tommasi I (1991) Catalytic properties of phenol carboxylase of *Pseudomonas aeruginosa* (strain from Venice lagoon). Eur J Biochem 197:473
49. Platen H, Schink B (1987) Methanogenic degradation of acetone by an enrichment culture. Arch Microbiol 149:136–141

50. Schnell S, Bak F, Pfenng N (1989) Anaerobic degradation of aniline and dihydroxybenzenes by newly isolated sulfate-reducing bacteria and description of desulfobacterium aniline. Arch Microbiol 152:556–563
51. Aresta M, Dibenedetto A (2002) Development of environmentally friendly syntheses: use of enzymes and biomimetic systems for the direct carboxylation of organic substrates. Rev Mol Biotechnol 90:113–128
52. Dibenedetto A, Lo Noce R, Pastore C, Aresta M, Fragale C (2006) First in vitro use of the phenylphosphate carboxylase enzyme in supercritical CO_2 for the selective carboxylation of phenol to 4-hydroxybenzoic acid. Environ Chem Lett 3:145–148
53. Ding B, Schmeling S, Fuchs G (2008) Anaerobic metabolism of catechol by the denitrifying bacterium *Thauera aromatica* – a result of promiscuous enzymes and regulators. J Bacteriol 190:1620–1630
54. Zhang X, Wiegel J (1994) Reversible conversion of 4-hydroxybenzoate and phenol by *Clostridium hydroxybenzoicum*. Appl Environ Microbiol 60:4182–4185
55. He Z, Wiegel J (1995) Purification and characterization of an oxygen-sensitive reversible 4-hydroxybenzoate decarboxylase from *Clostridium hydroxybenzoicum*. Eur J Biochem 229:77–82
56. He Z, Wiegel J (1996) Purification and characterization of an oxygen-sensitive reversible 3,4-dihydroxybenzoate decarboxylase from *Clostridium hydroxybenzoicum*. J Bacteriol 178:3539–4343
57. Miyazaki M, Shibue M, Ogino K, Nakamura H, Maeda H (2001) Enzymatic synthesis of pyruvic acid from acetaldehyde and carbon dioxide. Chem Commun 1800–1801
58. Wieser M, Yoshida T, Nagasawa T (2001) Carbon dioxide fixation by reversible pyrrole-2-carboxylate decarboxylase and its application. J Mol Catal B Enzym 11:179–184
59. Omura H, Wieser M, Nagasawa T (1998) Pyrrole-2-carboxylate decarboxylase from Bacillus megaterium PYR2910, an organic-acid-requiring enzyme. Eur J Biochem 253:480–484
60. Wuensch C, Glueck SM, Gross J, Koszelewski D, Schober M, Faber K (2012) Regioselective enzymatic carboxylation of phenols and hydroxystyrene derivatives. Org Lett 14 (8):1974–1977
61. Yoshida T, Fujita K, Nagasawa T (2002) Novel reversible indole-3-carboxylate decarboxylase catalyzing nonoxidative decarboxylation. Biosci Biotechnol Biochem 66(11):2388–2394
62. Rodriguez H, Landete JM, Curiel JA, de las Rivas B, Mancheno JM, Munoz R (2008) Characterization of the p-coumaric acid decarboxylase from *Lactobacillus plantarum* CECT 748(T). J Agric Food Chem 56:3068–3072
63. Cavin J-F, Barthelmebs L, Divies C (1997) Molecular characterization of an inducible p-coumaric acid decarboxylase from *Lactobacillus plantarum*: gene cloning, transcriptional analysis, overexpression in *Escherichia coli*, purification, and characterization. Appl Environ Microbiol 63:1939–1944
64. Gu W, Li X, Huang J, Duan Y, Meng Z, Zhang K-Q, Yang J (2011) Cloning, sequencing, and overexpression in *Escherichia coli* of the *Enterobacter* sp. Px6-4 gene for ferulic acid decarboxylase. Appl Microbiol Biotechnol 89:1797–1805
65. Prim N, Pastor FIJ, Diaz P (2003) Biochemical studies on cloned *Bacillus* sp. BP-7 phenolic acid decarboxylase PadA. Appl Microbiol Biotechnol 63:51–56
66. Goto M, Hayashi H, Miyahara I, Hirotsu K, Yoshida M, Oikawa T (2006) Crystal structures of nonoxidative Zn-dependent 2,6-dihydroxybenzoate (γ-resorcylate) decarboxylase from *Rhizobium* sp. strain Mtp-10005. J Biol Chem 281:34365–34373
67. Matte A, Grosse S, Bergeron H, Abokitse K, Lau PCK (2010) Structural analysis of *Bacillus pumilus* phenolic acid decarboxylase, a lipocalin-fold enzyme. Acta Crystallogr F66:1407–1414
68. Gu W, Yang J, Lou Z, Liang L, Sun Y, Huang J, Li X, Cao Y, Meng Z, Zhang K-Q (2011) Structural basis of enzymatic activity for the ferulic acid decarboxylase (FADase) from Enterobacter sp. p x6–4. PLoS One 6(1):e16262. doi:10.1371/journal.pone.0016262

69. Rodriguez H, Angulo I, de las Rivas B, Campillo N, Paez JA, Munoz R, Mancheno JM (2010) p-Coumaric acid decarboxylase from *Lactobacillus plantarum*: structural insights into the active site and decarboxylation catalytic mechanism. Proteins 78:1662–1676
70. Swaving J, de Bont JAM (1998) Microbial transformation of epoxides. Enzym Microb Technol 22(1):19–26
71. Allen JR, Ensign SA (1996) Carboxylation of epoxides to beta-keto acids in cell extracts of *Xanthobacter* strain Py2. J Bacteriol 178(5):1469–1472
72. Volbeda A, Fontecilla-Camps JC (2005) Structural bases for the catalytic mechanism of Ni-containing carbon monoxide dehydrogenases. Dalton Trans 3443–3450
73. Park SW, Taeksun S, Kim SY, Kim E, Oh J, Eom C, Kim YM (2007) Carbon monoxide dehydrogenase in mycobacteria possesses a nitric oxide dehydrogenase activity. Biochem Biophys Res Commun 362:449–453
74. Dobbek H, Gremer L, Meyer O, Huber R (1999) Crystal structure and mechanism of CO dehydrogenase, a molybdo iron-sulfur flavoprotein containing *S*-selanylcysteine. Proc Natl Acad Sci USA 96(16):8884–8889
75. Kato N, Sahm H, Wagner F (1979) Steady-state kinetics of formaldehyde dehydrogenase and formate dehydrogenase from a methanol-utilizing yeast, *Candida boidinii*. Biochim Biophys Acta Enzymol 566:12–20
76. Ljungdahl LG, Wood HG (1969) Total synthesis of acetate from CO_2 by heterotrophic bacteria. Annu Rev Microbiol 23:515–538
77. Lu Y, Jiang ZY, Xu SW, Wu H (2006) Efficient conversion of CO_2 to formic acid by formate dehydrogenase immobilized in a novel alginate-silica hybrid gel. Catal Today 115:263–268
78. Miyatani R, Amao Y (2002) Bio-CO_2 fixation with formate dehydrogenase from *Saccharomyces cerevisiae* and water-soluble zinc porphyrin by visible light. Biotechnol Lett 24:1931–1934
79. Parkinson BA, Weaver PF (1984) Photoelectrochemical pumping of enzymatic CO_2 reduction. Nature 309:148–149
80. Reda T, Plugge CM, Abram NJ, Hirst J (2008) Reversible interconversion of carbon dioxide and formate by an electroactive enzyme. Proc Natl Acad Sci USA 105:10654–10658
81. Ruschig U, Müller U, Willnow P, Höpner T (1976) CO_2 reduction to formate by NADH catalyzed by formate dehydrogenase from *Pseudomonas oxalaticus*. Eur J Biochem 70:325–330
82. Boyington JC, Gladyshev VN, Khangulov SV, Stadtman TC, Sun PD (1997) Crystal structure of formate dehydrogenase H: catalysis involving Mo, molybdopterin, selenocysteine, and an Fe4S4 cluster. Science 275:1305–1308
83. Kraulis K (1991) *MOLSCRIPT*: a program to produce both detailed and schematic plots of protein structures. J Appl Crystallogr 24:946–950
84. Bacon DJ, Anderson WF (1988) A fast algorithm for rendering space-filling molecule pictures. J Mol Graph 6:219–220
85. Merritt EA, Murphy MEP (1994) *Raster3D* Version 2.0. A program for photorealistic molecular graphics. Acta Crystallogr D50:869–873
86. Tsuru D (1979) Formaldehyde dehydrogenase from *Pseudomonas putida*. Purification and some properties. J Biochem 85(5):1165–1172
87. Tanaka N, Kusakabe Y, Ito K, Yoshimoto T, Nakamura KT (2002) Crystal structure of formaldehyde dehydrogenase from *Pseudomonas putida*: the structural origin of the tightly bound cofactor in nicotinoprotein dehydrogenases. J Mol Biol 324:519–533
88. Negelein E, Wulff HJ (1937) Diphosphopyridinproteid ackohol, acetaldehyd. Biochem Z 293:351–389
89. Theorell H, McKEE JS (1961) Mechanism of action of liver alcohol dehydrogenase. Nature 192(4797):47–50
90. Jörnvall H, Harris JI (1970) Horse liver alcohol dehydrogenase. On the primary structure of the ethanol-active isoenzyme. Eur J Biochem 13(3):565–576

91. Brändén CI, Eklund H, Nordström B, Boiwe T, Söderlund G, Zeppezauer E, Ohlsson I, Akeson A (1973) Structure of liver alcohol dehydrogenase at 2.9-Angstrom resolution. Proc Natl Acad Sci USA 70(8):2439–2442
92. Raaijmakers H, Macieira S, Dias JM, Teixeira S, Bursakov S, Huber R, Moura JJG, Moura I, Romão MJ (2002) Gene sequence and the 1.8 Å crystal structure of the tungsten-containing formate dehydrogenase from Desulfovibrio gigas. Structure 10:1261–1272
93. Kletzin A, Adams MWW (1996) Tungsten in biological systems. FEMS Microbiol Rev 18:5–63
94. de Bok FAM, Hagedoorn P-L, Silva PJ, Hagen WR, Schiltz E, Fritsche K, Stams AJM (2003) Two W-containing formate dehydrogenases (CO_2-reductases)involved in syntrophic propionate oxidation by *Syntrophobacter fumaroxidans*. Eur J Biochem 270:2476–2485
95. Aresta M (2010) CO_2 enzymatic carboxylation and reduction to methanol. *International Scientific Forum on CO_2 chemistry and biochemistry*, CO_2 Challenge Forum, Lyon
96. Kima S, Kimb MK, Leeb SH, Yoonc S, Junga K-D (2014) Conversion of CO_2- to formate in an electro-enzymatic cell using *Candida boidinii* formate dehydrogenase. J Mol Catal B Enzym 102:9–15
97. Jiang Z, Xu S, Wu H (2003) Novel conversion of carbon dioxide to methanol catalyzed by sol-gel immobilized dehydrogenases. International conference on carbon dioxide utilization ICCDU VII, Seoul-Korea
98. Xu S, Lu Y, Li J, Jiang Z, Wu H (2006) Efficient conversion of CO_2 to methanol catalyzed by three dehydrogenases Co-encapsulated in an alginate-silica (ALG–SiO_2) hybrid gel. Ind Eng Chem Res 45:4567–4573
99. Dibenedetto A, Stufano P, Baran T, Macyk W, Aresta M (2011) Hybrid technologies for an enhanced carbon recycling based on enzymatic CO_2 reduction to methanol in water. International conference on carbon dioxide utilization ICCDU XI, Dijon, FR
100. Dibenedetto A, Angelini A, Aresta M, Macyk W, Baran T (2013) Nanomaterials as photocatalysts for the CO_2 reduction to methanol in water. International conference on carbon dioxide utilization, ICCDU XII, Alexandria, VA
101. Dibenedetto A, Baran T, Macyk W, Aresta M (2013) Photonanomaterials for CO_2 reduction to methanol. 245th ACS national meeting, New Orleans
102. Dibenedetto A, Stufano P, Angelini A, Fragale C, Aresta M, Costa M (2012) Hybrid technologies for an enhanced carbon recycling based on enzymatic CO_2 reduction to methanol in water: chemical and photochemical NADH regeneration. ChemSusChem 5:373–378
103. Cazelles R, Drone J, Fajula F, Ersen O, Moldovan S, Galarneau A (2013) Reduction of CO_2 to methanol by a polyenzymatic system encapsulated in phospholipids–silica nanocapsules. New J Chem 37:3721–3730
104. Song J, Kim Y, Lim M, Lee H, Lee JI, Shin W (2011) Microbes as electrochemical CO_2 conversion catalysts. ChemSusChem 4:587–590
105. Aresta M, Dibenedetto A, Baran T, Angelini A, Łabuz P, Macyk W (2014) An integrated photocatalytic-enzymatic system for the reduction of CO_2 to methanol in bio-glycerol-water. Beilstein J Org Chem 10:2556–2565
106. Aresta M, Dibenedetto A, Baran T, Macyk W (2013) Patent application MI2013A001135

Chapter 10
Thermodynamics and Applications of CO_2 Hydrates

Abstract Gas hydrates are clathrate solid crystalline compounds consisting of a lattice formed by water molecules and entrapped gas molecules inside. They are stable under high pressure and low temperature. CO_2 hydrates, specifically, are composed of CO_2 as the guest molecule and water as the host molecule. CO_2 hydrates have a number of applications including CO_2 capture, cold storage, CO_2 sequestration, and, lately, the direct displacement of methane hydrates with CO_2 to simultaneously produce methane and sequester CO_2. This chapter provides a comprehensive overview of the fundamentals of CO_2 hydrates. The first section gives a general introduction and some basic concepts of gas hydrates. Section 10.2 shifts the focus to the microscopic perspective, looking into how gas hydrates form, the three structures of gas hydrates, and the characteristics of CO_2 hydrates. From there onward, the text focuses specifically on CO_2 hydrates. The physical properties of CO_2 hydrates are considered in Sect. 10.3. Section 10.4 deals with the phase equilibrium of CO_2 hydrate. Experimental methods and the phase diagram are shown in this section. The last section covers the applications of CO_2 hydrates, including the formation and dissociation of CO_2 hydrates, ocean sequestration, and the CH_4 replacement in hydrates by CO_2, which is an attractive potential method to produce natural gas.

10.1 Introduction

Hydrate is a term related to a substance that contains water of crystallization. In inorganic chemistry, hydrates refer to inorganic salts that have water molecules crystallized with salt compounds in a definite ratio. An example is copper sulfate, which turns from gray-white (anhydrous form) to blue (pentahydrate) upon hydration. Such hydration can easily be carried out by dissolving anhydrous (water free) copper sulfate in water and crystallizing. Laboratory research reveals that five water molecules occur in a copper sulfate crystal unit, and four of them are attached to the copper ion by coordination bonds, whereas the fifth is supposed to be held to sulfate

Co-authored by Junjie Zheng, Ponnivalavan Babu, and Praveen Linga, Department of Chemical and Biomolecular Engineering, National University of Singapore, Singapore with the Singapore Ministry of Education, MOE's Tier 1 (R-279-000-386-112) financial support.

M. Aresta et al., *Reaction Mechanisms in Carbon Dioxide Conversion*,
DOI 10.1007/978-3-662-46831-9_10

by a hydrogen bond. This chemical compound can be described by the chemical formula $CuSO_4{\cdot}5H_2O$.

Another class of hydrates is gas hydrates. Unlike the stoichiometric inorganic salt hydrates, gas hydrates are nonstoichiometric clathrate solid crystalline substances consisting of a lattice formed by water molecules (host) and entrapped gas molecules (guest). They are stable under high pressure and low temperature. The hydrate formation reaction can be described by the following general equation:

$$G + N_{\mathrm{H}}\,\mathrm{H_2O} = G \cdot N_{\mathrm{H}}\mathrm{H_2O}$$

where G is the gas molecule and N_{H} is the hydration number, which means the number of water molecules per gas molecule. Hydrates can form under these three conditions:

1. The presence of a sufficient amount of water
2. The presence of hydrate former (guest molecules)
3. The right combination of pressure and temperature (high pressure and low temperature are favorable for hydrate formation)

It should be noted that hydrate formation is not a chemical reaction because no chemical bond is formed or broken during the reaction. Officially, clathrate hydrates are not chemical compounds but one phase of gas–water mixture. As discussed in detail in Sect. 10.2, there are basically three types of structures of gas hydrates. Different structures comprise different types, shapes, sizes, and numbers of cavities. The polyhedral cavities are spherical-like cages formed by hydrogen-bonded water molecules, and a guest molecule with suitable size can fit into one cavity but not be bonded with the lattice, which is quite different than inorganic salt hydrates. Typically, the guest molecules could be low-molecular-mass gases, such as H_2, Ar, Kr, N_2, O_2, CH_4, Xe, H_2S, CO_2, and some relatively heavier hydrocarbons. Gas hydrates can be regarded as highly efficient molecular-level gas storage. Under ideal conditions, each volume of hydrate is capable of containing as much as 184 volumes of gas at standard temperature and pressure.

It was in 1810 that clathrate hydrates were first discovered by Sir Humphry Davy (1811), who found that chlorine (known as oxymuriatic gas at that time) in water could freeze more readily than pure water. Such a frozen substance was what is now known as the chlorine hydrates. During the century following the discovery of clathrate hydrates, many more species were found to form gas hydrates. In the 1930s, research on gas hydrates was stimulated upon the confirmation that gas hydrates are the cause of blockage in gas and oil pipelines, which is a major problem in the oil and gas industries and may cause significant financial loss. Since then, considerable investment has been made into research areas such as phase equilibria, thermodynamic models, and time-dependent kinetic behavior of gas hydrates. Since the 1960s, after they were found to be commonly located in the Earth's permafrost and deep sea regions, gas hydrates are emerging as a potential energy resource. The majority of natural gas hydrates contain CH_4 in overwhelming abundance. Though this resource has not been systematically mapped and evaluated, and current estimates vary widely, there is a consensus that the quantity of natural gas hydrates is vast and the carbon content in natural gas hydrates is far more than the carbon content in all the conventional fossil fuel resources combined. Thus, a number of gas hydrates research and development programs have emerged, including basic scientific research and field-based studies, related to hydrates exploration and recovery. At the same time, there are ongoing studies of the environmental and safety aspects of hydrates, for example, the impact of methane hydrate on climate change. A schematic snapshot of significance of gas hydrates or clathrate hydrates is shown in Fig. 10.1.

CO_2 hydrates are composed of CO_2 as the guest molecules and water as the host molecules. The first discovery of CO_2 hydrates dates back to 1882, when Wroblewski [1, 2] discovered a white snow-like substance while studying carbonic acid. He estimated that the composition of CO_2 hydrates was approximately $CO_2{\cdot}8H_2O$, whereas the hydration number of typical CO_2 hydrates is now known to be between 5.75 and 7.67 [3]. CO_2 hydrates have a number of applications including CO_2 capture and CO_2 sequestration, and because of the higher stability of CO_2 hydrates compared with that of CH_4 hydrates, displacing CH_4 directly in the methane hydrates with CO_2 to produce methane and sequester CO_2 simultaneously has recently been proposed.

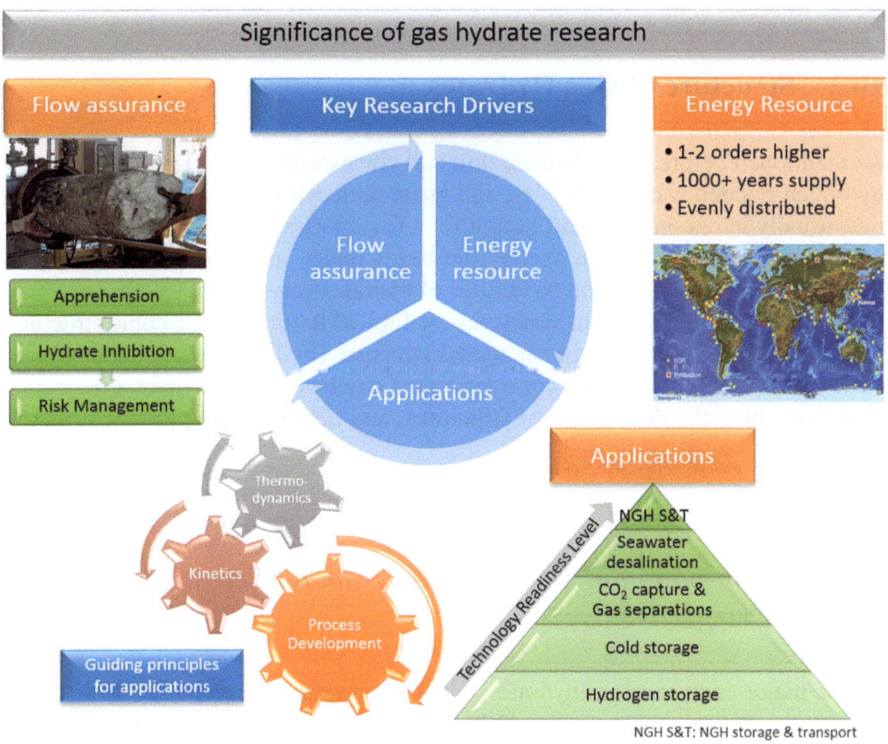

Fig. 10.1 Significance of gas hydrate or clathrate hydrate research

The endothermic effect of dissociation of CO_2 hydrates makes it possible for them to be used in refrigeration systems [4–6]. Another point making CO_2 hydrates important is the high possibility of the occurrence of CO_2 hydrates on other planets, especially Mars. Because CO_2 is an abundant component of the atmosphere on Mars, and with the prevailing temperature and pressure conditions, it is suggested that CO_2 hydrates, dry ice, liquid CO_2, and carbonated groundwater are common phases on Mars. The formation and transformation of these phases may play a significant role in the evolution of Martian geology. All the above factors lead to an increasing interest in the study of CO_2 hydrates.

This chapter provides a comprehensive overview of the fundamentals and applications of CO_2 hydrates. Section 10.2 focuses on the microscopic perspective, looking into how gas hydrates form, the three structures of gas hydrates, and the characteristics of CO_2 hydrates. From there onwards, the text focuses specifically on CO_2 hydrates. The physical properties of CO_2 hydrates are considered in Sect. 10.3. Section 10.4 deals with the phase equilibrium of CO_2 hydrate. Experimental methods and the phase diagram are showed in this section. The last section covers the applications of CO_2 hydrates, including the formation and dissociation of CO_2 hydrates, ocean sequestration, the CH_4 replacement in hydrates by CO_2, and the use of CO_2 hydrates in the refrigeration process.

10.2 Structure of Gas Hydrates

As mentioned before, a gas hydrate unit cell consists of several cavities formed by hydrogen-bonded water molecules, and guest molecules are trapped in those cavities, usually in one-in-one form except for a few cases such as hydrogen molecules whereby more than one can occupy the cages under certain experimental conditions. Although the formation conditions of hydrates vary widely for different guest molecules, the typical crystal structures are limited to three types: cubic structure I (sI), cubic structure II (sII), or hexagonal structure H (sH). This section details the structure of gas hydrates, first considering the cavities formed by water molecules, then describing the three types of structures, and finally shifting the focus to CO_2 hydrates.

10.2.1 Formation of Gas Hydrates from a Microscopic Perspective

As the major content of gas hydrates, water plays a crucial role in the formation of the cavities through the extensive H-bonding, which may link each of them to four others, two through the two lone pairs on oxygen and two through the two polar hydrogen atoms. The four surrounding molecules are arranged in a tetrahedral manner. The hydrogen bonding has an energy of ca. 7.5 kcal/mol and, because of such extra energy, water has a higher boiling point, melting point, and viscosity than other Group 16 element hydrides in which hydrogen bonds are absent or much weaker. Because it is much more significant than the van der Waals interaction, only the hydrogen bond is considered between adjacent molecules when hydrates form or dissociate.

Connected by hydrogen bonds, water molecules can form many kinds of three-dimensional frameworks, which include cavities (polyhedrons) to accommodate guest molecules. For simplicity, the polygons comprising those cavities are first to be considered here. Computer simulation work shows that the most likely natural structure in water at many temperatures is the pentamer, followed by hexamers and squares [7]. Molecular dynamics studies suggest that cyclic pentamer comprising many hydrate cavities is the only stable five-member cluster above 230 K [8]. More generally, it was claimed that closed rings were often more stable than open chains of the same cluster number, because of the extra energy of the hydrogen bond [9]. From another perspective, regarding planar rings, the O–O–O angle of pentagons is 108°, which is very similar to the normal water angle (104.5°) and the tetrahedral angle (109.5°), leading to the least strain on the bonds. Furthermore, the O–O–O angles of squares (90°) and the hexagons (120°) deviate more but are still acceptable. However, if the number of edges becomes larger than six or less than four, the strains are so high that such polygons occur infrequently in water structures.

Based on the three available types of polygons mentioned above, five types of polyhedra are mainly found in the three structures of gas hydrates. As shown in Fig. 10.2, they are pentagonal dodecahedron (5^{12}), tetrakaidecahedron ($5^{12}6^2$), hexakaidecahedron ($5^{12}6^4$), irregular dodecahedron ($4^35^66^3$), and icosahedron ($5^{12}6^8$). The formalism X^n is used to describe the structure of each cavity, where X denotes the type of cage face (4–square, 5–pentagon, and 6–hexagon), and n represents the number of such kinds of face in that cavity. The three structures of gas hydrates comprise different cavities: 5^{12} and $5^{12}6^2$ cavities form sI, 5^{12} and $5^{12}6^4$ cavities form sII, and sH is formed by 5^{12}, $4^35^66^3$, and $5^{12}6^8$ cavities.

The pentagonal dodecahedron (5^{12}) has 12 pentagonal faces with equal edge lengths and equal angles. It is the most common basic building block present in almost all gas hydrate structures. It can be seen from Fig. 10.2 that each 5^{12} cavity contains 20 water molecules, as each vertex represents an oxygen atom. However, when it comes to the gas hydrate unit cell, the average number of water molecules per guest is not 20 anymore because those cavities are connected by sharing faces, edges, or vertices. Also note that there are 30 bonds (edges) displayed in the 5^{12} cavity: as mentioned before, only one hydrogen atom lies between the adjacent two oxygen atoms, thus 30 hydrogen atoms are shown in this figure, whereas the other 10 hydrogen atoms protrude from the cavity, potentially making attachment to other cavities or molecules. It is suggested that this 5^{12} cavity structure is the most unstrained polyhedra compared with other clathrate clusters, and therefore geometrically favored by nature [11]. The tetrakaidecahedron ($5^{12}6^2$) has 12 pentagonal and 2 hexagonal faces. Each $5^{12}6^2$ cavity contains 24 water molecules. This $5^{12}6^2$

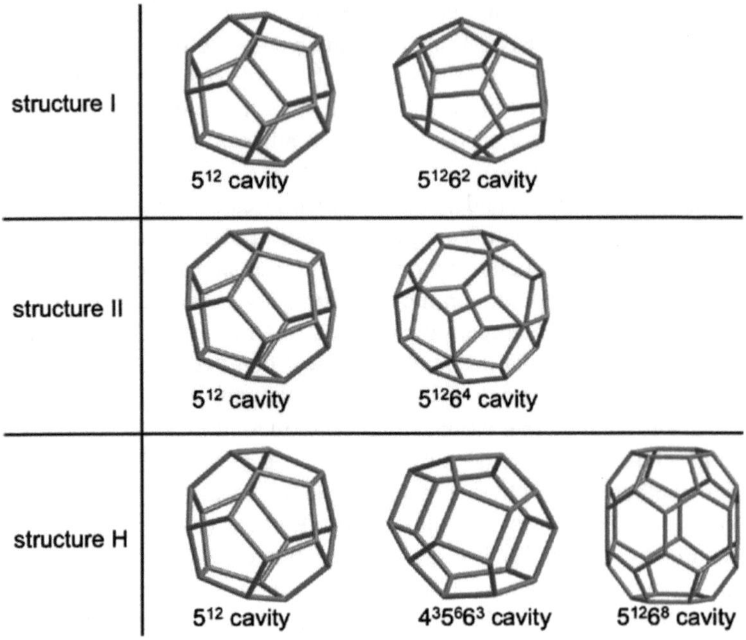

Fig. 10.2 Five building polyhedra of gas hydrate structures: pentagonal dodecahedron (5^{12}), tetrakaidecahedron ($5^{12}6^2$), hexakaidecahedron ($5^{12}6^4$), irregular dodecahedron ($4^35^66^3$), and icosahedron ($5^{12}6^8$) [10]. Reprinted by permission from Springer Science and Business Media

cavity is the least spherical cavity in sI and sII. The hexakaidecahedron ($5^{12}6^4$) has 12 pentagonal faces and 4 hexagonal faces. Those 4 hexagonal faces distribute symmetrically and each of them is surrounded by 6 pentagonal faces with sharing edges. Each $5^{12}6^4$ cavity contains 28 water molecules. This cavity is the most spherical cavity of the five types. The irregular dodecahedron ($4^35^66^3$) has 3 square faces, 6 pentagonal faces, and 3 hexagonal faces. Each $4^35^66^3$ cavity contains 20 water molecules. Because of the high proportion of square and hexagonal faces, the $4^35^66^3$ cavity becomes the most strained one, which may indicate a slow kinetics of formation. The icosahedron ($5^{12}6^8$) has 12 pentagonal faces and 8 hexagonal faces, 6 of which form a girdle and the other 2 as the top and bottom faces. Each $5^{12}6^8$ cavity contains 36 water molecules. It is clear that the $5^{12}6^8$ cavity is the largest of the five types and allows a larger occupant. It is also the most non-spherical one of all five cavities.

10.2.2 Crystal Structures of Gas Hydrates

All gas hydrates have repetitive crystal units, and a unit cell consists of several cavities sharing faces, edges, and/or vertices, with guest molecules trapped in those

Table 10.1 Characteristics of the three structures of gas hydrates

	I		II		H		
Ideal unit cell formula	$2(5^{12})\cdot 6(5^{12}6^{2})\cdot 46H_2O$		$16(5^{12})\cdot 8(5^{12}6^{4})\cdot 136H_2O$		$3(5^{12})\cdot 2(4^{3}5^{6}6^{3})\cdot 1(5^{12}6^{8})\cdot 34H_2O$		
Crystal system	Cubic		Cubic		Hexagonal		
Cavities	Small	Large	Small	Large	Small	Medium	Large
	5^{12}	$5^{12}6^{2}$	5^{12}	$5^{12}6^{4}$	5^{12}	$4^{3}5^{6}6^{3}$	$5^{12}6^{8}$
No. of cavities per unit cell	2	6	16	8	3	2	1
Average cavity diameter (Å)	7.9	8.66	7.82	9.46	7.88	8.08	11.58
Effective cavity diameter (Å)[a]	5.1	5.86	5.02	6.66	5.08	5.28	8.78

[a]Effective cavity diameter is obtained by subtracting the van der Waals diameter of the water molecule (2.8 Å) from the average cavity diameter

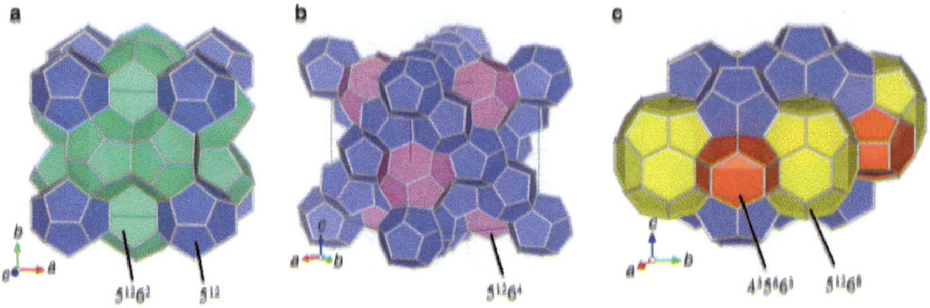

Fig. 10.3 Three structures of gas hydrates: (**a**) sI, (**b**) sII, and (**c**) sH [13]. Reprinted by permission from Macmillan Publishers Ltd: NATURE COMMUNICATIONS, copyright (2011)

cavities [12]. The most common unit cell structures found in gas hydrates are named as structure I (sI), structure II (sII), and structure H (sH). Some characteristics of the three structures are showed in Table 10.1. Figure 10.3 shows the appearance of these structures.

Structure I comprises two 5^{12} cavities and six $5^{12}6^2$ cavities, formed by 46 water molecules. It is clear from Fig. 10.3a that one 5^{12} cavity is located at each of the eight vertices. Considering the surrounding unit cells, each of the vertices only contains 1/8 of a 5^{12} cavity. By adding the 5^{12} cavity at the center of the structure, the total number of 5^{12} cavities becomes two per unit cell. For the $5^{12}6^2$ cavities, each of the six faces contains two halves of $5^{12}6^2$ cavities and the total number becomes six. Similar calculation can be done on other structures.

Structure II (see Fig. 10.3b) is a face-centered cubic lattice, consisting of 16 5^{12} cavities and 8 $5^{12}6^4$ cavities. Each unit cell of sII contains 136 water molecules. This structure can be regarded as completely defined by 5^{12} cavities (which share the pentagonal faces), whereas the residual voids form the $5^{12}6^4$ cavities (which share all the hexagonal faces). The two kinds of cavities are connected by sharing pentagonal faces.

Structure H (see Fig. 10.3c) is a hexagonal rather than cubic cell, consisting of 3 5^{12} cavities, 2 $4^35^66^3$ cavities and 1 $5^{12}6^8$ cavity, in total 34 water molecules. Considering a single unit cell, four $5^{12}6^8$ cavities are located at the vertices of a rhombus. Looking at the middle level plane, each of the six hexagonal faces on the girdle of $5^{12}6^8$ cavity is shared with a $4^35^66^3$ cavity, resulting in six $4^35^66^3$ cavities surrounding a $5^{12}6^8$ cavity. $4^35^66^3$ cavities are connected only through square faces with each other. All pentagonal faces of these two types of cavity are shared with 5^{12} cavities.

Apart from the framework formed by water molecules, the guest molecules are also very important to the unit cell structure. It is believed that the hydrate structure is stabilized mainly by the repulsive van der Waals interactions between guest molecules and cage walls. Without the support of guest molecules, the lattice structure may easily collapse to an ice crystal structure or liquid water. Typically, every cavity contains only one guest molecule, although at very high pressure some low-molecular-mass gas may multiply occupy the large cavities. The volume of cavities vary with temperature, pressure, and the size and shape of the guest molecules. Average and effective diameters of cavities are shown in Table 10.1. Both sI and sII have 5^{12} cavities, with effective diameters of 5.1 Å and 5.02 Å, respectively. Though the difference is small, it reflects on the size of guest molecules. For example, oxygen (with diameter of 4.2 Å) occupies the 5^{12} cavity of structure II, whereas methane (with diameter of 4.36 Å), though only slightly larger, occupies the 5^{12} cavity of structure I. As for the large cages in sI and sII ($5^{12}6^2$ and $5^{12}6^4$ cavities), the larger diameter makes them able to contain molecules of larger size. For some guest molecules with sizes between the small and large cages, they only occupy the large cages but leave the small ones vacant. For sH, the size difference between the small and large cavities is so significant that two kinds of guest molecules with different sizes are required to stabilize the structure. For example, methane and neohexane can form sH hydrates, in which methane occupies the small (5^{12}) and medium ($4^35^66^3$) cavities whereas neohexane enters the large cavities ($5^{12}6^8$).

In order to describe the ratio of water molecules to gas molecules in hydrates, hydration number is defined as the number of water molecules per guest. Take ideal N_2 hydrates as an example; N_2 occupies all the small and large cages of sII, resulting in the formula $24N_2{\cdot}136H_2O$, i.e., 5(2/3) H_2O per N_2. If the guest molecules only occupy the 8 large cages, such as C_3H_8, then the hydration number becomes 17. Similarly, CO_2 forms sI; the ideal occupancy (complete occupancy of CO_2 in small and large cages) yields a hydration number of 5(3/4) H_2O per CO_2. It is noted that, for the case of CO_2, it has been experimentally shown that CO_2 can occupy almost all of the large cages (>90%) although it occupies a fraction of the small cages, typically in the range of 20–60% depending upon the formation experimental conditions. As for sH, the concept of hydration number is only applicable for multi-guests. It is interesting that, under ideal condition, the mole fractions of water in the three structures are all around 85%. However, it is unlikely that all cages could be occupied in simple hydrates by guest molecules (hydrates with only one kind of guest). For example, the real formula of sI hydrates usually ranges between $G{\cdot}5(3/4)H_2O$ to $G{\cdot}19H_2O$, which depends on the guest composition, temperature, and pressure. Such variation is called the non-stoichiometry of gas hydrates, which distinguishes them from the stoichiometric salt hydrates in inorganic chemistry.

10.2.3 *Characteristics of CO_2 Hydrates*

Different from most gas hydrate formers, CO_2 is not hydrophobic. At 273 K, the solubility of CO_2 in water varies between 0.08 and 1.46 mol/L with the partial pressure of CO_2 between 0.1 and 3.4 MPa [14]. The dissolution of CO_2 in water produces species including dissolved unhydrated CO_2, hydrated CO_2, H_2CO_3, HCO_3^-, and CO_3^{2-}.

Carbon dioxide is known to form regular cubic sI as simple hydrates, although there is also some evidence which indicates a metastable sII CO_2 hydrate phase at temperatures near the ice melting point [15]. Stabilization is the major consideration when determining which structure gas hydrates form. For CO_2 hydrates, the diameter ratios of guest molecule to host cavities play a dominant role in stabilizing the structure. The ratios of molecular diameter of CO_2 to cavity diameter are shown in Table 10.2, with the diameter of CO_2 selected as 5.12 Å. For simple (single guest) hydrate formers, usually a size ratio higher than 0.76 is required to ensure the cavity stability. CO_2 forms regular cubic sI as simple hydrates. As can be seen in Table 10.2, for large cavity in sII, the size ratio is 0.769, indicating less cavity stability and therefore difficulty in forming a stable sII hydrate for CO_2 as a single guest. It is noted that there is evidence in literature that, for mixed hydrates with a hydrate former such as propane occupying the large cage of sII and thus making it stable, CO_2 has been shown to occupy the large cages in such a resultant mixed sII hydrates. For the large cavities the size ratio in sI is 0.843, although it is 0.769 for sII, indicating less significant cavity stability. Hence as a simple hydrate former, CO_2 forms sI.

Representing the hydrates composition, the hydration number of sI hydrates can be obtained from the equation

$$N_{\mathrm{H}} = \frac{46}{2\theta_{\mathrm{s}} + 6\theta_{\mathrm{l}}}$$

where N_{H} is the hydration number, θ_{s} is the degree of occupancy of small cages, and θ_{l} is the degree of occupancy of large cages.

Hydration number can be obtained experimentally through macroscopic methods using the enthalpies of formation/dissociation [16] or gas uptake/release during formation/dissociation [17]. Another class of approaches is microscopic methods, including X-ray and neutron diffraction and Raman and NMR spectroscopy. For CO_2 hydrates, data from a neutron diffraction study [18] indicate that, at a pressure of 6.2 MPa and temperature between 230 and 263 K, more than 95% of the large cages can be filled and the small cage occupancy is between 60% and 80%, yielding a hydration number between 6.05 and 6.67.

10.3 Physical Properties of CO_2 Hydrates

CO_2 hydrates are snow-like substance which can exist below 283 K at a wide range of pressures. Because CO_2 sequestration into the deep ocean has become a hot potential approach to dispose of CO_2, the physical properties of CO_2 hydrates are important to the feasibility study and modeling of CO_2 deposition. For example, the density of CO_2 hydrates is needed to determine whether they sink or float at certain depths of the ocean, which is a major problem when modeling the CO_2 dispersion. This section details the mechanical, thermal, and other physical properties of CO_2

Table 10.2 Diameter ratio of CO_2 molecule to host cavities

	I		II		H		
Ideal unit cell formula	$2(5^{12})\cdot 6(5^{12}6^{2})\cdot 46H_2O$		$16(5^{12})\cdot 8(5^{12}6^{4})\cdot 136H_2O$		$3(5^{12})\cdot 2(4^{3}5^{6}6^{3})\cdot 1(5^{12}6^{8})\cdot 34H_2O$		
Cavities	Small	Large	Small	Large	Small	Medium	Large
	5^{12}	$5^{12}6^{2}$	5^{12}	$5^{12}6^{4}$	5^{12}	$4^{3}5^{6}6^{3}$	$5^{12}6^{8}$
Effective cavity diameter (Å)[a]	5.1	5.86	5.02	6.66	5.08	5.28	8.78
Diameter ratio of CO_2 molecule to host cavities	1.004	0.874	1.020	0.769	1.008	0.970	0.583

[a]Effective cavity diameter is obtained by subtracting the van der Waals diameter of the water molecule (2.8 Å) from the average cavity diameter

hydrates. Because there is a lack of data for CO_2 hydrates, for some parts, the properties of some other common gas hydrates are shown, which may indicate the properties of CO_2 hydrates based on the similarity between gas hydrates.

10.3.1 Mechanical Properties

Generally, the mechanical strength of pure gas hydrates is much higher than that of ice (Ih). For example, Durham et al. [19] measured the creep resistance of methane hydrates and found it more than 20 times stronger than ice. This may be because of the much lower diffusion rate of water in hydrates compared to ice, limiting the motion of crystalline defects. However, when it comes to the hydrates-bearing sample, the mechanical strength may be much smaller than that of pure hydrates because of the liquid water or ice in the sample. It is believed that the strength increases with increasing hydrate saturation (the proportion of the hydrate volume to pore volume in the sample).

The elastic properties of hydrates are important to understanding the sonic and seismic velocity field data obtained from the natural hydrates-bearing sediments. Data on the mechanical properties of CO_2 hydrates are limited. Table 10.3 shows the elastic properties of ice, CH_4 hydrates, and CO_2 hydrates. It should be noted that these properties may vary for different guests and occupancies. For example, Kiefte et al. [21] measured the compressional velocity of methane, propane, and hydrogen sulfide hydrates as 3.3, 3.7, and 3.35 km/s, respectively.

10.3.2 Thermal Properties

Thermal transport in the bulk medium is one of most important factors that affect the phase transformation of gas hydrates. Thermal properties are needed to estimate the heat transfer in hydrates sediments. Thermal properties of gas hydrates vary a

Table 10.3 Elastic properties of ice, CO_2, and CH_4 hydrates

	Ice (Ih) [20]	CH_4 hydrates [20]	CO_2 hydrates
Conditions	253–268 K, 22.4–32.8 MPa	258–288 K, 27.6–62.1 MPa	
Poisson's ratio	0.3301	0.31403	No data
Bulk modulus (GPa)	9.097	8.762	No data
Shear modulus (GPa)	3.488	3.574	No data
Compressional velocity, Vp (m/s)	3 870.1	3 778	No data
Shear velocity, V_S (m/s)	1 949	1 963.6	No data

Table 10.4 Thermal conductivity of some samples

	Ice [22]	Water [22]	CH_4 hydrates [23]	CO_2 hydrates [24]
Temperature (K)	268	273	263	263
Thermal conductivity (W/m K)	2.3	0.56	0.49	0.49
Heat or enthalpy of dissociation and formation (kJ/mol)			52.7–56.9	57.7–63.6 (±1.8)
Thermal diffusivity (m^2/s)		1.33×10^{-7}	3.1×10^{-7}	No data

Table 10.5 Constant pressure heat capacity (J/g K) of ice, methane CO_2 and ethane hydrates

T (K)	Ice [25]	CH_4 hydrate [16]	C_2H_6 hydrate [16]	CO_2 hydrates
240	1.863	1.885	1.864	No data available Most probably it will be less than liquid water and possibly similar to ice.
250	1.935	1.939	1.936	
260	2.009	2.003	2.025	
270	2.083	2.077		

little between different guest molecules, and thus the data from other gas hydrates can only provide an approximation for CO_2 hydrates.

The thermal conductivity of gas hydrates is dependent on temperature, but has no pressure dependence. Table 10.4 shows the thermal conductivities of ice, water, CO_2 hydrates, and methane hydrates.

It is clear that the thermal conductivity of gas hydrates is much less than that of ice, but similar to liquid water. Furthermore, when it comes to hydrate/gas/water or hydrate/gas/water/sediment systems, the thermal properties are usually determined as the average values of the properties of the components by considering their saturation (volumetric fraction) in the sample. Because of the paucity of data of CO_2 hydrates, the heat capacities of ice, methane and ethane hydrates are shown in Table 10.5. Considering the similarity between CO_2 hydrates and other gas hydrates, the heat capacity of CO_2 hydrates is certainly less than that of liquid water and may be similar to that of ice.

Another thermal property is thermal expansion, which is important when estimating the density of CO_2 hydrates. Udachin et al. [26] combined his results with others' work and obtained the lattice parameter for CO_2 hydrates as a polynomial function of temperature as follows:

$$a = 11.81945 - 9.08711 \times 10^{-5}T + 4.59676 \times 10^{-6}T^2 - 8.35548 \times 10^{-9}T^3$$

where a represents the lattice parameter with unit Å and T is the temperature with unit K.

This function suggests that the lattice parameter, and thereby the volume, of CO_2 hydrates grows monotonically with the temperature within the range in which hydrates exist. It is worth noting that the thermal expansion of the CO_2 hydrates is much greater than that of ice.

10.3.3 Other Physical Properties

One of the advantages of gas hydrates is the significant volume reduction compared with the original gaseous phase. For example, if all cavities of sI are occupied, one volume of hydrate may contain 184 volumes of gas at standard temperature and pressure. Specifically, it is estimated that the maximum density of CO_2 hydrates is 1.13 g/cm^3, assuming the ideal (full) occupancy of CO_2 molecules in the cavities. On the other hand, if no guest molecule occupies the cages of sI, the density would be 0.796 g/cm^3 [27]. In actual conditions, however, the occupancy must be less than the ideal situation. By using microscopic observation techniques, such as X-ray and neutron diffraction, it is possible to obtain the cage occupancies (which can be used to calculate the hydrate composition) and lattice parameters (which depend on temperature and can be used to calculate the volume of hydrate unit cells). With these data, one can estimate the density of CO_2 hydrates. For example, in a Udachin et al. study [26], the lattice parameter of CO_2 hydrates was measured as 11.893 Å at 173 K, and the composition was $7.42CO_2{\cdot}46H_2O$, yielding the density as 1.14 g/cm^3, and when the temperature increased to 277 K, the lattice parameter became 11.970 Å, resulting in a density of 1.12 g/cm^3. The main uncertainty of the density estimated this way arises from the cage occupancy, which may be affected by the pressure. Generally, the density of CO_2 hydrates formed under deep sea conditions is slightly higher than that of liquid CO_2 (around 1.05 g/cm^3) under the same conditions.

10.4 Phase Equilibrium of CO_2 Hydrate

10.4.1 Experimental Methods to Study Hydrate Phase Equilibria

Knowledge of CO_2 hydrate formation conditions is very important for rational and design of processes in CO_2 capture and sequestration. Several studies have focused on determination of incipient CO_2 hydrate formation conditions. The condition at which an infinitesimal amount of hydrate phase is present in equilibrium with liquid phases is referred to as incipient hydrate formation condition [28]. Using different experimental methods, macroscopic hydrate phase equilibrium can be determined by employing a traditional stirred tank reactor with vigorous mixing. The four different methods are:

1. Temperature search method
2. Pressure search method
3. P-T cycle method
4. Calorimetric method

All these methods were used to determine phase equilibrium above the ice point, i.e., L_W–H–V line.

10.4.1.1 Temperature Search Method

This method operates at isobaric condition (constant pressure). The crystallizer with a gas–liquid system is set to pressures at which the hydrate equilibrium temperature needs to be determined. Initially the pressure of the system drops because of dissolution of the gas into the liquid caused by mixing. The pressure of the system is maintained constant by a supply of gas from the reservoir or fluid withdrawal. The system temperature is decreased to induce hydrate formation. The pressure of the system drops because of cooling and hydrate formation. After hydrate formation, the temperature of the system is increased slowly until an infinitesimal amount of hydrate crystals exists in equilibrium with the vapor and liquid phases. During heating, the system pressure is maintained constant by fluid withdrawal. This point is recorded as equilibrium condition. The equilibrium point is determined by visual observation.

10.4.1.2 Pressure Search Method

This method operates at isothermal condition (constant temperature). The crystallizer was precooled to the temperature at which the equilibrium pressure needs to be determined. The system was then set to a pressure higher than the equilibrium condition. The crystallizer content was mixed and the pressure of the crystallizer drops because of gas dissolution and hydrate formation. As the hydrates form, the temperature of the liquid–gas system increases and is restored back by external cooling. After hydrate formation, the pressure is reduced gradually until infinitesimal amount of hydrate crystals exists in equilibrium with the vapor and liquid phases. The equilibrium pressure is determined by visual observation of the hydrate crystal disappearance.

10.4.1.3 Isochoric P–T Cycle

This method operates at isochoric condition (constant volume). A P–T cycle method is illustrated in Fig. 10.4. The crystallizer content is mixed after pressurizing with the gas.

Fig. 10.4 Pressure–temperature trace for phase equilibrium measurement

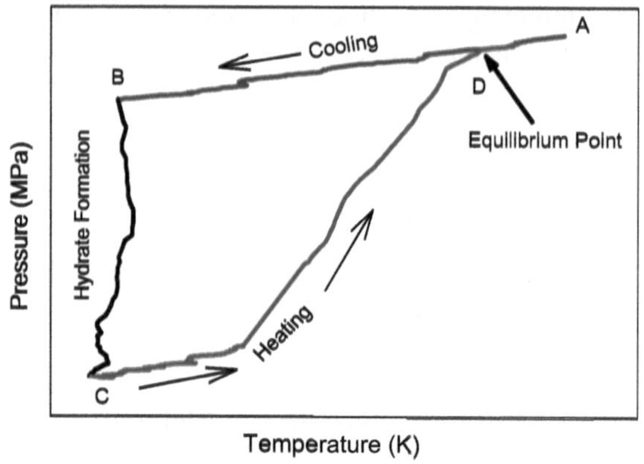

The system was allowed to reach a steady state. At this time, the system is far enough from the hydrate stable region. Upon pressure and temperature stabilization, the temperature of the system is lowered and the pressure drops because of slight cooling. At point B, a catastrophic hydrate formation leads to a sudden decrease in pressure. The temperature is then slowly increased to dissociate the formed hydrates. During the dissociation of the hydrate crystals inside the hydrate stable region, the increase in temperature increases the pressure gradually whereas, outside the hydrate stable region, a slight increase in pressure is observed. On the pressure–temperature diagram, the equilibrium point is the point at which the heating curve (hydrate dissociation trace) intersects with the cooling curve (Point D) or a change in slope of the dissociation trace is observed. This method is an alternative to visual observation of the hydrate crystal disappearance.

10.4.1.4 Calorimetric Method

As an alternative to these classical PVT methods, a calorimetric method can be used to determine the hydrate phase equilibrium [29–31]. A high pressure differential scanning calorimetry (DSC) is employed. The high pressure cell in the DSC is loaded with the solution and then pressurized with the hydrate forming gas. The cell is then cooled to a lower temperature to form hydrate. An exothermic peak is observed in the DSC thermograph because of hydrate formation. The cell is then heated at a constant rate to dissociate the hydrate crystal. DSC thermography shows an endothermic peak during hydrate dissociation. The onset temperature of the endothermic peak is taken as the equilibrium temperature at the given pressure. The PVT and DSC results are in good agreement with each other [31].

10.4.2 Pressure–Temperature Phase Diagram of CO_2+H_2O System

Equilibrium hydrate formation condition for CO_2 has been extensively investigated and available in the literature [32–44]. A hydrate phase diagram of the CO_2+H_2O system is presented in Fig. 10.5. As can be seen in the figure, equilibrium hydrate formation conditions above and below freezing point of water is presented.

The CO_2 vapor pressure curve is also presented. The lower quadruple point ($Q1 = I–L_W–H–V$) is located at the intersection of I–H–V and L_W–H–V lines. The upper quadruple point ($Q2 = L_W–H–V–L_{CO2}$) is located at the intersection of the L_W–H–V and L_W–H–L_{CO2} lines. The H–V–L_{CO2} and L_W–V–L_{CO2} originate at point Q2. Lines I–H–V, L_W–H–V, and L_W–H–L_{CO2} bound the hydrate formation region. Hydrates can form at higher pressures and lower temperatures to the left of the region enclosed by these three lines. Hydrates are unstable if we move towards the right from these three lines. Above the H–V–L_{CO2} line the hydrates form from liquid CO_2 whereas below this line gas CO_2 takes part in hydrate formation.

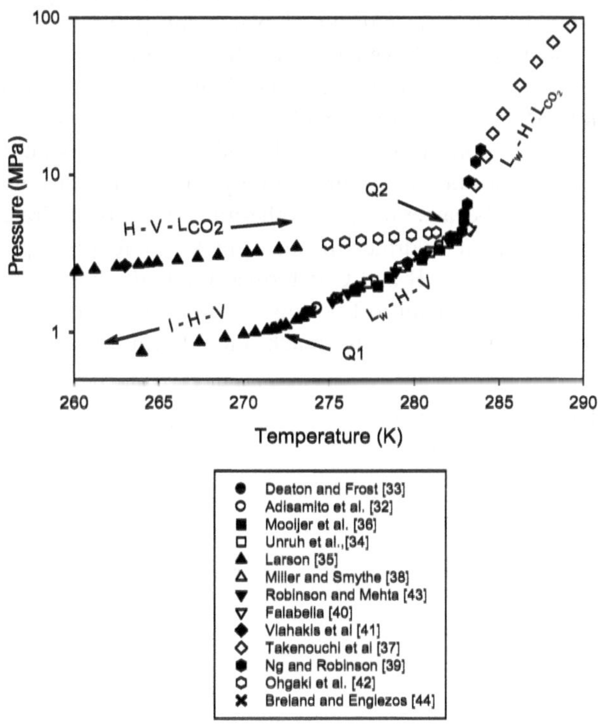

Fig. 10.5 Phase diagram of CO_2+H_2O system

10.5 Applications of CO_2 Hydrates

10.5.1 Formation of CO_2 Hydrate

Following the thermodynamics (phase equilibrium) of CO_2 hydrate, which is time-independent, it is also important to know the time-dependent phenomenon of hydrate, namely, how hydrates form and dissociate. Note that the study of such phenomena is much more challenging than that of the thermodynamic properties.

Figure 10.6 shows typical crystallization kinetics of the formation of CO_2 hydrate. The gas was contacted with the known amount of water in a semi-batch manner, i.e., pressure and temperature of the system is maintained constant. The pressure and temperature of the crystallizer is set at the hydrate stable region. The hydrate crystallization kinetics curve has three different phases. First the dissolution phase where the gas dissolves into the liquid. The moles of gas uptake at Point A corresponds to the three-phase hydrate equilibrium pressure at the experimental temperature. Further gas dissolves and the gas uptake plateaus off. Although the conditions are favorable for hydrate formation, hydrate crystals do not form immediately. This ability to maintain a non-equilibrium state is called metastability. The region between points A and B is characterized by the nucleation process. Nucleation is the restructuring of water and guest molecules into stable nuclei with critical size, followed by growth on these nuclei. The period during which the temperature and pressure conditions are within the hydrate stable zone but no hydrate is detected is usually called the induction period (in Fig. 10.6 this period begins at point B and ends when significant pressure drop occurs). At point C, the metastable state is ruptured and hydrate starts to form. A temperature spike can be observed because of the exothermic nature of hydrate formation. Along with the formation of hydrate, the gas uptake increases rapidly to point C. The region between points B and C is termed the hydrate growth phase. After some time the gas uptake plateaus off because of no further hydrate formation. The time taken for the hydrate growth to reach the plateau varies according to the crystallizer

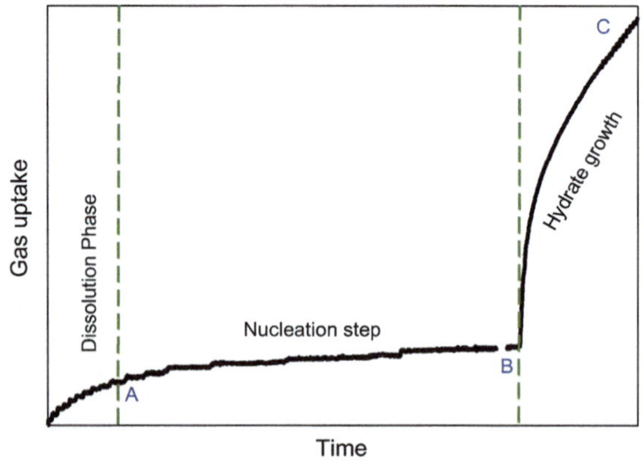

Fig. 10.6 Crystallization kinetics of hydrate formation

configurations. Usually it takes few hours in a stirred tank reactor for the hydrate growth to cease because of mass transfer limitation.

The formation of hydrate nuclei in a CO_2–water mixture is in the form of heterogeneous nucleation. Unlike homogeneous nucleation, heterogeneous nucleation occurs with a foreign body (e.g., inorganic crystals, clays and sands in natural waters, and any impurities in laboratory waters) or at a surface such as fluid interface and container wall. Foreign bodies may serve as catalysts during the nucleation process, and the interface could decrease the Gibbs free energy of nucleation.

At the molecular level, several mechanisms of hydrate formation are proposed but no agreement has been achieved. For example, the labile cluster nucleation hypothesis suggests that a "prehydrate" structure is first formed by liquid water molecules surrounding a dissolved solute molecule, with the essential correct coordination number. Then those clusters of dissolved species combine to form unit cells by sharing faces. When the cluster agglomerate achieves a critical radius, growth starts. During the growth process, for the first step, a temporal water cluster containing a guest is transported to the surface of the growing crystal. The following steps involve the adsorption and stabilization of the cluster onto the crystal. Generally, the growth occurs at the interface because of the high concentrations of guest and host molecules there. It should be noted that nucleation as well as induction time is stochastic and therefore unpredictable. The induction time is proved to depend on the characteristics of the apparatus, history of water, gas composition, presence of foreign particles, and so on.

10.5.2 Dissociation of CO_2 Hydrate

In contrast to hydrate formation, hydrate dissociation is an endothermic process in which energy is consumed to break the hydrogen bonds between water molecules and the van der Waals interaction between the guest and water molecules, with the production of water and gas. At around 273 K, the measured dissociation heat of CO_2 hydrate varies between 57.66 and 65.22 kJ/mol, with the hydration number of 6.21–7.23 [45–47]. For hydrate with single guest, the amount of heat absorbed by dissociation is equal to that released by formation. As an endothermic process, the hydrate dissociation is typically dominated by heat transfer but not intrinsic kinetics.

There are several approaches to dissociate hydrates: depressurization, thermal stimulation, thermodynamic inhibitor injection, or a combination of these methods. By depressurization or thermal stimulation, hydrate can be moved across the phase equilibrium line to the unstable zone. However, inhibitor injection does not aim to change the temperature or pressure conditions of the hydrate but switch the phase equilibrium line to a lower temperature and a higher pressure, leaving the hydrate in the unstable zone.

Based on its formation and dissociation, CO_2 hydrate has many potential applications. As described in detail later, ocean sequestration and CH_4/CO_2 replacement are based on the formation of CO_2 hydrate, and the application in refrigeration systems makes use of the dissociation of CO_2 hydrate.

10.5.3 CO_2 Capture and Sequestration

It is widely known that CO_2 is one of the major greenhouse gases responsible for global warming. Carbon capture and sequestration (CCS) has been considered as a means of reducing the emission of CO_2 into the atmosphere from fossil fuel use. CO_2 hydrates can play an important role both in CO_2 capture and sequestration.

The main CO_2 source comes from flue gas (N_2+CO_2) generated by power plants. Other feed gases for CO_2 separation may be the shifted synthesis gas (H_2+CO_2) and natural gas or biogas (CH_4+CO_2). There are several methods to separate CO_2 from gas mixtures, including chemical absorption, cryogenic fractionation, selective adsorption, and membrane separation. Apart from these, CO_2 hydrate separation is a relatively novel technology for CO_2 capture. This technology, called hydrate-based CO_2 capture (HBCC), is based on selective partitioning of CO_2 between the gaseous and hydrate phase. Specifically, the hydrate formation pressure at 273 K for CO_2 (1.22 MPa) is lower than N_2 (16.3 MPa), H_2 (213 MPa), O_2 (11.1 MPa), and CH_4 (2.65 MPa) [48]. Consequently, compared to the other major components in the mixture, CO_2 is easier to form hydrates and therefore can be separated by hydrate formation. Figure 10.7 illustrates the simplest idea of

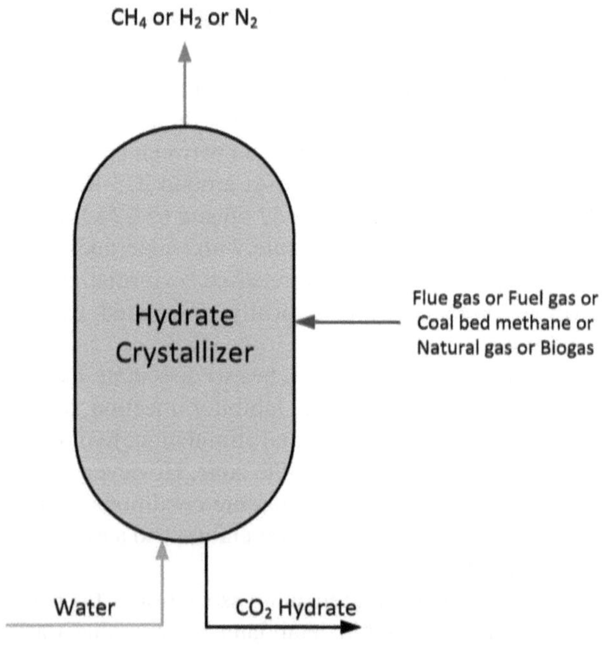

Fig. 10.7 Capture of CO_2 from gas mixture

HBCC by a single stage. For improvement, this may be combined with a second stage or some other separation methods to increase the efficiency of separation. Some additives such as THF (tetrahydrofuran) can also be added during the separation process to lower the equilibrium pressure for economic and safety concern. HBCC is of high capacity and easy to operate. It is free from contamination because only water is the major raw material. At the same time, CO_2 can be stored in hydrates after separation, making it is easier for transport and disposal. Further improvements can be made to increase the energy efficiency and lower the cost of this separation scheme.

Another main source of CO_2 is from the fuel gas generated by power plants such as integrated gasification combined cycle (IGCC) power plants. A typical pre-treated fuel gas exiting the IGCC power plant consists of 40% CO_2 in hydrogen in the pressure range of 2–7 MPa [49]. This pressure range is suitable for hydrate formation and preferential incorporation of CO_2 over H_2 in hydrate phase makes hydrate-based gas separation (HBGS) feasible for CO_2 capture from fuel gas mixture. Moreover, the CO_2 capacity of the HBGS process is very high compared to that of commercial physical solvents such as Selexol and Rectisol. Extensive studies have been reported demonstrating the HBGS process employing different reactor configurations and different promoters under various experimental conditions [50–54]. To align the operating pressure in line with the fuel gas mixture pressure range, several promoters such as propane [55–59], tetrahydrofuran [60–62], cyclopentane (CP) [29, 63–65], tetra-*n*-butyl ammonium bromide (TBAB) [52, 60, 66–68], tetra-*n*-butyl ammonium nitrate ($TBANO_3$) [69, 70], and tetra-*n*-butyl ammonium fluoride (TBAF) [71] are employed. Traditional stirred tank reactors were not suitable because of slow kinetics of hydrate formation. The slow kinetics is mainly because of the mass transfer limitation caused by agglomeration of hydrate crystals at the liquid–gas interface. To overcome the mass transfer limitation, a fixed bed reactor configuration was employed with different porous media. Silica sand, PU foam, glass beads, and silica gel were employed as porous media [51, 55, 56, 72]. Enhanced kinetics was reported in silica sand and PU foam compared to silica gel, suggesting that interstitial interconnectivity of pore space is important for hydrate formation [56, 57].

Following CO_2 capture, CO_2 sequestration is essential to the reduction of CO_2 emission as it aims to store CO_2 for a long term. There are three main technologies to store CO_2 – geological sequestration, terrestrial sequestration, and ocean sequestration. Geological sequestration uses underground geological formations such as depleted oil and gas reservoirs, saline aquifers, and coal seams to store CO_2. The storage in depleted oil and gas reservoirs is considered to be economically favorable but the storage potential is limited. For the saline aquifers and coal seams storage, the risk of CO_2 leakage may be high and further study is needed. Terrestrial sequestration uses plants to absorb CO_2. As a natural process of the carbon cycle, absorption of CO_2 by the global biosphere equals roughly one-third of the global carbon emission from human activities. The major objectives of this approach are to protect the current ecosystem and enhance the CO_2 absorption process. The idea of ocean sequestration is easy to understand, considering that the ocean stores a huge amount of CO_2 by absorption. It is estimated that 38 100 GtC (gigatonnes of carbon) is stored in the ocean whereas there is only 610 GtC in the global ecosystems. However, the kinetics of ocean absorption is slow and ocean acidification is another concern. Although many researchers focus on the mechanisms of CO_2 absorption in the ocean, another potential way is to store CO_2 at the deep ocean floor in terms of CO_2 hydrates, which is thermodynamically feasible.

Figure 10.8 shows the phase diagram of CO_2 in seawater (with a salinity of 34) and the ocean temperature profile. It is clear that CO_2 can form hydrate below a certain depth in the sea. Several scenarios have been proposed for CO_2 sequestration in the ocean: (1) sequestering CO_2 on the bottom of the sea (deeper than 3 500 m), (2) injecting liquid CO_2 deeper than 1 000 m of the sea, and (3) injecting CO_2 gas into the shallow sea (within the hydrate stability zone). However, all these schemes have economic and environmental challenges. The cost is much higher for deeper injection because of the long pipeline and high pressure. The shallow injection, though less expensive, may not be effective for long-term sequestration of CO_2 because of the high chance of CO_2 escaping to the atmosphere from the sea. The high CO_2 concentration and low pH near the injection area may also pose a risk to the ocean ecosystem. Considering the environmental issue, it is suggested to sequester CO_2 in the ocean sediments, where CO_2 hydrates can be fixed. Such geological trapping makes it more promising to reduce the leakage of CO_2. Furthermore, to overcome the economic barrier, it is proposed to inject CO_2 into the natural gas hydrates reservoirs to produce natural gas while sequestering CO_2. This is introduced in the following section. Overall, the ocean sequestration of CO_2 is

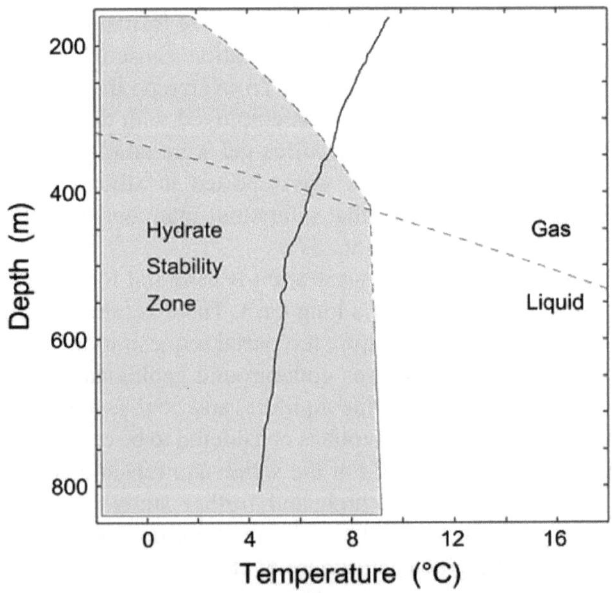

Fig. 10.8 The ocean temperature profile and CO_2 phase diagram for seawater [73]. Reprinted with permission from Copyright (2002) American Chemical Society

still in its early stage of development and the evaluation of technical feasibility and environmental impacts is ongoing.

10.5.4 Replacement of CH_4 by CO_2 in Naturally Occurring Hydrates

Natural gas (mainly CH_4) has become a more and more attractive fuel because it burns more cleanly than other fossil fuels. According to the statistical review of world energy by BP, the world consumed about 3.348 trillion m^3 of natural gas in 2013, and the amount is growing. This huge and growing demand for natural gas attracts attention to unconventional natural gas resources. Natural gas hydrate, found widely distributed under the deep ocean and in permafrost regions, is a very promising potential energy resource. Although with great uncertainty, it is estimated that totally 200–120 000 trillion m^3 of natural gas (STP) exists in the form of gas hydrate. Even the most conservative estimate significantly surpasses the amount of conventional fossil fuel reserves. Three exploitation methods are proposed to extract natural gas from hydrates, including (1) depressurization, (2) thermal stimulation, and (3) inhibitor injection. Another scheme combines the recovery of natural gas with the sequestration of CO_2, that is, to use CO_2 to replace CH_4 trapped in hydrates. This idea becomes attractive because it can simultaneously sequester CO_2 and produce natural gas, alleviating the total cost and making the most benefits, both economically and environmentally.

The feasibility of the replacement scheme is based on the thermodynamics of CH_4 and CO_2 hydrates equilibrium. Figure 10.9 illustrates the hydrate formation conditions of CO_2 and CH_4. In certain regions (A and B in the figure), CO_2 can form hydrate at lower pressure and higher temperature than CH_4. Moreover, the heat released from the exothermic formation of CO_2 hydrate is more than the dissociation heat of CH_4 hydrate. CH_4 hydrate can also be heated by the sensible heat of injected CO_2. Consequently, when CO_2 (as gas in region A or as liquid in region B) is injected into CH_4 hydrate-bearing sediments, it can assist the dissociation of CH_4 hydrate and simultaneously form CO_2 hydrate. Another benefit of the replacement reaction is that the formation of CO_2 hydrate could preserve the stability of sediment during the recovery process.

For real conditions, the pressure and temperature profiles in the permafrost seem to lie close to region A and B. However, under the deep sea floor, the pressure usually exceeds 10 MPa, and the equilibrium temperature of CO_2 hydrate is lower than that of CH_4 hydrate at the same pressure, which means CO_2 finds it harder to form hydrate than CH_4. Hence, the liquid CO_2 injected may not be able to form hydrate itself while it dissociates CH_4 hydrate. If CO_2 is not stored as hydrate but in liquid form, the stability of the reservoir may not be maintained, resulting in the risk of escape of CH_4 and CO_2. This problem could be avoided if there is a permeability

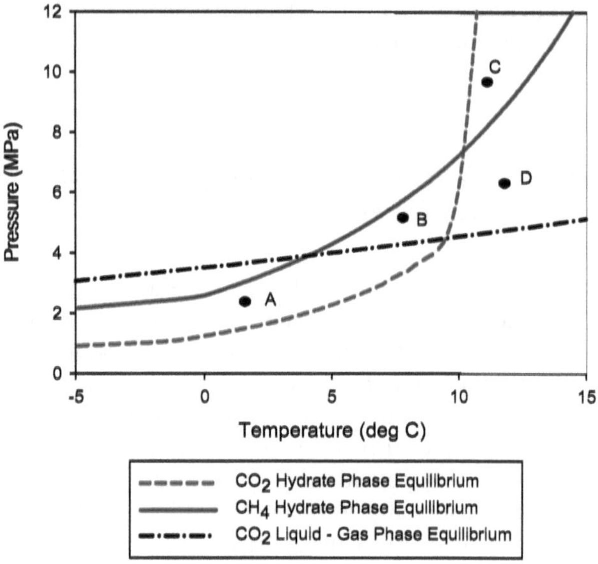

Fig. 10.9 Hydrate forming conditions of CO_2 and CH_4

barrier such as a lithological stratum to prevent CO_2 and dissociated CH_4 from being released out to the sea bed.

Nevertheless, it should be noted that the actual behavior of hydrate formation and dissociation depends on not only the pressure and temperature of the location but also many other factors, such as the properties of the in situ methane hydrate, the amount of injected CO_2, and the characteristics of the pore media. More research needs to be done on the kinetics and mechanism of the interchange of guests, the effects of the above factors, and the equilibrium conditions of gas hydrate with liquid CO_2 [74]. Overall, the replacement of CH_4 in hydrates with CO_2 is a potential method for both recovery of natural gas and long-term storage of CO_2. For the first consideration, this method should be used for selected reservoirs with suitable in situ pressure and temperature conditions that are favorable for the exchange process.

10.5.5 CO_2 Hydrates in Refrigeration Processes

The key basis of the refrigeration process is that a space can be cooled by the endothermic phase change of refrigerant. The latent heat of evaporation of some common refrigerants is around 200 kJ/kg. Because of its high latent heat of dissociation (374 $kJ/kg_{hydrate}$), CO_2 hydrate has a great potential to be used in refrigeration processes.

Scheme 10.1 shows a conceptual design of a clathrate hydrate-based refrigeration system. Hydrate is formed in the forming reactor at relatively high temperature (exceeding that of the warm environment) and pressure, releasing heat to the warm environment. Then the hydrate slurry is pumped to the dissociating reactor, the slurry pump working as an expansion valve which conveys hydrate slurry to a state with a lower pressure and temperature (lower than that of the refrigerated space). The dissociation of hydrate in the dissociating reactor absorbs heat from the environment and therefore cools the space. After dissociation, the separated gas and water are compressed to the forming reactor by a multiphase compressor to continue this cycle.

Although there is a lack of data for the CO_2 hydrate-based refrigeration system, Ogawa et al. [75] simulated the refrigeration system shown in Scheme 10.1 using cyclopentane and CH_2F_2 as working medium (hydrate formers), and revealed that the coefficient of performance (COP, ratio of cooling capacity to electrical energy consumed) of the system could be as high as 8 when the hydrate formation and dissociation temperatures were 298.5 K and 280.5 K, respectively. Such performance is really promising as the COP for conventional refrigeration systems is around 4.

Because the formation temperature of CO_2 hydrate is still well below that of the normal environment (say 303–313 K), there is a limitation for the use of it as a direct refrigerant for air conditioning. However, it is possible to use CO_2 hydrate as a secondary refrigerant. In a cooling system with secondary refrigeration, the secondary refrigerant is used to transfer heat from the space being cooled to a heat exchanger where the heat is absorbed by a primary refrigerant. A secondary refrigerant is often used in large refrigeration systems to transport cooling capacity from the plant room to the distributed points of use. Such a system has many advantages and the main points include: (1) the amount of primary refrigerant, usually traditional environment-unfriendly, can be minimized, because it only fills the pipelines in the plant room – meanwhile the cold is distributed by the secondary refrigerant, which is an environment-friendly fluid; (2) the primary refrigerant remains in the plant room and enables close supervision and control, allowing the use of some hazardous but efficient refrigerants (e.g., NH_3); (3) the overall efficiency could increase or at least be equal to the single refrigeration system. Currently, ice slurry is one of the common secondary refrigerants. CO_2 hydrate slurry becomes more promising not only because its latent heat of dissociation (374 $kJ/kg_{hydrate}$) is higher than the latent heat of melting of ice (333 kJ/kg_{ice}), but also the formation conditions of CO_2 hydrate (278–293 K) are more adaptable to air conditioning applications.

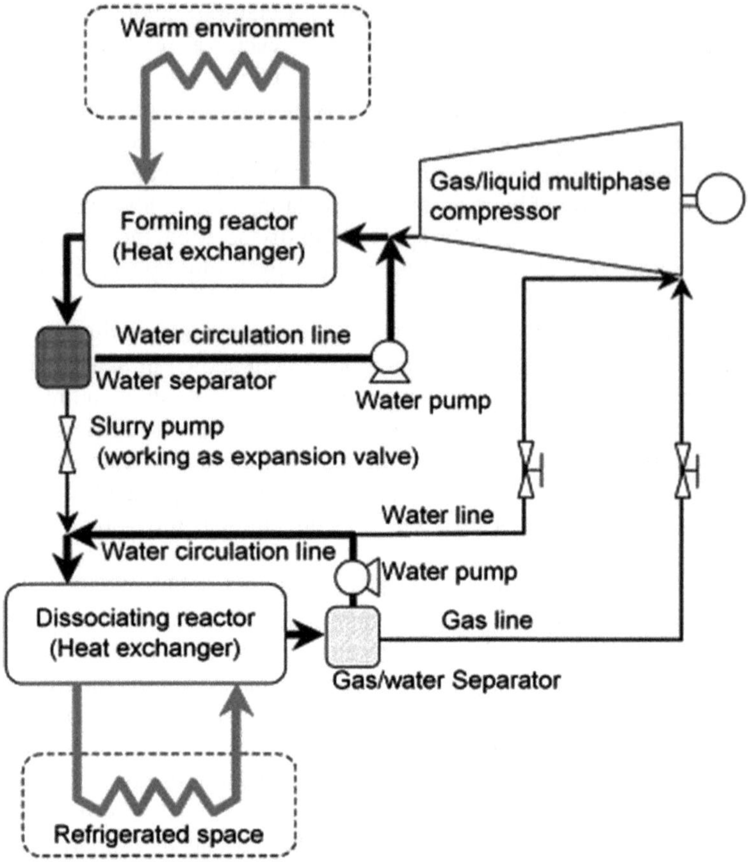

Scheme 10.1 Conceptual design of a clathrate hydrate based refrigeration system [75]. Reprinted with permission from Elsevier

Another application of the cooling capacity of CO_2 hydrate is cold storage. As a means to alleviate peak loading of electricity, cold storage uses spare electricity (usually during the night) to form cold storage media (e.g., ice), which is then used for cooling during the peak periods with high electricity demand. Compared to ice, CO_2 hydrate has a higher cold storage density (which would reduce the size of equipment) and can work at a temperature closer to environmental temperature, leading to a higher COP. CO_2 hydrate is a promising candidate for refrigeration and air conditioning use. The study of CO_2 hydrate as a refrigerant and cold storage medium is ongoing. Attempts are being made to decrease the equilibrium pressure, increase dissociation enthalpy, and enhance formation process and gas solubility.

References

1. Wroblewski S (1882) On the combination of carbonic acid and water. C R Acad Sci Paris 94: 212–213
2. Wroblewski S (1882) The composition of hydrated carbonic acid. J Chem Soc 42:1026
3. Teng H, Yamasaki A, Shindo Y (1996) Stability of the hydrate layer formed on the surface of a CO_2 droplet in high-pressure, low-temperature water. Chem Eng Sci 51(22):4979–4986
4. Shi XJ, Zhang P (2013) A comparative study of different methods for the generation of tetra-n-butyl ammonium bromide clathrate hydrate slurry in a cold storage air-conditioning system. Appl Energy 112:1393–1402
5. Zhang P, Ma ZW (2012) An overview of fundamental studies and applications of phase change material slurries to secondary loop refrigeration and air conditioning systems. Renew Sustain Energy Rev 16(7):5021–5058
6. Zhang P, Shi XJ, Ma ZW (2013) Solid fraction determination in cold storage by tetra-n-butyl ammonium bromide clathrate hydrate slurry. Int J Refrig 36(3):809–819
7. Stillinger FH, Rahman A (1974) Improved simulation of liquid water by molecular dynamics. J Chem Phys 60(4):1545–1557
8. Plummer PM, Chen T (1987) Investigation of structure and stability of small clusters: molecular dynamics studies of water pentamers. J Chem Phys 86(12):7149–7155
9. Frank HS (1970) The structure of ordinary water new data and interpretations are yielding new insights into this fascinating substance. Science 169(3946):635–641
10. Rauh F, Mizaikoff B (2012) Spectroscopic methods in gas hydrate research. Anal Bioanal Chem 402(1):163–173
11. Holland PM, Castleman A Jr (1980) A model for the formation and stabilization of charged water clathrates. J Chem Phys 72(11):5984–5990
12. Sloan ED Jr, Koh CA (2008) Clathrate hydrates of the natural gases, 3rd edn. CRC Press, Boca Raton, FL
13. Momma K, Ikeda T, Nishikubo K, Takahashi N, Honma C, Takada M, Furukawa Y, Nagase T, Kudoh Y (2011) New silica clathrate minerals that are isostructural with natural gas hydrates. Nat Commun 2:196
14. Houghton G, McLean A, Ritchie P (1957) Compressibility, fugacity, and water-solubility of carbon dioxide in the region 0–36 atm. and 0–100 C. Chem Eng Sci 6(3):132–137
15. Fleyfel F, Devlin JP (1991) Carbon dioxide clathrate hydrate epitaxial growth: spectroscopic evidence for formation of the simple type-II carbon dioxide hydrate. J Phys Chem 95(9): 3811–3815
16. Handa Y (1986) Calorimetric determinations of the compositions, enthalpies of dissociation, and heat capacities in the range 85 to 270 K for clathrate hydrates of xenon and krypton. J Chem Thermodyn 18(9):891–902
17. Circone S, Kirby SH, Stern LA (2005) Direct measurement of methane hydrate composition along the hydrate equilibrium boundary. J Phys Chem B 109(19):9468–9475
18. Henning RW, Schultz AJ, Thieu V, Halpern Y (2000) Neutron diffraction studies of CO2 clathrate hydrate: formation from deuterated ice. J Phys Chem A 104(21):5066–5071
19. Durham WB, Stern LA, Kirby SH (2003) Ductile flow of methane hydrate. Can J Phys 81(1–2):373–380
20. Helgerud MB, Circone S, Stern L, Kirby S, Lorenson TD (2002) Conference, Yokohama, p 716
21. Kiefte H, Clouter MJ, Gagnon RE (1985) Determination of acoustic velocities of clathrate hydrates by Brillouin spectroscopy. J Phys Chem 89(14):3103–3108
22. Kaye GWC, Laby TH (2007) Tables of physical and chemical constants, 16th edn. National Physical Laboratory, Middlesex, UK
23. Huang D, Fan S (2005) Measuring and modeling thermal conductivity of gas hydrate-bearing sand. J Geophys Res Solid Earth (1978–2012) 110:B01311. doi:10.1029/2004JB003314

24. Jung J, Espinoza DN, Santamarina JC (2010) Properties and phenomena relevant to CH_4–CO_2 replacement in hydrate-bearing sediments. J Geophys Res Solid Earth (1978–2012) 115: B10102. doi:10.1029/2009JB000812
25. Giauque W, Stout J (1936) The entropy of water and the third law of thermodynamics. The heat capacity of ice from 15 to 273°K. J Am Chem Soc 58(7):1144–1150
26. Udachin KA, Ratcliffe CI, Ripmeester JA (2001) Structure, composition, and thermal expansion of CO_2 hydrate from single crystal X-ray diffraction measurements. J Phys Chem B 105(19):4200–4204
27. Uchida T (1998) Physical property measurements on CO_2 clathrate hydrates. Review of crystallography, hydration number, and mechanical properties. Waste Manag 17(5):343–352
28. Englezos P (1993) Clathrate hydrates. Ind Eng Chem Res 32(7):1251–1274
29. Zhang J, Yedlapalli P, Lee JW (2009) Thermodynamic analysis of hydrate-based pre-combustion capture of CO_2. Chem Eng Sci 64(22):4732–4736
30. Zhang J, Lee JW (2008) Equilibrium of hydrogen + cyclopentane and carbon dioxide + cyclopentane binary hydrates. J Chem Eng Data 54(2):659–661
31. Dalmazzone D, Kharrat M, Lachet V, Fouconnier B, Clausse D (2002) DSC and PVT measurements. J Therm Anal Calorim 70(2):493–505
32. Adisasmito S, Frank RJ, Sloan ED (1991) Hydrates of carbon dioxide and methane mixtures. J Chem Eng Data 36(1):68–71
33. Deaton WM, Frost EM (1946) Gas hydrates and their relation to the operation of natural gas pipelines. US Bureau of Mines Monograph 8:101
34. Unruh CH, Katz DL (1949) Gas hydrates of carbon dioxide-methane mixtures. J Petrol Technol 1(04):83–86
35. Larson SD (1955) Phase studies of the two component carbon dioxide-water system involving the carbon dioxide hydrate. University of Illinois, Urbana, IL
36. Mooijer-van den Heuvel MM, Witteman R, Peters CJ (2001) Phase behaviour of gas hydrates of carbon dioxide in the presence of tetrahydropyran, cyclobutanone, cyclohexane and methylcyclohexane. Fluid Phase Equilib 182(1–2):97–110
37. Takenouchi S, Kennedy GC (1964) The binary system H_2O-CO_2 at high temperatures and pressures. Am J Sci 262(9):1055–1074
38. Miller SL, Smythe WD (1970) Carbon dioxide clathrate in the Martian ice cap. Science 170(3957):531–533
39. Ng H-J, Robinson DB (1985) Hydrate formation in systems containing methane, ethane, propane, carbon dioxide or hydrogen sulfide in the presence of methanol. Fluid Phase Equilib 21(1):145–155
40. Falabella BJ (1975) A study of natural gas hydrates. Ph.D Dissertation, University of Massachusetts, Ann Arbor, MI
41. Vlahakis J, Chen H, Suwandi M, Barduhn A (1972) The growth rate of ice crystals: properties of carbon dioxide hydrate, a review of properties of 51 gas hydrates. Syracuse University Research and Development Report 830
42. Ohgaki K, Makihara Y, Takano K (1993) Formation of CO_2 hydrate in pure and sea waters. J Chem Eng Jpn 26(5):558–564
43. Robinson D, Metha B (1971) Hydrates in the propane-carbon dioxide-water system. J Can Petrol Technol 10(01)
44. Breland E, Englezos P (1996) Equilibrium hydrate formation data for carbon dioxide in aqueous glycerol solutions. J Chem Eng Data 41(1):11–13
45. Anderson GK (2003) Enthalpy of dissociation and hydration number of carbon dioxide hydrate from the Clapeyron equation. J Chem Thermodyn 35(7):1171–1183
46. Kang S-P, Lee H, Ryu B-J (2001) Enthalpies of dissociation of clathrate hydrates of carbon dioxide, nitrogen, (carbon dioxide + nitrogen), and (carbon dioxide + nitrogen + tetrahydrofuran). J Chem Thermodyn 33(5):513–521
47. Yoon J-H, Yamamoto Y, Komai T, Haneda H, Kawamura T (2003) Rigorous approach to the prediction of the heat of dissociation of gas hydrates. Ind Eng Chem Res 42(5):1111–1114

48. Wang Y, Lang X, Fan S (2013) Hydrate capture CO_2 from shifted synthesis gas, flue gas and sour natural gas or biogas. J Energy Chem 22(1):39–47
49. Metz B, Davidson O, De Coninck H, Loos M, Meyer L (2005) IPCC 2005: IPCC special report on carbon dioxide capture and storage (Prepared by Working Group III of the Intergovernmental Panel on Climate Change). Cambridge University Press, Cambridge
50. Adeyemo A, Kumar R, Linga P, Ripmeester J, Englezos P (2010) Capture of carbon dioxide from flue or fuel gas mixtures by clathrate crystallization in a silica gel column. Int J Greenhouse Gas Control 4(3):478–485
51. Babu P, Kumar R, Linga P (2013) Pre-combustion capture of carbon dioxide in a fixed bed reactor using the clathrate hydrate process. Energy 50:364–373
52. Li X-S, Xia Z-M, Chen Z-Y, Yan K-F, Li G, Wu H-J (2010) Gas hydrate formation process for capture of carbon dioxide from fuel gas mixture. Ind Eng Chem Res 49(22):11614–11619
53. Linga P, Kumar R, Englezos P (2007) Gas hydrate formation from hydrogen/carbon dioxide and nitrogen/carbon dioxide gas mixtures. Chem Eng Sci 62(16):4268–4276
54. Linga P, Kumar R, Englezos P (2007) The clathrate hydrate process for post and pre-combustion capture of carbon dioxide. J Hazard Mater 149(3):625–629
55. Babu P, Kumar R, Linga P (2013) Medium pressure hydrate based gas separation (HBGS) process for pre-combustion capture of carbon dioxide employing a novel fixed bed reactor. Int J Greenhouse Gas Control 17:206–214
56. Babu P, Kumar R, Linga P (2013) A new porous material to enhance the kinetics of clathrate process: application to precombustion carbon dioxide capture. Environ Sci Technol 47(22): 13191–13198
57. Babu P, Kumar R, Linga P (2014) Unusual behavior of propane as a co-guest during hydrate formation in silica sand: potential application to seawater desalination and carbon dioxide capture. Chem Eng Sci 117:342–351
58. Babu P, Yang T, Veluswamy HP, Kumar R, Linga P (2013) Hydrate phase equilibrium of ternary gas mixtures containing carbon dioxide, hydrogen and propane. J Chem Thermodyn 61:58–63
59. Kumar R, Linga P, Ripmeester JA, Englezos P (2009) Two-stage clathrate hydrate/membrane process for precombustion capture of carbon dioxide and hydrogen. J Environ Eng 135 (6):411–417
60. Babu P, Ho CY, Kumar R, Linga P (2014) Enhanced kinetics for the clathrate process in a fixed bed reactor in the presence of liquid promoters for pre-combustion carbon dioxide capture. Energy 70:664–673
61. Lee HJ, Lee JD, Linga P, Englezos P, Kim YS, Lee MS, Kim YD (2010) Gas hydrate formation process for pre-combustion capture of carbon dioxide. Energy 35(6):2729–2733
62. Park S, Lee S, Lee Y, Lee Y, Seo Y (2013) Hydrate-based pre-combustion capture of carbon dioxide in the presence of a thermodynamic promoter and porous silica gels. Int J Greenhouse Gas Control 14:193–199
63. Ho LC, Babu P, Kumar R, Linga P (2013) HBGS (hydrate based gas separation) process for carbon dioxide capture employing an unstirred reactor with cyclopentane. Energy 63:252–259
64. Li X-S, Xu C-G, Chen Z-Y, Wu H-J (2011) Hydrate-based pre-combustion carbon dioxide capture process in the system with tetra-n-butyl ammonium bromide solution in the presence of cyclopentane. Energy 36(3):1394–1403
65. Lim Y-A, Babu P, Kumar R, Linga P (2013) Morphology of carbon dioxide–hydrogen–cyclopentane hydrates with or without sodium dodecyl sulfate. Cryst Growth Des 13(5): 2047–2059
66. Gholinezhad J, Chapoy A, Tohidi B (2011) Separation and capture of carbon dioxide from CO_2/H_2 syngas mixture using semi-clathrate hydrates. Chem Eng Res Des 89(9):1747–1751
67. Kim SM, Lee JD, Lee HJ, Lee EK, Kim Y (2011) Gas hydrate formation method to capture the carbon dioxide for pre-combustion process in IGCC plant. Int J Hydrogen Energy 36(1): 1115–1121

68. Li X-S, Xia Z-M, Chen Z-Y, Wu H-J (2011) Precombustion capture of carbon dioxide and hydrogen with a one-stage hydrate/membrane process in the presence of tetra-n-butylammonium bromide (TBAB). Energy Fuels 25(3):1302–1309
69. Babu P, Datta S, Kumar R, Linga P (2014) Impact of experimental pressure and temperature on semiclathrate hydrate formation for pre-combustion capture of CO_2 using tetra-n-butyl ammonium nitrate. Energy 78:458–464
70. Babu P, Yao M, Datta S, Kumar R, Linga P (2014) Thermodynamic and kinetic verification of tetra-n-butyl ammonium nitrate (TBANO3) as a promoter for the clathrate process applicable to precombustion carbon dioxide capture. Environ Sci Technol 48(6):3550–3558
71. Park S, Lee S, Lee Y, Seo Y (2013) CO_2 Capture from simulated fuel gas mixtures using semiclathrate hydrates formed by quaternary ammonium salts. Environ Sci Technol 47(13): 7571–7577
72. Yang M, Song Y, Jiang L, Zhao Y, Ruan X, Zhang Y, Wang S (2014) Hydrate-based technology for CO_2 capture from fossil fuel power plants. Appl Energy 116:26–40
73. Brewer PG, Peltzer ET, Friederich G, Rehder G (2002) Experimental determination of the fate of rising CO_2 droplets in seawater. Environ Sci Technol 36(24):5441–5446
74. Chong ZR, Yang SHB, Babu P, Linga P, Li X-S (2015) Review of natural gas hydrates as an energy resource: prospects and challenges. Appl Energy. DOI:10.1016/j.apenergy.2014.12.061. http://sciencedirect.com/science/article/pii/S030626191401318X
75. Ogawa T, Ito T, Watanabe K, Tahara K-i, Hiraoka R, J-i O, Ohmura R, Mori YH (2006) Development of a novel hydrate-based refrigeration system: a preliminary overview. Appl Thermal Eng 26(17):2157–2167

Index

A
Acceptor, 9
Acetogenic bacteria, 361
Acrylic acid, 152
Acyl-CoA carboxylation pathways, 351–353
Adiabatic excitation, 8
Adventitious water, 103
Alcohol dehydrogenase (ADH) enzyme, 358
Alcoholysis of urea, 208
Alkylative/arylative carboxylation, 157
Alkyl esters of acrylic acids, 151
Alkylmagnesium amides, 109
Alkyne cyclotrimerization, 162
Al_2O_3-modified ceria, 197
Amine-catalyzed mechanism, 103
Ammonium carbamates, 74
Amphoteric oxide, 9
Amphoteric reactivity, 11–12
Anionic clusters, 2
Antibonding, 4, 9
Antisymmetric stretching, 21
Antropogenic emissions, 1
Applications of CO_2 hydrates, 390–398
Ar matrix, 8
Aromatic amines, 74
Atomic orbitals (AOs), 3

B
Back-biting and elimination of cyclic carbonate, 223
Bacteria and methanogenic *Archaea,* 351
Band gap, 338
Bent equilibrium structures, 7
Bent excited valence state, 8
Bimetallic Al–O–Al catalysts, 219
Bimolecular decay of CO_2^-, 18
Bio-carboxylation of epoxides, 353–357
Biological CO_2 fixation, 348
$Bn_2NC(O)OH$, 75
Bonding character, 5
Bonding combinations, 3

C
Calorimetric method, 388
Calvin–Benson–Bassham-cycle, 349
Carbamic acid, 75
Carbon capture and sequestration (CCS), 392
Carbon dioxide chemical utilization (CCU), 1
Carbon dioxide radical anion, CO_2^-, 12–19
Carbon dioxide radical cation, CO_2^+, 19–20
α-Carbonia, 12
Carbonic anhydrase (CA), 91
Carbon monoxide dehydrogenase (CODH), 358–359
Carbon-recycling, 340
Carboxylation of alkynes, 156–163
Carboxylation of allenes, 164–169
Carboxylation of conjugated dienes, 169–176
Carboxylation of cyclic ethers, 216–223
Carboxylation of diols and polyols, 210–212
Carboxylation of epoxides, 217
Carboxylation of isoprene, 172
Carboxylation of methylenecyclopropane, 118
Carboxylation of N-tosylaziridines, 121
Carboxylation of olefins, 147–156
Carboxylation of quadricyclane, 117

M. Aresta et al., *Reaction Mechanisms in Carbon Dioxide Conversion*,
DOI 10.1007/978-3-662-46831-9

Carboxylation of styrenes, 155
Carboxylation reactions, 353–358
Carboxylative cyclization of enynes, 156
Catalytic systems for CO_2 hydrogenation, 280–285
Catalytic systems for CO_2-OCM, 239–241
Catalytic systems for CO_2-ODE to styrene, 256–258
Catalytic systems for CO_2-ODH of alkanes, 244–248
Catalytic systems for the DRM process, 267–271
CCS. *See* Carbon capture and sequestration (CCS)
CCU. *See* Carbon dioxide chemical utilization (CCU)
C-cycle, 1
C-electrophile, 10–11
C-footprint, 187
Chemical water traps, 200
Chemico-physical techniques, 187
Chemiluminescence, 7
Chemoautotrophic organisms, 348
Chlorine hydrates, 374
Climate change, 1
^{13}C-NMR, 26
CO($^1\Sigma$), 24
CO_2-binding constants, 324
$CoCl(NO)_2[PhP(OCH_2CH_2)_2N\text{-}CO_2H]$, 75
CODH. *See* Carbon monoxide dehydrogenase (CODH)
C-O distance, 2
CO_2 economy, 340
Coefficient of performance (COP), 397
Co-encapsulated enzymes, 365
C-O equilibrium distances, 7
CO_2 as Lewis acid, 9
CO_2 hydrates, 375
CO_2 hydrates in refrigeration processes, 396–398
CO_2 hydrates on other planets, 376
CO_2 hydrogenation to methanol and DME, 278–280
CO_2 insertion into activated C–H bonds, 129
CO_2 insertion into C–C bonds, 117–120
CO_2 insertion into C–N bonds, 120–127
CO_2 insertion into M–C bonds, 95–96
CO_2 insertion into M–N bonds, 102–115
CO_2 insertion into M–OH bonds, 91–94
CO_2 insertion into M–OR bonds, 97–98
CO_2 insertion into M–P bonds, 115–117
CO_2 insertion into M-phosphido bond, 116
CO_2 insertion into the N–Si bond, 114
Co(*n*-Pr-salen)(CO_2)K(THF), 39
Collisions of alkali-metal atoms, 13
Collisions of electrons, 13
Compressional velocity, 384
Computational studies, 96
Conformational isomers, 326
Conversion of CO_2 and H_2 into HCO_2H, 88
Conversion of CO_2 into methanol, 364
Cooligomerization of butadiene with CO_2, 173
Co-operative effect, 329
Co-ordination of CO_2, 36
CO_2 photodissociation, 23
Copolymerization of aziridines, 224
CO_2^- radical anion, 2
CO_2^- radicals, 14
CO_2 (dry) reforming of methane, 265–277
Corrins, 335
Corroles, 335
Coupling of two reactions, 187
Coupling terminal alkynes with CO_2, 158
C-protonation, 12
Crossover zone, 26
Crown-ether, 81
Crystalline "super-hard" phase, 11
Crystallization kinetics of hydrate, 390
Crystal structures, 376
Cubic structure I (sI), 376
Cubic structure II (sII), 376
C_{2v}, 3
Cyclic carbamate, 126
Cycloaddition of allene with CO2, 168
Cycloaddition of CO_2 to aziridines, 120
Cycloaddition of CO_2 with aziridines, 123

D

Decay of the CO_2^- radical, 14
Degenerate π-orbitals, 5
Dehydrogenation of hydrocarbons (DH), 237
DFT calculations, 153, 191, 324
$D_{\infty}\eta$, 3
$\Delta H°$ and $\Delta G°$ for the synthesis of organic carbonates, 186
Dialkylcarbonates, 191
Dicarboxylate, 147
1,4-Dicarboxylation of 1,3 dienes, 170
Dicyclohexylcarbodiimide, 200
Diethylcarbonate (DEC), 183
Diffusion electrodes, 315
Dihydroxo-carbene, 328
3,4-Dihydroxybenzoate, 354
Dimeric alkylzinc amides, 109

Dimethyl-carbonate (DMC), 183
Dimethylether, 237
4,6-Dimethyl-2-pyrone, 167
Direct carboxylation of alcohols, 185
Discovery of CO_2 hydrates, 375
Dissociation of CO_2 hydrate, 391–392
Dissociative ionization, 20
DME, 279–280
Donor, 9
Doubly degenerate, 3
Dual-face electrode, 340

E
Efficiency of PV, 333
Elastic properties of hydrates, 384
Electric quadrupole, 2
Electrocatalysis, 320, 347
Electrochemical reductive conversion of CO_2, 365
Electrode deterioration, 320
Electron delocalization, 12
Electronically excited oxygen atoms, 23
Electronic ground state, 2
Electron scattering, 13
Electron-transfer processes, 321
Electrophilic attack on coordinated CO_2, 61
Electroreduction of CO_2, 312
End-on (η^1-OCO) coordination, 10
End-on protonation, 10
Energy profiles, 194
Energy thresholds, 17, 23
Enolate moiety, 350
Enthalpy effect, 186
Enthalpy of reaction, 186
Entropic effect, 186
Enzymatic approach to convert CO_2, 347
Enzymatic reduction of CO_2, 366
Enzyme-catalyzed processes, 347
Excitation energy, 7
Excited electronic configuration, 6

F
fac-(bpy)$(CO)_3$Re(OCHO), 87
Faradic efficiency, 335
Fermi level pinning, 338
Fluorescence, 7
Fluxional behavior, 54
Fluxionality, CO_2 Molecule, 51–54
Formaldehyde dehydrogenase (F_{ald}DH) enzyme, 358
Formate dehydrogenases (F_{ate}DH), 358
Formation of metallacycles, 146
Formation of polyurethanes, 223–225
Fourier transform IR (FTIR) studies, 36
Free Gibbs energy of formation, 312
Frost diagram, 312
Frustrated Lewis pairs, 40

G
Gas hydrates, 374
Glycerol carbonate, 211
Green sulfur photosynthetic bacterium, 350
Guest (entrapped gas molecules), 374

H
H-bonded pyridinium-to-pyridine dimers, 330
η^1-C coordination, 41
η^2-C,O coordination, 39
η^1(C)-CO_2 Ru(diphos)$_2$(CO_2), 90
Head-to-head recombination, 18
Head-to-tail homocoupling, 144
Head-to-tail recombination, 18
Hemicarbonate, 190
Heterogeneous catalysis, 194–199, 347
Heterogeneous nucleation, 391
Heterogenized catalysts, 188
Hexagonal structure H (sH), 376
Hexakaidecahedron ($5^{12}6^4$), 378
Highest occupied molecular orbital (HOMO), 8
Highest unoccupied molecular orbital (HUMO), 35
Highly excited vibrational states, 8
High resolution absorption spectrum, 8
High temperature catalyzed reactions, 237
η^1-O-coordinated CO_2, 43
Homogeneous nucleation, 391
Host (water molecules), 374
Hybrid bioinspired systems, 366
Hybrid system, 334
Hydration–dehydration of CO_2, 91
Hydration number of sI hydrates, 381, 382
β-Hydride elimination from a metalla-lactone complex, 154
Hydrides, 71
Hydrocarboxylation of allenes, 165
Hydrocarboxylation of styrenes, 155

Hydrogenation of CO_2, 277–294
Hydrogenation of CO_2 to formic acid, 63
Hydrogenation of $HCO_3^-{}_{(aq)}$, 88
Hydrogen carbonate, 315
Hydroxycarbonyl cation $HOCO^+$, 9
4-Hydroxy-3-methylbenzoate, 354
Hydroxymethylene, 329
3-Hydroxypropionate/4-hydroxybutyrate pathway, 351
Hyperfine coupling coefficients, 13
Hyperfine interaction, 26

I
Icosahedron ($5^{12}6^8$), 378
Incorporation of CO_2 into the cyclopropane, 119
Increase of the catalyst activity, 195
Indole-3-carboxylase, 356
Infrared data of transition metal complexes, 47–49
Insertion into amides of non-metallic elements, 112–115
Insertion into C–H bonds, 129–130
Insertion into main group and post-transition metal amides, 110–112
Insertion into M–H bonds, 85–91
Insertion into M–M bonds, 127–128
Insertion into M–O_2 bonds, 98–102
Insertion into Si–H bonds, 128–129
Insertion into the P–N bond, 113
Insertion into transition metal amides, 102–108
Insertion of CO_2 in the Zn–N bond, 110
Insertion of CO_2 into a Ru–H bond, 62
Insertion of CO_2 into the Si-N bond, 112–113
"2+2" Interaction, 86
Interaction of CO_2 with C–C multiple bonds, 143–176
Interstellar lines, 9
Intramolecular transfer, 191
Inverse isotope dependence, 86
Ionic hydrides, 71
Ionic metal hydroxides, 72
$[(IPr)Ni]_2(\mu\text{-}CO)(\mu\text{-}\eta^2,\eta^2\text{-}CO_2)$, 43
$Ir(CO_2)(Cl)(dmpe)_2$, 41
Irregular dodecahedron ($4^35^66^3$), 378
IR spectra of CO_2-TM complexes, 47
IR spectroscopy, 20–22
Isobaric condition, 387
Isochoric P–T Cycle, 387–388
Isocitrate dehydrogenase, 350
Isothermal condition, 387
Isotope-labelled CO_2, 36
Isotopic labelled CO_2, 47
Isotopologues, 26
Isourea protonation, 203

K
α-Ketoglutarate, ferredoxin oxidoreductase, 350
Kinetic modeling for the DRM process, 276–277
Kinetics and reaction mechanism for CO_2 exchange, 110
Kinetics studies, 76
Kolbe–Schmitt reaction, 353
Kolbe–Schmitt synthesis, 129

L
Latent heat of evaporation, 396
Lewis base, 9
$Li^+CO_2^-(C_s)$, 16
$Li^+CO_2^-(C_{2v})$, 16
L-Lactic acid, 355
Lowering of the selectivity, 195
Lowest excited states, 6
Lowest excited triplet state, 8
Lowest unoccupied molecular orbital (LUMO), 8, 35
Low-temperature solid-inert-gas matrices, 55–57

M
Macrocyclic N-ligands, 315
Matrix isolation, 15, 55
Maximum density of CO_2 hydrates, 385
$M^+CO_2^-(C_{2v})$, 16
M-C(O)OH, 89, 91
Mechanical properties of CO_2 hydrates, 384
Mechanical strength of pure gas hydrates, 384
Mechanism of conversion of Ti-CO_2 into O=Ti (CO), 56
Mechanism of the coupling reaction of CO_2, 148
Mechanism the formation of the metal carbamate, 103
Mechanistic studies, 117
Metal–amido bond, 102, 103
Metallacarboxylic species, 89
Metalla dihydroxycarbene, 91
Metalla-lactone structure, 151
Metal organic frameworks (MOF), 46
Metastable sII CO_2 hydrate phase, 381

Methyl acrylate, 151
Methyl-Co-species, 350–351
Microscopic perspective, 376
Miniaturization, 340
Mixed oxides, 197
$M[N(SiMe_3)_2]_2$, 114
$M^+(CO_2)_n$ adducts, 57
M–OCHO, 85
Mo-CODH enzymes, 358
Modes of bonding of CO_2, 35
Modes of bonding of the formate ion, 87
Molecular-level gas storage, 374
Molecular orbital (MO), 3
Multielectron process, 312
Multinuclear complexes, 43–46
Multiple ionization, 20
Multireference configuration, 7

N

NADH-dependent FDH, 363
NADH-independent FDH enzymes, 363
Natural photosynthesis, 321
$Nb(OR)_4(OCO_2)R$, 189–190
n-$Bu_2Sn(OR)_2$, 188
n-Dibutyldialkoxostannanes, 188
Neutral clusters, 2
New concept photobioreactors, 340
$(PC_{y3})_2Ni(CO_2)$, 37
Ni-carbonyl-species, 351
Nickelacycle carboxylate, 149
Nickelacyclopentadiene, 162
$Ni(cyclam)Cl_2$, 40
Ni-CODH/ACS (acetyl CoA synthetase), 358
Ni-CODH enzymes, 358
Ni-C(O)OH, 89
NMR data, transition metal complexes, 49–51
Nodal planes, 4
Non-stoichiometric clathrate solid, 374
Normal and abnormal insertion, 85
Normal vibration modes, 21
N-substituted phenylaziridines, 126
Natural population analysis (NPA), 12
Nucleation, 390

O

$O(^1D)$, 24
$O(^3P)$, 24
OCM. *See* Oxidative coupling of methane (OCM)
OCO bond angle, 3, 36
2-OH-benzoic acid, 353
4-OH-benzoic acid, 353
One-electron transfer, 312, 324
One-electron transfer to CO_2, 321–322
One-step pathway for CO_2-ODE, 259
O-nucleophile, 9–10
O-protonation, 12
Organic carbonates, 183–185
Organic promoters, 201–208
Overpotential, 323
Overvoltage, 320
Oxazolidinones, 212, 224
Oxidation state of Ni, 37
Oxidative carbonylation of methanol, 184
Oxidative carboxylation of olefins, 213–216
Oxidative coupling of methane (OCM), 237, 239–243
Oxidative coupling with CO_2, 144
Oxidative dehydrogenation of alkanes, 243–254
Oxidative dehydrogenation of ethylbenzene, 254–265
Oxygen anion mobility, 239
Oxygen containing fuels, 237

P

Pd-C(CO)OH2, 328
Pentagonal dodecahedron (5^{12}), 377, 378
Peroxocarbonates, 98
Pervaporation membrane, 201
Phase equilibria of CO_2, 374
Phase equilibrium of CO_2 hydrates, 376, 386–389
$[(PhCH_2)_2NCO_2H]_2$, 104
Phenolic acid decarboxylases, 357
Phosphite dehydrogenase (PTDH), 365
Phosphocarbamates, 112
Photoautotrophic organisms, 348
Photocatalysis, 347
Photocatalytic C–H activation, 130
Photocatalytic reduction, 366
Photochemical (PC), 321
Photoelectrochemical (PEC), 321
Photoreduction processes, 17
Photosynthesis, 1
Phthalocyanines, 335
Physical properties of CO_2 hydrates, 382–386
$(C_3H_5N_2)_3(C_3H_4N_2)[PMo_{11}CoO_{38}(CO_2)]$ $4H_2O$, 46
Point group $D_{\infty}\eta$, 2
Polyaniline, 339
Polycarbonates, 221–223
Polydentate P-ligand, 328
Polyhedral cavities, 374

Pressure–temperature phase diagram of CO_2 $+H_2O$, 389
Production of acetic acid, 361-362
Proteobacteria, 350
Proton affinity, 9
PTDH. *See* Phosphite dehydrogenase (PTDH)
P-type semiconductors, 239
Pulsed ESR, 17
PV, 321
PV-generated H_2, 333
Pyridine as an organic catalyst, 330
Pyridine-N-carboxylic acid, 330
Pyrone, 161
Pyrrole-2-carboxylate decarboxylase enzyme, 355–356
Pyruvate decarboxylase, 354
Pyruvate, ferredoxin oxidoreductase, 350

Q
Quantum efficiency, 334

R
Rate constants, 72, 324
Rate determining step, 77
Rate of carbamate formation, 77
γ-Rays, 13
Reaction mechanism for CO_2 hydrogenation, 285–294
Reaction mechanism for CO_2-OCM, 241–243
Reaction mechanism for CO_2-ODE to styrene, 258–265
Reaction mechanism for CO_2-ODH of alkanes, 248–254
Reaction mechanism for DME synthesis, 293–294
Reaction mechanism for methanol synthesis, 287–293
Reaction mechanism for RWGS, 285–287
Reaction mechanism for the carboxylation of epoxides, 219
Reaction mechanism for the DRM process, 271–276
Reaction of alcohols with urea, 185
Reaction of co-ordinated CO_2 with electrophiles, 59–63
Reaction of CO_2 with amines, 74–82
Reaction of CO_2 with carbanions, 73–74
Reaction of CO_2 with hydroxyl and alkoxyl species, 72–73
Reaction of CO_2 with $R_3Si–H$, 128
Reaction of CO_2 with the hydride ion, 71–72
Reaction of ionic hydrides with CO_2, 71
Reactions of co-ordinated CO_2, 58–59
Reactions of co-ordinated CO_2 with nucleophiles, 63–64
Reduce emission, 1
Reduction of CO_2 to carbon monoxide/ formate, 362–364
Reduction reactions, 358–366
Reductive acetyl-CoA (Wood–Ljungdahl) pathway, 350–351
Reductive TCA, 350–351
Reforming of methane (MDR), 237
Refrigeration systems, 376
Regeneration of the cofactor NADH, 366
Regularity of the alternate insertion, 222
Replacement of CH_4 by CO_2, 395–396
Reverse water gas shift reaction (RWGS), 278
Reversible binding, 324
Reversible interconversion, 191
$Rh(diars)_2(Cl)(CO_2)$, 41
$(diphos)Rh(O_2C\text{-}C_6H_5)$, 95
Rh-catalyzed cooligomerization of butadiene, 175
Ring-opening carboxylative polymerization, 216
Ring opening of epoxides, 217
Ring opening of vinylaziridines, 127
Ring stress energy, 216
Rotational structure, 8
$Ru(bpy)_2(CO)(COOH)^{2+}$, 90
$[Ru(diphos)_2(COOH)]^+$, 90
RubisCO, 349
$RuCl_2$(1,3,5-triaz-7-phosphaadaman tane), 89
$RuCO_2H$, 327
$RuC(O)OH_2$, 327
RWGS. *See* Rese water gas shift reaction (RWGS)

S
Secondary refrigerant, 397
Second-order reaction, 72
Semiconducting materials, 366
Sequential "one-electron plus one-proton", 326–332
Single-metal catalysts, 194–195
Single-step, third-order reaction, 78
Singlet ground state ($^1\Sigma_g$+), 24
Sn-tethered complexes, 194
Solar driven cyclic interconversion, 59
Solar-driven water splitting, 340
Solar energy, 321

Solid–liquid junction, 339
Solid matrix, 13
Solubility of CO_2, 315
Solubility of CO_2 in organic solvents, 319
Solubility of CO_2 (g) in water, 319
Sonic and seismic velocity, 384
Space issue, 340
Spherical-like cages, 374
Spontaneous release of formic acid, 89
Structure of gas hydrates, 376–382
Substituted allenes, 164
Sustainability, 187
Symmetric stretching mode, 21
Synthesis of oxalates, 322

T
$(^{Ad}ArO)_3tacnU(\eta^1\text{-}OCO)$, 43
Tetrakaidecahedron ($5^{12}6^2$), 378
Theoretical calculations, 36
Thermal conductivity of gas hydrates, 385
Thermal expansion, 385
Thermal properties of CO_2 hydrates, 384–386
Thermodynamic and kinetic issues, 185–187
Thermodynamic barriers, 185
Thermodynamic data, 88
Thermodynamic models, 374
Thermodynamic potential, 320
Time-dependent kinetic behavior, 374
Time-to-market, 340
TOF, 327
Trans-esterification, 184
Transfer of "1e^{-}" to CO_2, 311
Transition state, 86, 98
Transition states, 62
trans-$Mo(CO_2)_2HN$
(CH$_2$CH$_2$PMe$_2$)$_2$(PMe$_3$), 38
trans-$Ru(dmpe)_2(O_2C\text{-}CH_3)_2$, 96
trans-$Ru(dmpe)_2(H)CH_3$, 95
trans-$Ru(dmpe)_2(O_2C\text{-}CH_3)CH_3$, 96
Triplet-singlet transition, 8
Triplet state $^3B_2$, 26
Two-electron transfer, 324
Two-step pathway, EB dehydrogenation, 262, 263

U
Unfavorable thermodynamics, 186
Unimolecular transition structure, 154
Unoccupied molecular orbitals, 4
Unpaired electron spin density, 17
Unperturbed electronic structure of CO_2^-, 16
Unsaturated carboxylic acid, 164
Upper (bound) singlet state $^1B_2$, 25
Urea as an active form of CO_2, 208–210
UV spectrum, 22–26

V
Variable temperature NMR spectroscopy, 26
Vibrational structure, 5
5-Vinyloxazolidinones, 127
VT-multinuclear NMR, 51

W
Walsh diagram, 7, 35
Water removal, 200–201
Water trapping, 187
W-shape, 16

X
XANES, 324
X-ray diffraction (XRD) structure of solid complexes, 36

Z
Zwitterion mechanism, 78

MIX
Papier aus verantwortungsvollen Quellen
Paper from responsible sources
FSC® C105338

If you have any concerns about our products,
you can contact us on
ProductSafety@springernature.com

In case Publisher is established outside the EU,
the EU authorized representative is:
Springer Nature Customer Service Center GmbH
Europaplatz 3, 69115 Heidelberg, Germany

Printed by Libri Plureos GmbH
in Hamburg, Germany